MW01625458

HANDBOOK OF MATHEMATICAL ECONOMICS

VOLUME III

EHEHEHEHEHEHEHEHEHEHEHEHEHEHEHEHEHEHEHE

HANDBOOKS IN ECONOMICS

1

Editors

KENNETH J. ARROW

MICHAEL D. INTRILIGATOR

NORTH-HOLLAND
AMSTERDAM • NEW YORK • OXFORD • TOKYO

EHEHEHEHEHEHEHEHEHEHEHEHEHEHEHEHEHEHEHEHEHE

HANDBOOK OF MATHEMATICAL ECONOMICS

VOLUME III

Edited by

KENNETH J. ARROW
Stanford University

and

MICHAEL D. INTRILIGATOR
University of California, Los Angeles

NORTH-HOLLAND
AMSTERDAM • NEW YORK • OXFORD • TOKYO

ELSEVIER SCIENCE PUBLISHERS B.V.
Sara Burgerhartstraat 25
P.O. Box 211, 1000 AE Amsterdam, The Netherlands

Distributors for the United States and Canada:
ELSEVIER SCIENCE PUBLISHING COMPANY, INC.
655 Avenue of the Americas
New York, N.Y. 10010, U.S.A.

ISBN : 0 444 86128 9
ISBN set: 0 444 86054 1

First edition: 1986
Second impression: 1989

Library of Congress Cataloging in Publication Data
(Revised for volume III)
Main entry under title:
Handbook of mathematical economics.
(Handbooks in economics; bk. 1)
Includes bibliographies and index.
1. Economics, Mathematical–Addresses, essays, lectures. I. Arrow, Kenneth Joseph, 1921- II. Intriligator, Michael D. III. Series.
HB135.H357 330'.01'51 81-2820
ISBN 0-444-86054-1 (U.S.: set)

© ELSEVIER SCIENCE PUBLISHERS B.V., 1986

All rights reserved. No part of this publication may be reproduced, stored in a retrieval system, or transmitted, in any form or by any means, electronic, mechanical, photocopying, recording or otherwise, without the prior written permission of the publisher, Elsevier Science Publishers B.V. / Physical Sciences and Engineering Division, P.O. Box 1991, 1000 BZ Amsterdam, The Netherlands.

Special regulations for readers in the U.S.A. - This publication has been registered with the Copyright Clearance Center Inc. (CCC), Salem, Massachusetts. Information van be obtained from the CCC about conditions underwhich photocopies of parts of this publication may be made in the U.S.A. All other copyright questions, including photocopying outside of the U.S.A., should be referred to the copyright owner, Elsevier Science Publishers B.V., unless otherwise specified.

No responsibility is assumed by the publisher for any injury and/or damage to persons or property as a matter of products liability, negligence or otherwise, or from any use or operation of any methods, products, instructions or ideas contained in the material herein.

PRINTED IN THE NETHERLANDS

INTRODUCTION TO THE SERIES

The aim of the *Handbooks in Economics* series is to produce Handbooks for various branches of economics, each of which is a definitive source, reference, and teaching supplement for use by professional researchers and advanced graduate students. Each Handbook provides self-contained surveys of the current state of a branch of economics in the form of chapters prepared by leading specialists on various aspects of this branch of economics. These surveys summarize not only received results but also newer developments, from recent journal articles and discussion papers. Some original material is also included, but the main goal is to provide comprehensive and accessible surveys. The Handbooks are intended to provide not only useful reference volumes for professional collections but also possible supplementary readings for advanced courses for graduate students in economics.

CONTENTS OF THE HANDBOOK

VOLUME I

VOLUME II

Part 2 – MATHEMATICAL APPROACHES TO MICROECONOMIC THEORY

Part 3 – MATHEMATICAL APPROACHES TO COMPETITIVE EQUILIBRIUM

Part 5 – MATHEMATICAL APPROACHES TO ECONOMIC ORGANIZATION AND PLANNING

PREFACE TO THE HANDBOOK

The field of mathematical economics

Mathematical economics includes various applications of mathematical concepts and techniques to economics, particularly economic theory. This branch of economics traces its origins back to the early nineteenth century, as noted in the historical introduction, but it has developed extremely rapidly in recent decades and is continuing to do so. Many economists have discovered that the language and tools of mathematics are extremely productive in the further development of economic theory. Simultaneously, many mathematicians have discovered that mathematical economic theory provides an important and interesting area of application of their mathematical skills and that economics has given rise to some important new mathematical problems, such as game theory.

Purpose

The *Handbook of Mathematical Economics* aims to provide a definitive source, reference, and teaching supplement for the field of mathematical economics. It surveys, as of the late 1970's, the state of the art of mathematical economics. Bearing in mind that this field is constantly developing, the Editors believe that now is an opportune time to take stock, summarizing both received results and newer developments. Thus all authors were invited to review and to appraise the current status and recent developments in their presentations. In addition to its use as a reference, the Editors hope that this Handbook will assist researchers and students working in one branch of mathematical economics to become acquainted with other branches of this field. Each of the chapters can be read independently.

Organization

The Handbook includes 29 chapters on various topics in mathematical economics, arranged into five parts: *Part 1* treats *Mathematical Methods in Economics*, including reviews of the concepts and techniques that have been most useful for the mathematical development of economic theory. *Part 2* elaborates on *Mathematical Approaches to Microeconomic Theory*, including consumer, producer, oligopoly, and duality theory. *Part 3* treats *Mathematical Approaches to Competi-*

tive Equilibrium, including such aspects of competitive equilibrium as existence, stability, uncertainty, the computation of equilibrium prices, and the core of an economy. *Part 4* covers *Mathematical Approaches to Welfare Economics*, including social choice theory, optimal taxation, and optimal economic growth. *Part 5* treats *Mathematical Approaches to Economic Organization and Planning*, including organization design and decentralization.

Level

All of the topics presented are treated at an advanced level, suitable for use by economists and mathematicians working in the field or by advanced graduate students in both economics and mathematics.

Acknowledgements

Our principal acknowledgements are to the authors of chapters in the *Handbook of Mathematical Economics*, who not only prepared their own chapters but also provided advice on the organization and content of the Handbook and reviewed other chapters.

KENNETH J. ARROW
Stanford University
MICHAEL D. INTRILIGATOR
University of California, Los Angeles

CONTENTS OF VOLUME III

Part 5 – MATHEMATICAL APPROACHES TO ECONOMIC ORGANIZATION AND PLANNING

PART 4

MATHEMATICAL APPROACHES TO WELFARE ECONOMICS

Chapter 22

SOCIAL CHOICE THEORY *

AMARTYA SEN

All Souls College, Oxford

1. Social welfare functions

1.1. Distant origins

The origins of social choice theory can be traced to two rather distinct sources, and it so happens that the theory is nearly in a position to celebrate the bicentenary of each of its two origins. One source is the study of normative analysis in terms of personal welfare (extensively explored in modern welfare economics), and the origins of this, through utilitarianism, can certainly be traced at least to Jeremy Bentham (1789). The other is the mathematical theory of elections and committee decisions, which is comfortably traced to Borda (1781) and Condorcet (1785). The influences of these two different origins will become clear as the modern developments in social choice theory are reviewed.

No approach to welfare economics has received as much support over the years as utilitarianism. If $U_i(\cdot)$ is the utility function of person i defined for each person $i=1,\ldots,n$, over the set X of alternative social states, then on the utilitarian approach any state x is at least as good as another y, denoted xRy, if and only if $\sum_{i=1}^{n} U_i(x) \geq \sum_{i=1}^{n} U_i(y)$.

It is clear that utilitarianism uses cardinality and interpersonal comparability of personal utilities. Both these practices received severe reprimand in the 1930's,[1]

* The first version of this paper was written during 1978–79. While the paper has been now revised, I have not tried to bring it "up to date" regarding more recent publications. (There are some references to later publications, but most of these were in fact available in pre-print form earlier.) My greatest debt is to Kenneth Arrow, Michael Dummett and Peter Hammond for extremely helpful comments and suggestions on the earlier version of this paper. I have also benefited greatly from the comments of Brian Barry, Charles Blackorby, Julian Blau, Graciela Chichilnisky, Peter Coughlin, Bhaskar Dutta, Alan Feldman, Wulf Gaertner, Louis Gevers, Geoffrey Heal, Michael Intriligator, Jocelyn Kynch, Tapas Majumdar, John Muellbauer, Prasanta Pattanaik, Robert Pollak, Ariel Rubinstein, Maurice Salles, David Schmeidler, Margaret Sjöberg, Steven Slutsky, Kotaro Suzumura, and H. P. Young.

[1] The most influential attack came from Robbins (1932).

Handbook of Mathematical Economics, vol. III, edited by K.J. Arrow and M.D. Intriligator
© 1986, Elsevier Science Publishers B.V. (North-Holland)

with the rebuke drawing sustenance from a single-minded concern with basing utility information on non-verbal behaviour only, dealing with choices in the absence of risk. It thus appeared that social welfare must be based on just the n-tuple of ordinal, interpersonally non-comparable, individual utilities. This informational restriction would, of course, make the traditional utilitarian approach – and a great many other procedures – unworkable.

This "informational crisis" is important to bear in mind in understanding the form that the origin of modern social choice theory took. In fact, with the binary relation of preference replacing the utility function as the primitive of consumer theory, it made sense to characterize the exercise as one of deriving a social preference ordering R from the n-tuple of individual orderings $\{R_i\}$ of social states.

The other source, dealing primarily with election methods, had in any case the tradition of concentrating on the information given by an n-tuple of individual orderings – reliant on an informational framework that was much less ambitious than utilitarianism. Borda, Condorcet, Dodgson (Lewis Carroll), Nanson and others had pursued various results of voting, and had discussed the superiority of some voting systems over others.[2] Economists did not, however, take much notice of this literature, or of the problem studied in them, until the "informational crisis" sent them searching for other methods.

The union produced modern social choice theory. The big bang that characterized the beginning took the form of an "impossibility theorem", viz. Arrow's (1950, 1951) "General Possibility Theorem". It appeared that some conditions that look mild – and are indeed satisfied comfortably by utilitarianism when translated into its cardinal interpersonally comparable framework (see Section 6) – cannot be fulfilled by *any* rule whatsoever that has to base the social ordering on n-tuples of individual orderings. This theorem, which had a profound impact on the way modern social choice theory developed, will be discussed in Section 2.[3]

1.2. The Bergson – Samuelson social welfare function

The concept of a *social welfare function* was first introduced by Bergson (1938). This was defined in a very general form indeed: as a real-valued function $W(\cdot)$, determining social welfare, "the value of which is understood to depend on all the variables that might be considered as affecting welfare" (p. 417). If the relevant information about the social states in set X can be obtained, then such a social welfare function – swf for short – might as well be thought to be a real-valued

[2] For an excellent account of the literature, see Black (1948, 1958).

[3] Another important contribution to the early development of modern social choice theory was Kenneth May's axiomatization of the majority rule [see May (1952, 1953)].

function defined on X. If the issue of numerical representation is not emphasized, this really amounts to an ordering R of X.

While the idea of a social welfare function came from Bergson, the uses to which such a swf can be put were definitively investigated by Samuelson (1947). His exercises made use of many criteria that a swf may be required to satisfy.[4] One of them is the old Pareto criterion. This can be defined in many forms, and since the differences will turn out to be of some significance, we might as well seize this opportunity of distinguishing between them (though not all these versions were, in fact, used by Samuelson).

Let P and I be the asymmetric and symmetric factors of the social preference relation R ("at least as good as"), standing respectively for "strictly better than" and "indifferent to". And let the corresponding individual preference relation and its asymmetric and symmetric factors for any person i be R_i, P_i and I_i, respectively. The different versions of the Pareto Principle may now be stated. The following are all defined with the universal quantifier $\forall x, y \in X$ ("for all x, y in X"):

Condition P (weak Pareto principle)
$(\forall i\colon xP_iy) \Rightarrow xPy$.

Condition P° (Pareto indifferent rule)
$(\forall i\colon xI_iy) \Rightarrow xIy$.

*Condition P** (strong Pareto principle)
$(\forall i\colon xR_iy) \Rightarrow xRy$. And if to the antecedent is added $\exists i\colon xP_iy$, then the consequence is xPy.

It is obvious that Condition P* implies both Conditions P and P°, but is not implied by them even jointly.

If a swf satisfies Condition P*, we shall call it a *Pareto-inclusive* swf. It may be remarked that, given the form in which Bergson defined a swf, it may or may not be possible to check whether it is Pareto-inclusive or not, since there is no obligation to specify the individual preferences in defining a Bergson $W(\cdot)$. However, from the motivating discussion of Bergson (1938, 1948) and more so from the operations chosen by Samuelson to demonstrate the use of such a swf, it appears that the intention is to take note of individual preferences at least to the extent of being Pareto-inclusive.[5]

[4] For excellent examples of application and use of Bergson–Samuelson and social welfare functions, see Dasgupta and Heal (1979), Atkinson and Stiglitz (1981), and Dasgupta (1982).

[5] In using utility for such social criteria (Pareto optimality, equality, justice, etc.), one source of ambiguity is the possibility of defining them *either* in terms of ex post utilities, *or* in terms of ex ante utilities. On this see Starr (1973). Also Hammond (1983).

Sometimes a Bergson–Samuelson social welfare function is described as "individualistic". There is an ambiguity in this expression which is worth clarifying since it has been the source of some confusion. A swf can be individualistic in the sense of reflecting the preferences of all the individuals in the society taken together *when such preferences do not conflict*, in ranking any pair of social states. In this sense, an individualistic swf is simply a Pareto-inclusive swf. There is a second interpretation, which makes social welfare W a function of the vector of individual utilities u irrespective of the non-utility characteristic of the social states from which the utilities emanate: $W = W(\boldsymbol{u})$; see Samuelson (1947, pp. 228–229, 246), Graaff (1957, pp. 48–54), and Bergson (1948, p. 418), among others. This is a version of a condition of "neutrality" (sometimes called "welfarism"), which relates closely to Arrow's result (Section 2.1), and which will be further examined in Sections 6 and 9. In effect, it asserts the neutrality of the social ranking towards non-utility features, which can then affect the social ranking *only through* their influence on individual utilities, or preferences. It is easily checked that neither does Pareto-inclusiveness imply this condition of neutrality, nor the converse, and these two interpretations of individualism are, thus, completely independent of each other.

Finally, none of the conditions that Samuelson imposed on a swf for his exercises happened to specify how the social ordering might alter if different n-tuples of individual orderings were considered. If any n-tuple of individual preference orderings is called a "profile", then his exercises – and those considered by Bergson – were all "single profile" problems (see Section 9).

1.3. The Arrow social welfare function

Arrow (1951) defined a social welfare function – henceforth SWF (to be distinguished from the Bergson–Samuelson swf) – as a functional relation specifying one social ordering R for any given n-tuple of individual orderings $\{R_i\}$, one ordering for each person,

$$R = f(\{R_i\}). \tag{1.1}$$

Note that if a Bergson–Samuelson swf is defined as a social ordering R, then an Arrow SWF is a function the *value* of which would be a Bergson–Samuelson swf. Arrow's exercise, in this sense, is concerned with the way of arriving at a Bergson–Samuelson swf. Alternatively, if the Bergson–Samuelson swf is taken as a function $W(\cdot)$, defined over a particular profile of individual ordinal utilities, then a Bergson–Samuelson swf fits into the form (1.1). The Arrow exercise can, then, be seen as a way of extending the set of single-profile formulations into one consistent multiple-profile function, specifying correspondences between the respective values of R (or parts thereof) for different n-tuples $\{R_i\}$.

Arrow proceeded to impose a variety of conditions that a reasonable SWF could be expected to satisfy. One of them deals specifically with the multiple-profile characteristics of a SWF: the independence of irrelevant alternatives. For stating this condition, Arrow used the notion of a choice function for the society, $C(\cdot)$, which was defined with respect to the binary relation R, satisfying what is sometimes called the "Condorcet condition" [Condorcet (1785)].[6] For all subsets S of X,

$$C(S) = [x | x \in S \;\&\; \forall y \in S\colon xRy]. \tag{1.2}$$

Condition I (independence of irrelevant alternatives)
For any two n-tuples $\{R_i\}$ and $\{R_i'\}$ in the domain of f, and for any $S \subseteq X$, with the choice functions $C(\cdot)$, and $C'(\cdot)$ corresponding to $f(\{R_i\})$ and $f(\{R_i'\})$, respectively,

$$\left[\forall i\colon \left(\forall x, y \in S\colon xR_i y \Leftrightarrow xR_i' y\right)\right] \Rightarrow C(S) = C'(S).$$

This condition requires that as long as individual preferences remain the same over a subset S of X, the social choice from that subset should also remain the same.

The property of independence can also be considered in purely relational terms as well, without invoking a choice function at all.[7]

Condition I^2 (pairwise relational independence)
The restriction of the social preference relation over any pair $\{x, y\}$ is a function of the n-tuple of restrictions of individual preferences over that pair,

$$R|^{\{x,y\}} = f^{x,y}(\{R_i|^{\{x,y\}}\}). \tag{1.3}$$

[6] The "Condorcet condition" is sometimes defined specifically for the majority relation only, which relates to Condorcet's (1785) own original concern. On these issues, see Black (1958), Fishburn (1973a), and Young (1977).

[7] Given the binary specification of the choice function for society, as in (1.2), it is easily checked that this relational independence condition I^2 is exactly equivalent to Arrow's choice-functional independence condition I. In the proofs that will be presented in Section 2.1, Condition I^2 will be used, because it simplifies matters, and makes Arrow's theorem entirely relation-theoretic. However, Conditions I and I^2 are not *generally* equivalent. When the choice function for the society cannot be represented by a binary relation (to be investigated in Section 4 below), Condition I can be used without implying Condition I^2, and vice versa. Indeed, in a purely relation-theoretic framework with the use of Condition I^2, it need not even be assumed that a choice function for the society exists. Similarly, in a purely choice-oriented framework, the relation-theoretic notions (including Condition I^2) can be entirely dispensed with. Finally, it is possible to define the binary relation of social preference just in terms of choices over pairs, i.e. xRy if and only if $x \in C(\{x, y\})$, which will make Condition I^2 strictly weaker than Arrow's choice-functional independence condition I. This avenue, which will be explored in Section 4.2, will be useful in interpreting some recent results on collective rationality, and in demonstrating that Arrow had proved a more general result than he had claimed.

2. Arrow's impossibility theorem

2.1. The general possibility theorem

Arrow's General Possibility Theorem asserts the inconsistency of some mild-looking conditions imposed on social welfare functions, viz. Conditions P and I as defined in the last section and the following two additional ones.[8]

Condition U (unrestricted domain)
The domain of the SWF, i.e. $f(\cdot)$ defined by (1.1), includes all (logically) possible n-tuples of individual orderings of X.

Condition D (non-dictatorship)
There is no individual i such that for all preference n-tuples in the domain of $f(\cdot)$, for each ordered pair $x, y \in X$, $xP_i y \Rightarrow xPy$.

Denoting the set of individuals in the society as H, and the cardinality of the set X of social states as $\#X$, the General Possibility Theorem can be stated thus.

General possibility theorem (GPT)
If H is finite and $\#X \geq 3$, then there is no SWF satisfying Conditions U, I, P and D.

This result has been the prime mover in getting the discipline of social choice theory started, and though recently the focus has somewhat shifted from impossibility theorems to other issues, there is no doubt that Arrow's formulation of the social choice problem in presenting the GPT laid the foundations of social choice theory. In seeking a demonstration of the GPT, Condition I^2 can be used rather than Condition I to get a fully relation-theoretic statement, which can be used with or without the further assumption of binary choice.

Pair relational general possibility theorem (GPT*)
If H is finite and $\#X \geq 3$, then there is no SWF satisfying Conditions U, I^2, P and D.

In establishing the General Possibility Theorem, it is convenient to go via two lemmas that are of interest in themselves. We shall call a set G of *persons* a "group" G (but – beware – no "group theory" is involved!). We define a group G of persons "decisive" over the ordered pair $\{x, y\}$, denoted $\overline{D}_G(x, y)$, if and only if xPy whenever $xP_i y$ for all i in G. Group G is "almost decisive" over that

[8] This version of the GPT was presented in the 2nd edition of Arrow's book [Arrow (1963, pp. 96–100)]. An error in Arrow's (1950, 1951) original presentation was noted and rectified by Blau (1957).

ordered pair, denoted $D_G(x, y)$, if and only if xPy whenever xP_iy for all i in G and yP_ix for all i not in G, i.e. for all i in $H-G$. Obviously decisiveness implies almost decisiveness, but in general not vice versa.

The first lemma applies not merely to SWFs, but to a broader class of transformations from individual preferences to social preferences, relaxing the requirement of full transitivity of social preference relation R.

Transitivity
R is transitive on X if and only if $\forall x, y, z \in X$, $(xRy \; \& \; yRz) \Rightarrow xRz$.

Quasi-transitivity
R is quasi-transitive on X if and only if $\forall x, y, z \in X$, $(xPy \; \& \; yPz) \Rightarrow xPz$.

Acyclicity
R is acyclic on X if and only if there is no cycle of strict preference: that is, no subset $(x_1, x_2, \ldots, x_k)$ of X such that $x_1Px_2, x_2Px_3, \ldots, x_{k-1}Px_k$, and x_kPx_1.

Obviously, in this framework, transitivity implies quasi-transitivity, but not vice versa, and quasi-transitivity implies acyclicity but not vice versa. Where the line of "collective rationality" is to be drawn depends partly on what use is to be made of the social preference relation R. While we have for the moment kept aside the question of whether or not to base social choice entirely on a binary relation – as given by (1.2) – it is relevant to note that for a reflexive and complete preference relation, acyclicity is the *necessary and sufficient* condition for the choice set $C(S)$, as defined by (1.2), to be non-empty for every non-empty, finite subset S of X.[9] We may call a choice function that has a non-empty $C(S)$ for every non-empty, finite $S \subseteq X$, a "finitely complete choice function".

If it is required that the binary relation of social preference should provide a minimally sufficient basis for a finitely complete choice function, then it is natural to confine the range of the function $f(\cdot)$ to preference relations that are reflexive, complete and *acyclic*. Such a function will be called a *social decision function* SDF. If the range is further restricted to reflexive, complete and *quasi-transitive* preference relations, then $f(\cdot)$ will be called a *quasi-transitive social decision function* QSDF (a transferred epithet to be sure, but it need not cause any confusion). If the range is further restricted to preference relations that are reflexive, complete and *transitive*, then we are back to the case of Arrow's social welfare functions SWF. It is trivial that a SWF is a QSDF, and a QSDF is a SDF, but in general not vice versa.

[9]For infinite sets, acyclicity would require supplementation by other conditions for guaranteeing the existence of a best element. This supplementation has been investigated in different ways. See Herzberger (1973), Smith (1974), Bergstrom (1975), Suzumura (1976a), Birchenhall (1977), Mukherji (1977), and Walker (1977). See also Aizerman, Zavalishin and Piatnitsky (1976), and Aizerman and Malishevski (1980).

Field expansion lemma
For any quasi-transitive social decision function (QSDF) satisfying Conditions U, I^2 and P, with $\#X \geq 3$, if any group is almost decisive over any ordered pair of social states, then it is decisive over every ordered pair of social states,

$$[\exists x, y \in X\colon D_G(x, y)] \Rightarrow [\forall a, b \in X\colon \bar{D}_G(a, b)].$$

To see clearly how this works, it may be useful to consider the case of four distinct states x, y, a and b. Let the preference ordering – in strict decreasing order – of every i in G be a, x, y, b and let everyone not in G strictly prefer a to x, y to b, and y to x, leaving the ordering of a and b completely unspecified. By the weak Pareto principle aPx and yPb. Further, since $D_G(x, y)$, clearly xPy. Thus, by quasi-transitivity, aPb. By Condition I^2, this must depend only on the individual orderings of a and b, of which – in fact – only the orderings of those in G have been specified. Hence $\bar{D}_G(a, b)$.

By virtue of the Field Expansion Lemma, there is no difference between a group being almost decisive over some ordered pair and being decisive over every ordered pair. Let such a group be called a decisive group.

Group contraction lemma
For any social welfare function (SWF) satisfying Conditions U, I^2 and P, with $\#X \geq 3$, if any group G, with $\#G > 1$, is decisive, then so is some proper subset of that group.

To prove this, partition a decisive group G into two non-empty proper subsets G_1 and G_2, respectively. Let the preference orderings of the three groups be the following in strict descending order, over some triple $\{x, y, z\}$: $G_1\colon x, y, z$; $G_2\colon y, z, x$; $H - G\colon z, x, y$. By the decisiveness of G, it follows that yPz. Clearly, either xPz, or zRx, by the completeness of R. Hence it follows from yPz that xPz or yPx, by the transitivity of R. Hence either G_1 is almost decisive over $\{x, z\}$, or G_2 is almost decisive over $\{y, x\}$. By the Field Expansion Lemma, either G_1 or G_2 is, thus, decisive.

Now Arrow's General Possibility Theorem (GPT).

*Proof of GPT**
By the weak Pareto principle, the group of all persons H is decisive. By the Group Contraction Lemma, we can go on persistently eliminating some members in each contraction, still leaving the rest decisive. Since H is finite, this would lead ultimately to some individual being a dictator.

Proof of GPT
By (2), Condition I implies I^2. Hence GPT* entails GPT.

2.2. *Variants*

In the original version of the General Possibility Theorem, Arrow (1950, 1951) had not used the Pareto principle, and had used instead a pair of conditions which he had called "citizens' sovereignty" and "positive association". The former is a requirement of "non-imposition", asserting that social preference should not be imposed from outside irrespective of the preferences of the members of the community, while the latter is, in fact, a weak condition of "monotonicity" [see Murakami (1968)], requiring non-negative response of social preference to individual preferences. Let $R = f(\{R_i\})$ and $R' = f(\{R'_i\})$.

Condition NI (non-imposition)
For no pair of social states $\{x, y\}$ it is true that xRy for every possible n-tuple $\{R_i\}$ in the domain of $f(\cdot)$.

Condition M (weak monotonicity)
For any two n-tuples $\{R_i\}$ and $\{R'_i\}$, for a given social state x, if for all individuals i, for all states y, $xI_i y \Rightarrow xR'_i y$, and $xP_i y \Rightarrow xP'_i y$, and for all states a and b both distinct from x, $aR_i b \Leftrightarrow aR'_i b$, then $xPy \Rightarrow xP'y$.

The original version of the impossibility theorem [Arrow (1950, 1951)] was concerned with showing the irreconcilability of Conditions U, I, M, NI and D for any SWF. In fact, for a SWF, Conditions U, I, M and NI together imply the weak Pareto principle, and thus the earlier version would be a corollary of the GPT presented in Arrow (1963), and thus of the GPT*.

There was, however, another difference in the original presentation of Arrow (1950, 1951). A weaker domain condition was used, requiring only that the domain of $f(\cdot)$ must include all n-tuples of individual orderings consistent with ordering a particular triple $\{x, y, z\}$ in any way whatsoever (but not necessarily ordering the whole X in any way). This proved insufficient for the impossibility result, and required strengthening as Blau (1957) showed, and hence the domain requirement was tightened to Condition U.[10] An alternative way of obtaining the impossibility is found in leaving the domain condition in its weaker form, while strengthening the non-dictatorship condition by ruling out *local* dictators over the specified triple $\{x, y, z\}$ on which individual preferences could be freely varied according to the weaker version of the domain condition.[11]

[10] The logical problem was absent in one of the earlier versions of Arrow's theorem [viz. Arrow (1952)], which did not, however, go into "field expansion" beyond a triple. Blau's contributions (1957, 1971, 1972, 1976) have brought out the "neutrality" implications of Arrow's framework for social choice by clarifying the full "field expansion" consequences of that framework.

[11] See Murakami (1961, 1968) and Pattanaik (1971).

A great many other variations in the theme of Arrow's impossibility theorem have been explored in the literature. Some of the variations will come up in the discussion of specific issues in later sections, and here I shall confine myself to a few remarks only. Recently Kelly (1978) has provided an excellent account of the main lines of development since Arrow's pioneering contribution.[12]

First, some versions of the result use neither the Pareto principle nor any condition of non-negative responsiveness. Consider the following requirement, which rules out "reverse dictators".

Non-suppression (NS)
There is no individual i such that for every preference n-tuple in the domain of $f(\cdot)$, for each ordered pair $x, y \in X$, $xP_i y \Rightarrow yPx$.

Conditions U, I, NI, NS and D are inconsistent for a social welfare function [see Wilson (1972b) and Binmore (1975); for related results, see Murakami (1968), Hansson (1969a, 1969b), Wilson (1972a), Fishburn (1974a), Binmore (1976), Monjardet (1979), and Kim and Roush (1980a)]. In fact, given unrestricted domain and independence of irrelevant alternatives, the possibilities that are open are (i) dictatorship, (ii) reverse dictatorship, and (iii) collective impotence. Either one person's strict preferences are fully reflected in social rankings of all pairs (positively or negatively), or not even everyone put together can influence social preference over some pair. The weak Pareto principle eliminates collective impotence as well as reverse dictatorship, leaving us only with the possibility of dictatorship (as in the version of the GPT presented in the last subsection).

Second, when the set of individuals is infinitely large, Arrow's conditions are mutually consistent, even though the permitted decision procedures are not very attractive [see Fishburn (1970b), Hansson (1972, 1976), Kirman and Sondermann (1972), Brown (1974), Schmitz (1977), and Armstrong (1980)]. There is, however, no "approximate" consistency for "very large" communities, and the impossibility result continues to hold exactly for all finite communities no matter how large, so that the practical relevance of the consistency possibility may not be very great. Furthermore, the Field Expansion Lemma and the Group Contraction Lemma both continue to hold for infinitely large communities and decisive sets can be endlessly curtailed, effectively disenfranchising nearly everybody [leading to such "limit" concepts as the existence of "invisible dictators", to use Kirman and Sondermann's (1972) description].

Third, McManus (1975, 1978, 1982, 1983) has investigated important issues of inter-taste consistency and inter-profile welfare comparisons, continuity condi-

[12] See also Murakami (1968), Pattanaik (1971), Fishburn (1973a), Brams (1976), and Plott (1976). There is also an important Russian book, viz. Mirkin (1974), with an English translation (1979). On closely related matters, see also Blin (1973), Brams (1976), Gottinger and Leinfellner (1978), Pattanaik (1978), Laffont (1979), Mueller (1979), Feldman (1980a), Kim and Roush (1980a), Moulin (1983), Suzumura (1983a), and Peleg (1984).

tions imposed on social welfare evaluation, and related matters [see also Inada (1964a) on an earlier study with a bearing on these issues]. He has provided both impossibility results and positive possibility theorems involving various combinations of these conditions. He has also provided reasons for not requiring the "independence" conditions, making positive possibilities that much easier.

Fourth, Chichilnisky (1976, 1980a, 1982a, 1982b) has established a set of important impossibility results without the use of the "independence" condition. For a class of social aggregation problems satisfying unanimity (a weak version of the Pareto principle) and anonymity, she shows the absence of continuous rules of transforming n-tuples of individual preferences into social preferences. Continuity too is a condition of "inter-profile consistency", but of a very different sort from "independence". She has investigated various general properties of individual and social choice [Chichilnisky (1979, 1980a, 1980b, 1981, 1983)], and also explored the possibilities of generalizing her original impossibility results by systematic relaxation of specific restrictions, such as non-satiation, preference ordinality, etc. [Chichilnisky (1980c, 1982a, 1983)].

Fifth, the formulation of social choice problems can be broadened by bringing in lotteries on alternatives [see Zeckhauser (1969), Shepsle (1970), Niemi and Weisberg (1972), Fishburn (1972b, 1973a, 1975b), Intriligator (1973, 1979), Nitzan (1975), Barbera (1979), Kalai and Megiddo (1980), Machina and Parks (1981), Coughlin and Nitzan (1981, 1983), and Heiner and Pattanaik (1983)]. This opens up new possibilities. If the problem is reformulated as demanding a lottery over *social preferences* (rather than over the alternatives to be chosen), based on n-tuples of individual orderings of social states, then Arrow-like impossibilities re-emerge in the form of arbitrary distribution of power (the exclusion of which would appear to be reasonable); see Barbera and Sonnenschein (1978), Bandyopadhyay, Deb and Pattanaik (1979), McLenan (1980), and Heiner and Pattanaik (1983).

Sixth, another variation that has been recently investigated is the eschewal of the assumption of completeness of the social preference. Arrow's impossibility result can be adapted for such an extended framework with only a little loss of power [Barthelemy (1983) and Weymark (1983)]. These analyses are, in fact, closely related to results dealing with admitting social intransitivity (see Section 3 below), since intransitivity can be given the particular form of dropping completeness.

Finally, many variations of the way of setting up the problem of social choice will be examined in some detail in the following sections: admitting non-transitive social preference (Section 3); admitting non-binary social choice (Section 4); seeking the acceptable rather than the best (Section 5); enriching the input of utility information (Section 6); restricting the domain of social choice procedures (Section 8); and weakening the independence condition and enriching the use of non-utility information (Section 9). While the focus very often will not be on the specific issue of avoiding Arrow's impossibility result, the implications of these different approaches for that problem will be, inter alia, clarified.

3. Non-transitive social preference

3.1. Quasi-transitivity

There has been speculation for some time as to whether the impossibility results of the type pioneered by Arrow could be avoided by weakening the requirement of collective rationality. There have been broadly two approaches to this question. One retains the Arrovian focus on a social preference relation R, but weakens the consistency requirement of R from the full dose of transitivity to milder conditions. The other dispenses with the notion of social preference as such and formulates the problem in choice functional terms. In this section the use of the first approach is discussed, while the second approach will be taken up in Section 4.

In establishing Arrow's theorem, two lemmas were used in the last section. The Field Expansion Lemma requires no more than quasi-transitivity of social preference, while the Group Contraction Lemma cannot be derived from quasi-transitivity alone, and was, in fact, established by using full transitivity of social preference. The latter result is crucial to deriving dictatorship from Arrow's Conditions U, P and I (or I^2), and if that result is nullified by relaxing the requirement of consistency of social preference to quasi-transitivity only, the Arrow impossibility result will fail to hold. On the other hand, quasi-transitivity is more than sufficient for generating a finitely complete choice function from a reflexive and complete social preference relation. Thus the avoidance of the Arrow impossibility result can be shown to exist strictly within the limits of Arrow's search for a preference-based social choice procedure satisfying Conditions U, P, I and D [see Sen (1969, 1970a) and Schick (1969)]. A simple example of such a procedure is a social decision function that yields the "Pareto-extension rule", with x being socially preferred to y if and only if everyone prefers x to y, while x and y being socially "indifferent" if either they are Pareto-indifferent or Pareto-non-comparable [Sen (1969); for an axiomatic examination of the Pareto-extension rule, see Pollak (1979)]. The unattractiveness of such a social decision procedure (despite its providing a formal route to escape the Arrow impossibility) led to the question as to whether or not the Arrow conditions were in an important sense "too weak" rather than "too strong" [Sen (1969)].

The Pareto-extension rule gives everyone a "veto", and if anyone prefers x to y strictly, he can guarantee that x is socially at least as good as y. Allan Gibbard showed in an unpublished paper [discussed in Sen (1970a)] that the existence of a veto is a necessary result of resolving the Arrow problem through weakening the transitivity of social preference to quasi-transitivity. Define a person i as "semi-decisive" over some ordered pair $\{x, y\}$ if $xP_i y$ implies xRy. A person has a veto if and only if he is semi-decisive over every ordered pair. A SDF is called oligarchic if and only if there is a unique group G of persons such that G is decisive and every member of G has a veto.

Quasi-transitive oligarchy theorem
If H is finite and $\#X \geq 3$, then any QSDF satisfying Conditions U, P and I[2] must be oligarchic.

Just like the Field Expansion Lemma, which continues to hold, it is possible to establish a "Veto-Field Expansion Lemma" asserting that any person who is almost semi-decisive over some ordered pair must be semi-decisive over all ordered pairs, i.e. must have a veto. (Almost-semi-decisiveness of i over x, y is defined as the requirement that $xP_i y$ *and*, for all $j \neq i$, $yP_j x$ must together imply xRy.) Now take a smallest decisive group G of persons, which must exist by the weak Pareto principle and the finiteness of H. Split G into any unit set $\{i\}$ consisting of one person i and the rest $G-\{i\}$. Assume the following preference orderings (shown in strict descending order) over a triple x, y, z: $G-\{i\}: x, y, z$; $\{i\}: y, z, x$; and $H-G: z, x, y$. By the decisiveness of G, we have yPz. By G being a smallest decisive group, $G-\{i\}$ cannot be decisive. But if xPz, then it will be almost decisive over this ordered pair, and thus by the Field Expansion Lemma, must be decisive. So zRx. If we now have xPy, this together with yPz and zRx will contradict quasi-transitivity. Hence yRx. But then i is almost semi-decisive over some ordered pair, and thus by the Veto-Field Expansion Lemma has a veto. This can be shown for every member of G. The proof is completed by noting that no group other than a superset of G can be decisive since every member of G has a veto.[13]

The replacement of transitivity by quasi-transitivity has translated the possibility of dictatorship to oligarchy with veto powers, and while the existence of vetoers may be less unattractive than that of a dictator, it is unappetizing enough not to provide a grand resolution of the Arrow problem.

In fact, even the dictatorship result reappears if the conditions imposed are supplemented by the requirement of "positive responsiveness" – a stricter version of the weak monotonicity condition (M) defined earlier. Positive responsiveness is defined below in a framework that incorporates independence of irrelevant alternatives. Denote $R = f(\{R_i\})$, and $R' = f(\{R'_i\})$.

Condition PR (positive responsiveness)
For any $x, y \in X$, if for all i, $(xP_i y \Rightarrow xP'_i y$ & $xI_i y \Rightarrow xR'_i y)$, and for some i, $(xI_i y$ & $xP'_i y)$ or $(yP_i x$ & $xR'_i y)$, then $xRy \Rightarrow xP'y$.

[13] This theorem was first established by Gibbard, and in different ways by Schwartz (1972), Mas-Colell and Sonnenschein (1972), and Guha (1972). Guha noted a hierarchy of oligarchies with a stricter version of the Pareto principle such that indifference by an oligarchic group would lead to a fresh oligarchy among the rest.

Quasi-transitive positive-responsive dictatorship theorem
If H is finite and $\#X \geq 3$, then there is no QSDF satisfying Conditions U, I^2, P, D and PR.

This theorem, established by Mas-Colell and Sonnenschein (1972), shows that transitivity can be weakened to quasi-transitivity of social preference if a stricter version of the monotonicity requirement is imposed.

3.2. *Acyclicity*

Quasi-transitivity may also be thought to be too demanding a condition, especially since acyclicity – a weaker requirement than quasi-transitivity – is sufficient for generating a finitely complete choice function based on the binary relation of social preference. Mas-Colell and Sonnenschein (1972) have a veto-result with *acyclicity* as such. (It can, in fact, be shown that the result goes through even with the weaker condition of "triple acyclicity", i.e. no cycles over triples.[14])

Triple-acyclic positive-responsive vetoer theorem
For H finite, $\#H \geq 4$, and $\#X \geq 3$, any SDF (even with the requirement of acyclicity relaxed to triple acyclicity) satisfying Conditions U, I^2, P and PR, must yield someone with veto.

An alternative way of generating the vetoer result is to use the weaker monotonicity condition M (essentially, non-negative responsiveness), but marry it with a requirement of neutrality towards the nature of social states. Combining neutrality with independence (in the form of I^2) and monotonicity (in the weak form) yields the following:

Condition NIM (neutrality, independence cum monotonicity)
For any $x, y, a, b \in X$, if for all i, $xP_i y \Rightarrow aP_i' b$, and $xI_i y \Rightarrow aR_i' b$, then $xPy \Rightarrow aP'b$.

The following theorem was established by Blau and Deb (1977).

Acyclic neutral monotonicity vetoer theorem
If $\#X \geq \#H$, with a finite H, then any SDF satisfying Conditions U and NIM must yield someone with a veto.[15]

[14] See Blair, Bordes, Kelly and Suzumura (1976).

[15] See also Schwartz (1974). The cycle involved in the proof is that of the $(n-1)$-majority rule. On related matters, see Dummett and Farquharsen (1961), Murakami (1968), Craven (1971), Pattanaik (1971), Fishburn (1973a), Ferejohn and Grether (1974), Deb (1976), Blau and Brown (1978), Nakamura (1978), Peleg (1978, 1979b), and Suzumura (1983a).

To establish this, suppose – to the contrary – there is no vetoer. So there is no one who is semi-decisive over all pairs. By the neutrality and monotonicity properties of NIM, there is thus no one who is almost semi-decisive over any pair. (If someone were, then by monotonicity he will be semi-decisive over that pair, and by neutrality a vetoer.) So everyone loses over any pair if unanimously opposed by others. With this in mind, consider the following n-tuple of preference orderings (in descending order) over a subset $\{x_1, x_2, \ldots, x_n\}$ of X, for the n individuals $1, \ldots, n$.

1: $x_1, x_2, \ldots, x_{n-1}, x_n,$

2: $x_2, x_3, \ldots, x_n, x_1,$

⋮

n: $x_n, x_1, \ldots, x_{n-2}, x_{n-1}.$

Clearly, x_1Px_2, x_2Px_3, ..., $x_{n-1}Px_n$, and x_nPx_1. This violation of acyclicity shows the falsity of the contrary hypothesis.

Thus, even acyclicity does not help very much in delivering us from the Arrow problem. A weaker consistency condition combined with other properties leads to a weakening – rather than elimination – of the dictatorship result, in the form of the existence of vetoers. And acyclicity *is* necessary for binary choice using the Condorcet condition.

Recently, Blair and Pollak (1982, 1983) and Kelsey (1982, 1983a, 1983b) have established various extensions of these impossibility results. Blair and Pollak have shown in particular that even without neutrality, some of the sting of the veto power remains in the form of an individual being semi-decisive over $(m-n+1)$ $(m-1)$ pairs of states, where m and n are respectively the numbers of states and individuals. Given the individuals, when larger and larger sets of states – without bound – are considered, the proportion of pairs over which the individual is semi-decisive approaches unity [Blair and Pollak (1982)]. Kelsey (1982, 1983a, 1983b) has established similar – though weaker – arbitrariness of power (semi-decisive or *anti*-semi-decisive) over a large proportion of pairs of states – approaching $\frac{1}{2}$ as more and more states are considered – without neutrality and even without the Pareto principle.

3.3. *Semi-transitivity, interval order and generalizations*

I turn now to a somewhat different question. From quasi-transitivity to move to acyclicity is an act of weakening. What about the act of strengthening in going from just quasi-transitivity to semi-orders (and similar structures) without moving all the way to full transitivity? Would the Arrow impossibility result hold with full force in such "intermediate" ground? The answer seems to be: yes, in a lot of that intermediate ground, and some areas outside it.

A semi-order satisfies the two following properties:[16]

Semitransitivity
For any $x, y, z, a \in X$, if xPy and yPz, then xPa *or* aPz.

Interval order property
For any $x, y, a, b \in X$, if xPy and aPb, then xPb *or* aPy.

Each of these properties implies quasi-transitivity for a complete R. Arrow's impossibility result can be established with either of these less demanding properties, and with still weaker structures, and recently Blair and Pollak (1979) and Blau (1979) have provided elegant proofs of these – and further – extensions. [For earlier contributions to this question, see Blau (1959), Schwartz (1974), Brown (1975b), and Wilson (1975).]

General possibility theorem for semi-transitivity
If H is finite and $\#X \geq 4$, then there is no SDF satisfying Conditions U, I^2, P and D, and yielding semi-transitive social preference.

In establishing this theorem, it may be first noted that since semi-transitivity implies quasi-transitivity, the Field Expansion Lemma still holds. The Group Contraction Lemma can also be re-established. Let G be a decisive group, which is partitioned into two non-empty subsets G_1 and G_2. The following preference orderings are postulated:

$$G_1: \quad x, y, z, a,$$

$$G_2: \quad a, x, y, z,$$

$$H - G: \quad z, a, x, y.$$

By the decisiveness of G, xPy and yPz. By the semi-transitivity of R, xPa *or* aPz. In the first case, G_1 is almost decisive over $\{x, a\}$; in the second case, G_2 is almost decisive over $\{a, z\}$. By the Field Expansion Lemma, therefore, some proper subset of G is, thus, decisive. This establishes the Group Contraction Lemma. The rest of the proof is the same as with the GPT, presented in Section 2.

General possibility theorem for interval order property
If H is finite and $\#X \geq 4$, then there is no SDF satisfying Conditions U, I^2, P and D, and yielding social interval orders.

[16]For discussions of the properties of semi-orders, see Luce (1956), Scott and Suppes (1958), Fishburn (1970a, 1975a), Chipman et al. (1971), Jamison and Lau (1973, 1977), Sjöberg (1975), and Schwartz (1976).

In this case the following preference orderings are considered:

G_1: $x, y, a, b,$

G_2: $a, b, x, y,$

$H-G$: $y, a, b, x.$

By the decisiveness of G, xPy and aPb. By the interval order property, xPb *or* aPy. In the first case G_1 is decisive; in the second, G_2. The rest of the proof is unaltered.

Since a semi-order is both semi-transitive and an interval order, clearly it is, a fortiori, adequate to sustain the Arrow impossibility result fully. While for an ordering even one strict preference "filters through" one indifference $PI \Rightarrow P$ and $IP \Rightarrow P$, i.e. $(xPy \;\&\; yIx) \Rightarrow xPz$ and $(xIy \;\&\; yPz) \Rightarrow xPz$, for a semi-order it is only the combined force of *two* strict preferences that is guaranteed to filter through one indifference, i.e. $P^2I \Rightarrow P$, $IP^2 \Rightarrow P$, and $PIP \Rightarrow P$. Generalizing, let s-and-t-order only guarantee $P^sIP^t \Rightarrow P$. The Arrow impossibility result translates intact to this case in general, provided $\#X \geq s+t+2$. Since an s-and-t-order need not be quasi-transitive, it is first established that for a SDF satisfying Conditions U, I^2 and P, and yielding an s-and-t-order must lead to quasi-transitivity of social preference. Then the proof can follow a variant of the Group Contraction Lemma for s-and-t-order (in the same way as the proofs for semi-transitivity and interval orders), and then the final result, much like GPT* and GPT.

In the case of orderings, originally studied by Arrow, $s+t$ is 1, and it works for $\#X \geq 3$. In case of semi-orders, $s+t$ is 2, and it works for $\#X \geq 4$. In the general finite case, s and t can be any positive integer or zero, and it works if $\#X \geq s+t+2$. For an infinite X, the range of the SDF may be confined to the doubly infinite *union* of sets of all s-and-t-orders.

3.4. *Prefilters, filters and ultrafilters*

Let Ω be the class of decisive sets of individuals – a subset of the power set of H. Since this is considered pair by pair and since no distinction is made between decisiveness over one pair and that over another, the structure studied has features of independence and neutrality. Consider the following properties:

(1) $H \in \Omega,$

(2) $[G \in \Omega \;\&\; G \subseteq J] \Rightarrow J \in \Omega,$

(3) $[G_1, G_2, \ldots, G_k \in \Omega \text{ for } k \text{ finite}] \Rightarrow \bigcap_j G_j \neq \emptyset,$

(4) $[G, J \in \Omega] \Rightarrow G \cap J \in \Omega$,

(5) $[G \notin \Omega] \Rightarrow H - G \in \Omega$.[17]

Ω is a *prefilter* if and only if it satisfies (1), (2) and (3). It is a *filter* if and only if it, additionally, also satisfies (4).[18] It is an *ultrafilter* if and only if it satisfies all these conditions, i.e. (1) to (5).

Brown (1973, 1974, 1975a), Hansson (1972, 1976) and others have studied the properties of the class of decisive groups as a function of the regularity properties of individual and social preferences.[19] Consider the transformation function f: $\{R_i\} \to R$. Each R_i and each R are taken to be reflexive and complete and, in addition, they are required to satisfy some regularity condition of consistency (the same for R_i as for R). It has been shown that for $f(\cdot)$ satisfying Conditions U, P and I:

(I) acyclicity implies that Ω is a prefilter;

(II) quasi-transitivity implies that Ω is a filter;

(III) semi-order properties imply that Ω is an ultrafilter;

(IV) transitivity implies that Ω is an ultrafilter.[20]

These results can be used to derive the various dictatorship and veto results studied in the earlier subsections. In particular, in Arrow's case of full transitivity, Ω is an ultrafilter. If non-dictatorship were to hold, then each unit set of persons must be non-decisive, and thus by (5) in the community with n people, all sets with $n-1$ people would be decisive. But this class of decisive sets has an empty intersection, thereby contradicting (3), and also (4). The proof extends readily to semi-orders, given result (III).

In the case of acyclicity, Ω is a prefilter, and by virtue of (3), there is a group of persons – Brown calls it a "collegium" – such that every member of it belongs to every decisive set of persons.[21] With quasi-transitivity Ω is a filter, and by (4) the collegium would be decisive and thus define the oligarchy.

[17] These relations can be seen as features of "simple games"; see von Neumann and Morgenstern (1947), Guilbaud (1952), Monjardet (1967, 1979, 1983), Bloomfield (1971, 1976), Wilson (1971, 1972a), Nakamura (1975, 1978, 1979), Salles (1976), and Peleg (1978, 1983, 1984).

[18] In fact, given the other conditions, (3) will now be automatically fulfilled.

[19] See also Ferejohn (1977), Jain (1977a), and Monjardet (1979, 1983).

[20] See Brown (1973, 1974, 1975a, 1975b), Hansson (1976), Blau (1979), and Blair and Pollak (1979). See also Chichilnisky (1982b).

[21] Ferejohn (1977) points out that this does not in itself imply that every member of the collegium has a veto, since the social decisions induced by the prefilter may have to be supplemented by other procedures when some members of the collegium are indifferent. The gap can, however, be closed by further use of the neutrality property.

4. Non-binary social choice

4.1. Cycles and transitive closures

Arrow formulated the problem of social choice in relational terms with the social welfare function determining a binary relation of social preference – in fact, an ordering. It has often been taken for granted that Arrow's impossibility result relates crucially to having a binary choice function for the society, i.e. on the choice function satisfying the so-called Condorcet condition (1.2) presented in Section 1. It will be argued presently that this is not the case (see Section 4.2), but for the moment let us not dispute this and examine instead what types of escape routes can emerge if a non-binary formulation of social choice is chosen.

In a great many contributions in recent years a non-binary formulation of the social choice problem has been preferred [see, particularly, Hansson (1969a), Schwartz (1970, 1972), Fishburn (1971, 1973a, 1974a), Campbell (1972, 1976), Plott (1972, 1973, 1976), Bordes (1976)].[22] And it has been found that the non-binary formulation can cope better with at least some of the problems that arise with strict preference cycles. Whether this leads to an escape from Arrow's impossibility problem is, thus, an interesting issue.

Take the classic example of the "paradox of voting", with three persons having the following strict orders: (1) x, y, z, (2) y, z, x, and (3) z, x, y. The majority rule[23] leads to xPy, yPz, and zPx, a strict preference cycle. Faced with the choice over $\{x, y, z\}$, it is tempting to conclude that there is "nothing in it", and any state is as good as any other. This converts a set with strict preference cycle into an indifference class. This can be done through several alternative procedures using transitive closures, and here we concentrate on two, which we may call, respectively, Weak Closure Maximality and Strong Closure Maximality. While the two methods lead to the same result in this simple case, they differ in other choice situations, as we shall presently discuss. But before defining these procedures, it is useful to remind ourselves of the definitions of "transitive closure" and "maximality".

If B is a binary relation, then its transitive closure B^* is defined in the following way: xB^*y if and only if there is a sequence $z_1Bz_2, z_2Bz_3, \ldots, z_{k-1}Bz_k$, with $z_1 = x$, and $z_k = y$.[24] If B is a binary relation, then the maximal subset of a set S is the undominated subset of S with respect to the asymmetric factors B^A

[22] See also the formulation of the social choice problem as simple games in Monjardet (1967, 1979, 1983), Wilson (1971), Bloomfield (1971), Nakamura (1975, 1978), Peleg (1978, 1979b, 1983, 1984), Salles (1976), Salles and Wendell (1978), and others.

[23] The majority relation R is defined thus: xRy if and only if the number for whom xR_iy holds is at least as large as the number holding yR_ix. The strict majority relation P is the asymmetric factor of R.

[24] B^* is often called "the ancentral" of B [see Quine (1940) and Herzberger (1973)], a term that goes back to Whitehead and Russell and the concept at least to Frege.

of B, $xB^A y$ being defined as xBy and *not* yBx,

$$M(S, B) = [x | x \in S \ \& \ \text{not} \ \exists \ y \in S\colon \ yB^A x]. \tag{4.1}$$

The choice $C(S)$ from any subset S is identified in the following way under the two procedures, respectively (R^* is the transitive closure of R, and P^* that of P, the asymmetric factor of R):

Weak closure maximality: $C(S) = M(S, R^*)$.

Strong closure maximality: $C(S) = M(S, P^*)$.

To illustrate with the case of the paradox of voting, over $\{x, y, z\}$, the weak transitive closure R^* of the majority relation makes aR^*b hold for every pair $a, b \in \{x, y, z\}$, and also the strong transitive closure P^* makes aP^*b hold for every pair $a, b \in \{x, y, z\}$. Thus neither $aR^{*A}b$ nor $aP^{*A}b$ hold for any $a, b \in \{x, y, z\}$. Hence $C(\{x, y, z\}) = \{x, y, z\}$ for both the weak and strong closure methods in this special case.

But the two methods are not in general equivalent, consider the following binary relation with P being – as before – the asymmetric factor of R: xPy, yPz, zRx, and xRz. Clearly the transitive closure of R defines the following relations: aR^*b for all $a, b \in \{x, y, z\}$, while the transitive closure of P defines: xP^*y, xP^*z, and yP^*z. Hence $M(S, R^*) = \{x, y, z\}$, while $M(S, P^*) = \{x\}$. Indeed, in general, $M(S, P^*) \subseteq M(S, R^*)$.[25]

These closure methods have been directly used or indirectly entailed in several contributions to the resolution of the Arrow dilemma through non-binary choice procedures, in particular, Schwartz (1970, 1972), Bloomfield (1971), Campbell (1972, 1976, 1980), and Bordes (1976).[26] It can be seen that the Schwartz rule amounts to the uniform use of strong closure maximality for all social choices. In contrast, Bloomfield (1971), Campbell (1972, 1976), and Bordes (1976) use Weak Closure Maximality for social choice.[27]

In what sense do these solutions resolve the Arrow paradox? Instead of demanding a social welfare function it is possible to demand a "social choice function",[28] $g(S, \{R_i\})$ which specifies a non-empty subset $g(S, \{R_i\}) \subseteq S$, for every non-empty, finite $S \subseteq X$. This is essentially equivalent to making the value of the function $f(\{R_i\})$ a finitely complete choice function $C(\cdot)$ for the society, and not – as with social welfare functions or social decision functions – a social

[25] Deb (1977, proposition 1).

[26] In fact, Campbell used a version of the Weak Maximality Closure which is consistent with binary choice at the expense of weakening the independence condition. The trade-off between binary choice and independence is examined in Section 9.

[27] Deb has helpfully analysed the relations between these two closure methods.

[28] See Fishburn (1973a).

preference relation R,

$$C(\cdot)=f(\{R_i\}). \tag{4.2}$$

For such a function $f(\cdot)$, which we may call a functional collective choice rule, FCCR, the Arrow conditions can be readily translated in several distinct ways. The translation that has been typically used (the limitations of which will be discussed later), takes the form of restricting choices over pairs only.

Condition $\hat{U}$ (unrestricted domain)
The domain of $f(\cdot)$ includes all logically possible n-tuples of individual orderings of X.

Condition $\hat{P}$ (pair-choice Pareto principle)
For all $x, y \in X$, $(\forall i\colon\ xP_i y) \Rightarrow \{x\} = C(\{x, y\})$.

Condition $\hat{D}$ (pair-choice non-dictatorship)
There is no individual i such that for all n-tuples in the domain of $f(\cdot)$, for each ordered pair $x, y \in X$, $xP_i y \Rightarrow \{x\} = C(\{x, y\})$.

The non-dictatorship condition can, in fact, be strengthened to a non-vetoer condition, and further extended to a condition of full "anonymity".

Condition $\hat{A}$ (anonymity)
If $\{R_i\}$ is a permutation of $\{R_i'\}$, then $f(\{R_i\}) = f(\{R_i'\})$.

These conditions can now be combined with Arrow's independence of irrelevant alternatives (Condition I), which was already defined in choice-functional terms. To tighten up the real possibility result further, the conditions of "positive responsiveness" (PR) and "neutrality, independence cum monotonicity" (NIM) can be similarly translated from relational to choice-functional terms, PR and $\widehat{\text{NIM}}$, constraining choices over pairs corresponding to the binary relations.

Choice-functional positive possibility theorem
For $\#H \geq 2$, there is a FCCR satisfying Conditions Û, I, P̂, D̂, Â, and $\widehat{\text{NIM}}$.[29]

The theorem is established by considering a particular example, e.g. the procedure generated by Weak Closure Maximality *or* by Strong Closure Maximality, applied to the majority rule relation R. The same operations can also be applied to other Pareto-inclusive, non-dictatorial, non-acyclic relations, of which there are plenty.

[29] The case of $\#H = 1$ is not covered for the simple reason that in a one-person community the Pareto principle conflicts with non-dictatorship, which – I hope – would give food for thought to this lonely individual.

The satisfaction of these conditions are obvious enough, with the possible exception of Condition I. In order to satisfy that condition, it is important to define the transitive closure R^* of R over the subset S from which the choice is being made, i.e. in the definition of R^* given above, all the elements $z_1, z_2, \ldots, z_k$ must belong to S. To avoid ambiguity R^* used for the choice over S, derived from preferences over S, may be denoted R^*_S, and the Weak and Strong Closure Maximality procedures can be clearly seen as consisting in identifying $M(S, R^*_S)$ and $M(S, P^*_S)$, respectively. It is obvious that Condition I will be fulfilled. But since R^*_S and P^*_S will vary with S even as far as the restrictions over some given $T \subseteq S$ is concerned, the choice function covering different subsets will not in general be representable by one binary relation. For example, in the case of the "paradox of voting", either procedure applied to the majority relation will identify the following choices: $C(\{x, y, z\}) = \{x, y, z\}$, $C(\{x, y\}) = \{x\}$, $C(\{y, z\}) = \{y\}$, $C(\{z, x\}) = \{z\}$. This choice function is, of course, defiantly non-binary.[30]

Since the above theorem is not too challenging the conditions may be tightened by demanding other conditions as well, and all the advocates of this class of solutions have offered other desirable conditions that the chosen rules will satisfy. Whether these good qualities are adequate for what may be called a satisfactory resolution of the Arrow problem will be discussed in Section 4.3, but before that an interpretative analysis of the nature of the Arrow problem from the non-binary perspective should be useful.

4.2. *The unimportance of binariness in Arrow's impossibility*

Consider the distinction between (i) using the social aggregation procedure to yield a binary relation of social preference, and (ii) using that binary relation of social preference to determine the choice function. Arrow (1950, 1951, 1963) did, in fact, endorse both, but while (ii) does play a crucial motivational part in the Arrow exercise (since he identified the meaning of the social preference relation in terms of the choice function), it has no role whatsoever in the genesis of the impossibility result. Indeed, the choice-theoretic interpretation of the social preference relation remains, strictly speaking, a separate issue that need not be brought into the impossibility theorem at all once the independence condition has been redefined to pairwise relational independence (Condition I^2), leading to GPT* (see Section 2.1).

It is no less important to recognize that choice-theoretic interpretations of the Arrow result can themselves take several *different* forms. Perhaps the simplest is

[30] On the factorization of necessary and sufficient conditions for binariness – or "normality" or "basic binariness" – of a choice function, see Sen (1971) and Herzberger (1973).

to give the social preference relation R the interpretation of being the "base relation" $\bar{R}_C$ of the choice function $C(\cdot)$, defined for choices over pairs only,

$$\forall x, y \in X: x\bar{R}_C y \quad \text{if and only if } x \in C(\{x, y\}). \tag{4.3}$$

Using this interpretation of R has the effect of not telling us anything whatsoever about how choices should be made from sets larger than pairs. In the choice-functional formulation of $f(\cdot)$ given in (4.2), all we need do is to replace the requirement of $C(\cdot)$ being finitely complete by the requirement that $C(\cdot)$ be complete over all pairs, and that for all $x, y, z \in X$, if $x \in C(\{x, y\})$ and $y \in C(\{y, z\})$, then $x \in C(\{x, z\})$. Such a FCCR cannot satisfy Conditions Û, P̂, D̂, and the condition of independence I weakened to I_C^2 to apply to choice over pairs only. (The restriction of R over a pair $\{x, y\}$ is denoted $R|^{\{x,y\}}$.)

Condition I_C^2 (pairwise choice independence)
For any pair of social states $x, y \in X$, $C(\{x, y\}) = f^{\{x,y\}}(\{R_i|^{\{x,y\}}\})$.

Base-relational general possibility theorem
If H is finite and $\#X \geq 3$, then there is no FCCR satisfying Conditions Û, I_C^2, P̂ and D̂, with $\bar{R}_C$ transitive.

The important point here is not the assertion that this theorem is valid, which it is, but that this is indeed Arrow's own theorem with an interpretational twist. The same proof suffices.

Recently, binariness of choice has been subjected to severe criticism, and Fishburn (1971), among others, has forcefully argued that "social choice from among more than two feasible alternatives should not be based on social choice from two alternative subsets" (p. 133). This contrasts sharply with Arrow's (1951) view that "one of the consequences of the assumptions of rational choice is that the choice in any environment can be determined by a knowledge of the choices in two-element environments" (p. 16). This is, of course, a question of much interest on its own, but Arrow's impossibility theorem does not depend on the answer to this question, and can be established without making any statement whatsoever on how choices over sets larger than pairs be made.

This recognition raises one *immediate* question: how can such procedures as the use of Weak Closure Maximality or Strong Closure Maximality provide any escape from the Arrow impossibility since these procedures doctor choices only over subsets *larger* than pairs? If the Arrow result is about choice over pairs and the escape routes under examination leave that completely untouched, then how can escape conceivably take place? In fact, the escape routes must be seen not as methods of avoiding Arrow's impossibility problem with its concentration on choices over pairs, but as methods of softening its *implications* for choices over sets larger than pairs. The distinction can be brought out by considering the contrast between two issues raised by pairwise inconsistency of choice. Suppose x is chosen in the choice over the pair $\{x, y\}$, and y in the choice over the pair $\{y, z\}$, but in choosing over the pair $\{x, z\}$, x is rejected and z chosen. This can be regarded as unsatisfactory for two rather different reasons. First, the choices over the pairs themselves may appear to be contrary, even uncanny. (Cf. "between Bermuda and Honolulu, I will choose Bermuda; between Honolulu and Pago Pago, it must be Honolulu; and between Pago Pago and Bermuda, I think Pago Pago".) Second, it augurs badly for the choice over the triple $\{x, y, z\}$.

The escape routes under examination are concerned *exclusively* with the second issue. This is no mean task, and thus the methods used deserve to be examined seriously as choice procedures for larger sets; this will be done in the next subsection. But this leaves the first issue *completely* untouched.

Before turning to questions of choice over more than two-alternative sets, a possible source of misunderstanding should be cleared up. In an important paper, Blau (1971) has shown that in Arrow's framework for social choice, what he calls "binary" independence is exactly equivalent to "m-ary" independence for any $m < \#X$, and that all these independence conditions are equivalent to Arrow's demanding condition.[31] Doesn't this indicate, it might be asked, that to focus on pairwise choice independence $\mathrm{I}^2_{\mathrm{C}}$ is equivalent to focussing on Arrow's own Condition I dealing also with choices over larger sets? Does it, then, make any difference at all whether we look at choices over pairs only, or over larger sets of social states?

To sort out this ambiguity it is useful to distinguish between what we may call "m-ary relational independence" and "m-ary choice independence".

Condition I^m (m-ary relational independence)
For any $S \subseteq X$ such that $\#S = m$, for $f(\cdot)$ given by (1.1), $R|^S = f^S(\{R_i|^S\})$.

Condition I^m_C (m-ary choice independence)
For any $S \subseteq X$ such that $\#S = m$, for $f(\cdot)$ given by (4.2), $C(S) = f^S(\{R_i|^S\})$.

Blau (1971) simply observed that for Arrow's social welfare function, i.e. given (1.1), what he called "binary" independence (i.e. our pairwise relational independence I^2) implies "m-ary" relational independence, and proceeded to prove the converse – a deep result – that "m-ary" relational independence for any $m < \#X$ also implies "binary" (i.e. pairwise relational) independence. This is a relational theorem – important on its own – but establishes nothing whatever about the correspondence between pairwise *choice* independence and the class of m-ary *choice* independence, *unless* choice is defined in binary terms, e.g. in the form of the Condorcet condition (1.2). So once the binary property of choice is eschewed,

[31] See also Murakami (1968), Fishburn (1974a), Binmore (1975), and d'Aspremont and Gevers (1977).

i.e. (1.2) denied and (1.1) replaced by (4.2), the equivalence result of Blau becomes unavailable. Thus, I_C^2 can be asserted without commitment to I_C^m for $m > 2$, and vice versa. Base-relational GPT stands as a theorem about the impossibility of consistent social choice over pairs *without* affirming or denying that social choice be binary.

4.3. Consistency of social choice

A FCCR generates a choice function $C(\cdot)$. To be able to choose from any non-empty finite subset S of X, $C(\cdot)$ is taken to be finitely complete. In addition, conditions of "consistency" of choice would have to be considered. Consistency conditions of choice used in the literature can be classified *or* factorized into requirements of two essentially different types, viz. *contraction* consistency and *expansion* consistency [Sen (1970a, 1977a)]. The former deals with requirements of the kind that insist that something chosen from a set must – under certain conditions to be specified – continue to be chosen when the menu offered is *contracted*. The latter, on the other hand, insists that something chosen from a set must – under circumstances to be specified – continue to be chosen when the menu offered is *expanded*.

The most used contraction consistency condition is called Property α (also called the "Chernoff condition"), while the natural complement of that condition is a requirement of expansion consistency which is called Property γ [Sen (1971)]. The set of definitions that follows are specified for all $x, y \in X$ and all $S, T \subseteq X$.

Property α (standard contraction consistency)
$[x \in C(S)\ \&\ x \in T \subseteq S] \Rightarrow x \in C(T)$.

Property γ (standard expansion consistency)
$[x \in C(S_j)$ for all S_j in any class of subsets of $X] \Rightarrow x \in C(\cup_j S_j)$.

The two together make the choice function essentially binary in the sense that its informational content can be exactly captured by a binary relation R defined on X. The "Condorcet condition" defined in Section 1.3 had specified how a choice function may be constructed from a binary relation R, and this is restated below with $\hat{C}(S, R)$ standing for the choice set of S as constructed from the relation R,

$$\hat{C}(S, R) = [x | x \in S\ \&\ \forall y \in S:\ xRy]. \tag{1.2'}$$

Consider now the opposite problem of constructing a binary relation of preference from a choice function.[32] There are at least two distinct natural claimants to this role, viz. the "revealed preference relation" R_C given by choices over all subsets of X containing the pair that is being ranked in any particular case, and the "base relation" $\bar{R}_C$ given by the choice exactly over that pair, already defined in (4.3).

Revealed preference relation
$xR_C y$ if and only if $\exists S$: $[x \in C(S)$ & $y \in S]$.

Base relation
$x\bar{R}_C y$ if and only if $x \in C(\{x, y\})$.

It is obvious that $x\bar{R}_C y \Rightarrow xR_C y$, but in general not vice versa, and that Property α does imply the converse, i.e. guarantees $R_C = \bar{R}_C$.

A choice function $C(\cdot)$ is "binary" (or "normal", or "rationalizable") if and only if the revealed preference relation R_C generated by it is adequate to generate back the choice function $C(\cdot)$ itself [using (1.2′)]. $C(\cdot)$ is "basic binary" if and only if the base relation $\bar{R}_C$ generated by it can generate back $C(\cdot)$ through (1.2′).

Binariness of a choice function
$C(S) = \hat{C}(S, R_C)$ for all $S \subseteq X$.

Basic binariness of a choice function
$C(S) = \hat{C}(S, \bar{R}_C)$ for all $S \subseteq X$.

Binariness lemma
A finitely complete choice function is binary if and only if it is basic binary, and also, if and only if it satisfies Properties α and γ.[33]

There are some alternative conditions of expansion consistency. A few are considered here.

Property β
$[x, y \in C(S)$ & $S \subseteq T] \Rightarrow [y \in C(T) \Rightarrow x \in C(T)]$.

[32] The word "preference" has some ambiguity in the individual context, since it can have *at least* two primitive meanings, viz. the reflection of choice behaviour and the reflection of well-being (or utility). To identify the two would provide a very limited model of behaviour [see Sen (1977c) and Schick (1978)]. A similar problem may arise for the concept of "social preference" as well, since it can be defined either in terms of characterisation of social *choice* or the concept of social *welfare*. Here the first meaning is taken as the primitive.

[33] See Sen (1971) and Herzberger (1973). As Kanger (1975) points out, binariness in this sense is a very limited interpretation of "choice based on preference", and more generally the chosen elements from a set A can be made to depend on a binary relation P_V that depends on the specification of a "background" set V. Binariness, as defined here, corresponds to taking $V = A$. Kanger (1975) provides a rich analysis of the more general case of "choice based on preference".

If both x and y are chosen in S, a subset of T, then one of them (say, y) can't be chosen in T without the other (i.e. x) being also chosen. This condition can be *strengthened* to Property β^+ by relaxing the antecedent in such a way that x being chosen in S in the presence of y (whether or not y is chosen in S) should entail the same consequent, i.e. y mustn't be chosen in T without x being also chosen.

Property β^+
$[x \in C(S)$ & $y \in S \subseteq T] \Rightarrow [y \in C(T) \Rightarrow x \in C(T)]$.

And Property β can be *weakened* through replacing the consequent by demanding only that y be not chosen *exclusively* in T, whether or not x is among the chosen elements of T.

Property δ
$[x, y \in C(S)$ & $S \subseteq T] \Rightarrow \{y\} \neq C(T)$.

Finally, a weakening of Property α to Weak α requires only that a state x chosen from any set S and belonging to a subset T of S must be chosen from T *if* it is not rejected in the choice over any other subset of S.

Property weak α
$[x \in T \subseteq S$ & for all $Y \subseteq S$ such that $Y \neq T$: $x \in C(Y)] \Rightarrow x \in C(T)$.

The following lemmas, among others, are useful in establishing possibility results for social choice.

Sundry choice-functional lemmas
For any finitely complete choice function $C(\cdot)$:

(1) $[\alpha$ & $\beta] \Leftrightarrow [C(\cdot)$ is binary and $R_C = \bar{R}_C$ is transitive$]$;

(2) $[\alpha$ & γ & $\delta] \Rightarrow [C(\cdot)$ is binary and $R_C = \bar{R}_C$ is quasi-transitive$]$;

(3) $\beta^+ \Rightarrow [\beta$ & γ & $\delta]$;

(4) $\beta^+ \Leftrightarrow R_C$ is transitive;

(5) [Weak α & $\beta] \Rightarrow \bar{R}_C$ is transitive;

(6) [Weak α & $\delta] \Rightarrow \bar{R}_C$ is quasi-transitive;

(7) $\alpha \Rightarrow \bar{R}_C$ is acyclic, i.e. there is no strict $\bar{P}_C$-cycle;

(8) Weak $\alpha \Rightarrow \bar{R}_C$ is triple-acyclic, i.e. there is no strict $\bar{P}_C$-cycle over any triple.

Since (1)–(4) and (7) have been proved elsewhere[34] and have been widely used, I shall concentrate here on establishing (5), (6) and (8). First, consider (5) and take $x\bar{R}_C y$ and $y\bar{R}_C z$, but the contrary supposition, i.e. *not* $x\bar{R}_C z$. Since $C(\cdot)$ is finitely complete, $\{z\} = C(\{x, z\})$. If $z \in C(\{x, y, z\})$, then by Weak α, clearly $\{y, z\} = C(\{y, z\})$, and thus by β, $y \in C(\{x, y, z\})$. Then by Weak α, $\{x, y\} = C(\{x, y\})$, and thus by β, $x \in C(\{x, y, z\})$. So by Weak α, $x \in C(\{x, z\})$, which is a contradiction. So $z \notin C(\{x, y, z\})$, and therefore, either x or y or both are in $C(\{x, y, z\})$. These possibilities in turn lead to the same contradiction with $\{z\} = C(\{x, z\})$, as shown above. Hence $x\bar{R}_C z$.

Taking up (6) next, consider $x\bar{P}_C y$ & $y\bar{P}_C z$. If, contrary to the quasi-transitivity of $\bar{R}_C$, *not* $x\bar{P}_C z$, then $z\bar{R}_C x$. If now $z \in C(\{x, y, z\})$, then by Weak α, $z \in C(\{y, z\})$, which is a contradiction of $y\bar{P}_C z$. Hence $z \notin C(\{x, y, z\})$. The hypothesis that $y \in C(\{x, y, z\})$ would imply, by Weak α, $y \in C(\{x, y\})$, which is a contradiction of $x\bar{P}_C y$. So by the finite completeness of $C(\cdot)$, we must have $\{x\} = C(\{x, y, z\})$, and by Weak α, also $x \in C(\{x, z\})$. So $x\bar{I}_C z$, and thus by Property δ, $z \in C(\{x, y, z\})$ or $y \in C(\{x, y, z\})$, which has already been proved impossible. Thus $x\bar{P}_C z$.

Finally (8). Suppose not. Consider a cycle over the triple $\{x, y, z\}$: $x\bar{P}_C y$, $y\bar{P}_C z$, and $z\bar{P}_C x$. Note that x cannot belong to $C(\{x, y, z\})$, since this would contradict $x \notin C(\{z, x\})$ given Weak α and $x \in C(\{x, y\})$. For similar reasons, nor can y or z. Since $C(\cdot)$ is finitely complete, the contrary supposition is, therefore, unsustainable, and hence Weak α does imply triple acyclicity of the base relation.

These lemmas permit us to translate the theorems obtained earlier into corresponding choice-functional results.

Choice-functional general possibility theorem (CFGPT)
If H is finite and $\#X \geq 3$, then there is no FCCR satisfying Conditions $\hat{\text{U}}$, I^2_{C}, $\hat{\text{P}}$ and $\hat{\text{D}}$, and generating choice functions satisfying Weak α and β.

In view of lemma (6) above, this reduces to the "Base-Relational General Possibility Theorem" discussed in Section 4.2. An immediate corollary is that no FCCR satisfies these conditions and generates choice functions fulfilling α and β.

[34] For proofs of (1) and (2), see Sen (1971); of (3) and (4), Bordes (1976); and of (7), Blair, Bordes, Kelly and Suzumura (1976). For related results, see Arrow (1959), Chipman, Hurwicz, Richter and Sonnenschein (1971), Hansson (1969a), Sen (1969, 1970a, 1971, 1977a), Schwartz (1970, 1972, 1974, 1976), Batra and Pattanaik (1972b), Herzberger (1973), Fishburn (1973a, 1974b, 1974c), Aizerman, Zavalishin and Piatnitsky (1976), Blair, Bordes, Kelly and Suzumura (1976), Bordes (1976, 1979), Parks (1976b), Richelson (1977, 1978), Suzumura (1976a, 1983a), Ferejohn and Grether (1977a, 1977b), Kelly (1978), Sertel and Van der Bellen (1979, 1980), Aizerman and Malishevski (1980), Grether and Plott (1982), and Matsumoto (1982), among other contributions.

While this *corollary* is very often taken to be the choice-functional "translation" of Arrow's result, it follows from the discussion in the last subsection that this interpretation is unduly restrictive, and CFGPT is a finer version.

If attention is shifted from the base relation $\overline{R}_C$ to the revealed preference relation R_C, then the picture is much more "positive".

Choice-functional positive possibility theorem with transitive social preference
For $\#H \geq 2$, there is a FCCR satisfying Conditions $\hat{\mathrm{U}}$, I, $\hat{\mathrm{P}}$, $\hat{\mathrm{D}}$, $\hat{\mathrm{A}}$ and $\widehat{\mathrm{NIM}}$, generating choice functions that meet expansion consistency properties δ, γ, β and β^+, and inducing transitive revealed preference relations R_C.

An example used for the "Choice-Functional Positive Possibility Theorem" (presented in Section 4.1) will do for this also. Bordes (1976) has demonstrated that his procedure of basing choice on Weak Closure Maximality applied to the majority relation satisfies β^+, and thus by lemma (4) above, must yield a transitive R_C, and by lemma (3) above, must fulfill Properties δ, γ and β also.

But even weak doses of contraction consistency creates problems when added to some expansion consistency, and also when used on its own.

Choice-functional oligarchy theorem
If H is finite and $\#X \geq 3$, then any FCCR satisfying Conditions $\hat{\mathrm{U}}$, I_C^2 and $\hat{\mathrm{P}}$, and generating choice functions satisfying Weak α and δ must be oligarchic.

This follows directly from lemma (6) above and Gibbard's oligarchy theorem (presented in Section 3.1, called "Quasi-transitive Oligarchy Theorem"), given the re-interpretation outlined in the last subsection.

Choice-functional positive-responsive dictatorship theorem
If H is finite and $\#X \geq 3$, then there is no FCCR satisfying Conditions $\hat{\mathrm{U}}$, I_C^2, $\hat{\mathrm{P}}$, $\hat{\mathrm{D}}$ and $\widehat{\mathrm{PR}}$, and yielding choice functions that fulfill Weak α and δ.

This follows from Mas-Colell and Sonnenschein's (1972) "Quasi-transitive Positive-Responsive Dictatorship Theorem" (presented in Section 3.1), in view of lemma (6) above, given the base relation interpretation. It is more general than the Choice-Functional General Possibility Theorem (CFGPT) established above in using a weaker condition of expansion consistency of social choice (δ rather than β), but is less general in having to use the additional requirement of positive responsiveness $\widehat{\mathrm{PR}}$.

Similarly, using the "Acyclic Positive-Responsive Vetoer Theorem" [Mas-Colell and Sonnenschein (1972)], its "triple-acyclic" extension, and the "Acyclic Neutral Monotonic Vetoer Theorem" [Blau and Deb (1977)], the following results can be immediately derived from lemmas (6), (7) and (8) above, through the "base relation" interpretation of the relational framework (outlined in the last subsection).[35]

Choice-functional vetoer theorems

For a finite H, a FCCR satisfying Conditions $\hat{\mathrm{U}}$, I^2_C and $\hat{\mathrm{P}}$, must have a vetoer, if:

either $\#X \geq 4$, the FCCR must satisfy $\widehat{\mathrm{PR}}$, and the choice function must fulfill Weak α;

or $\#X > \#H$, the FCCR must satisfy $\widehat{\mathrm{NIM}}$, and the choice function must fulfill α.

Similar translations can be made for the results on "prefilters, filters and ultrafilters" obtained by Brown (1973, 1974, 1975a), Hansson (1972, 1976), Monjardet (1979, 1983), and others.

These results bring out a contrast between the respective effects of contraction and expansion consistency. The former raises rather serious problems, even when used on its own without any expansion consistency requirement, whereas the latter seems typically satisfiable unless coupled with some contraction consistency. Property β^+ is a strong condition of expansion consistency (subsuming the other conditions β, γ and δ, all operating in that direction), but it causes in the present context no problem at all. Property α, however, has a wrecking impact, and so has even Weak α.

One reason for the contrast lies in the way the regularity conditions are defined in this framework. The Pareto principle, the non-dictatorship condition, positive responsiveness, neutrality, independence cum monotonicity, and veto conditions, are all defined in terms of pairwise relations, which translate in a natural way into conditions on choice over pairs. Inconsistencies thus generated over pairs rule out consistent choice over larger sets when contraction consistency is insisted on. But expansion consistency does not carry over these inconsistencies to larger sets, since it can be met by arbitrarily enlarging the choice set.[36] Thus the contrast is, to a great extent, presentational.

If this interpretation of the contrast is accepted, it is natural to expect that the impossibility results can be regenerated in a non-binary framework *without* being dependent on contraction consistency if the regularity conditions are defined not

[35] See Sen (1977a). These theorems can, of course, also be derived independently of the earlier relational results. See Blair, Bordes, Kelly and Suzumura (1976) and Kelly (1978). See also Ferejohn and Grether (1977b), Bordes (1979), Suzumura (1983a), Grether and Plott (1982), and Matsumoto (1982).

[36] It is worth noting in this context that "contraction consistency" conditions are various requirements of retaining the *inclusion* of elements in the choice set as the menu is contracted, and are equivalent to the corresponding conditions on retaining the exclusion of elements from the choice set as the menu is *expanded*. Similarly, expansion consistency conditions are requirements of the *inclusion* of chosen elements as the menu is *expanded* and of the *exclusion* of unchosen elements as the menu is *contracted* [see Sen (1977a, pp. 65–68)].

for choices over pairs but for choices over subsets of any size. For example, the weak Pareto principle can be redefined in the following non-pair-choice form.

Condition $\hat{\hat{P}}$ (general-choice Pareto principle)
For all $x, y \in X$, $[\forall i: xP_i y] \Rightarrow [\forall S \subseteq X: x \in S \Rightarrow y \notin C(S)]$.

It is easily checked that this Condition $\hat{\hat{\mathrm{P}}}$ conflicts directly with such rules as Weak Closure Maximality and Strong Closure Maximality based on the majority relation, through which escape from preference cycles have been often sought (see Section 4.1). Consider the following set of strict preference orderings of three persons over four states [suggested by Ferejohn and Grether (1977a)]: (1) x, y, z, w, (2) y, z, w, x, (3) z, w, x, y. This leads to the majority relation strict cycle, xPy, yPz, zPw, wPx. Either of the two "closure" methods would now suggest $\{x, y, z, w\} = C(\{x, y, z, w\})$. But w is Pareto inferior to z, and its choice thus violates Condition $\hat{\hat{\mathrm{P}}}$.[37]

Through this procedure of defining the regularity conditions that deal directly with choices other than pairs, inconsistencies can be precipitated without relying on contraction consistency. While only a few such results have been formally derived,[38] it is quite clear that pair-choice inconsistencies can quite generally be translated in a natural way into choices over larger sets of social states.

It was argued in the last subsection that binariness of choice was unimportant for the impossibility results, and that it was sufficient to concentrate on choices over pairs only (without saying anything about choices over sets larger than pairs). The last analysis shows that while that is indeed the case, it is also possible to obtain – alternatively – the impossibility results by redefining all the conditions in terms of choices over sets larger than pairs without saying anything about choices over pairs as such. This can be done in a variety of different ways [see Fishburn (1974a), Matsumoto (1982), Grether and Plott (1982), and Sen (1982)]. The impossibility results following from Arrow's work are robust enough to surface in widely different formulations of the problem of consistency of social choice.

4.4. *Path independence*

Among the various justifications considered by Arrow for the condition of transitivity of social preference is the argument that it "will insure" the "indepen-

[37]As Suzumura (1983a) has noted, this particular case can be effectively dealt with by using the transitive closure over the *Pareto optimal subset* of the set of states, but there are other difficulties that are less easy to deal with in the "general-choice" interpretation of the Arrow conditions.

[38]See Hansson (1969a, 1973), Sen (1970a, section 6.3), Batra and Pattanaik (1972b), Fishburn (1973a, 1974a), Ferejohn and Grether (1974, 1977a, 1977b), Binmore (1975), Bandopadhyay (1983), and Suzumura (1983a).

dence of the final choice from the path to it" [Arrow (1963, p. 120)]. It is clear, however, that binariness of social choice with a transitive social preference relation is an overly strong condition for path independence, as Plott (1973) has noted. He has provided a characterization of path independence, and there have been a number of important contributions on possibility results using some variant or other of path independence as the condition of consistency of social choice.

The commonest characterisation of path independence is the following, defined for any class of subsets $S_j \subseteq X$:

Property PI (path independence)
$C(\cup_j S_j) = C(\cup_j C(S_j))$.

This implies that no matter how a set is split up for "divide and choose", the final outcome must be the same.

Weaker conditions of path independence have also been studied by Parks (1971), Plott (1973), Schwartz (1974, 1976), Suzumura (1976a), and Ferejohn and Grether (1977a, 1977b), among others. Here two complementary conditions are noted, which together make up Property *PI*.

*Property PI** (upper path independence)
$C(\cup_j S_j) \subseteq C(\cup_j C(S_j))$.

*Property *PI* (lower path independence)
$C(\cup_j S_j) \supseteq C(\cup_j C(S_j))$.

Obviously, $PI \Leftrightarrow (PI^* \;\&\; {}^*PI)$.

Many interesting results have been derived using some version or other of path independence.[39] A few of these are noted here, mainly aimed at their use in possibility results of the Arrow type.

Path independence lemma
For any finitely complete choice function $C(\cdot)$:

(1) $PI^* \Leftrightarrow \alpha \Rightarrow [\bar{R}_C = R_C \text{ acyclic}]$;

(2) $PI \Rightarrow [\bar{R}_C = R_C \text{ is quasi-transitive}]$;

(3) [Binariness of choice function & R_C quasi-transitive] $\Leftrightarrow [PI \;\&\; \gamma]$;

[39]See Plott (1973), Parks (1971, 1976b), Schwartz (1974), Blair (1975), Blair, Bordes, Kelly and Suzumura (1976), Bordes (1976, 1979), Suzumura (1976b, 1983a), Ferejohn and Grether (1977a, 1977b), Kelly (1978), Schofield (1978), Kalai and Megiddo (1980), Machina and Parks (1981), and Matsumoto (1982).

(4) $^*PI$ and β are independent of each other;

(5) $\beta^+ \Rightarrow {}^*PI$.

The following results follow from these lemmas and from results presented in the last subsection.

Path-independent positive possibility theorem
For $\#H \geq 2$, there is a FCCR satisfying Conditions $\hat{\text{U}}$, I, $\hat{\text{P}}$ and $\hat{\text{D}}$, and also generating path-independent choice functions.

Path-independent dictatorship theorem
For a finite H and $\#X \geq 3$, there is no FCCR satisfying Conditions $\hat{\text{U}}$, I^2_C, $\hat{\text{P}}$, $\hat{\text{D}}$ and $\widehat{\text{PR}}$, and generating path-independent choice functions.

Path-independent oligarchy theorem
For a finite H and $\#X \geq 3$, there is no non-oligarchic FCCR satisfying Conditions $\hat{\text{U}}$, I^2_C and $\hat{\text{P}}$, and generating path-independent choice functions.

Upper path-independent vetoer theorems
For a finite H, a FCCR satisfying Conditions $\hat{\text{U}}$, I^2_C and $\hat{\text{P}}$, and generating upper path-independent choice functions (fulfilling PI^*), must have a vetoer, if:
either $\#H \geq 4$, $\#X \geq 3$, and the FCCR must satisfy $\widehat{\text{PR}}$;
or $\#X \geq \#H$ and the FCCR must satisfy $\widehat{\text{NIM}}$.

Lower path-independent possibility theorem with transitive social preference
For $\#H \geq 2$, there is a FCCR satisfying Conditions $\hat{\text{U}}$, I, $\hat{\text{P}}$, $\hat{\text{D}}$, $\hat{\text{A}}$ and $\widehat{\text{NIM}}$, and generating lower path-independent choice functions (fulfilling $^*PI$), and inducing transitive revealed preference relation R_C.

There is some ray of hope in the last, which is an extension of theorems noted by Bordes (1976) and Ferejohn and Grether (1977a), and is established by the use of the "closure" methods. Choices based on Weak Closure Maximality applied to the majority relation satisfy β^+, and thus reveal a transitive social preference (without the choice function being binary), and the choices over pairs permit the fulfillment of such pair-choice conditions as $\hat{\text{P}}$, $\hat{\text{D}}$ and $\widehat{\text{NIM}}$. Ferejohn and Grether (1977a) have argued forcefully in favour of the view that lower path independence, which they call "Weak Path Independence", is the proper reflection of Arrow's (1963, p. 120) justification for path independence. They note, however, that such choice procedures can go against the general-choice version of the Pareto principle (called $\hat{\hat{\text{P}}}$ in the last subsection). Indeed, the possibility of generating impossibility results by redefining the regularity conditions for choices over sets larger than pairs remains real (as discussed in the last subsection).

Finally, upper path independence PI^*, which is equivalent to the contraction consistency condition α [see Path Independence Lemma (1)[40]], will immediately translate the inconsistencies of choices over pairs into inconsistencies for choices over larger sets. This range of issues was extensively discussed in the last subsection.

5. Efficiency and fairness

5.1. Good quality?

While the exercises outlined so far deal with the problem of social choice in rather comprehensive terms, there are some approaches that aim to do no more than separate out a subset of the set X of social states for special commendation. The specified subset is seen as good, but there is no claim that they represent the "best" alternatives, all *equally* choosable. There is no attempt to give an answer to the overall problem of social choice, and the exercise is quite different from the specification of a social preference over X (as with social welfare functions or social decision functions), as well as from the identification of a choice function specifying in each non-empty finite subset S of X, the *optimal* subset $C(S)$ of S (as with social choice functions or FCCRs generating finitely complete choice functions). This general approach, which we may call the "good quality" approach, has been extensively used in the context of such concepts as Pareto optimality, the core, equitability and fairness.

Is this, in any sense, a "superior" approach? In presenting his analysis of "fairness" based on "equity" and "efficiency", Varian (1974) makes the following critical comment on "standard" social choice theory.

> "Social decision theory views the specification of the social welfare function as a problem in aggregating individual preferences. Its chief results are of the form 'There are no reasonable ways to aggregate individual preferences.' ...Social decision theory asks for too much out of the process in that it asks for an entire *ordering* of the various social states (allocations in this case). The original question asked only for a 'good' allocation; there was no requirement to rank all allocations. The fairness criterion in fact limits itself to answering the original question. It is limited in that it gives no indication

[40] This important result was first established by Parks (1971) in an unpublished paper, and analysed and further studied by Plott (1973). See also Blair (1975).

of the merits of two nonfair allocations, but by restricting itself in this way it allows for a reasonable solution to the original problem." (pp. 64–65)[41]

While Varian addresses his criticisms to the search for a social *ordering* – and thus to social welfare functions only – the reasoning applies equally well to social decision functions and social choice functions, for they too seek a complete solution of the problem of social decision (or of social choice).

Efficiency provides a classic example of a "Good Quality" approach. Some binary relation of dominance D is used, e.g. in the case of "technical efficiency" the vector dominance of output (with inputs taken as negative outputs), and the maximal set with respect to that dominance relation is declared as "efficient". Taking the dominance relation D in the "weak" (reflexive) form, let D^{A} be its asymmetric factor,

$$E(S) = [x | x \in S \text{ \& for } no\; y \in S\text{: } y D^{\mathrm{A}} x]. \tag{5.1}$$

Since the dominance relation D is a quasi-ordering (weak *partial* ordering), the maximal set $E(S)$ is not to be interpreted as a choice set of "best" elements. A complete social ranking R of which the dominance relation D is a sub-relation can order the efficient points in any way whatsoever.[42]

While "technical efficiency" is a common concept in the resource allocation literature, in welfare economics the more common notion of efficiency is that of so-called Pareto optimality, where the dominance relation D is that of weak dominance of utility ranking, or weak unanimity of individual preference: xDy if and only if $\forall i$: $xR_i y$.[43] It is sometimes referred to – more sensibly – also as "Pareto efficiency" or "economic efficiency". Often – merciless to the reader – also as "efficiency".

The notion of "the core", which is thoroughly studied elsewhere in this Handbook, extends the approach of Pareto efficiency in one particular direction, viz. that of *equilibrium* (of all groups as well as of individuals).[44] A different extension, with a clearer ethical relevance, relates to supplementing Pareto efficiency, which pays no attention to the equity of distributions, with explicit

[41]Another important line of criticism of the formulation of the choice problem in social choice theory deals with the way the "menu" is specified. Braybrooke (1978) emphasizes the limitation of a fixed menu which gives no room for "issue processing and transformation of issues". The question relates also to the long-standing debate on the relevance of log-rolling in the formulation of social choice problems, the importance of which has been emphasized by Buchanan and Tullock (1962) and extensively discussed in the literature on political processes. For a defence of the formulation used in social choice theory, see Arrow (1963, pp. 108–109) and Wilson (1969, 1971).

[42]See Arrow (1963, pp. 64–68), extending a result of Szpilrajn (1930).

[43]For a powerful use of Pareto optimality as the basis for public decisions, see Buchanan and Tullock (1962).

[44]For motivational discussions, see Hahn (1973) and Dasgupta and Heal (1979).

criteria of equitability, yielding tests of "fairness". This approach is looked at in the next subsection.

5.2. *Envy, equity and fairness*

If no individual prefers the bundle of good enjoyed by another person to his own, then that allocation is called *equitable*. If an allocation is both Pareto optimal and equitable, then it is called *fair*.[45] This concept, introduced in the modern literature by Foley (1967), has been extensively explored recently.[46] Much of the literature has been concerned with problems of existence and consistency.

Some rather negative conclusions have been established even with the "standard" assumptions of production and exchange – convex production possibilities, convex preferences, self-seeking choice, no externalities.[47] In economies with production, fair allocations need not exist [see Pazner and Schmeidler (1974) and Varian (1974)]. Pareto optimal equilibrium conditions may require that the more productive should work harder and be paid more. And the leisure-loving more productive may, in this situation, envy the unhurried less productive, while the income-loving less productive may envy the opulent more productive. Furthermore, without production, i.e. in a purely exchange economy, fair allocations may not exist if individual preferences are not all convex.

Even with convex preferences and even in the context of pure exchange, equitable endowment allocations can lead to non-equitable competitive equilibria through trade which happens to be profitable for all [see Feldman and Kirman (1974) and Goldman and Sussangkarn (1978)]. It is possible to construct examples in which the diversity of tastes guarantees that none of the parties envies the commodity basket of any other before trade, but after a mutually advantageous trade, at least one person envies the basket that another ends up with. Pareto improvements may, thus, conflict with the preservation of equitability, and this type of conflict between equity and Pareto efficiency may arise with great generality.

[45] There seems to be some non-uniformity of terminology in the literature. Sometimes "fair" is defined simply as "equitable", e.g. in Pazner and Schmeidler (1974) and Feldman and Kirman (1974).

[46] See Kolm (1969, 1972), Schmeidler and Yaari (1970), Schmeidler and Vind (1972), Pazner and Schmeidler (1972, 1974, 1978), Feldman and Kirman (1974), Varian (1974, 1975, 1976a, 1976b), Daniel (1975), Crawford (1977, 1979), Gärdenfors (1975), Allingham (1976), Pazner (1977), Svensson (1977, 1980), Goldman and Sussangkarn (1978), Archibald and Donaldson (1979), Crawford and Heller (1979), Feldman and Weiman (1979), Sobel (1979), Champsaur and Laroque (1981), and Suzumura (1983a), among others.

[47] Note that the concept of "envy" used in these models is one of "preferring" the position of another, and not – as in another interpretation of envy – "suffering from" the superior position of another. It is only in the former sense that envy can be present *without* "externality"!

Partly under the influence of such negative results, but also for their own interest, other rival concepts of fairness have also been explored. In the case of "wealth fairness", the concept of equity is reduced to non-envy of the "complete" position of any other person including his commodity bundle, his leisure, as well as his production. This criterion is formulated in such a way that "if it is impossible for agent i to produce what j produces", then the equity condition is "vacuously satisfied for these two agents" [Varian (1974, p. 73)]. Fair allocations, in this sense, do exist under standard assumptions, but the criterion may appear to be morally quite arbitrary, and many would share Pazner's (1977) inability "to see any possible moral justification for this concept in the case of innate (or, more generally, exogenous) productivity differentials" (p. 459).

Another concept is "income-fairness" where the object of envy is another person's income, not his commodity bundle.[48] This leads to the requirement that, at efficiency prices corresponding to the allocation, there must be equalisation of potential income, i.e. equalisation of the value of each person's commodity-cum-leisure bundle [see Pazner and Schmeidler (1972) and Varian (1974, 1975, 1976a), Feldman and Weiman (1979)].[49] While income-fair allocations do not involve logical problems of existence under standard assumptions, it is remarkably exacting in terms of its institutional implications.[50]

Other variants of the concept of fairness make the criteria typically a good deal less exacting, but in the process also make the "good quality" rather ad hoc. An "egalitarian-equivalent" allocation is one in which the distribution of personal utilities could have been generated by an equal division of some – not necessarily feasible – vector of goods [see Pazner and Schmeidler (1978) and Crawford (1979), among others]. Egalitarian-equivalence is consistent with Pareto efficiency even in situations in which fair allocations may not exist (e.g. with production under standard assumptions, or in a pure exchange economy with some non-convexity of preference). But the use of equal distribution of some purely *hypothetical* commodity vector to identify a good quality of an *actual* distribution may appear to be quite arbitrary. It also goes thoroughly against the rationale of Arrow's condition of "independence of irrelevant alternatives".[51] One may have to pay dearly for one's dislike of some particular good of little importance in the *actual* basket if it looms large in some *hypothetical* basket.

[48] It may be argued that the envy of someone's *income* may be a more cogent basis of judging relative advantage than the envy of someone's *commodity bundle*. Cf. the old story of the father–son conversation: "Dad, I wish I had the money to buy an elephant." "Why, son, what will you do with an elephant?" "Don't be daft – why should I buy an elephant with that money?"

[49] See also Archibald and Donaldson's (1979) criterion of economic equality in terms of identical sets of bundles to choose from (what they call "identical choice sets", not to be confused with "choice sets" as defined here).

[50] Sufficiency conditions for envy-equitable, Pareto-optimal allocations under competitive equilibrium have been investigated by Champsaur and Laroque (1981).

[51] For various motivations underlying the independence condition, see Arrow (1951, 1963), Ray (1973), Mayston (1974, 1975), and Plott (1976).

As a final example, consider Daniel's (1975) criterion of a "just" allocation as one satisfying (i) Pareto optimality, and (ii) being "balanced" in the sense that "the number of people who envy a person is equal to the number of people that he envies" (p. 102). Daniel establishes the existence of such "just" allocations under standard assumptions, but it is not altogether clear whether being "balanced", in this sense, can be described as a "good" quality. A situation in which everyone envies everybody else – hardly a "nice" society – is clearly "balanced".

There are other variants of the fairness criteria – many of these have been critically surveyed by Pazner (1977) – but the tension between avoiding extremely exacting requirements and eschewing arbitrary discrimination is widely observed. This raises some general questions about the extent to which these procedures have been able to provide a more satisfactory approach to social decisions than traditional social choice theory has offered. This issue is taken up in the next subsection.

5.3. *Good quality approaches vs. traditional social choice formulations*

There is little doubt that the "good quality" approaches have provided a worthwhile field for investigation. The ambitiousness of the traditional social choice formulations in seeking a social ordering, or a finitely complete choice function (specifying the optimal subsets for each choice problem), causes not a little problem, and here the Good Quality approaches have some potential advantage. On the other hand, it is difficult to agree on a particular quality as especially good (irrespective of other qualities), and partitioning the set of possibilities into good and bad subsets based on any of these qualities suffers from some arbitrariness.

As it happens the more comprehensive good qualities, taking into account consideration of both Pareto efficiency and equitability, have also raised serious problems of *existence* of good subsets (e.g., satisfying "fairness" based on the requirement of Pareto optimality of equitable allocations without envy). Even when "existence" has been guaranteed in principle, the practical relevance of the partitioning has been constrained by the fact that one side of the partition has been occupied only by allocations that are truly demanding (e.g. Pareto-efficient *equal* distributions of income – no less!). While Varian (1974) may be right to criticise traditional social choice approaches by arguing that "there was no requirement to rank all allocations", still an approach that "gives no indications of the merits of two nonfair allocations" (p. 65) may not take us a great distance when fair allocations don't exist, or require conditions so exacting that they are unlikely to be practically achievable in the near future. The traditional social choice approach, in contrast, can offer more, since it discriminates more – even between the *bad* and the *worse*!

The chief contribution of the "fairness" literature has rested elsewhere. First, it has shown the relevance of informational parameters that the traditional social choice approaches have tended to ignore in the single-minded concern with individual orderings of complete social states. Comparisons of different persons' positions within a state have been brought into the calculation, enlarging the informational basis of social judgments.[52]

Second, in raising rather concrete questions regarding states of affairs, the fairness literature has pushed social choice theory in the direction of more structure. Criteria such as "unrestricted domain", or "independence", or "non-dictatorship", are very general requirements of good social choice procedures, while requirements of "fairness" or "equity" make the demands more specific. There is some obvious gain in this extension.

6. Social welfare functionals

6.1. *Invariance requirements: Measurability and comparability of utilities*

The informational base of the traditional social choice approaches can be enriched by making the social preference relation R, or choice function $C(\cdot)$, not a function of the n-tuple of individual orderings $\{R_i\}$, but the n-tuple of individual utility functions $\{U_i(\cdot)\}$. Such formulations are, of course, not new, and indeed the classical utilitarian characterization of social welfare (in the works of, say, Edgeworth, Marshall, Pigou, or Ramsey) is only a special case of such a form.[53] However, the difficulty with this way of formalizing the functional relations arises from the fact that given the measurability and comparability assumptions of individual utilities, the utility function has to be represented not by one n-tuple of individual utilities, but by a set of n-tuples of individual utilities which are informationally identical (for the given assumptions of measurability and comparability). This problem is met in the approach of *social welfare functionals* through imposing a class of *invariance requirements* [Sen (1970a, chapters 7–9, 1974, 1977a) and Roberts (1977, 1980b)], which demand the same outcome for each of the n-tuples of utility functions that could reflect the same underlying reality.

A social welfare functional SWFL specifies exactly one social ordering R over the set X of social states for any given n-tuple $\{U_i(\cdot)\}$ of personal utility functions, each defined over X, one for each person $i=1,\ldots,n$. The invariance

[52]As Varian (1974) has argued, "[traditional] social decision theory does not put enough into the aggregating process" (p.65). See also Svensson (1977).

[53]Harsanyi's (1955) well-known axiomatic derivation of utilitarianism also uses a general form of this kind. See also Kolm's (1969, 1972) extensive studies of justice and equity.

requirement takes the general form of specifying that for any two n-tuples in the same comparability-set $\bar{L}$, reflecting the assumptions of measurability and interpersonal comparability of individual utilities, the social ordering generated must be the same,

$$R = F(\{U_i\}). \tag{6.1}$$

Invariance requirement
For any two n-tuples $\{U_i\}$ and $\{U_i^*\}$ belonging to the same comparability-set $\bar{L}$, $F(\{U_i\}) = F(\{U_i^*\})$.

The specification of the measurability–comparability assumptions takes the form of characterizing $\bar{L}$. Depending on the assumption of measurability, each person i has a family L_i of (essentially equivalent) utility functions: each a positive, monotonic transformation of any other in the family in case of ordinality; each a positive, affine transformation of any other in the family in case of cardinality; each a positive, homogeneous linear transformation of any other in the family in case of ratio-scale measure; etc. The Cartesian product of the n-tuple of families of utility functions $\{L_i\}$ is the measurability-set $L = \prod_{i=1}^{n} L_i$, specifying all possible n-tuples of individual utility functions consistent with the measurability assumption.

If there is no interpersonal comparability at all, then there is no further restriction, and $\bar{L} = L$. If, however, interpersonal comparability of any type is permitted, then $\bar{L} \subset L$.[54] For example, with full comparability, if a transformation $\psi(\cdot)$ permitted by the measurability assumption is applied to one person's utility function in moving from one n-tuple $\{U_i\}$ to another $\{U_i^*\}$, then the same transformation $\psi(\cdot)$ must have been applied to everyone's utility function as a necessary and sufficient condition for $\{U_i\}$ and $\{U_i^*\}$ to belong to the same comparability-set $\bar{L}$. Some distinguished cases of measurability–comparability assumptions are considered below [see Sen (1970a, 1974), Hammond (1976a, 1977b), Maskin (1978, 1979b), d'Aspremont and Gevers (1977), Deschamps and Gevers (1978), Gevers (1979), Blackorby and Donaldson (1979), and Roberts (1980a, 1980b)].

Alternative measurability – comparability frameworks
For any utility n-tuple $\{U_i^*\}$ belonging to $\bar{L}$, it is required that $\bar{L}$ must consist of exactly all n-tuples $\{U_i\}$ such that for some n-tuple of transformations $\{\psi_i\}$

[54] For various interpretations of interpersonal comparisons, see Vickrey (1945), Little (1950), Harsanyi (1955), Arrow (1963), Suppes (1966), Sen (1970a, 1973, 1979a), Jeffrey (1971), Rawls (1971), Waldner (1972), Hammond (1977a), and Borglin (1982).

satisfying the following alternative restrictions, $U_i = \psi_i(U_i^*)$ for all i:

ordinal non-comparability (ONC): each ψ_i is a positive, monotonic transformation;

cardinal non-comparability (CNC): each ψ_i is a positive affine transformation, $\psi_i(\cdot) = a_i + b_i \cdot (\cdot)$, with $b_i > 0$;

ratio-scale non-comparability (RNC): each ψ_i is a positive, homogeneous linear transformation, $\psi_i(\cdot) = b_i \cdot (\cdot)$, with $b_i > 0$;

ordinal level comparability (OLC):[55] for all i, $\psi_i(\cdot) = \psi(\cdot)$, a positive, monotonic transformation;

cardinal full comparability (CFC): for all i, $\psi_i(\cdot) = \psi(\cdot)$, a positive, affine transformation, $\psi(\cdot) = a + b \cdot (\cdot)$, with $b > 0$;

ratio-scale full comparability (RFC): for all i, $\psi_i(\cdot) = \psi(\cdot)$, a positive, homogeneous, linear transformation, $\psi(\cdot) = b \cdot (\cdot)$, with $b > 0$;[56]

cardinal unit comparability (CUC): each ψ_i is a positive, affine transformation, $\psi_i(\cdot) = a_i + b \cdot (\cdot)$, with $b > 0$, the same for all i;

cardinal level comparability (CLC): each ψ_i is a positive, affine transformation, $\psi_i(\cdot) = a_i + b_i(\cdot)$, with $b_i > 0$, and there is a positive, monotonic transformation $\phi(\cdot)$ such that $U_i(x) = \phi(U_i^*(x))$, for all $x \in X$, for all i;

cardinal unit and level comparability (CULC):[57] each ψ_i is a positive, affine transformation, $\psi_i(\cdot) = a_i + b \cdot (\cdot)$, with $b > 0$, the same for all i, and there is a positive, monotonic transformation $\phi(\cdot)$ such that $U_i(x) = \phi(U_i^*(x))$, for all $x \in X$, for all i.[58]

The invariance restriction applied to these respective cases will be denoted as ON, CN, RN, OL, CF, RF, CU, CL, and CUL, respectively. For example, ON is the invariance restriction for the case of ordinal non-comparability ONC. Note also that the less the precision of information, the wider the set $\bar{L}$, and the more

[55] This can, in fact, be called "ordinal full comparability" as well, since ordinal *intra*personal comparisons can be fully extended here to *inter*personal comparisons.

[56] Utility values have to be confined to being non-negative in this case, to avoid perversity; see footnote 74 in Section 6.7 below.

[57] This is a somewhat wider class of $\bar{L}$ than under cardinal full comparability, thereby inducing a more demanding invariance restriction than under the latter, and represents less usable information than with cardinal full comparability. The difference will depend on X and the actual utility configurations. Gevers' (1979) case of "almost co-cardinal" (ACC*) corresponds to CULC except for requiring that the common monotonic $\phi(\cdot)$ function should apply not necessarily to the whole of X but to each pair of utility vectors separately. ACC* is in this sense still more demanding than CULC, requiring invariance over a wider class, and thus represents *less* informational availability.

[58] Other cases can be correspondingly specified, e.g. ratio-scale level comparability.

demanding is the invariance restriction. With less information more signals are indistinguishable.

It will be convenient later to consider comparability cases that are not fully specified, e.g. levels being comparable whether or not anything else is. Let $\bar{L}(L)$ and $\bar{L}(U)$ be comparability sets with ordinal level comparability and cardinal unit comparability respectively.

Level-plus comparability (L^+C) is defined at $\bar{L} \subseteq \bar{L}(L)$, and *unit-plus comparability* (U^+C) as $\bar{L} \subseteq \bar{L}(U)$, respectively, in each case. The invariance restriction applied to these measurability–comparability frameworks will be denoted as L^+ and U^+, respectively.

6.2. *Arrow's impossibility result and richer utility information*

For a SWFL the Arrow conditions can be readily redefined.

Condition $\tilde{U}$
The domain of $F(\cdot)$ includes all logically possible n-tuples of utility functions $\{U_i\}$, defined over X.

Condition $\tilde{I}^2$
For any pair of social states $x, y \in X$, $R|^{\{x,y\}} = F^{\{x,y\}}(\{U_i(x), U_i(y)\})$, so that if $U_i(a) = U_i^*(a)$ for all i, for $a = x, y$, then $xF(\{U_i\})y$ if and only if $xF(\{U_i^*\})y$.

Condition $\tilde{P}$
For any pair $x, y \in X$, $[\forall i\colon U_i(x) > U_i(y)] \Rightarrow xPy$.

Condition $\tilde{D}$
There is no individual i such that for all $x, y \in X$ and for all $\{U_i\}$ in the domain of $F(\cdot)$, $U_i(x) > U_i(y) \Rightarrow xPy$.

Since Arrow (1963) dealt with the case of ordinal non-comparability, the General Possibility Theorem translated to SWFLs yields the following:

Arrow's theorem for SWFL
For a finite H and $\#X \geq 3$, there is no SWFL satisfying Conditions $\tilde{\mathrm{U}}$, $\tilde{\mathrm{I}}^2$, $\tilde{\mathrm{P}}$, $\tilde{\mathrm{D}}$, and the invariance restriction ON.

This is established by noting that with ON, a SWFL is, in fact, a SWF, and observing that in this case Conditions $\tilde{\mathrm{U}}$, $\tilde{\mathrm{I}}^2$, $\tilde{\mathrm{P}}$ and $\tilde{\mathrm{D}}$ imply U, I^2, P and D applied to the SWF to which the SWFL is reduced.

In fact, the impossibility result extends readily to the case of *cardinal* non-comparability as well [Sen (1970a, theorem 8*2)].

Arrow's theorem extended to cardinal non-comparable utilities
For a finite H and $\#X \geq 3$, there is no SWFL satisfying Conditions $\tilde{\mathrm{U}}$, $\tilde{\mathrm{I}}^2$, $\tilde{\mathrm{P}}$, $\tilde{\mathrm{D}}$, and the invariance restriction CN.

This is established by taking any two n-tuples of utility functions $\{U_i\}$ and $\{U_i^*\}$ such that each individual ranks the set X in the same way in the two cases. For every pair $x, y \in X$, by exploiting the two degrees of freedom in an affine transformation, an n-tuple of positive, affine transformations $\{\psi_i\}$ applied to $\{U_i^*\}$ yields $U_i'(z) = \psi_i(U_i^*(z)) = U_i(z)$, for $z = x, y$, for all i. By the independence condition $\tilde{\mathrm{I}}^2$, $xF(\{U_i\})\,y$ if and only if $xF(\{U_i'\})y$, and by CN, $xF(\{U_i'\})y$ if and only if $xF(\{U_i^*\})y$. Since this holds pair by pair, clearly $F(\{U_i\}) = F(\{U_i^*\})$, so the SWFL is, in fact, a SWF. The rest of the proof is the same as with Arrow's original theorem.

A similar impossibility result can be obtained by replacing the pair-relational independence condition $\tilde{\mathrm{I}}^2$ by the m-ary relational independence condition $\tilde{\mathrm{I}}^m$, since Blau's (1971) result about the equivalence of pair-relational independence and m-ary relational independence for social welfare functions can be extended to social welfare functionals as well [see d'Aspremont and Gevers (1977)].[59]

While cardinality without interpersonal comparability does not change matters as far as the Arrow impossibility result is concerned,[60] interpersonal comparability without cardinality does, however, make a real difference. With ordinal level comparability, Conditions $\tilde{\mathrm{U}}$, $\tilde{\mathrm{I}}^2$, $\tilde{\mathrm{P}}$ and $\tilde{\mathrm{D}}$ are perfectly consistent, and an example of these conditions being fulfilled along with the invariance restriction OL is provided by the so-called Rawlsian "maximin" criterion (interpreted in terms of individual utilities). The *stronger* Pareto principle $\tilde{\mathrm{P}}^*$, which is violated by maximin, can also be satisfied, if we use the lexicographic version of the "maximin" rule [Sen (1970a) and Rawls (1971)], often called "leximin".[61] Let

[59]Kalai and Schmeidler (1977) have presented another impossibility result permitting cardinal utility with a weakened dictatorship condition, but involving some additional requirements, most notably continuity. For other impossibility results with cardinality, see Schwartz (1970), DeMeyer and Plott (1971), Fishburn (1972b), and Chichilnisky (1980c).

[60]If, however, "independence" is not required, then various possibilities exist, notably the Nash bargaining solution. On "Nash social welfare functions", see Nash (1950), Luce and Raiffa (1957), Sen (1970a), Kalai and Smordinsky (1975), Harsanyi (1977b), Kaneko and Nakamura (1979), Kaneko (1980), Kim and Roush (1980a), Coughlin and Nitzan (1981), and Binmore (1981).

[61]*Strong Pareto principle* ($\tilde{\mathrm{P}}^*$): $\forall x, y \in X$, $[\forall i\colon xR_i y \;\&\; \exists i\colon xP_i y] \Rightarrow xPy$, and $[\forall i\colon xI_i y] \Rightarrow xIy$.

For general discussions of the Rawlsian approach, see Rawls (1971, 1982), Sen (1970a, 1976b, 1977b), Arrow (1973, 1977), Barry (1973), Phelps (1973, 1977), Dasgupta (1974), Daniels (1975), Barry and Rae (1975), and Yaari (1981), among others. See also the literature on axiomatic derivation of maximin and leximin, discussed below.

$r(x)$ be the rth worst-off person in state x; in case of more than one person having the same utility level, rank them in any arbitrary strict order.

Leximin
For any $x, y \in X$, if there is k, $1 \leq k \leq n$, such that $U_{k(x)}(x) > U_{k(y)}(y)$, and for all $r < k$, $U_{r(x)}(x) = U_{r(y)}(y)$, then xPy. If, on the other hand, for all r, $1 \leq r \leq n$, $U_{r(x)}(x) = U_{r(y)}(y)$, then xIy.

Leximin satisfies Conditions $\tilde{\mathrm{U}}$, $\tilde{\mathrm{I}}^2$, $\tilde{\mathrm{P}}^*$, $\tilde{\mathrm{D}}$ and OL (and obviously "level plus" L^+ invariance restrictions – with utility information richer than OL such as CL, CUL, CF, RF, etc.). It also satisfies several other conditions that have been proposed in the literature, such as Anonymity, Neutrality, Separability, Suppes's (1966) "grading principle of justice", and several "equity" criteria including Hammond's (1976a) demanding Axiom E.

Condition $\tilde{A}$ (anonymity)
If $\{U_i\}$ is a re-ordering (permutation) of $\{U_i^*\}$, then $F(\{U_i\}) = F(\{U_i^*\})$.

Condition $\tilde{N}$ (neutrality)
If $\mu(\cdot)$ is a permutation function applied to X, and $\mu[R]$ is the ordering R modified by the same permutation $\mu(\cdot)$, and if for all i, $U_i(x) = U_i^*(\mu(x))$ for all $x \in X$, then $F(\{U_i^*\}) = \mu[F(\{U_i\})]$.

Condition SE (separability)
If the set H of individuals partitions into two proper subsets H_1 and H_2 such that for all i in H_1, $U_i(x) = U_i^*(x)$ for all x in X, and for all i in H_2, $U_i(x) = U_i(y)$ and $U_i^*(x) = U_i^*(y)$, for all x, y in X, then $F(\{U_i\}) = F(\{U_i^*\})$.

Condition S (Suppes principle)
If $\rho(\cdot)$ is a permutation function applied to the set H of individuals, and if for any $x, y \in X$, $U_i(x) \geq U_{\rho(i)}(y)$ for all i, then xRy. If *additionally*, for some i, $U_i(x) > U_{\rho(i)}(y)$, then xPy.

Condition HE (Hammond's equity axiom)
For any $x, y \in X$, if for some pair $g, h \in H$, $U_g(y) > U_g(x) > U_h(x) > U_h(y)$, and for all $i \neq g, h$, $U_i(x) = U_i(y)$, then xRy.

Anonymity states that permuting the utility functions among the people does not affect the social ordering. Neutrality asserts that permuting the social states in individual orderings permutes the social states in the social ordering in exactly the same way. Separability says that if the utility numbers for all states remain unchanged for all *non-indifferent* individuals, then the social ordering should not change either. The Suppes principle extends the Pareto principle by using dominance in an *anonymous* way. First, dealing with weak ranking, if each person in x is at least as well off as the corresponding person in y, then xRy. If, additionally, someone in x is strictly better off than the corresponding person in y, then xPy. Hammond's equity principle demands that if person h is worse off than person g in both x and in y and if h prefers x to y while g prefers y to x, with all other persons indifferent between x and y, then xRy.

Both maximin and leximin can be seen as incorporating the dictatorship of a particular "rank", viz. the rank of being worst-off. While ordinal level comparability provides an adequate informational base for escaping Arrow's impossibility, it is interesting to enquire whether the escape must take the form of rules that incorporate dictatorship of *some* rank (e.g. of the worst-off, the best-off, the kth worst-off). Certainly the Arrow conditions imposed on a SWFL satisfying invariance for ordinal level comparability push us in that direction, and all other possible rules – typically rather odd ones – can be weeded out by strengthening the condition of non-dictatorship to anonymity [see Gevers (1979) and Roberts (1980a)]. With anonymity, in the presence of the other conditions, "rank" remains an invariant and usable signal (personal identity does not), and the absence of cardinality and of comparability of units makes rank effectively the only such invariant signal. This permits the translation of the Arrow-type reasoning about personal decisiveness to a corresponding reasoning about rank decisiveness, moving from the decisiveness of all ranks put together (guaranteed by the weak Pareto principle) to the decisiveness of some particular rank (as under the "Group Contraction Lemma").

Rank dictatorship theorem

For a finite H and $\#X \geq 3$, a SWFL satisfying Conditions $\tilde{\mathrm{U}}$, $\tilde{\mathrm{I}}^2$, $\tilde{\mathrm{P}}$, $\tilde{\mathrm{A}}$, and the invariance restriction OL, must be rank-dictatorial, i.e. there will be a rank k such that for all $x, y \in X, U_{k(x)}(x) > U_{k(y)}(y) \Rightarrow xPy$.[62]

Leximin implies not only the dictatorship of the worst-off, but a whole hierarchy of dictatorial powers so that each rank has dictatorial power *when* the lower ranks are all "indifferent". Leximax defines the opposite hierarchy, with the best-off being the unconditional dictatorial rank, and the other ranks enjoying dictatorial powers conditional on the higher ranks being indifferent. The definition of leximax is the same as that of leximin but for the change that the condition refers to $r > k$ in place of $r < k$. The rank dictatorship result can be modified to precipitate either leximin or leximax [see d'Aspremont and Gevers (1977)], by demanding separability and replacing the weak Pareto principle by the strong Pareto principle $\tilde{\mathrm{P}}^*$ (corresponding to P^*, as $\tilde{\mathrm{P}}$ does to P).

[62] For this and related results, see Roberts (1977, 1980a, 1980b) and Gevers (1979). Also Deschamps and Gevers (1979).

Leximin – leximax theorem

For a finite H and $\#X \geq 3$, a SWFL satisfying Conditions $\tilde{\mathrm{U}}$, $\tilde{\mathrm{I}}^2$, $\tilde{\mathrm{P}}^*$, $\tilde{\mathrm{A}}$, SE, and the invariance restriction OL, must be leximin or leximax.

6.3. *Axiomatic derivation of leximin*

The Suppes principle, which like the Pareto principle builds on dominance of utilities (but does this in an anonymous way and is thus remarkably more extensive than the Pareto principle), can be stated in many different forms. Two weakenings are considered next, before proceeding to the interesting subject of the axiomatic derivation of the leximin rule. One weakening confines the "anonymous" comparisons to permutations between exactly two persons only, and the other concentrates on indifference only (correspondingly to the Pareto indifference rule).

Condition S_2 (2-person Suppes principle)

For any $x, y \in X$, if for any two persons $g, h \in H$, *either* $U_j(x) \geq U_j(y)$ for $j = g, h$, *or* $U_g(x) \geq U_h(y)$ and $U_h(x) \geq U_g(y)$, while for all $i \neq g, h$, $U_i(x) = U_i(y)$, then xRy. If, furthermore, at least one of the two inequalities $\geq$ holds strictly $>$, then xPy.

Condition S° (Suppes indifference rule)

For any $x, y \in X$, if for some permutation function $\rho(\cdot)$ applied to the set H of individuals $U_i(x) = U_{\rho(i)}(y)$ for all i, then xIy.

Condition S_2° (2-person Suppes indifference rule)

For any $x, y \in X$, if for two persons $g, h \in H$, $U_g(x) = U_h(y)$ and $U_h(x) = U_g(y)$, while for all $i \neq g, h$, $U_i(x) = U_i(y)$, then xIy. Also the Pareto indifference rule holds.

Hammond's equity condition can also be weakened to what d'Aspremont and Gevers (1977) have called "minimal equity", to derive leximin axiomatically.

Condition ME (minimal equity)

The SWFL is not the leximax principle.

Finally, since the Blau (1971) result on the equivalence of pair-relational independence with m-ary relational independence holds (as has already been remarked), we might as well simply take the general relational independence.

Condition $\tilde{I}$ (relational independence)
For any subset $S \subseteq X$, if for all i, for all $x \in S$, $U_i(x) = U_i^*(x)$, then $F(\{U_i\})|^S = F(\{U_i^*\})|^S$.

Leximin has been neatly axiomatized by Hammond (1976a), Strasnick (1976, 1978), and d'Aspremont and Gevers (1977), and further by Maskin (1979b), Deschamps and Gevers (1978, 1979), Roberts (1977, 1980a, 1980b), Arrow (1977), Sen (1977b), and Gevers (1979). The main results can be put in the form of a rather comprehensive theorem. In this theorem – and indeed throughout Section 6 – it is assumed that $\#X \geq 3$ and that H is finite.

Leximin derivation theorem
A SWFL satisfying unrestricted domain $\tilde{\mathrm{U}}$ and independence of irrelevant alternatives $\tilde{\mathrm{I}}$ must be leximin if it satisfies invariance for level-plus comparability L^+, and one of the following set of conditions:

[1] $\tilde{\mathrm{P}}^*$, $\tilde{\mathrm{A}}$, SE, ME and OL;

[2] S, SE, ME and OL;

[3] $\tilde{\mathrm{P}}^*$, $\tilde{\mathrm{A}}$ and HE;

[4] $\tilde{\mathrm{P}}^*$, S° and HE;

[5] $\tilde{\mathrm{P}}$, S_2° and HE;

[6] S and HE;

[7] S_2 and HE.

The last set, viz. [7], is taken up first. One way of establishing the result is through a reduction technique used in Sen (1976b, 1977b). It reduces the problem of getting leximin for n-person judgments to getting leximin for 2-person judgements (with the rest indifferent).[63]

First define leximin-k as the leximin principle applied to ranking any pair of states over which there are exactly k non-indifferent persons. One of the unappealing features of leximin is that it permits the interest of one person (if relatively badly off) to override the interests of a great many others, possibly a billion of them. This possibility can be eliminated by confining the application of leximin to cases of a small number of non-indifferent persons. But it can be shown that such a programme of constraining leximin would be hopeless for a

[63] Hammond (1979b) has shown that this 2-to-n person correspondence of principle applies not merely to leximin, but to a whole class of principles. This also yields an alternative way of deriving leximin from set [7] by establishing first S from S_2 in the presence of $\tilde{\mathrm{U}}$ and $\tilde{\mathrm{I}}$. Ulph (1978) has extended the correspondence.

SWFL satisfying unrestricted domain and independence because of the following result:

Leximin from Inch to Ell
For any SWFL satisfying Conditions Ũ and Ĩ, leximin-2 implies leximin.

The proof of this proposition, which will not be presented here, can build on showing first, that leximin-2 implies leximin-1, and then that leximin-1,...,leximin-$(r-1)$ together imply leximin-r [Sen (1977b, theorem 8)]. In view of this result, the leximin derivation by route [7] can be done via the following lemma:

Leximin-2 derivation
A SWFL satisfying invariance for level-plus comparability L^+, and fulfilling Conditions Ũ, Ĩ, S_2 and HE, must satisfy leximin-2.

In proving this proposition, the "Paretian" comparisons subsumed by leximin-2 cause no problems, since they are subsumed by S_2 as well. So we need be concerned with only the non-Paretian comparisons. Take first the case of two rank-ties: $U_1(x)=U_2(y)$, and $U_2(x)=U_1(y)$. Again, directly from S_2, it follows that xIy, which is what leximin-2 requires. Similarly, with exactly one rank-tie, say, $U_1(x)=U_2(y)$, there is again an immediate application of S_2 ranking x and y entirely by the ranking of $U_2(x)$ and $U_1(y)$, and this corresponds exactly to leximin-2. That leaves only the case of two non-indifferent, non-rank-ties. But again if the two inequalities point in the *same* direction, say $U_1(x)>U_2(y)$ and $U_2(x)>U_1(y)$, the xPy by S_2, which is exactly what leximin-2 demands. Thus, the only case that is not immediate is one in which there are two inequalities pointing in opposite directions.

Without loss of generality, consider $U_1(x)>U_2(y)$ and $U_2(x)<U_1(y)$. Noting that $U_2(x)\neq U_2(y)$, since 2 is non-indifferent, again without loss of generality, take $U_2(x)>U_2(y)$. To establish leximin-2, we have to show that xPy. Consider a third state z, and an n-tuple $\{U_i^*\}$ such that $U_i(a)=U_i^*(a)$ for all i and $a=x, y$; $U_i^*(z)=U_i^*(x)=U_i^*(y)$ for all $i\neq 1,2$; $U_1^*(z)>U_2^*(z)>U_2^*(y)$; and $U_i^*(a)>U_1^*(z)$ for $a=x, y$, and all i, other than the particular combination $a=y$ and $i=2$. It follows from Hammond's Equity Axiom HE, that zR^*y. From the 2-person Suppes principle S_2, we have xP^*z. Hence by transitivity of R, xP^*y. By independence xPy. This establishes leximin-2.

Due to the above result, leximin in its full force follows from the *same* axioms in view of Leximin from Inch to Ell.

Obtaining leximin from the alternative set [5] is similarly done, since in the presence of the other conditions, P* and S_2° imply S_2. Sets [3], [4] and [6] are similarly covered since each of these sets implies S_2. None of these combinations of conditions relies on the measurability–comparability framework to be restricted to ordinal level comparability. If that restriction is imposed, then leximin can be axiomatized on the basis of the Leximin–leximax Theorem. This covers the combinations given by [1] and [2]. Anonymity and the strong Pareto principle follow from the Suppes relation S. Minimal Equity ME eliminates leximax. That leaves only leximin.

6.4. *Strong neutrality and strong anonymity*

It was mentioned earlier that Leximin satisfies the conditions of neutrality and anonymity. In fact, it satisfies a stronger version of each condition. So do utilitarianism and many other procedures. Before proceeding further it is useful to consider these stronger versions of neutrality and anonymity.

Condition SN (strong neutrality)
For any two pairs of social states $\{x, y\}$ and $\{a, b\}$, and any two n-tuples of utility functions $\{U_i\}$ and $\{U_i^*\}$, if for all i, $U_i(x) = U_i^*(a)$ and $U_i(y) = U_i^*(b)$, then $x F(\{U_i\}) y$ if and only if $a F(\{U_i^*\}) b$.

Condition SA (strong anonymity)
If for any pair of utility n-tuple $\{U_i\}$ and $\{U_i^*\}$, there is a permutation function $\rho(\cdot)$ over the set H of persons such that for some x, for all i, $U_i(x) = U_{\rho(i)}^*(x)$, and for all $y \neq x$, for all i, $U_i(y) = U_i^*(y)$, then $F(\{U_i\}) = F(\{U_i^*\})$.

Strong neutrality implies neutrality $\tilde{\text{N}}$ and independence $\tilde{\text{I}}^2$, and is indeed equivalent to the combination of the two. It permits neutrality to be applied pair by pair, and asserts that the utility information regarding any two social states is all that is needed for ranking that pair. Strong anonymity asks for invariance not merely when utility functions are permuted between the persons, but also when the utility *values* for any particular state x are permuted between the persons without doing anything to the utility values for other states. Clearly, such permutations can alter the list of preference orderings embedded in an n-tuple of utility functions, and ordering-based rules such as the Method of Majority Decision, while satisfying anonymity (and strong neutrality), do not in general fulfil strong anonymity.

Given strong neutrality, social welfare W can be seen as a function of the individual utility vectors $\boldsymbol{u}$, bringing us back to a classic formulation of the Bergson–Samuelson social welfare function,[64]

$$W = W(\boldsymbol{u}). \tag{6.2}$$

With strong anonymity added to this, the function $W(\cdot)$ is symmetric.

[64] See Samuelson (1947, pp. 228–229, 246), Bergson (1948, p. 418), and Graaff (1957, pp. 48–54).

For SWFLs satisfying unrestricted domain and independence of irrelevant alternatives, the Pareto indifference rule $\tilde{P}^{\circ}$ implies strong neutrality, and the Suppes indifference rule S° implies both strong neutrality and strong anonymity.[65]

Strong neutrality theorem
For any SWFL fulfilling Conditions U and I^2, $P^{\circ} \Leftrightarrow SN$.

Strong anonymity theorem
For any SWFL fulfilling Conditions U and I^2, $S^{\circ} \Leftrightarrow (SN \,\&\, SA)$.

For proofs, see Sen (1977b).

6.5. *Utilitarianism: Harsanyi's theorems*

Harsanyi's (1955) axiomatic treatment of utilitarianism provided a classic contrast to the ordering-based social welfare judgments in Arrow's social welfare function and related structures. A richer base of utility information permitted Harsanyi to consider the class of weighted sum of individual utilities – a class that could not have been accommodated within social welfare functions, or for that matter in structures permitting only ordinal level comparability.

Harsanyi (1955) established two – essentially independent – results about utilitarianism. One, which I shall call Harsanyi's "Impersonal Choice Utilitarianism", requires any individual's social welfare function – reflecting his ethical judgments – to be based on what his preferences about the social states would have been if he had an equal chance of being in the position of anyone in the society.[66] With consistent choice the von Neumann–Morgenstern (1947) postulates are assumed to be fulfilled. Then the social welfare from a state can be seen as the "utility" of an *as if* lottery, having a probability $1/n$ of being anyone

[65] Neutrality in a milder form – involving only strict (antisymmetric) individual orderings – played an important part in Arrow's (1951) impossibility theorem, and this was explicitly noted by Blau (1957). (See the Field Expansion Lemma in Section 2 above.) The first explicit version of the Strong Neutrality Theorem (applied to social decision functions with quasi-transitive social preference) was presented by Guha (1972) and Blau (1976). The theorem as presented here, dealing with the wider informational framework of SWFLs, is due to d'Aspremont and Gevers (1977). Roberts (1980b) provides an alternative derivation with the weak Pareto principle $\tilde{P}$ rather than $\tilde{P}^{\circ}$ through the use of a continuity axiom. The Strong Anonymity Theorem figures in various forms in Hammond (1976a, 1979b), d'Aspremont and Gevers (1977), Roberts (1977, 1980b), and Sen (1977b).

[66] On this way of characterizing social welfare, see also Vickrey (1945). For a critique of the moral acceptability of the approach, see Diamond (1967), and the controversy on that and related issues in Harsanyi (1975, 1977a) and Sen (1976b, 1977d). For other types of critiques, see McClennen (1978) and Blackorby, Donaldson and Weymark (1980). The broader ethical issue of "impersonal choice" as the basis of moral judgments – going well beyond the status of the utilitarian form – has been illuminatingly discussed by Harsanyi (1958) in his model of "ethics in terms of hypothetical imperatives". See also Harsanyi (1977b, 1979).

in that state. If $W_i(x)$ is the utility of the "prize" i (i.e. of being person i, in state x) in the von Neumann–Morgenstern scale, then clearly

$$W(x) = \frac{1}{n} \sum_{i=1}^{n} W_i(x) \quad \text{for all} \quad x \in X. \tag{6.3}$$

For a given population size, (6.3) is not essentially different from the straightforward utilitarian formula for social welfare.

The other result, which I shall call Harsanyi's "Utility Sum Theorem" has less of a moral basis, but is analytically more assertive. If in a given situation, (a) the family of individual utility functions of each person i is cardinal, given by a class of positive affine transformations, (b) the social welfare function is also cardinal, given by a class of positive affine transformations, and (c) the Pareto indifference rule is assumed, i.e. $U_i(x) = U_i(y)$ for all i must imply $W(x) = W(y)$, then social welfare must be a linear weighted sum of individual utilities,

$$W(x) = \sum_{i=1}^{n} a_i U_i(x) \quad \text{for all} \quad x \in X. \tag{6.4}$$

In recent discussion on utilitarianism, it is Impersonal Choice Utilitarianism, (6.3), that has received most attention [see, for example, Arrow (1973)]. This is a theorem about utilitarianism in a rather limited sense in that the von Neumann–Morgenstern cardinal scaling of utilities covers *both* W_i and W within *one* integrated system of numbering, and the individual utility numbers W_i do not have any independent meaning other than the value associated with each "prize", in predicting choices over lotteries. There is no *independent* concept of individual utilities of which social welfare is shown to be the sum, and as such the result asserts a good deal less than classical utilitarianism does.

Consider, for example, the case in which a person's ethical judgments – and his "impersonal" choices – are based on maximizing the sum of independently measured,[67] ratio-scale comparable (RF) individual utilities (uniformly non-negative) raised to the power t (a constant),

$$W = \frac{1}{t} \sum_{i=1}^{n} (U_i(x))^t \quad \text{for all } x.^{68} \tag{6.5}$$

With $t < 1$ social welfare is strictly concave on (and thus non-utilitarian in terms

[67]See Krantz, Luce, Suppes and Tversky (1971).

[68]Mirrlees (1971) uses this formulation of social welfare [but see also Mirrlees (1982)]. This formulation is axiomatically analysed and discussed by Roberts (1977, 1980b), and Blackorby and Donaldson (1977, 1979).

of) the independently measured utilities U_i. It would, however, appear to be utilitarian within the von Neumann–Morgenstern scaling system, since that scaling would allow $W_i = (n/t)(U_i(x))^t$, the whole scaling being unique up to positive affine transformations of these. Since the only role of W_i is to predict the person's choices under uncertainty, this is a rather superficial form of utilitarianism. As it happens (6.5) permits a whole class of non-utilitarian rules (for all cases other than $t=1$),[69] and by making t go to minus infinity "Rawlsian" maximin or leximin can also be covered,[70] for the independently scaled utilities.

Harsanyi's Utility Sum Theorem does not, however, suffer from this problem, and is in this sense a good deal more assertive. But it is primarily a "representation theorem". It deals only with single-profile exercises and does not claim that the constants a_i in (6.4) will remain the same when the individual utility functions change (i.e. when a *family* L_i of positive affine transformations alters).[71] Not only, therefore, does it not establish that all the a_i must equal each other as under the utilitarian formula (indeed for the axioms specified they can even be negative), but it does not even require that the set of a_i will be invariant with respect to changes in individual utility characteristics (as opposed to representational change *within* a given positive affine family).

The upshot of this discussion is that there is need for an axiomatic derivation of utilitarianism despite Harsanyi's theorems. What is needed is an axiomatization that (1) permits independent formulation of individual utilities, and (2) which has the invariance property of having the set of a_i determined independently of the utility functions to be aggregated (and in particular having $a_i = 1$). Such axiomatic results have recently been presented, and will be taken up in the next subsection. But before closing the discussion on Harsanyi's framework, it is worth asserting unequivocally that the failure to provide a fully-fledged axiomatic derivation of utilitarianism does not render Harsanyi's results useless. Indeed, far from it. The representation theorem is of much interest in itself, and Harsanyi's framework of impersonal choice has proved to be one of the most fruitful ones in social ethics.

6.6. *Utilitarianism: Axiomatic derivations*

Define a utilitarian SWFL as one which for any *n*-tuple of individual utility functions, for any $x, y \in X$, declares xRy if and only if $\sum_{i=1}^{n} U_i(x) \geq \sum_{i=1}^{n} U_i(y)$.[72] The following theorem, established by d'Aspremont and Gevers (1977, theorem 3), uses the invariance requirement for cardinal *unit* comparability CU in

[69] Note that $U_i(\cdot)$ and $(U_i(\cdot))^t$ cannot belong to the same positive affine class unless of course $t=1$.

[70] Cf. Atkinson (1970), Arrow (1973), and Hammond (1975).

[71] This issue has been illuminatingly discussed by Nader-Isfahani (1979).

[72] Yaari (1978) defines "the utilitarian form" less restrictively, using a weighted-sum *formula*, with the weights being endogenously determined. One set of assumptions is shown to lead to the equivalence of Rawlsian and utilitarian SWFLs. Yaari, thus, provides an axiomatic (and also intuitive) analysis of a much wider class of rules than utilitarianism, as it is normally defined.

addition to other conditions to eliminate rules rival to utilitarianism. As in Sections 6.2–6.4, it is assumed that H is finite and $\#X \geq 3$.

Utilitarianism derived with unit comparability
A SWFL satisfying Conditions $\tilde{\mathrm{U}}$, $\tilde{\mathrm{I}}$, $\tilde{\mathrm{P}}^*$, $\tilde{\mathrm{A}}$ and CU must be utilitarian.

It is first checked that a utilitarian SWFL must indeed satisfy these conditions. This is immediate for $\tilde{\mathrm{U}}$, $\tilde{\mathrm{I}}$, $\tilde{\mathrm{P}}^*$ and $\tilde{\mathrm{A}}$. Regarding CU, it need only be noted that translating anyone's utility function by adding a constant (positive or negative) to it must leave all the differences $[U_i(x)-U_i(y)]$ unaffected. And multiplying each U_i by the same constant leaves the *relative* differences unchanged. So we need be concerned only with establishing that these conditions together do not permit any other kind of a SWFL.

It follows from the Strong Neutrality Theorem that the SWFL in question must be strongly neutral. Since given unrestricted domain, independence and anonymity, the Pareto indifference rule implies Suppes indifference rule, the SWFL must also be strongly anonymous by the Strong Anonymity Theorem. So in ranking any pair $x, y \in Y$, we need be concerned only with the utility vectors for x and y, and we can permute the utility values among the individuals for any state without changing the social ranking.

Take, first, a case in which the individual utility sums for x and y are equal; we have to show xIy. Permute the utility numbers among the persons in each state separately in such a way that we have the utility order in line with the individual numbers: $U_n(a) \geq U_{n-1}(a) \geq \cdots \geq U_2(a) \geq U_1(a)$, for $a = x, y$. Now deduct from *each* $U_i(a)$ the minimal of the two values $\{U_i(x), U_i(y)\}$. (Note that this is a permitted transformation under CU, being a translation of individual origins, which can be freely done.) After the deductions permute the individual utilities again in each state to get them in line with individual numbers: $U_n^1(a) \geq U_{n-1}^1(a) \geq \cdots \geq U_2^1(a) \geq U_1^1(a)$. This yields $\{U_i^1\}$. By repeating this process, for some r, we shall get $U_i^r(a) = 0$, for all i and for $a = x, y$. By the Pareto principle, xIy for this utility n-tuple $\{U_i^r\}$, and by CU this must be the case for all $\{U_i\}$ in $\bar{L}$. Hence xIy.

If, instead, we started with the individual utility sum being larger for x than for y, then we would have reached $U_i^r(y) \geq 0$, for all i, with $U_i^r(x) > 0$ for some i. So by the strong Pareto principle, xPy. And this establishes that the SWFL is indeed utilitarian.

Various other axiomatizations of utilitarianism have also been presented [see Deschamps and Gevers (1978, 1979), Maskin (1978), Blackorby and Donaldson (1977, 1979), Roberts (1980b), Myerson (1983), Blackorby, Donaldson and Weymark (1984)], without making the levels non-comparable as in d'Aspremont and Gevers' (1977) method.[73]

[73] For a very different route to the axiomatization of utilitarianism, see Ng (1975). See also Danielson (1974) and Mirrlees (1982).

Maskin's axiomatisation supplements the imposed conditions by separability (Condition SE) and a requirement of continuity, to wit, that $W(\cdot)$ in (6.2) be continuous.

Utilitarianism derived with separability and continuity
A SWFL satisfying Conditions $\tilde{\mathrm{U}}$, $\tilde{\mathrm{I}}$, $\tilde{\mathrm{P}}^*$, $\tilde{\mathrm{A}}$, SE, continuity, and the invariance requirement for cardinal full comparability CF, must be utilitarian.

It follows from the application of Debreu's (1960) theorem on additive separability, that due to $\tilde{\mathrm{U}}$, $\tilde{\mathrm{I}}$, $\tilde{\mathrm{P}}^*$ and SE, it must be the case that there exist continuous functions $v_i(\cdot)$ such that xRy if and only if $\sum_{i=1}^{n} v_i(U_i(x)) \geq \sum_{i=1}^{n} v_i(U_i(y))$. By anonymity, for all i, $v_i(\cdot) = v(\cdot)$. Maskin completes the proof by demonstrating (with the help of the invariance requirement CF, and continuity, in addition to $\tilde{\mathrm{U}}$, $\tilde{\mathrm{I}}$ and $\tilde{\mathrm{P}}^*$) that $v(\cdot)$ must be a positive affine transformation. That establishes that the SWFL is utilitarian.

Deschamps and Gevers (1978) have proved a theorem that provides another route to axiomatic derivation of utilitarianism – strictly speaking a slightly weakened version of it. A SWFL will be called "utilitarian-type" if it yields a utilitarian strict preference for all cases in which the utility sums to be compared are different; it may or may not declare two equal-utility-sum states as indifferent.

Joint characterization theorem
A SWFL satisfying Conditions $\tilde{\mathrm{U}}$, $\tilde{\mathrm{I}}$, $\tilde{\mathrm{P}}^*$, $\tilde{\mathrm{A}}$, SE, ME, and the invariance condition CF, must be either leximin or of the utilitarian-type.

We know from the Leximin Derivation Theorem, in particular case [1] of it, that these conditions with the additional requirement of invariance for ordinal level comparability OL will lead to leximin. By broadening the utility informational framework to cardinal full comparability, the only additional rules that are admitted must be of the utilitarian-type. If now leximin is excluded by some axiom, and there are many "mild" axioms that will do this, the class of utilitarian-type rules would have been axiomatized. The advantage of this route lies in the fact that it demands neither continuity, which may not be accepted to be an intuitively "basic" social welfare property (though satisfied by utilitarianism in particular), nor the informational limitation of CU, which renders an important parameter (viz. comparative utility levels) unavailable for use. On the other hand, the Joint Characterization Theorem delivers a little bit less, viz. utilitarian-type rules rather than the utilitarian rule, and also this route requires some additional exclusion, notably something to knock out leximin.

Myerson (1983) derives utilitarianism from Pareto optimality and a linearity condition, but – more importantly – shows that Pareto optimality, independence and a concavity condition together ensure that the social welfare rule must be *either* utilitarian *or* egalitarian – a remarkable elimination of all other rules.

6.7. *Other informational structures*

While ordinal non-comparability, cardinal non-comparability, ordinal level comparability, cardinal full comparability and cardinal unit comparability have been the informational assumptions that have been most used (as in the results discussed above), other alternative informational structures have also received some attention. Indeed, recently the various alternative possibilities have been fairly thoroughly investigated by Roberts (1977, 1980a, 1980b), Gevers (1979), Blackorby and Donaldson (1979), and Blackorby, Donaldson and Weymark (1984).

While space will not permit a discussion of the different possibilities, the particular case of ratio-scale full comparability must be briefly mentioned. Using axioms similar to those used to arrive at utilitarianism for cardinal full comparability, Roberts (1980b) has established that with ratio-scale full comparability, the SWFL must be of the more general class specified by (6.5) above, i.e. with constant elasticity (t) transforms of individual utilities being added to arrive at social welfare W. The value of t is unspecified. To obtain the special case of utilitarianism, viz. $t=1$, would require some additional restriction, e.g. the invariance requirement for cardinal full comparability, which is a good deal more restrictive than ratio-scale full comparability. An alternative route towards utilitarianism has been pointed out by Blackorby and Donaldson (1979) by considering negative as well as positive utility values and demanding that social welfare be quasi-concave on individual utilities.[74]

Another possibility that seems important is the case of "partial" comparability and "partial" measurability. For example, if $\bar{L}(1)$ and $\bar{L}(0)$ are comparability sets respectively for cardinal unit comparability and cardinal non-comparability, then a case of "partial unit comparability" is one in which the comparability set $\bar{L}$ lies somewhere in between the two, i.e. $\bar{L}(1) \subseteq \bar{L} \subseteq \bar{L}(0)$. The partial nature of the comparability assumption reflects a certain amount of "vagueness" about the way individual utility units can be compared with each other. It leads to quasi-orderings (reflexive, transitive, but not necessarily complete) for such rules as utilitarianism – the quasi-ordering getting monotonically

[74] Note also that with some ratio-preserving transformations that are commonly used in economic exercises, the consideration of negative utilities would cause problems. As Blackorby and Donaldson (1979) note, in the negative utility orthant, the Atkinson (1970) measure of inequality based on means of order r would react perversely to a Lorenz curve improvement. In these cases, the argument for imposing boundary conditions on utility functions guaranteeing non-negativity may well be strong – indeed overwhelming.

extended as the fuzziness diminishes, and a defined "degree" of partial comparability rises systematically from 0 to 1 [see Sen (1970a, chapter 7*; 1970c)]. Other partial comparability cases can also be considered, e.g. partial level comparability. Measurability parameters can also be taken to be partial, e.g. partial cardinality.[75]

An important contrast between the results dealing with such "partial" frameworks and the results of "pure" types reported earlier relates to the output of the aggregation exercise and correspondingly to the way the invariance requirement is defined. If the demand is for a *complete* social ordering, as with SWFLs, it is natural to require that if some rule leads to xRy for some n-tuple $\{U_i\}$ in $\bar{L}$, and to yPx for some other n-tuple $\{U_i^*\}$ in the *same* $\bar{L}$, then that rule is to be rejected altogether. This is what the invariance requirement specified in Section 6.1 – and used in most of the literature – does. This can be called "global" invariance requirement. The alternative – the "local" requirement – is less restrictive and works especially well for cases of partial comparability. It insists only on the social preference being a quasi-ordering, and in the case of an inconsistency over some pair – as in the example above – it leaves that pair unranked. Only those pairs that are consistently ranked by all $\{U_i\}$ in $\bar{L}$ are then ranked in the social preference. The contrast between the two approaches can be illustrated by remarking that with utilitarianism and ordinal non-comparability, the global approach will record an inconsistency, while the local approach would simply assert the Pareto quasi-ordering.[76] While the global approach has received a good deal more attention than the local one, there is much to be said for the wasteless use of available information that the local approach permits.[77] Since completeness of social ordering is a demanding requirement – as we have discussed earlier – a more thorough exploration of the local avenue might well be rewarding.[78]

7. Informational availability and manipulation

7.1. *Problem types*

Under the broad hat of "aggregation" in social choice theory rest problems of quite distinct types. Among various bases of classification, one concerns the

[75] See Sen (1970a, chapter 7*; 1979a), Blackorby (1975), Fine (1975a), Basu (1979), and Bezembinder and van Acker (1979). There are some similarities with Levi's (1974) treatment of "indeterminate probabilities".

[76] The contrast was explored in Sen (1970a), where the global approach was the one used in Chapters 8 and 8* and the local approach in Chapters 7 and 7*.

[77] It is also possible to relax the requirement of consistency of social preference from transitivity to quasi-transitivity or acyclicity, and to consider non-binary formulations of social choice, in line with the procedures considered in Sections 3 and 4.

[78] Another important problem concerns combining an n-tuple of "extended orderings" (including each person's interpersonal comparisons). The problem was first investigated by Suppes (1966), and it has received attention from Sen (1970a, chapter 9*; 1977b), Hammond (1976a), Roberts (1977, 1980b), Kelly (1978, chapter 8), Mizutani (1978), Suzumura (1983a), Gaertner (1983), and others.

interpretation of individual preferences R_i (or utilities U_i). These could reflect a person's conception of his *own* well-being, or – alternatively – his idea of what is good for the *society* [see Harsanyi (1955), Suppes (1966), Sen (1977c)]. To assert the distinction is not to deny that a person's conception of his own well-being may well take note of the welfare of the others in the society, but still the questions "what is best for the society?" and "what is best for me?" are different ones, even though they are clearly interrelated.

At the risk of oversimplification we may distinguish between an exercise of "interest-aggregation" – wherein different people's personal interests are aggregated – and that of "judgment-aggregation" – wherein different persons' judgments about what is good for the community are aggregated.[79] The typical formulation of the problem of the "fair division" of a cake among a group of cake-loving individuals illustrates the former.[80] On the other hand, Borda's famous method of aggregating different views on the "merits" of a candidate to membership of the Academy of Sciences (later denounced – effectively – by a new member called Napoleon Bonaparte) was clearly addressed to the problem of aggregation of judgments.

In an interest-aggregation exercise, the informational base of the individual orderings of the social states is particularly limiting, and it can be sensibly supplemented by additional information about rankings of different persons' positions in a given state (see Section 5) or by straightforward interpersonal comparisons of well-being and of gains and losses from change (see Section 6). This is the typical framework for economic planning [see, for example, Dobb (1955), Malinvaud and Bacharach (1967), Chakravarty (1969), Arrow and Kurz (1970), Heal (1973), Dasgupta and Heal (1979), Dasgupta (1982), and Majumdar (1983)]. Even when one person does the personal exercise of finding out what his "ethical preferences" should be [see Harsanyi (1955)], he may have to go well beyond just the n-tuple of individual orderings, bringing in interpersonal comparisons of utility, perhaps placing himself in the position of others [see Vickrey (1945), Harsanyi (1955), Rawls (1958), Suppes (1966), and Arrow (1963, pp. 14–15)].[81] On the other hand, in judgment-aggregation exercises, especially in such institutional contexts as committee decisions, or elections, it may be very difficult to have room for anything other than mechanically recording people's preference rankings (or declared preference rankings). There the exercise may have to make do with the n-tuple of individual orderings only. If this – admittedly oversimplified – dichotomy is accepted, then it may well be the case that the

[79] The distinction is explored in Sen (1977a). There could, of course, be *mixed* cases in which the aggregation exercise takes into account *both* judgments *and* interests of the people involved; for an example, see Graaff (1977). See also Bose (1975).

[80] See Luce and Raiffa (1957, section 14.9) for a discussion of fair mechanisms for cake division. A different type of norm and a different class of ideas on fairness can be found in various concepts of "exploitation", on which see Roemer (1982).

[81] Recent contributions include Kern (1978) and Leinfellner (1978), among others.

Arrovian informational format is more relevant for some exercises – typically aggregation of judgments – while the richer informational structures analysed in Sections 5 and 6 are more relevant for others – typically aggregation of interests.

Even when the Arrovian informational base of n-tuples of individual orderings is taken as appropriate, and institutional mechanisms are geared to this informational format, there remains the important problem of getting hold of the "true" orderings of social states by the individuals.[82] If the procedure for collecting this information is some type of voting mechanism, then the problem of guaranteeing "sincere voting" arises. This problem of "strategy-proof" voting procedures has been much investigated recently, and in the rest of this section this question is examined.

7.2. *Manipulability and dominant strategies*

That the characterisation of social choice in terms of social welfare functions abstracts from the "game aspects" of the problem was noted by Arrow (1951), conjecturing that "once a machinery for making social choices from individual tastes is established, individuals will find it profitable, from a rational point of view, to misrepresent their tastes by their actions" (p. 7).[83] A firmer conjecture about the potential manipulability of social choice mechanisms – with a persuasive defence – was presented by Vickrey (1960).[84] And Dummett and Farquharson (1961) made a universalized conjecture: "It seems unlikely that there is *any* voting procedure in which it can *never* be advantageous for any voter to vote 'strategically', i.e., non-sincerely" (p. 34, italics added).[85] The recent investigation of the manipulability of voting mechanisms – starting with the contributions of Murakami (1968), Gibbard (1973), Pattanaik (1973) and Satterthwaite (1975) – has essentially confirmed these pessimistic conjectures.

A voting scheme picks one social state x from a given set X of social states for any logically possible n-tuple of *reported* preference orderings (or ballots, for short) of X. A voting scheme is "manipulable" (not "strategy-proof") if and only if for some n-tuple of true individual preference orderings, there is at least one person k who can improve the outcome for himself by reporting a preference

[82] Much insight has been gained recently by experimental studies of behaviour and response [see Plott (1979), V. L. Smith (1979), Ordeshook (1980), and other recent contributions].

[83] For an early conjecture of the manipulability result, see Hoag and Hallett (1926, pp. 396–397). I am indebted to Duff Spafford for this interesting reference.

[84] Majumdar (1956) presented reasons for expecting widespread manipulability of "issues", i.e. the possibility of gain from sponsoring unfavoured alternatives for strategic reasons. This type of manipulability has not yet been analysed as much as it seems to deserve. See, however, Luce and Raiffa (1957, section 14.8) and Pattanaik (1978, chapter 9), and on related problems of agenda manipulation, Campbell (1979) and Plott and Levine (1978).

[85] See also Farquharson (1956, 1969).

ordering different from his true one when others report true preferences. More formally, a voting scheme $V(\{R_i\}) = x$ is *manipulable* if and only if for some $\{R_i\}$, some k, and some R_i^*, $V(\{R_i^*\})P_kV(\{R_i\})$ when $R_i^* = R_i$ for all $i \neq k$. A voting scheme is dictatorial if and only if there is a person i such that whichever element of the *range* of $V(\cdot)$ he ranks highest in his ballot is invariably the element that is chosen by the voting mechanism.

Gibbard–Satterthwaite manipulability theorem
Every non-dictatorial voting scheme with at least three distinct outcomes is manipulable.

Gibbard (1973) establishes this theorem as a corollary of another one dealing with "game forms" in general, of which voting schemes are special cases. A game form does not restrict the strategies to be chosen by the individuals to the orderings of social states, i.e. to ballots, and each person i's strategy set S_i can be any set of signals. A game form specifies an outcome x from a given set Y for every n-tuple of strategy choice $(s_1, \dots, s_n)$ with $s_i \in S_i$ for all i (that is, a game form is a mapping from the Cartesian product of strategy sets of individuals to the set Y of outcomes). A voting scheme is a game form such that the strategy set S_i of each person is a set of declared orderings (ballots) of a set X of social states including the set Y of outcomes. A game form is "straightforward"[86] if for each person i and for any preference ordering of the outcomes that he might have, he has a *dominant* strategy, i.e. a best strategy with respect to his ordering of the outcomes *irrespective* of what the strategies of others might be. A game form is dictatorial if there is a person k such that for every outcome x, there is a strategy $s_k(x)$ for k such that if k chooses $s_k(x)$, then the outcome must be x, no matter what others choose. (It is readily checked that a dictatorial voting scheme must be a dictatorial game form.) Gibbard's theorem about game forms in general – rather than about voting schemes in particular – is the following:

Gibbard's non-dominance theorem about game forms
No non-dictatorial game form with at least three possible outcomes can be straightforward, i.e. in every non-dictatorial game form, there is at least one person who does not have a dominant strategy for some preference ordering of the outcomes.

The existence of dominant strategies for everyone for every possible preference n-tuple would, of course, be a pretty demanding requirement, so the Non-dominance Theorem is not really counter-intuitive. But, as Gibbard notes, the Manipu-

[86] This concept, like many others in this part of the literature, was introduced by Farquharson (1956, 1969). One of the other notions introduced by Farquharson, viz. "sophisticated voting" (based on successive elimination of dominated strategies), has been very fruitfully investigated recently by Brams (1975), Pattanaik (1978), Moulin (1979, 1983), and others.

lability Theorem follows immediately from this Non-dominance Theorem. If the voting scheme were non-manipulable, then everyone must have a dominant strategy, viz. recording his true preference irrespective of what others do. Since Gibbard establishes – most elegantly – the Non-dominance Theorem, he obtains the Manipulability Theorem directly from it.[87]

The analytical connection between the Gibbard–Satterthwaite theorem and the Arrow theorem has been widely noted. It is possible to define social preference R with respect to a voting scheme such that for a voting scheme to be strategy-proof, that social preference relation R has to be determined by an Arrovian social welfare function satisfying pair-relational independence I^2. The demand for a non-manipulable, non-dictatorial voting scheme with at least three outcomes can then be translated as the demand for a social welfare function satisfying Arrow's conditions U, P, I and D. Since the latter demand cannot be met, neither can the former.[88]

This close correspondence between impossibility results on the existence of reasonable social decision procedures and impossibility results about manipulability of voting schemes applies also to extensions and variations of the Arrow impossibility result. This has been investigated for cases involving many variations, such as non-transitive social preferences, non-binary social choice, probabilistic social preference, cardinal individual utilities, infinite set of voters, restricted domain of social welfare functions, etc., and a number of striking correspondence results have been established.[89]

7.3. *Manipulability with multiple outcomes and with counterthreats*

The Gibbard–Satterthwaite Manipulability Theorem is constrained by two rather limiting features of the chosen characterization of manipulability. First, the voting schemes (and more generally the game forms) are characterized as having a unique outcome x for any combination of ballots (more generally, strategies). Second, the formulation of the optimum choice of strategy does not give any room to strategic responses by others and considerations of "counterthreats" are not brought in.

[87] For other proofs of the Gibbard–Satterthwaite manipulability theorem, and related matters, see Satterthwaite (1975), Gärdenfors (1976), Jain (1977b), Pattanaik (1978, chapter 5), Schmeidler and Sonnenschein (1978), Batteau and Blin (1979), Chichilnisky and Heal (1979), Dasgupta, Hammond and Maskin (1979), Barbera (1980b), Batteau, Blin and Monjardet (1981), Moulin (1983), and Peleg (1984).

[88] See Gibbard (1973), Satterthwaite (1975), Schmeidler and Sonnenschein (1978), and Pattanaik (1978). In fact, the correspondence applies not merely to voting schemes but also to the more general case of game forms, and Gibbard established his Non-dominance Theorem by using the Arrow impossibility result.

[89] The literature is quite vast. For good accounts of the main results, see Kelly (1978), Pattanaik (1978), Kim and Roush (1980a), Moulin (1983), and Peleg (1984).

Gibbard's (1973) investigation of manipulability was, in fact, paralleled contemporaneously by a similar exploration by Pattanaik (1973), who did not however insist that the voting scheme must yield a unique outcome x, but rather that it could specify a non-empty subset $C(X)$, the choice set of X. Pattanaik's results were indeed much less negative – in fact, he established some positive possibility theorems requiring that voting schemes should be able to specify a subset $C(X)$, rather than invariably a single state, and he used "maximin" behaviour in choosing over subsets of outcomes.[90]

Once non-unique outcomes are admitted, there is need for supplementing the voting mechanism by specification of (i) rules about how to break ties, and (ii) characterization of how the people involved would behave faced with uncertainty about the final outcome (from the subset specified by the voting mechanism). Gibbard (1977) has extended the manipulability result to the case of a pure lottery mechanism in selecting a Pareto efficient final outcome, and a behaviour pattern that relies entirely on expected utility maximization. A similar result about the impossibility of a non-dictatorial and non-manipulable mechanism can also be arrived at by a much weaker requirement on behaviour under risk provided the mechanism satisfies a rather stringent condition of "positive responsiveness" of the subset $C(X)$ to individual preferences [see Barbera (1977a)].[91]

Since "positive responsiveness" of mechanisms as well as individual behaviour based entirely on expected utility maximization are both demanding assumptions, the investigation has been continued into cases with less stringent specification. A variety of impossibility results have emerged under alternative combinations of requirements, with specific attempts to make the restriction on behaviour as weak as possible [see especially Barbera (1977b), Pattanaik (1978), MacIntyre and Pattanaik (1981), and Peleg (1982)]. The main message to emerge from all this literature is that while the original manipulability result does need substantial revision when the voting mechanism is not required to yield unique outcomes, the pessimism about finding non-dictatorial and non-manipulable mechanisms remains well-grounded.

The same general message emerges from the investigation of manipulability defined more stringently by taking into account counterthreats. Various alternative ways of characterizing behaviour in the presence of response of others has led to different formulations of "strategy-proofness",[92] but in each case the pessimism about non-manipulable and non-dictatorial voting mechanisms seems to

[90] See also Gärdenfors (1976, 1979), Gardner (1977), Kelly (1977), Pattanaik (1978), and Sengupta (1980a).

[91] See also Barbera and Sonnenschein (1978), Kelly (1978), Pattanaik (1978), Barbera (1979, 1980a), Feldman (1979, 1980b), Dutta (1980b), and Sengupta (1980a).

[92] See Pattanaik's (1978) distinctions between Types II, III and IV of strategy-proofness (Chapter 6). Type I is strategy-proofness or manipulability in the absence of any response, or counterthreats, by others. See also Pattanaik and Sengupta (1980).

re-emerge in the reformulated format [see, especially, Pattanaik (1976b, 1976c, 1978)]. The impossibility of reasonable voting procedures that would be non-manipulable seems to survive a good deal of variation in the requirement of reasonableness of such procedures and in the characterization of non-manipulability.[93]

7.4. *Equilibrium, consistency and implementation*

The focus on "honest" revelation of preferences in the literature surveyed above has come under serious scrutiny in recent years. If the object of the exercise is effectiveness in the sense of getting an appropriate outcome (rather than having the moral glory of everyone being perfectly honest in reporting their preferences), then the thing to investigate is the existence of an effective mechanism rather than a strategyproof one. If, for example, a non-strategy-proof voting mechanism yields an equilibrium of dishonest behaviour that produces the same outcome as honest revelation of preferences would have, then the mechanism could well be regarded as successful in terms of effectiveness.

Various alternative ways of characterizing an "equilibrium" have been considered. Obviously, there is no advantage in asking for a *dominant* strategy equilibrium, for the Gibbard Non-dominance Theorem is exactly concerned with this case.[94] Perhaps less obviously (but obviously enough), there is not much point in asking for a voting mechanism that yields truthful ballots as Nash equilibria. If such a mechanism were to exist, then everyone's honest strategy would be his best strategy given the honest strategy choice of others, and these latter could be any set of strategies at all. Thus a mechanism that guarantees that any n-tuple of honest strategies must be a Nash equilibrium, would also guarantee that honest strategies must be dominant strategies. Since the latter requirement would lead to impossibility, so would the former.

Hence, in this framework, if the solution concept is based on Nash equilibrium, then one must admit dishonest strategies as well, and be content with Nash equilibria such that they yield the same outcome as the true preferences would.

[93] See also Sengupta (1978a).

[94] Gibbard's (1973) "Non-dominance Theorem" is, of course, not a result concerned with honesty as such. It translates into a theorem about manipulability only because with ballots as strategies, an honest strategy has to be a dominant one. For general game forms, truth may not require dominance in this sense. For example, in seeking implementation rules for optimum allocation of public goods the strategies in the Groves–Ledyard mechanism in the form of declaration of "the increment (or decrement) of each public good the consumer would like to add (or subtract) to the amount requested by others" [Groves and Ledyard (1977, p. 796)] must make the truthfulness of such strategies dependent on the declaration of others. The absence of non-dominant strategy equilibrium as identified by Gibbard's Non-dominance Theorem, in this more general context, implies nothing about an equilibrium of sincere strategies. See also, Green and Laffont (1979), Dasgupta, Hammond and Maskin (1979), Laffont and Maskin (1981), Chichilnisky and Heal (1981), for various aspects of incentive compatibility.

Peleg (1978) calls a voting mechanism to be "exactly consistent" if and only if "for each profile of true preferences of individuals, it possesses a Nash equilibrium point which yields the *same* social choice as that corresponding to the profile of true preferences" (p. 153).[95] It turns out that a very wide class of voting mechanisms are exactly consistent in this sense, as demonstrated by Dutta and Pattanaik (1978). Voting mechanisms that are exactly consistent and furthermore not distorted by manipulation of preferences by *coalitions* are called by Peleg "exactly and strongly consistent". The score here is much more divided, and the existence of exactly and strongly consistent voting mechanisms depends on the minimal number of persons in a coalition that makes it a "winning" group, compared with the numbers of persons and social states [see Peleg (1978)].[96] If there are at least as many states as there are persons, then at least one person must have a veto.

It is possible to broaden the format of the problem by permitting the use of a game form G *different from* the function F used for making the normative judgment (e.g. the social welfare function or social choice function). The problem can then be formulated as that of finding a game form G such that for any preference situation, the best social state as judged by F would be yielded by G, as an equilibrium outcome [see Pattanaik (1978), Maskin (1978, 1979a), and Roberts (1979)]. The parallel literature on public goods and "revelation of preferences" has been concerned with variants of this type of formulation.[97]

Gibbard (1978) has presented, in this type of format, an impossibility result which parallels Arrow's theorem. Gibbard permits the normative judgment to be based on richer information than Arrow-type social welfare functions, and indeed in effect takes a social welfare functional SWFL satisfying unrestricted domain, the weak Pareto principle and the absence of a "weak dictator" (i.e. non-existence of a vetoer), and yielding a social ordering of judgments. However, the game form used for implementation defines a "social choice function" SCF that – by virtue of the combination of two postulated axioms – is made to relate social choice over each pair to individual preferences over that pair. Thus, despite the cardinality of the individual utility function, a condition much like pair-choice independence (Condition $\mathrm{I}^2_{\mathrm{C}}$) holds for the implementation mechanism (not necessarily for the SWFL). Assuming that there are at least four distinct social states, Gibbard

[95] See also Dummett and Farquharson's (1961) characterization of "majority games". Exact consistency is quite a mild requirement. There can be *many* Nash equilibria, only one of which might be "desirable" (in terms of true preferences), and most of which could be terrible. Contrast Hurwicz and Schmeidler's (1978) insistence on at least Pareto optimality of *all* Nash equilibria.

[96] See also Dutta and Pattanaik (1978), Pattanaik (1978), Maskin (1979a), Dutta (1980b, 1983), Pattanaik and Sengupta (1980), Peleg (1982), and Moulin (1983).

[97] Dasgupta, Hammond and Maskin (1979) have provided an extensive treatment of this class of problems. See also the literature cited there on the related problem of incentive compatibility, starting with the pioneering contribution of Hurwicz (1962).

demonstrates that it is impossible for such a social choice function to guarantee that *only* those elements that are optimal with respect to the SWFL will be chosen.

Note that the approach used here dissociates the discipline of social judgment from the act of marshalling individual utility information to pick what would be regarded as best points according to that procedure of social judgment.[98] The richer informational structures used in Section 6 – involving interpersonal comparisons – are in principle admissible for making social judgments, but in fact they can't be used in implementation because of the limitation of the signalling device of individual preferences pair by pair. These signals reflect individual choices "when players who are guided by their true utilities interact strategically" [Gibbard (1978, p. 158)]. Despite the different formulation of the exercise, the impossibility result turns out ultimately to be rather similar to Arrow's impossibility theorem in the version involving non-comparable cardinal utility (presented in Section 6.2). As Gibbard (1978, p. 163) puts it:

> "...if we take the conditions needed for a cardinal version of the Arrow theorem (Sen, 1970 [1970a here], p. 129), there are only two differences. One is that Arrow's non-dictatorship condition is weaker than the condition of No Weak Dictator given here. The other, more crucial difference is that Arrow has a strengthened version of the condition of Optimality. Optimality here requires that all members of the choice set of the SCF be best feasible alternatives; Arrow requires in addition that all best feasible alternatives be included in the choice set."

I end with some less discouraging remarks. First, various other types of "implementation" problems can be and have been considered, e.g. implementing social choice *correspondences* rather than specifically social choice *functions*, and some of these offer positive possibilities for both Nash equilibrium and strong equilibrium [see Moulin and Peleg (1982), Peleg (1982), and Moulin (1983)]. Second, procedures such as "voting by veto" [Mueller (1978)], while unattractive in some respects, do offer scope for true revelation of preferences [Barbera and Dutta (1982)] as well as for exact and strong consistency [Dutta (1983)]; see also Moulin (1983). Third, domain restriction can play an important part in making manipulability less of a problem. Indeed, the requirement of unrestricted domain is very limiting in many economic contexts, e.g. when people can be relied upon to prefer more to less. [On implementational possibilities with domain restriction, see particularly Dasgupta, Hammond and Maskin (1979).] In the more traditional format in which strategies take the form of preference rankings, the necessary and sufficient conditions for strategy-proofness have been identified by Maskin (1976a),

[98]A similar approach was used earlier by Campbell (1976), as Gibbard notes.

Kalai and Muller (1977), and Ritz (1981).[99] In the format of "consistency", with strategies restricted to preference rankings, but with various different solution concepts (such as Nash equilibrium, the core, exact and strong consistency, etc.), Dutta (1980b), Peleg (1982) and others have provided extensive investigations of the domain restrictions that are adequate for the purpose at hand.

Further, a consequence of combining unrestricted domain, the weak Pareto principle and – in the particular context of implementation – independence, is to produce a "neutrality" result (see the Field Expansion Lemma in Section 2, and the Strong Neutrality Theorem in Section 6). This has the effect of ruling out any essential use of non-utility information for implementable social welfare judgments. Since many public decision procedures are based on direct use of *non-utility* information (e.g. in providing social security to the hungry, the ill, or the unemployed) rather than on expressed utility information (e.g. basing social security on expressions of disutility from hunger or joblessness), the informational base for practical decision-taking is indeed a good deal wider than is allowed by the implementation mechanisms characterized in these exercises.

Finally, the assumption that each individual's choice depends exclusively on the pursuit of personal utility or preference in a strategic way, irrespective of other considerations, may not be very realistic [see Johansen (1976), Sen (1977c)]. Indeed, within the limits of such an assumption there is some difficulty in explaining why people are ready to take the trouble of voting at all in large elections.[100] The assumption is particularly galling when the social choice exercise is taken to be one of aggregation of judgments about what is best for society, rather than of aggregation of personal interests.

8. Domain restrictions

8.1. *Restricted preferences and voting outcomes*

After establishing the impossibility theorem, Arrow (1951) had proceeded to suggest an escape route through a domain condition that is called "single-peaked preferences" [see also Black (1948)]. If individual preferences happen to be single-peaked and if the number of voters happens to be odd, then the method of

[99] The famous Arrow–Black condition of "single-peaked preferences" turns out to be inadequate for strategy-proofness [see Blin and Satterthwaite (1976)]. See also Moulin (1980).

[100] On various aspects of this complex problem, see Downs (1957), Barry (1965, 1970), Olson (1965), Tullock (1968), and Riker and Ordeshook (1968, 1973), among others. The possibility of a single voter affecting the outcome is related to the probability of "ties", and this probability is very low for large communities. Chamberlain and Rothschild (1981) show that in an election with $2n+1$ voters, the probability that any one voter casts the decisive ballot is of the order $1/n$.

majority decision would yield transitive social preference [Arrow (1951, theorem 4)]. Roughly speaking, single-peakedness requires that the set of social states can be so arranged on a line that the utility curve ("intensity of preference") of everyone would be unimodal – either monotonically rising, or monotonically falling, or rising up to a maximum and falling thereafter. Such a condition looks plausible if everyone votes according to some one characteristic, e.g. how "left-wing" the alternative is. If the states are lined up according to that characteristic (the more left on the line, the more left-wing the alternative), then the voters' preferences – under the postulated one-characteristic system of ranking – can be represented from left to right as rising uniformly ("the extreme right-wing hyenas"), falling uniformly ("the extreme left-wing creeps"), or rising up to the point of "optimum" left-wingness and falling thereafter (variants of "wishy–washy centrists").

Since transitivity is a property of triples, it is immediate that the required conditions can be weakened – without losing the result – by demanding single-peakedness over triples even if the set of all alternatives cannot be so arranged. Other extensions appear natural, e.g. having "single-caved preferences" [Vickrey (1960), Inada (1964b), and Ward (1965)]. Indeed, this type of condition leads to a generalized condition called "value restriction". To motivate this generalization, note that single-peakedness over a triple requires that if x, y, z is the order in which they are arranged, then for anyone i for whom xR_iy, it must be the case that yP_iz. But this condition is equivalent to demanding that for all i, *not* (xR_iy and zR_iy), and it can be seen as simply restricting y from being a "worst" alternative in anyone's preference order over that triple. Value restriction requires that for any triple x, y, z, there is at least one alternative, say y, and at least one "value" (viz., "worst", "best" or "medium") such that in no one's preference ordering does that alternative have that value. Since individuals who are indifferent over all three alternatives in a triple do not sway the majority voting outcome, indifference over the triple need not be ruled out, and such people are called "unconcerned" over that triple. Let $H(x, y, z)$ be the set of people who are *not* unconcerned ("concerned") over that triple.

Value restriction (VR)

Individual preferences are value restricted over X if for every triple in X, there is an alternative, say x, such that the following condition holds, denoting the other two alternatives as y and z: $[\forall i \in H(x, y, z)\colon xP_iy \text{ or } xP_iz]$ *or* $[\forall i \in H(x, y, z)\colon yP_ix \text{ or } zP_ix]$ *or* $[\forall i \in H(x, y, z)\colon (xP_iy \;\&\; xP_iz) \text{ or } (yP_ix \;\&\; zP_ix)]$.

Arrow's theorem about single-peaked preferences, suitably generalized, can cover all value restricted preferences [Sen (1966) and Majumdar (1969b)].

Value restriction SWF theorem
If individual preferences are value restricted over X and the number of concerned individuals for every triple is odd, then the majority rule is a SWF, yielding (transitive) orderings.

Since the arbitrary condition of oddness of number is a bit of a peculiar restriction, the following result is rather less ad hoc [Sen (1969)].

Value restriction SDF theorem
If individual preferences are value restricted, then the majority rule is a quasi-transitive social decision function QSDF.

As far as full transitivity is concerned, a sufficient condition is Extremal Restriction, requiring that if anyone i has an antisymmetric (strict) preference order over a triple, $xP_i y$ & $yP_i z$, then no one j should partially oppose this by preferring z to x without having *exactly* the opposite preference of i (that is, either *not* $zP_j x$, or $zP_j y$ & $yP_j x$).[101]

Extremal restriction (ER)
Individual preferences are extremal restricted over X, if for every triple $x, y, z \in X$, $(\exists i\colon xP_i y \;\&\; yP_i z) \Rightarrow [\forall j\colon zP_j x \Rightarrow (zP_j y \;\&\; yP_j x)]$.

Not only is ER sufficient for transitivity of majority decision, it is also necessary in an interesting sense. A domain restriction for some property of the range (e.g. that social preferences be all transitive) is necessary, in this sense, if every violation of the restriction leads to a list of preference orderings such that some assignment of these orderings over some number of individuals would lead to the violation of that property of the range (e.g. would lead to *intransitive* social preference). The following theorem was established by Sen and Pattanaik (1969), and an essentially equivalent result was proved by Inada (1969).[102]

Necessary and sufficient preference restriction for majority rule (SWF)
The necessary and sufficient restriction of preferences for the majority rule to be a SWF (in particular, to yield *transitive* social preference relations) is extremal restriction.

As far as the weaker demand of social decision functions are concerned, yielding *acyclic* social preference relations, the required restriction is less exacting.

[101] While ER was proposed in this form in Sen and Pattanaik (1969), Inada's (1969) "dichotomous preferences", "echoic preferences" and "antagonistic preferences" *together* cover exactly the same ground.

[102] See also Inada (1970), Kelly (1974a), Kaneko (1975), and Chichilnisky and Heal (1983).

It is necessary here to introduce a further condition, viz. one that demands that in every triple there is a limited agreement to the effect that some pair is weakly ranked by everyone in the same way.

Limited agreement (LA)
Individual preferences satisfy limited agreement over X if in every triple there is a pair, say (x, y), such that for all i, $xR_i y$.

Necessary and sufficient preference restriction for majority rule (SDF)
The necessary and sufficient restriction of preferences for the majority rule to be a SDF (in particular, to yield *acyclic* social preference relations) is that either extremal restriction, or value restriction, or limited agreement, be satisfied.[103]

Inada (1970) showed that the necessary and sufficient restrictions for a majority rule QSDF are also exactly the same.

With strict (antisymmetric) preferences, the necessary and sufficient condition in both cases is fulfilment of VR.[104]

Necessary and sufficient conditions for voting rules other than majority decision, e.g. multi-stage majority decision rule, non-minority rule, semi-strict majority rule, and other variants, have also received much attention,[105] but they will not be pursued here.

A limitation of the "restricted preference" approach, on which the above results are based, may now be noted. The restrictions considered *rule out* certain types of preferences and impose *no other* condition about the number of people holding one type of preference or another.[106] It is possible instead to investigate the domain conditions that have to be satisfied taking into account actual numbers of people holding different preference orderings. To this alternative approach, I turn in the next subsection.

8.2. Number-specific domain conditions

The credit for pioneering the approach of number-specific constraints should go to Nicholson (1965) and Tullock (1967). Tullock's sufficiency condition for

[103] Sen and Pattanaik (1969). See also Sen (1970a), Pattanaik (1971), Taylor (1971), and Fishburn (1972a, 1973a).

[104] Sen and Pattanaik (1969), Sen (1970a), and Pattanaik (1971).

[105] See Pattanaik (1971), Kelly (1971, 1974a, 1978), Davis, De Groot and Hinich (1972), Fishburn (1972a, 1973a), Blin (1973), Sloss (1973), Kramer (1973, 1977), Ferejohn and Grether (1974), Kuga and Nagatani (1974), Saposnik (1974, 1975a, 1975b), Salles (1975, 1976), Blin and Satterthwaite (1976), Deb (1976), Plott (1976), Schofield (1977a, 1983a, 1983b), Slutsky (1977), Peleg (1978, 1984), Chichilnisky and Heal (1983), Kim and Roush (1980a), and Blair and Muller (1983).

[106] In fact, the meaning of "necessity" is ambiguous in this context. For other characterizations, see Pattanaik (1971) and Kelly (1974a), among others.

transitive majority rule has been subsequently generalized – most powerfully by Grandmont (1978) – and it may be useful to consider Tullock's characterization in some detail. Consider a real plane E^2. For any voter i, let a_i, a point in E^2, represent his or her best alternative, all alternatives are ranked by i entirely on the basis of their distance from a_i. The indifference curves for everyone are, thus, circles with centre a_i, not necessarily the same for different individuals.[107] Tullock assumes that the sets of a_i, that is the "centres" (or best points), are symmetrically distributed over a rectangle with centre a^*. The majority relation must then be transitive.

The Tullock conditions are suitable for generalization in many different ways. First, the uniform distribution over a rectangle can be replaced by other distributions with similar effect, e.g. uniform distribution on the *boundary* of a rectangle with centre a^*, or on a disc (or on its boundary) with centre a^*. Second, instead of a plane, an m-dimensional characterization can be chosen, and the result correspondingly generalised [see Davis, DeGroot and Hinich (1972)]. The important point about Tullock's example is that every line through a^* cuts the distribution of voters (i.e. of a_i) into two parts of equal measure, and every line that does such an equal division goes through a^*. These properties have been generalized by Grandmont (1978).

For Grandmont, preference ordering R_a is defined by a, belonging to an open convex subset A of E^n. The family of the preference relations $(R_a)_{a \in A}$ satisfy a weak continuity property H.1, viz. the set $\{a \in A \mid xR_a y\}$ is closed in A. There is, in addition, the regularity condition H.2 that if a is a strictly convex combination of a' and a'', then R_a must be "intermediate" between $R_{a'}$ and $R_{a''}$, in the sense that $R_{a'}$ (resp. $R_{a''}$) must be a subrelation of R_a conditionally on $R_{a''}$ (resp. $R_{a'}$) holding over the relevant ordered pair.[108] Finally, Grandmont assumes that the distribution of individual preferences represented by the distribution of a_i satisfies the property that there exists some a^* in A such that every hyperplane through a^* produces equal proportions of a_i in the two closed half spaces, and every hyperplane with that equal division characteristic goes through a^* (condition M.1).

[107]It is tempting to think – and has been often suggested – that this is a generalization of Arrow's single-peaked preferences from a line to a plane. It is certainly true that both sets of conditions satisfy the condition that on any line from the most preferred point, the further away one goes the less one likes the alternative. On the other hand, while that is all that is required in the case of single-peaked preferences, with Tullock's condition two points at the *same* distance from the most preferred alternative must be indifferent, irrespective of the direction in which one moves (in the case of single-peaked preferences which permits movements in two opposite directions, no such requirement is imposed). Furthermore, the circular shape of indifference curves has to be supplemented by some assumption about uniform distribution of a_i, to get Tullock's result.

[108]That is, for any x, y, given $xR_{a'}y$, $(xR_{a''}y \Rightarrow xRy)$ & $(xP_{a''}y \Rightarrow xPy)$, and given $xR_{a''}y$, $(xR_{a'}y \Rightarrow xRy)$ & $(xP_{a'}y \Rightarrow xPy)$. Grandmont's own statement is somewhat different, but equivalent.

Grandmont's theorem on intermediate preferences
Conditions H.1, H.2 and M.1 imply that the majority preference relation must coincide with R_{a*}.

As an immediate corollary, the transitivity of the majority relation follows from the transitivity of R_{a*}, when individual preferences are taken to be transitive.

In interpreting Grandmont's result, it should be noted that unlike in Tullock's example (with circular indifference curves) the structure of individual preferences is given a great deal of latitude here, requiring only that the family of such preferences should satisfy a weak continuity property and the "intermediate preference" condition. On the other hand, condition M.1 retains the demanding numerical requirement that *every* hyperplane through a^* would split the voters in two equal halves. However, Grandmont shows that this condition can be relaxed substantially.[109]

Another interesting feature of Grandmont's result is that it produces a distribution of voters such that all preference orders except R_{a*} in effect either cancel each other out in a majority contest, or reinforce R_{a*} by pulling in opposite directions. This idea of preference order combinations neutralizing each other, leaving an intermediate ordering ruling the roost, has been explored by other writers as well [e.g. by Nicholson (1965), Plott (1967), Saposnik (1975a), Slutsky (1977, 1979), Gaertner and Heinecke (1978), and Matthews (1978)]. Saposnik (1975a) shows the sufficiency (and under special conditions, also the necessity) of "cyclical balance" in which the same number of individual preferences belong to the "clockwise cycle" ($xRyRz$, $yRzRx$, $zRxRy$) as the number belonging to the "counter-clockwise cycle" ($xRzRy$, $zRyRx$, $yRxRz$). Gaertner and Heinecke (1978) have analysed "cyclically mixed preferences", which is a generalization of Saposnik's notion of cyclical balance. They show that the majority decision relation is transitive if and only if it is cyclically mixed.

Using a somewhat similar approach, Slutsky (1977) has provided a complete characterization of preference profiles that lead to consistent majority decision. The technique of analysis involves showing the equivalence of actual preference profiles to some hypothetical ones that are easier to analyse. The "transitive strict preference" (TSP) equivalence is constructed by replacing preferences with indifference by a corresponding set of strict (i.e. antisymmetric) preferences. The profiles are reduced to the "equivalent irreducible society" by jettisoning groups of persons whose combined preferences would lead to indifference among all the

[109]Grandmont also demonstrates that individual preferences being single-peaked or single-caved implies that the family will satisfy H.1 and H.2, and furthermore with an odd number of voters a relaxed version of condition M.1 (viz., his condition M) will also be fulfilled. Thus this provides an alternative way of proving Arrow's theorem about single-peaked preferences, and the corresponding theorem about single-caved preferences.

alternatives for that group under majority decision.[110] With these translations from actual to hypothetical preference profiles, in the "equivalent irreducible society", agreement among the members of a winning coalition dominates the social ranking and gives it the required consistency.

These studies – and others – have substantially enriched our understanding of the consistency problems of majority rule and related decision procedures.

Finally, two general comments on the number-specific approach to domain restriction may be worth making. First, even the domain conditions in the "exclusion" form of "restricted preference" can be given number-specific interpretations, so that the line between the two approaches may be less sharply drawn than it may at first appear. For example, with single-peaked preferences and an odd number of concerned voters, when the *median* voter (in terms of the position of his "best" alternative) is identified, then it can be said that the number of people on "one side" of him is exactly equal to the number on the "other". Thus the single-peakedness characteristic can be translated into a condition requiring that such statements are well-defined (and true). Indeed, even Black's (1948) original theorem about single-peaked preferences took the form of asserting that the best alternative for the *median* voter will win [see also Black (1958, pp. 16–17)].[111]

Second, in order to make the exercise worthwhile, the number-specific conditions must have some intuitive meaning that helps the interpretation of the nature of the preference configurations. Otherwise, there is the danger of merely translating the formal requirement of transitivity (or acyclicity) of the majority relation into a more elaborately stated – but equivalent – number-specific form. When $N(x, y)$ is the number of people who prefer x to y, clearly a condition that asserts that for all x, y, z, $[N(x, y) \geq N(y, x)$ & $N(y, z) \geq N(z, y)] \Rightarrow [N(x, z) \geq N(z, x)]$, is a number-specific requirement for transitivity – irresistably necessary *and* sufficient, and obviously no less "general" than any other condition! The merit of the conditions proposed and the characterizations provided rests in their ability to capture patterns that have *independent* interest and interpretative value.

8.3. *Domain conditions for Arrovian social welfare functions*

While the majority rule is an interesting social choice procedures, it is by no means uniquely so. This leads to the interesting question as to what domain

[110] Gaertner and Heinecke (1978) undertake a similar "reduction".

[111] This, in fact, corresponds very closely to the form of Grandmont's Theorem on "intermediate preferences" discussed above, and it is for this reason that Grandmont (1978) could claim – correctly – that "the transitivity of the majority rule when preferences are single-peaked is indeed a particular case of the analysis of this paper" (p. 326). See also Fishburn (1972a), Denzau and Parks (1975), Saposnik (1975a), Hinich (1977), and Gaertner and Heinecke (1978).

conditions will be adequate when we are not confining our attention to majority rule only. In fact, the domain conditions for majority rule were arrived at in Sen and Pattanaik (1969) by arguments involving only certain characteristics of the majority rule, e.g. only strong neutrality and non-negative responsiveness in the case of value restriction. The method could be applied to various other types of choice procedures [see Pattanaik (1971)].[112] Salles (1975) posed a more general problem – using a game-theoretic framework developed by Wilson (1972a) and Bloomfield (1971) – by asking for necessary and sufficient domain restrictions for a SWF satisfying certain general conditions, including independence and "Pareto-transitivity" (xRy and y unanimously preferred to z must together imply xPz), and found the answer to be the fulfilment of either "value restriction" or a rather demanding condition which he called "cyclical indifference".[113]

A crucial question concerns the required domain restriction for a SWF satisfying Arrow's other conditions (viz., I, P and D), which can take the form of many rules other than majority decisions. For a particular class of restrictions, the necessary and sufficient conditions for this have recently been obtained by Maskin (1976a) and Kalai and Muller (1977).[114] Let $\mathscr{R}_X$ be the set of all orderings of X, the set of social states. The class of domain restrictions considered are characterized by specifying a subset $\mathscr{R}$ of $\mathscr{R}_X$, and restricting the domain of the SWF to $\mathscr{R}^n$, i.e. the SWF is required to specify a social ordering R for any n-tuple $\{R_i\}$ with each $R_i \in \mathscr{R}$.[115] Maskin and Kalai and Muller concentrate on strict (antisymmetric) orderings only.

The investigation is immensely simplified by a remarkable reduction result established by Maskin (1976a) and by Kalai and Muller (1977). It asserts that an n-person SWF (for any $n \geq 2$) satisfying I, P and D exists for a particular domain (in the class specified) if and only if such a 2-*person* SWF exists for that domain. This result permits an exact characterization, independently of n, of the permissible domain for SWFs satisfying Arrow's Conditions P, I and D [see Maskin (1976a, pp. 22–24) and Kalai and Muller (1977, pp. 462–463)], and this necessary and sufficient condition has been called "decomposability".[116]

[112] For various results related to domain conditions, see Craven (1971), Pattanaik (1971), Blin (1973), K. Fine (1973), Fishburn (1973a), Sloss (1973), Ferejohn and Grether (1974), Rosenthal (1975), Saposnik (1975b), Deb (1976), Kelly (1978), Salles and Wendell (1978), Slutsky (1979), Coughlin (1981), Brams and Fishburn (1983), and Chichilnisky and Heal (1983).

[113] Cyclical indifference requires that for any triple x, y, z, either all individual preferences are of the form $aI_i b$ & $bP_i c$, or all of the form $aP_i b$ & $bI_i c$, with $a, b, c \in \{x, y, z\}$, all distinct. See also Salles (1976).

[114] See also Kalai and Ritz (1980). See also Kaneko (1975), Nakamura (1978), Monjardet (1979), and Peleg (1982).

[115] Note that this restricts the permissible individual preferences rather than leaving them free but restricting permissible *combinations* of individual preferences. In this respect the Maskin–Kalai–Muller conditions are quite different from conditions such as "extremal restriction", "value restriction", or "single-peaked preferences", which – following Arrow's lead – investigate "similarity" (in a very broad sense) among the preferences of different individuals.

[116] See also Dasgupta, Hammond and Maskin (1979) and Kalai and Ritz (1980).

It appears that the domain restriction needed for a non-dictatorial and non-manipulable voting mechanism (discussed in Section 7) is also exactly the same, viz. decomposability [see Kalai and Muller (1977, pp. 467–468) and Maskin (1976b)]. This identity helps to highlight the exact correspondence of the Arrow impossibility theorem about SWFs satisfying U, P, I and D and the Gibbard–Satterthwaite impossibility theorem about strategy-proof, non-dictatorial voting mechanisms.[117]

Having said that, however, it is worth mentioning that the domain restrictions have somewhat different roles in the two problems. In the context of Arrow's impossibility, the domain restriction is a statement about what actual preferences people can, in fact, *have*. If a similar restriction is applied in the context of all preferences in the manipulability exercise, then one is restricting not merely the preferences that people can actually have, but also the *strategies* that they are permitted to adopt. Even if a restriction (e.g. decomposability or value restriction) were reasonable as a description of *actual* preferences, it does not follow at all that such a restrictions would make sense in confining people's *strategic* choices of ballots (i.e. *reported* preferences). If the restriction of domain in the manipulability exercise is applied to the true preferences *without* constraining the ballots in any way, then the relevant domain restrictions become a good deal more stringent.[118]

8.4. *Most unlikely?*

There is quite an extensive literature on the "probability" of transitivity of the majority relation and the existence of a majority winner.[119] The calculations are typically based on assuming that every preference pattern is as likely as any other, and they tend to lead to most discouraging results, especially for societies with many people and – much more importantly – in choice situations with many alternative states. For large communities choosing over a large set of social states, the "probability" of a majority winner seems minute.[120] But, it can be argued,

[117] See also Blin and Satterthwaite (1976), Chichilnisky and Heal (1979, 1983), Kim and Roush (1980a), Satterthwaite and Sonnenschein (1981), Moulin (1983), and Peleg (1984).

[118] See Dutta (1977), Pattanaik (1978), Sengupta and Dutta (1979), and Pattanaik and Sengupta (1980).

[119] See Guilbaud (1952), Riker (1961), Campbell and Tullock (1965, 1966), Williamson and Sargent (1967), Garman and Kamien (1968), Niemi and Weisberg (1968), DeMeyer and Plott (1970), Fishburn (1973a), Kelly (1974b, 1978), Gehrlein and Fishburn (1976, 1979), and Fishburn, Gehrlein and Maskin (1979).

[120] However, the probability that there is a majority winner is "substantially larger" than the probability that the majority preference relation be transitive. Indeed, the ratio of the latter to the former goes to zero rapidly as the number of voters is increased – a point that was established by Graaff (1965).

that this is an odd way of going about checking the actual probabilities, since individual preference n-tuples are results of social processes involving interconnections, and preferences are not formed in real societies by an equal-chance lottery mechanism. Given such interconnections, the plausible preference n-tuples can quite possibly be more conducive to consistent majority decision.

This is fair enough, but analyses of plausible preference patterns in many common circumstances have been hardly more encouraging. Kramer (1973) established an important result by taking a case in which the set X of alternatives can be seen as points on a multi-dimensional real space (e.g. commodity space or policy space). If individual preferences are representable by quasi-concave differentiable utility functions, even a very modest extent of heterogeneity of tastes would imply that value restriction (VR), limited agreement (LA) and extremal restriction (ER) will all be violated. Since these restrictions together constitute the necessary and sufficient conditions for a majority rule SDF in the approach of "restricted preferences" (see Section 8.1), the result seems damaging. In fact, similar problems can occur even without the assumption of quasi-concavity and even when the set X has no Euclidean metric properties at all, but has instead the structure of a differentiable manifold [see Chichilnisky (1976) and Schofield (1977a)]. Other decent burial grounds for majority rule have been found, and possibilities of total cycles involving *all* social states have been identified.[121]

Pessimism reigns. But it is not altogether clear whether so much pessimism is appropriate. Consider the distinction between the interest-aggregation exercise and the judgment-aggregation exercise (discussed in Section 7.1). The assumptions about individual preferences made in models such as those of Kramer (1973) and others are reasonable enough for the interest-aggregation exercise; this is indeed how individual utility functions over private and public goods are typically characterized.[122] But – as was argued earlier – for the interest-aggregation exercise the Arrow formulation of the problem may be informationally unduly restrictive, since it rules out the use of interpersonal comparisons of utility as well as the use of non-utility information except in very special circumstances.[123] In particular, the majority rule may be a very odd way of doing resource allocation

[121]See McKelvey (1975, 1976, 1979), Schofield (1977, 1978, 1980, 1983a, 1983b), and Rubinstein (1979, 1980b). The literature on majority decision on multi-dimensional space (with or without probabilistic voting, and with both "global" and "local" formulations) has developed vigorously in recent years, following Plott's (1967) pioneering formulation of the problem. For various distinct problems within this general approach, see Kramer (1973, 1977), Heal (1973, chapter 2), Kramer and Klevorick (1974), Nitzan (1975), Wagstaff (1976), Fishburn and Gehrlein (1977b), Hinich (1977), Kalai, Muller and Satterthwaite (1977), Slutsky (1977, 1979), Matthews (1979), Ordeshook (1980), Cohen and Matthews (1980), Coughlin (1981), and Couglin and Nitzan (1981), among others.

[122]A classical burial ground for majority rule is the cake division problem with strictly monotonic preferences, with each preferring any division with more cake for himself. It is easy to show that with three or more people when all divisions are considered, extensive majority cycles *will* occur in this case.

[123]See Sen (1973, 1977b), Hammond (1976b), and Gevers (1979).

or economic planning,[124] and richer informational structures may be needed (see Section 6). On the other hand, while the Arrow format might well be more appropriate for the exercise of aggregating judgments of different people as to what is good for society (e.g. whether "positive discrimination" should be pursued, whether tax systems should be more progressive, or whether multinational investments should be encouraged in developing countries), it is not at all clear that these preferences would have the characteristics on which the negative results were based. It is, therefore, possible to argue that while the negative results are of much analytical interest, they may not be altogether devasting either for the judgment-aggregation exercise, or for the exercise of aggregation of interests.

9. Independence, neutrality and liberty

9.1. *Independence and Bergson–Samuelson impossibilities*

In the preceding discussion various modifications of Arrow's social welfare functions SWF have been investigated, including – among other structures – social decision functions SDF (permitting non-transitive social preference), social choice functions SCF or functional collective choice rules FCCR (permitting non-binary social choice), and social welfare functionals SWFL (permitting the use of richer utility information). But the case of the Bergson–Samuelson social welfare function SWF, briefly outlined in Section 1, has not yet been further examined here. This lacuna is particularly important to fill since it has been repeatedly claimed that the Arrow impossibility theorem and related results do not affect the existence of Bergson–Samuelson social welfare functions in any way [see Little (1952) and Samuelson (1967a, 1967b, 1977)].

It has been pointed out that since the Bergson–Samuelson exercise is based on "individual tastes as being given", conditions of inter-profile consistency such as independence of irrelevant alternatives, are not to be imposed on the SWF in this case. And, it is argued, since the Arrow impossibility result is crucially dependent on the independence condition, the result can hardly affect the Bergson–Samuelson SWF. Indeed, "Arrow's work has no relevance to the traditional theory of welfare economics, which culminates in the Bergson–Samuelson formulations" [Little (1952, pp. 423–425)]. "For Bergson, one and only one of

[124]Under majority rule with self-seeking non-satiated preferences, social "improvements" can be persistently carried out by cutting the income of one person and dividing the loot among the rest (two or more), and this "improving" process can go on until the fall-guy has no income left! See Sen (1977a). For examples of more standard economic and political problems, see Downs (1957), Frey (1978, 1983), and Usher (1981).

the...possible patterns of individuals' orderings is needed" [Samuelson (1967a, pp. 48–49)]. Hence, "it is not true, as many used to believe, that Professor Kenneth Arrow of Stanford has proved the 'impossibility of a social welfare function'" [Samuelson (1967b, p. vii)].[125] In a formal sense that last statement, applied to Bergson–Samuelson SWF, is entirely correct, but it may be useful to examine why odd beliefs like this could flourish at such places as Stanford.

For a SWF to be impossible some restrictions, obviously, would have to be imposed on it. Until these are specified, "the impossibility of the traditional Bergson welfare function of economics", which Samuelson (1967a) rightly holds to be false (p. 42), is hardly worth commenting on. It does not appear to be Samuelson's intention to deny the need to fulfill the condition of unrestricted domain since the pattern of individuals' orderings "could be *any* one, but it is *only* one" (p. 49). Nor is the Pareto condition to be dispensed with since so many of the Bergson–Samuelson exercises seem to use this principle [see Samuelson (1947, chapter 8)]. Indeed, as Johansen (1970) pointed out in his illuminating examination of the relevance of Arrow's theorem for economic planning, "a Bergson welfare function is essentially nothing but such a social preference ordering which is positively associated with the individual preference orderings in the ... Paretian sense" (p. 42). If this is all that is required of a Bergson–Samuelson SWF, then the question of its existence would be quite trivial since the Pareto quasi-ordering – like any other quasi-ordering – can be completed into an ordering. But this completion can be done in so many different ways, and the question would arise as to whether a "reasonable" Bergson–Samuelson SWF should not fulfill some additional conditions. Arrow (1951) presumably thought that independence and non-dictatorship would be such conditions, while Samuelson does not find independence reasonable in this context since he does not wish to impose *any* inter-profile consistency condition.[126]

It appears, however, that the Bergson–Samuelson SWF has often been combined with the requirement of "strong neutrality" *within* a given profile of individual preferences, and the so-called "individualistic" version of SWF makes social welfare a function of the vector of individual utilities: $W = W(\boldsymbol{u})$, viz. as in (6.2) presented in Section 6.4 [see Samuelson (1947, pp. 228–229, 246), Bergson (1948, p. 418), and Graaff (1957, pp. 48–54)]. But as was noted in establishing

[125] Samuelson (1967a) denies that the Bergson–Samuelson SWF need not satisfy the independence condition: "my formulation builds it *from the beginning* into Axiom 1" (p. 47). But this appears to be the result of a misunderstanding, to wit: "if the ordering is transitive, it *automatically* satisfies the condition called 'independence of irrelevant alternatives'" (p. 43). Not at all so.

[126] Note that even when independence is dropped from the Arrow framework, impossibility results can be generated by other types of inter-profile conditions. For a novel and interesting example of an impossibility theorem *without* the use of the independence condition, see Chichilnisky (1976, 1982a, 1982b), who uses *continuity* as the inter-profile link. See also McManus (1975, 1978, 1982), and Ferejohn, Grether, Matthews and Packel (1980).

Arrow's impossibility theorem, one of the main uses of the independence condition (along with unrestricted domain and the Pareto principle) is precisely to precipitate a neutrality result (see Sections 2.1 and 6.4). In the "individualistic" case, it is handed on a plate. Even when the scope of the equation $W = W(\boldsymbol{u})$ is restricted in a way consistent with the absence of inter-profile conditions, it still follows that if in a *given* profile x and y have exactly the same utility characteristics as a and b respectively (for $x, y, a, b \in X$), then the social ordering of x vis-a-vis y must be the same as the social orderings of a vis-a-vis b, for that *given* profile. Given the use of non-comparable ordinal utilities in the traditional Bergson–Samuelson framework, this limited neutrality condition can be formulated in this way.

Condition SPN (single-profile neutrality)
For any given n-tuple $\{R_i\}$ of individual preference orderings, for any $x, y, a, b \in X$, not necessarily all distinct, if for all i, $xR_iy \Leftrightarrow aR_ib$ and $yR_ix \Leftrightarrow bR_ia$, then $xf(\{R_i\})y \Leftrightarrow af(\{R_i\})b$.

What is the effect of imposing single-profile neutrality on a social welfare function $f(\cdot)$ which is also required to satisfy weak Pareto principle P and have a domain with some diversity of preferences? It is that the social welfare function will be dictatorial in a "single-profile" sense, viz. there will be a person j such that all his strict preferences will be reflected in the social preference, *for that profile*. A variant of this result was first established by Parks (1976a), and others by Kemp and Ng (1976), Hammond (1976b), Pollak (1979), Roberts (1980c), and Rubinstein (1981). A "single-profile dictator" can, of course, have the *same* preference as everyone else. This won't then be a disturbing result and can indeed be a consequence of the Pareto Principle. Some of the authors establish their theorems with domains that have built-in "diversity" because of dealing with the space of income vectors, or of commodity distributions, or with directly-specified diversity.[127]

The proofs of these single-profile results go through easily enough – a good deal more easily than the proof of Arrow's impossibility theorem. Much of the effort in proving Arrow's theorem rests in establishing "neutrality", which is the main part of what we have called the Field Expansion Lemma (see Section 2.1), and this is simply *given here* by virtue of taking social welfare as a function of the vector of individual utilities (i.e. by assuming the social welfare function to be "individualistic").[128] The independence condition is *used* in Arrow's case to establish

[127]See Rubinstein (1981) on the logical correspondence between single- and multiple-profile results. Note also that single-profile impossibility results do involve *choosing* a profile from the domain (unrestricted, or restricted to a permissible class). The fact that such a choice is involved must not be confused with the *simultaneous* use of several profiles (as in proving Arrow's multiple-profile impossibility theorem; see Section 2.1).

[128]From the exchange between Kemp and Ng (1976, 1977) and Samuelson (1977), it would appear

this property, but since the property is given here, it does not have to be established. And since the end-product to be obtained is a single-profile dictatorship result (and not – as in Arrow's case – the much stronger multiple-profile dictatorship result), there is no further need for the inter-profile condition of independence.

9.2. *The Borda rule and the use of positional information*

With ordinal, non-comparable utilities, the single-profile neutrality condition took the coincidence of utility characteristics over two distinct pairs $\{x, y\}$ and $\{a, b\}$ to imply that the society should rank $\{x, y\}$ in the same way as $\{a, b\}$. In this description, no attention is paid as to whether or not there are other – "irrelevant" – alternatives in between x and y, or in between a and b. One way of avoiding the impossibility result is to enrich the description by taking note of the position of other alternatives (including "intermediate" states between any pair) in each person's preference. Neutrality can be redefined to demand that x and y be ranked in the same way as a and b if they occupy the same position vis-a-vis each other *and* vis-a-vis other – "irrelevant" – alternatives. Then the dictatorship consequence will be avoided.[129] Indeed, this relaxation will yield enough freedom to demand the fulfillment also of some other appealing conditions.[130] The merits of rules that take note of such positional information, are not, of course, confined to avoiding Arrow's impossibility result.[131] [Borda (1781), who put forward the first known formal rule based on such information, had presumably not lost any sleep on Arrow's paradox.]

The Borda rule can be seen as based on attaching a number to any alternative equal to the sum of its ranks in each person's preference ordering (e.g. in a 3-person, 3-state world, if x is first in one person's ordering and third in the other two persons', then the "Borda count" for x is $3+1+1=5$). The Borda rule ranks the states socially in the *inverse* order of these numbers. Recently, the Borda rule

---

that it is not – indeed never was – Samuelson's intention to insist on neutrality. It is certainly the case that Samuelson (1947) made critical comments on this "extreme assumption" (pp. 223–224), and while this did not stop him from dealing extensively with cases in which this condition is fulfilled (pp. 228–247), the traditions of economic theory do not, of course, permit one to deduce belief from extensive use.

[129] On issues raised by the case for relaxing independence, see Hansson (1973), Ray (1973), Mayston (1974, 1975, 1980), Karni and Schmeidler (1976), Osborne (1976), Packard and Heiner (1977), Kelly (1978), and Pattanaik (1978), among others.

[130] Indeed the eschewal of independence permits the use of a social welfare function based on Nash's (1950) solution of the bargaining problem; on this see Kaneko and Nakamura (1979) and Kaneko (1980). See also Luce and Raiffa (1957), Sen (1970a), DeMeyer and Plott (1971), Yaari (1978), and Mayston (1982).

[131] See, for example, Moon (1976), Rubinstein (1980b), Nitzan and Rubinstein (1981), and Mayston (1982).

has been nicely axiomatized involving a variable electorate [see Young (1974a, 1974b, 1975), and also Gärdenfors (1973), Smith (1973), Fine and Fine (1974), Fishburn and Gehrlein (1976), Hansson and Sahlquist (1976), Gardner (1977), Farkas and Nitzan (1979), and Nitzan and Rubinstein (1981)].

Gärdenfors (1973) and Fine and Fine (1974) have provided a thorough exploration of positional rules. These include "finite ranking rules", which are based on attaching weights according to the position occupied by an alternative in each person's ordering (the weights being non-decreasing function of ranks, applied in the same way to everyone's ordering, i.e. anonymously). The social ranking is made to reflect the ranking of the sum of weights on the different states. A special case of this is the Borda method. Another is utilitarianism *with* "utilities" taken to be reflected by positions. The *intersection* of all finite ranking rules yields a quasi-ordering exactly reflecting rank-dominance R^{D}, when $xR^{\mathrm{D}}y$, if and only if for some interpersonal permutation x occupies at least as high a position in each person's ordering as y does in the corresponding person's ordering.[132] The axiomatic structure of various positional rules analysed in recent contributions have enriched our understanding of the nature and operation of these important classes of decision procedures.[133]

Positional discrimination can also be combined with the use of ordinal level-comparable utilities, and the weights can be based on the rank of a "station" (x, i), i.e. that of being person i in state x, in an interpersonal order of the entire Cartesian product of X and H. While the general format will be that of ranking social states according to the sum of weights on all stations involving that state, the interpersonal rank-order rule IROR corresponds exactly to the Borda rule, in making the weight on each station equal its rank number from bottom upwards.[134] If, for example, the ranking of nine stations involving three states and three persons is given by the following: $(x,1)$, $(y,2)$, $(z,3)$, $(x,2)$, $(y,3)$, $(z,1)$, $(x,3)$, $(y,1)$, $(z,2)$, then the majority rule will yield a preference *cycle*, the Borda rule will yield universal indifference, but IROR will yield the strict ordering xPy & yPz. While this coincides with the Rawlsian maximin (defined on utilities), a conflict between the two can be brought about by switching the positions of $(x,3)$ and $(z,2)$, which would leave the IROR ranking unchanged, but exactly reverse the Rawlsian ordering to zPy & yPx. It is perhaps worth mentioning that the "Rank-Dictatorship Theorem" (with Arrow-like conditions married to the invariance restriction under ordinal level comparability OL), which was presented in Section 6.2, would not conflict with the possibility of interpersonal positional rules under OL because of the violation of the independence condition in these rules.

[132] Fishburn (1973a) has discussed such "permuted dominance" for strict orderings. Fine and Fine (1974) have provided extensive analysis – and axiomatic derivation – of rules of this type.

[133] As a contrast, see also Brams and Fishburn's (1978, 1983) definitive exploration of "approval voting", which is a flexible voting procedure *without* use of positional data.

[134] Sen (1977b, section 5), Mizutani (1978), and Gaertner (1983).

In the last case, i.e. with interpersonal positional rules, the positional information is used, as it were, to convert ordinal level comparability into some kind of a devised cardinal full comparability based on ranks in the extended ordering of $X \times H$. In the case of ordinary positional rules, including the Borda rule, the positional information is used, as it were, to convert non-comparable ordinal utility information into assumed cardinal full comparability by building on the ranks in each person's ordering taken separately. It is the arbitrariness of translating rank values into numerical weights that is typically found to be the weakest aspect of both these classes of rules. Indeed Arrow's (1951) defence of the condition of independence rested partly on the need to avoid such arbitrariness.

9.3. *Independence versus collective rationality*

There are two ways of defining the Borda rule depending on whether the Borda counts are based on the ranks in the total set X, or in the set S from which the choice $C(S)$ is to be made, with $S \subseteq X$. It can be easily checked that while the former, which may be called the "broad" Borda rule violates independence but yields a transitive social ordering, the latter, which may be called the "narrow" Borda rule, satisfies the independence condition but can yield non-binary choice functions. The "narrow" version has the merit of providing a social choice function – possibly non-binary – satisfying all of Arrow's conditions, viz. U, P, I and D. In this respect, the "narrow" Borda rule is a serious rival of social choice functions based on the transitive closures of the majority rule, investigated by Schwartz (1970, 1972), Bloomfield (1971), Campbell (1972, 1976), Bordes (1976), and Deb (1977), and discussed earlier in Section 4.1.

Just as modifying the Borda rule from its usual broad version to its narrow variant takes one across the independence-binariness line, similarly changing the majority closure methods from their usual "narrow" formulations to the corresponding broad variants will take one across the same line in the opposite direction. The broad version of the Weak Closure Method is defined by obtaining the transitive closure R^*_X of the majority relation over the *entire set* X, and then identifying the choice set $C(S)$ for any non-empty $S \subseteq X$ as the maximal set $M(S, R^*_X)$ of S.[135] This process identifies a social welfare function in the sense of Arrow; R^*_X is a (fully transitive) social ordering. It would, however, violate the

[135] Campbell (1976) presents, inter alia, the "broad" formulation of the Weak Closure Method as defining a "democratic preference function" satisfying the property of generating complete social *orderings* and fulfilling other requirements. Independence *is* violated, which, Campbell argues, "should be introduced, not as a normative restriction on the mapping of individual into social preference, but as a technical requirement the force of which is to ensure that a social preference function can be implemented by some iterative procedure" (p. 259).

independence of irrelevant alternatives, since the choice set $C(S)$ will depend on individual preferences over the whole X, *including* $X-S$, and not merely over the subset S. (A similar possibility exists through using the broad version of the Strict Closure Maximality, based on P_X^*.)

The contrast between the narrow and broad versions of the familiar rules of Borda and the majority closures bring out the fact that the Arrow impossibility result is largely built on the tension between independence and collective rationality over a large domain.[136] We can have independence *or* collective rationality (but *not* both) from these rules, depending on whether we choose the narrow or the broad version. There are other rules also of which narrow and broad versions could be contrasted with the same division between independent and non-binary procedures on one side and binary and non-independent procedures on the other. Indeed, all complete positional rules (including finite ranking rules) can be thus treated, as well as those variants of the majority rule that yield a complete but not necessarily transitive relation [e.g. R_{maj} defined by Dummett and Farquharson (1961) and extensively studied by Pattanaik (1971)].

9.4. *Neutrality and the use of non-utility information*

It was argued earlier – in Section 6 – that an escape from the Arrow impossibility result could be found by enriching the information that could be used for social choice. The General Possibility Theorem builds on combining poor utility information (in particular, no interpersonal comparisons) with an effective ban on the use of non-utility information (through the derived characteristic of neutrality). The same applies to the single-profile versions of the Arrow theorem, tailored for Bergson–Samuelson SWF (discussed in Section 9.1), in which poor utility information was combined with a ban on any essential use of non-utility information through the neutrality property of the SWF in the form $W=W(\boldsymbol{u})$. The informational enrichment that was explored in Section 6 concentrated on the improvement of the utility information, a possibility that was suggested by Arrow (1963) himself, and the approach of SWFL with invariance restrictions [Sen (1970a)] was based on that foundation. It is easy to recognize that the enrichment of the utility information would have the same eliminating effect on the impossibility result in the single-profile framework as well, and neutrality can indeed be combined with interpersonally comparable utilities fulfilling the Arrow conditions on an appropriately defined single-profile SWFL.

An alternative way of avoiding the impossibility problem in either framework rests in relaxing neutrality and permitting the use of non-utility information,

[136] See Hansson (1972), Fishburn (1974a), Binmore (1976), Hammond (1977b), and Sen (1977a).

instead of improving the utility information. Many acts of economic judgment for the society (e.g. planning exercises) are based on taking explicit note of non-utility information, e.g. data on hunger, or poverty, or inequality, or national income, or violation of acknowledged rights (such as personal liberty).[137] What is less clear is whether these non-utility data are used to get indirectly at utility information,[138] or whether selected non-utility data would have a status of its own *even when* utility information is as rich as it can be. Support for the latter position from diverse sources can be seen in Rawls's (1971) focus on "primary goods", the importance attached to description of work and social relations in such analyses as Marx's (1875, 1887) treatment of "exploitation" and "alienation", and in the wide use of principles like "equal pay for equal work" in recent normative discussions. The contrast will not be pursued further here; I have tried to do this elsewhere [Sen (1979b, 1979c)].[139]

It is, however, worth clarifying a distinction that seems to be sometimes confused, partly because of the ambiguous use of the characteristic of a social welfare function being "individualistic". Individualism could mean neutrality – indeed strong neutrality, so that two pairs of states $\{x, y\}$ and $\{a, b\}$ must be socially ranked in exactly the same way when the individual utility characteristics of x vis-a-vis y are exactly the same as those of a vis-a-vis b, respectively. Alternatively, it could mean that the Pareto Principle – indeed the Strong Pareto Principle – holds, and if x has at least as much utility as y in everyone's preference ordering, then x is socially at least as good as y, and if furthermore at least one of the individual inequalities is strict, then so is the social preference.[140] Some conceptualizations of the principle that "individuals' preferences are to 'count'" seem to cover both characteristics [see Samuelson (1947, pp. 223–224, 228, 236) and Graaff (1957, pp. 9–10)], but it is easily checked that the two requirements are completely independent of each other. A strongly neutral social welfare function must satisfy the Pareto indifference rule P^0, but need not fulfill the strong Pareto condition P^*, or for that matter the weak Pareto

[137] While the "ethical" measurement of inequality has been typically based on richer utility information [see Kolm (1969), Atkinson (1970), Sen (1973), Blackorby and Donaldson (1978)], the use of non-utility information has played a crucial role in the recent contributions to welfare-based national income comparisons and to the measurement of poverty [see Sen (1976c, 1976d, 1979d), Hamada and Takayama (1978), Blackorby and Donaldson (1978, 1980a, 1980b), Hammond (1978), Takayama (1979), Thon (1979), Kakwani (1980a, 1980b, 1981), Anand (1983), Chakravarty (1983a, 1983b), Graaff (1983), Kundu and Smith (1983), Foster (1984)].

[138] The problem of "recovering" utility information from non-utility data (such as incomes) may be a very complex one in practice, because of variations of other parameters of the individual utility functions [see Lindbeck (1983)].

[139] See also Williams (1973), Nozick (1974), Scanlon (1975), Dworkin (1978), Roemer (1982), and Sen and Williams (1982).

[140] "Individualist" as an adjective seems misleading for both, since utility is scarcely the only expression of one's individuality. "Individual rights" are typically formulated taking explicit note of non-utility information, and can also conflict with the Pareto principle (to be discussed in the next subsection).

condition P of Arrow, since it need not be strictly monotonic (positively responsive). Similarly, a Pareto-inclusive social welfare function may still use non-utility information for discrimination when the individual utility rankings do not coincide over either pair $\{x, y\}$ or $\{a, b\}$, but the individual utility rankings – individually divergent as they are – happen to be exact reflections of each other (x vis-a-vis a, and y vis-a-vis b) in the two cases.

It seems natural to argue that of the two interpretations of an "individualistic" social welfare function, neutrality is the more demanding. If social welfare *is* a function of the vector of individual utilities only, it seems difficult to argue that it need not be an *increasing* function. On the other hand, while the Pareto indifference rule or the strong Pareto principle does lead to strong neutrality in the presence of unrestricted domain and independence (see the Strong Neutrality Theorem in Section 6.4), both these additional conditions are quite demanding. In fact, traditionally the Pareto principle has appeared to be a very mild requirement indeed, but it is clear that it has remarkable cutting power in excluding various natural formulations of rights and liberties precisely because they make use of non-utility information. This problem is discussed in the next subsection.

9.5. *The impossibility of the Paretian liberal*

Various formulations of liberty have been based on identifying certain types of "self-regarding" choices as being in a person's "protected sphere", on which that person's wishes should rule, and this has provided a common theme of diverse libertarian writings from John Stuart Mill (1859) to Hayek (1960). For example, what a person wears, or what he or she reads, may in many circumstances be regarded as being in such a "protected" or "personal" sphere. When two social states x and y differ only in this respect, it may be argued that libertarianism should demand that the relevant person's preference over this pair must be reflected in the social preference. The requirements of liberty can be defined in a mild form by demanding that everyone has a non-empty protected sphere, and "minimal liberty" as a condition requires that at least two persons must have a non-empty protected sphere each [Sen (1970a, 1976a)]. Define a person as strongly decisive over $\{x, y\}$, if $xP_i y \Rightarrow xPy$, and $yP_i x \Rightarrow yPx$.

Condition ML (minimal liberty)
At least two persons are strongly decisive over one pair of social states each.[141]

[141] Whether these conditions are reasonable must depend, among other things, on what the set X of social states consist of. If the variations between them involve *only* such "non-personal" differences as British forces vacating Ulster, or wheat being stockpiled by the United Nations, clearly both L and ML would be very unreasonable conditions, and in particular their non-fulfilment would not imply anything about the absence of libertarian decision procedures, as commonly understood. So the usefulness of conditions of this type depends inter alia on the nature and the richness of the set X.

The impossibility of the Paretian liberal
There is no SDF satisfying unrestricted domain (U), the weak Pareto principle (P) and minimal liberty (ML).

For there to be two persons strongly decisive over one pair each, there have to be at least three distinct social states. Consider first the case in which person 1 is strongly decisive over $\{x, y\}$, while 2 is over $\{y, z\}$, with one state (viz. y) in common. Consider the following individual orderings of the three states, viz. for 1, zP_1x & xP_1y; for 2, yP_2z & zP_2x; and for all $i \neq 1,2$, zP_ix. By strong decisiveness of 1 and 2 over $\{x, y\}$ and $\{y, z\}$ respectively (i.e. by ML), it follows that xPy and yPz. But by the weak Pareto principle, zPx. Thus the social preference relation must violate acyclicity, and hence no SDF can satisfy these conditions. Now taking the case in which the two persons 1 and 2 are strongly decisive over $\{x, y\}$ and $\{a, b\}$, when all four are *distinct* states, consider the following strict orderings, in descending order:

1	2	$i \neq 1,2$ (partially specified)
b	y	$\binom{b}{x}\binom{y}{a}$
x	a	
y	b	
a	x	

By ML, xPy & aPb, and by the weak Pareto principle, yPa & bPx. This strict preference cycle shows the impossibility of any SDF satisfying U, ML and P.

It is an immediate corollary of this result that there is no social welfare function satisfying these conditions, since a SWF is also a SDF.

There are several interesting features of this theorem that are worth noting. First, it is a single-profile impossibility result – established by considering one profile – and applies immediately to a Bergson–Samuelson SWF. In particular, it makes no use of the inter-profile condition of independence.[142]

Second, it makes no use of the requirement of transitivity (or of quasi-transitivity) of social preference, just of acyclicity.

Third, it can be easily extended to social choice functions, or functional collective choice rules, by translating the pair-choice requirements to general *choice* constraints. Redefine a person being "strongly decisive" over x, y as the requirement that if xP_iy, then y cannot be chosen in the presence of x, and if yP_ix, then x cannot be chosen in the presence of y, and let this convert

[142] Note that both the Pareto principle and the libertarian ones have the characteristic of basing the ranking of a pair of states on individual preferences over that pair only, which can be seen, in some ways, as an "independence" property [see Blau (1975)]. But Arrow's independence condition is unnecessary for this result. Or any other condition of multiple-profile correspondence (except what results indirectly from the Pareto principle and minimal liberty).

conditions L and ML into $\hat{\hat{\text{L}}}$ and $\widehat{\widehat{\text{ML}}}$. Also, consider the Pareto principle in the general choice-functional form $\hat{\hat{\text{P}}}$ presented in Section 4.3.

Choice-functional impossibility of the Paretian libertarian
There is no functional collective choice rule FCCR satisfying Conditions $\hat{\text{U}}$, $\hat{\hat{\text{P}}}$ and $\widehat{\widehat{\text{ML}}}$.

While this translation is immediate [see Sen (1970a, pp. 81–82)], more complex choice functional variants of this result can be derived by making the conditions constrain choices over pairs only, but linking these choices with choices over larger sets by consistency conditions of social choice [see Batra and Pattanaik (1972b)].

Fourth, escape from this impossibility result can scarcely be found in enriching the utility information, unlike in the case of the Arrow impossibility result (Section 6). While some authors have considered the possibility that one's right to be able to do personal things without let or hindrance should be conditional on one's utility gain from this being large [see Ng (1971)], the libertarian approach is to assert these rights on grounds of the *nature* of the choice – that they are "personal" matters – and not on the basis of balancing the utility gains of the person concerned against the utility losses of the nosey.[143] Liberty is, however, one value among many, and it is possible to constrain libertarian rights by making them conditional on not violating some elementary requirements of utility-based justice such as Suppes' (1966) "grading principle of justice". However, it has been demonstrated by Kelly (1976a) that conditioning the libertarian requirements in this way leaves the impossibility result virtually unaffected; see also Austen-Smith (1980), Wriglesworth (1982b), and Suzumura (1983a).

Finally, the result is not based on ignoring non-utility information, as may arguably be the case with the Arrow result (see the Field Expansion Lemma in Section 2.1). Indeed, non-utility information is given an explicit role in the libertarian conditions. In fact, it can be argued that unlike in the case of the Arrow impossibility result, the basis of the impossibility of the Paretian libertarian rests not on *inadequate* information, but on *inconsistent* use of information. The Pareto principle insists on basing a class of social decisions *exclusively* on *utility* information, while the libertarian principles insist on giving crucial role to *non-utility* information in another class of social decisions, through the specification of protected spheres.[144] The impossibility result captures the tension between the two.

[143] Cf. John Stuart Mill (1859): "...there is no parity between the feeling of a person for his own opinion, and the feeling of another who is offended at his holding it; no more than between the desire of a thief to take a purse, and the desire of the right owner to keep it. And a person's taste is as much his own peculiar concern as his opinion or his purse" (p. 140). See also Riley (1983) on Mill.

[144] For general studies of the analytics of a system of rights, see Kanger (1972) and Lindahl (1977).

9.6. *Rights and principles*

Various extensions of the impossibility of the Paretian libertarian have been discussed, and other issues in the normative theory of rights have been explored in this context. Batra and Pattanaik (1972b) have been concerned with rights of groups intermediate between individuals and the whole community, e.g. in a federal country the rights of members of a state to do certain local things irrespective of the wishes of people in other states. The "impossibility of Paretian federalism" can be readily established on the same lines as the impossibility of Paretian libertarianism so long as the groups involved are pairwise disjoint. The proofs are virtually the same. Even when the groups are not disjoint, impossibilities can occur if the within-group decision mechanism is not unanimity but some other rule, e.g. the majority rule, unless the groups structure is severely restricted [see Stevens and Foster (1978) and Wriglesworth (1982a)].

Gibbard (1974) has noted that even in the absence of the Pareto principle an impossibility result could arise if individual rights are asserted not merely over a *non-empty* protected sphere but *generally* over pairs for which the states differ from each other in a respect "personal" to someone – other things given. A simple example brings out the nature of the conflict that is envisaged. In a 2-person community, let each person's right to wear a hat of any design be accepted, and the social preference is required to reflect a person's preference about his own hat other things given (in particular, given the other person's hat). Now assume that person 1 wants to wear a hat of the same design as the one worn by 2, while person 2 wants a hat of a different design from the one worn by 1. It is easy to see that an impossibility result can be constructed even without invoking the Pareto principle. To assert libertarian rights consistently, they would have to be formulated differently, e.g. by restricting the rights to "coherent" domains [see Suzumura (1978)], or by making the rights conditional on "independent individual preferences" [see Gibbard (1974), Hammond (1981)].[145] But even when the rights are internally consistent, the conflict with the Pareto principle can easily arise. (See the proof of the impossibility of the Paretian liberal, p. 1156.)

Other extensions have been presented, including a probabilistic version of the impossibility of the Paretian libertarian [see Bandopadhyay, Deb and Pattanaik (1979)] and its use in various game-theoretic contexts [see Aldrich (1977a, 1977b), Miller (1977), Breyer and Gardner (1980), and Gardner (1980)].

Various ways of resolving the impossibility of the Paretian libertarian and related results have been proposed in the literature. Some methods involve

[145] On this, see Ng (1971), Gibbard (1974), Farrell (1976), Sen (1976a), Kelly (1978), Suzumura (1978, 1983a), Hammond (1981), and Wriglesworth (1983a, 1983b).

constraining the libertarian rights, or the exercise thereof; for formal results as well as analyses of pros and cons of such procedures, see Gibbard (1974), Bernholz (1974), Blau (1975), Seidl (1975), Buchanan (1976), Campbell (1976), Kelly (1976a, 1976b, 1978), Ferejohn (1978), Karni (1978), Mueller (1979), Austen-Smith (1980), Breyer and Gardner (1980), Gardner (1980), Suzumura (1980, 1983a), Baigent (1981), Gaertner and Krüger (1981, 1983), Wriglesworth (1982b, 1983a, 1983b), and Basu (1984). Other ways involve constraining the Pareto principle, either by "amending" individual preferences, or by "counting" only a subrelation of a person's preference for the purpose of the Pareto judgment, taking note of the underlying motivation behind the preferences; for formal results, motivational analyses and assessment, see Sen (1970a, 1976a), Farrell (1976), Suzumura (1978, 1983a), Hammond (1981, 1982), Austen-Smith (1982), Rawls (1982), Wriglesworth (1982b, 1983a), and Coughlin (1983). Still others have explored domain restrictions that would avoid the impossibility in question [see Bergstrom (1970), Blau (1975), Fine (1975b), Seidl (1975), Breyer (1977), Breyer and Gigliotti (1980), and Nalebuff (1981)]. Some have argued in favour of limiting the scope of social choice theory through technical devices that would amount to a refusal to pronounce judgments on choices that are personal [see Ramachandra (1972) and Farrell (1976)]. Others have argued for incorporating rights not in the evaluation of states of affairs but as deontological constraints on action in an essentially non-consequentialist framework; for presentations and critiques of this approach, see Nozick (1973, 1974), Bernholz (1974, 1980), Rowley and Peacock (1975), Buchanan (1976), Aldrich (1977a, 1977b), Miller (1977), Perelli-Minetti (1977), Gärdenfors (1981), Sugden (1981), and Chapman (1983).

Constraints of space will not permit discussion of these various approaches here [see, however, Sen (1983)]. It should, however, be obvious that the interest of the "impossibility of the Paretian libertarian" and related results lies not so much in their value as paradoxes and brain-teasers, but as grounds for *re-examining* the usual formulations of individual and group rights and principles of decisions usually accepted, including such allegedly non-controversial rules as the Pareto principle. In the earlier sections of this paper I have tried to argue that a similar remark can be made about the much deeper impossibility result contained in Arrow's General Possibility Theorem.

10. A concluding remark

It was argued earlier in this paper that under the broad hat of social choice theory can be found quite a few different types of problems. Consider the following examples of "social choice" problems: choosing procedures for committee decisions; fixing electoral rules; choosing a constitution for a newly independent country; judging whether the government of a country has failed to serve the interests of the nation; choosing methods of assessing fiscal policies; doing central planning based on interests of the community; making systematic social welfare judgments; constructing ethically significant indicators of national prosperity, poverty or inequality. There are indeed things in common between these exercises, but also fundamental differences. In a broad sense they are all "social choice" problems, and all deal with methods of marshalling information, particularly those relating to the people involved, to arrive at correct social judgments or acceptable group decisions. But the nature of the possible informational inputs vary, as do the required outputs of judgments or decisions or the required means of settlement. The balance of moral and pragmatic considerations also varies with the nature of the exercise. There are other differences, e.g. whether the procedures can permit the use of discretion in interpreting individual utilities (e.g. in making social welfare judgments) or must be rather mechanical (e.g. electoral procedures).

The nature of the exercise affects the appropriate specification of the "social choice" format. This relates to distinctions between structures such as social welfare functions (Sections 1 and 2), social decision functions (Section 3), social choice functions or functional collective choice rules (Sections 4 and 7), or social welfare functionals (Section 6). It also affects the appropriateness of particular axioms within a given structure, e.g. whether the social welfare function should satisfy the independence condition (Sections 6, 7 and 9), or what types of interpersonal comparability – if any – should be used (Sections 5 and 6), or what domain conditions would make sense (Section 8).

The relevance of the various results presented and discussed in different sections of this paper depends on the *particular* nature of the exercise to which application may be sought. It is important to bear this in mind in understanding the rather bewildering collection of results that three decades of social choice theory have produced. They do not all deal with the same type of exercise. Between them they cover vastly different types of problems with only a very general "social choice" character in common. Indeed, the richness of the subject owes much to this diversity.

References

Aizerman, M. A. and A. V. Malishevski (1980), "Choice of best variants: Some aspects of general theory". Moscow: Institute of Control Sciences.

Aizerman, M. A., N. V. Zavalishin and E. S. Piatnitsky (1976), "Global functions of sets in theory of choice". Moscow: Institute of Control Sciences.

Aldrich, J. (1977a), "The dilemma of a Paretian liberal: Some consequences of Sen's theorem", Public Choice, 30:1–21.

Aldrich, J. (1977b), "Liberal games: Further comments on social choice and game theory", Public Choice, 30:29–34.
Allingham, M. (1976), "Fairness and utility", Economie Appliquée, 29:257–266.
Anand, S. (1983), Inequality and Poverty in Malaysia: Measurement and Decomposition. Oxford: Oxford University Press.
Archibald, G. C. and D. Donaldson (1979), "Notes on economic equality", Journal of Public Economics, 12:205–214.
Armstrong, T. E. (1980), "Arrow's theorem with restricted coalition algebras", Journal of Mathematical Economics, 7:55–75.
Arrow, K. J. (1950), "A difficulty in the concept of social welfare", Journal of Political Economy, 58:328–346.
Arrow, K. J. (1951), Social Choice and Individual Values. New York: Wiley.
Arrow, K. J. (1952), "Le principe de rationalité dans les décisions collectives", Economie Appliquée, 5:469–484.
Arrow, K. J. (1959), "Rational choice functions and orderings", Economica, 26:121–127.
Arrow, K. J. (1963), Social Choice and Individual Values, 2nd ed. New York: Wiley.
Arrow, K. J. (1967), "Public and private values", in: Sidney Hook, ed., Human Values and Economic Policy. New York: New York University Press.
Arrow, K. J. (1973), "Some ordinalist-utilitarian notes on Rawls' theory of justice", Journal of Philosophy, 70.
Arrow, K. J. (1977), "Extended sympathy and the possibility of social choice", American Economic Review, 67:219–225.
Arrow, K. J. and M. Kurz (1970), Public Investment, the Rate of Return and Optimal Fiscal Policy. Baltimore: Johns Hopkins University Press.
Atkinson, A. B. (1970), "On the measurement of inequality", Journal of Economic Theory, 2:244–263.
Atkinson, A. B. and J. E. Stiglitz (1981), Lectures on Public Economics. Maidenhead: McGraw-Hill.
Austen-Smith, D. (1980), "Fair rights", Economic Letters, 4:29–32.
Austen-Smith, D. (1982), "Restricted Pareto and rights", Journal of Economic Theory, 26:89–99.
Baigent, N. (1981), "Decomposition of minimal libertarianism", Economic Letters, 7:29–32.
Bandyopadhyay, T. (1983), "Expansion consistency, Pareto optimality and the power structure", mimeographed.
Bandyopadhyay, T. and R. Deb (1983), "Strategic voting for weakly binary group decision functions: The case of linear individual orders", in: Pattanaik and Salles (1983).
Bandyopadhyay, T., R. Deb and P. K. Pattanaik (1979), "The structure of coalition power under probabilistic group decision rules", mimeographed. Birmingham University.
Barbera, S. (1977a), "The manipulation of social choice mechanisms that do not leave 'too much' to chance", Econometrica, 45:1573–1588.
Barbera, S. (1977b), "Manipulation of social decision functions", Journal of Economic Theory, 15:226–278.
Barbera, S. (1979), "Majority and positional voting in probabilistic framework", Review of Economic Studies, 46:379–389.
Barbera, S. (1980a), "A note on group strategy-proof decision schemes", Econometrica, 47:637–640.
Barbera, S. (1980b), "Stable voting schemes", Journal of Economic Theory, 23:267–274.
Barbera, S. and B. Dutta (1982), "Implementation via protective equilibria", Journal of Mathematical Economics, 10:49–65.
Barbera, S. and H. Sonnenschein (1978), "Preference aggregation with randomized social orderings", Journal of Economic Theory, 18:244–254.
Barry, B. (1965), Political Argument. London: Routledge and Kegan Paul.
Barry, B. (1970), Sociologists, Economists and Democracy. London: Macmillan.
Barry, B. (1973), The Liberal Theory of Justice. Oxford: Clarendon Press.
Barry, B. and D. Rae (1975), "Political evaluation", in: F. I. Greenstein and N. W. Polsby, eds., Handbook of Political Science, Vol. I. Reading: Addison-Wesley.
Barthelemy, J. P. (1983), "Arrow's theorem: Unusual domains and extended codomains", in: Pattanaik and Salles (1983).

Basu, K. (1979), Revealed Preference of Governments. Cambridge: Cambridge University Press.
Basu, K. (1984), "The right to give up rights", Economica, 51:413–422.
Batra, R. N. and P. K. Pattanaik (1971), "Transitivity of social decisions under some more general group decision rules than majority voting", Review of Economic Studies, 38:295–306.
Batra, R. N. and P. K. Pattanaik (1972a), "Transitive multistage majority decisions with quasi-transitive individual preferences", Econometrica, 40:1121–1135.
Batra, R. M. and P. K. Pattanaik (1972b), "On some suggestions for having non-binary social choice functions", Theory and Decision, 3:1–11.
Batteau, P. and J.-M. Blin (1979), "Elements for new insights into the Gibbard–Satterthwaite theorem", in: Laffont (1979).
Batteau, P., J.-M. Blin and B. Monjardet (1981), "Stability of aggregation procedures, ultra-filters and simple games", Econometrica, 49:527–534.
Baumol, W. J. (1952), Welfare Economics and the Theory of the State. Cambridge: Harvard University Press.
Bentham, J. (1789), An Introduction to the Principles of Morals and Legislation. London: Payne. Also Oxford: Clarendon Press, 1907.
Bergson, A. (1938), "A reformulation of certain aspects of welfare economics", Quarterly Journal of Economics, 52:310–334.
Bergson, A. (1948), "Socialist economics", in: H. S. Ellis, ed., A Survey of Contemporary Economics. Homewood: Irwin.
Bergson, A. (1966), Essays in Normative Economics. Cambridge: Harvard University Press.
Bergstrom, T. C. (1970), "A 'Scandinavian consensus' solution for efficient income distribution among non-malevolent consumers", Journal of Economic Theory, 2:383–398.
Bergstrom, T. C. (1975), "Maximal elements of acyclic relations on compact sets", Journal of Economic Theory, 10:403–440.
Bernholz, P. (1974), "Is a Paretian liberal really impossible?", Public Choice, 20:99–108.
Bernholz, P. (1977), "Prisoner's dilemma, logrolling and cyclical group preferences", Public Choice, 29:73–84.
Bernholz, P. (1980), "A general social dilemma: Profitable exchange and intransitive group preference", Zeitschrift für Nationalökonomie, 40:1–23.
Bezembinder, Th. and P. van Acker (1980), "Intransitivity in individual and social choice", in: E. D. Lantermann and H. Feger, eds., Similarity and Choice. Bern: Huber, and New York: Wiley.
Bezembinder, Th. and P. van Acker (1979), "A note on Sen's partial comparability model". Department of Psychology, Katholieke Universiteit, Nijmegen, The Netherlands.
Binmore, K. (1975), "An example in group preference", Journal of Economic Theory, 10:377–385.
Binmore, K. (1976), "Social choice and parties", Review of Economic Studies, 43:459–464.
Binmore, K. (1981), "Nash bargaining theory", ICERD Discussion Paper. London School of Economics.
Birchenhall, C. R. (1977), "Conditions for the existence of maximal elements in compact sets", Journal of Economic Theory, 16:111–115.
Black, D. (1948), "On the rationale of group decision making", Journal of Political Economy, 56:23–34.
Black, D. (1958), The Theory of Committees and Elections. London: Cambridge University Press.
Blackorby, C. (1975), "Degrees of cardinality and aggregate partial orderings", Econometrica, 43:845–852.
Blackorby, C. and D. Donaldson (1977), "Utility vs. equity: Some plausible quasi-orderings", Journal of Public Economics, 7:365–381.
Blackorby, C. and D. Donaldson, (1978), "Measures of relative equality and their meaning in terms of social welfare", Journal of Economic Theory, 18:59–80.
Blackorby, C. and D. Donaldson (1979), "Interpersonal comparability of origin- or scale-independent utilities: Admissible social evaluation functionals", Discussion Paper No. 79-04. Vancouver: Department of Economics, University of British Columbia.
Blackorby, C. and D. Donaldson (1980a), "Ethical indices for the measurement of poverty", Econometrica, 48:1053–1060.

Blackorby, C. and D. Donaldson (1980b), "A theoretical treatment of indices of absolute equality", International Economic Review, 21:107–136.
Blackorby, C., D. Donaldson and J. A. Weymark (1980), "On John Harsanyi's defences of utilitarianism", Discussion Paper. Vancouver: University of British Columbia.
Blackorby, C., D. Donaldson and J. A. Weymark (1984), "Social choice with interpersonal utility comparisons: A diagrammatic introduction", International Economic Review, 25:327–356.
Blair, D. H. (1975), "Path-independent social choice functions: A further result", Econometrica, 43:173–174.
Blair, D. H. (1979), "On variable majority rule and Kramer's dynamic competitive processes", Review of Economic Studies, 46:667–673.
Blair, D. H., G. Bordes, J. S. Kelly and K. Suzumura (1976), "Impossibility theorems without collective rationality", Journal of Economic Theory, 13:361–379.
Blair, D. H. and R. Muller (1983), "Essential aggregation procedures on restricted domain of preferences", Journal of Economic Theory, 30:34–53.
Blair, D. H. and R. A. Pollak (1979), "Collective rationality and dictatorship: The scope of the Arrow theorem", Journal of Economic Theory, 21:186–194.
Blair, D. H. and R. A. Pollak (1982), "Acyclic collective choice rules", Econometrica, 50:931–943.
Blair, D. H. and R. A. Pollak (1983), "Polychromatic acyclic tours in coloured multigraphs", Mathematics of Operations Research, 8:471–476.
Blau, J. H. (1957), "The existence of a social welfare function", Econometrica, 25:302–313.
Blau, J. H. (1959), "Aggregation of preferences" (abstract), Econometrica, 27:283.
Blau, J. H. (1971), "Arrow's theorem with weak independence", Economica, 38:413–420.
Blau, J. H. (1972), "A direct proof of Arrow's theorem", Econometrica, 40:61–67.
Blau, J. H. (1975), "Liberal values and independence", Review of Economic Studies, 42:413–420.
Blau, J. H. (1976), "Neutrality, monotonicity and the right of veto: A comment", Econometrica, 44:603.
Blau, J. H. (1979), "Semiorders and collective choice", Journal of Economic Theory, 21:195–206.
Blau, J. H. and D. J. Brown (1978), "The structure of neutral monotonic social functions", mimeographed.
Blau, J. H. and R. Deb (1977), "Social decision functions and veto", Econometrica, 45:871–879.
Blin, J.-M. (1973), Patterns and Configurations in Economic Science. Dordrecht: Reidel.
Blin, J.-M. and M. Satterthwaite (1976), "Strategy-proofness and single-peakedness", Public Choice, 26:51–58.
Blin, J.-M. and M. Satterthwaite (1977), "On preferences, beliefs, and manipulation with voting situations", Econometrica, 45:881–888.
Bloomfield, S. (1971), "An axiomatic formulation of constitutional games", Technical Report No. 71–18. Stanford: Operations Research House, Stanford University.
Bloomfield, S. (1976), "A social choice interpretation of the von Neumann–Morgenstern game", Econometrica, 44:105–114.
Borda, J. C. (1781), "Mémoire sur les élections au scrutin", Memoires des l'Académie Royale des Sciences; English translation by A. de Grazia, Isis, 44 (1953).
Bordes, G. (1976), "Consistency, rationality and collective choice", Review of Economic Studies, 43:447–457.
Bordes, G. (1979), "Some more results on consistency, rationality and collective choice", in: Laffont (1979).
Borglin, A. (1982), "States and persons – On the interpretation of some fundamental concepts in the theory of justice as fairness", Journal of Public Economics, 18:85–104.
Bose, A. (1975), Marxian and Post-Marxian Political Economy. Harmondsworth: Penguin Books.
Brams, S. J. (1975), Game Theory and Politics. New York: Free Press.
Brams, S. J. (1976), Paradoxes in Politics. New York: Free Press.
Brams, S. J. and P. C. Fishburn (1978), "Approval voting", American Political Science Review, 72:831–847.
Brams, S. J. and P. C. Fishburn (1983), Approval Voting. Boston: Birkhäuser.
Braybrooke, D. (1978), "Policy-formation with issue processing and transformation of issues", in: Hooker, Leach and McClennen (1978).

Breyer, F. (1977), "The liberal paradox, decisiveness over issues, and domain restrictions", Zeitschrift für Nationalökonomie, 37:45–60.

Breyer, F. and R. Gardner (1980), "Liberal paradox, game equilibrium and Gibbard optimum", Public Choice, 35:469–481.

Breyer, F. and G. A. Gigliotti (1980), "Empathy and respect for the right of others", Zeitschrift für Nationalökonomie, 40:59–64.

Brown, D. J. (1973), "Acyclic choice", Cowles Foundation Discussion Paper. New Haven: Yale University.

Brown, D. J. (1974), "An approximate solution to Arrow's problem", Journal of Economic Theory, 9:375–383.

Brown, D. J. (1975a), "Aggregation of preferences", Quarterly Journal of Economics, 89:456–469.

Brown, D. J. (1975b), "Collective rationality", Cowles Foundation Discussion Paper 393. New Haven: Yale University.

Buchanan, J. M. (1976), "An ambiguity in Sen's alleged proof of the impossibility of a Paretian libertarian", mimeographed. Virginia Polytechnic.

Buchanan, J. M. and G. Tullock (1962), The Calculus of Consent. Ann Arbor: University of Michigan Press.

Camacho, A. (1974), "Societies and social decision functions", in: Leinfellner and Kohler (1974).

Camacho, A. and J. C. Sonsteile (1974), "Cardinal welfare, individualistic ethics and interpersonal comparisons of utilities: A note", Journal of Political Economy, 82:60–61.

Campbell, C. D. and G. Tullock (1965), "A measure of the importance of cyclical majorities", Economic Journal, 75:853–857.

Campbell, C. D. and G. Tullock (1966), "The paradox of voting: A possible method of calculation", American Political Science Review, 60:684–685.

Campbell, D. E. (1972), "A collective choice rule satisfying Arrow's five conditions in practice", in: Theory and Application of Collective Choice Rule. Toronto: Institute for Quantitative Analysis of Social and Economic Policy, University of Toronto.

Campbell, D. E. (1976), "Democratic preference functions", Journal of Economic Theory, 12:259–272.

Campbell, D. E. (1978), "Realization of choice functions", Econometrica, 46:171–180.

Campbell, D. E. (1979), "Manipulation of social choice rules by strategic nomination of candidates", Theory and Decision, 10:247–264.

Campbell, D. E. (1980), "Algorithms for social choice functions", Review of Economic Studies, 47:617–627.

Chakravarty, S. (1969), Capital and Development Planning. Cambridge: MIT Press.

Chakravarty, S. R. (1983a), "Ethically flexible measures of poverty", Canadian Journal of Economics, 16:74–85.

Chakravarty, S. R. (1983b), "Measures of poverty based on representative income gap", Sankhyā: The Indian Journal of Statistics, 45 B:69–74.

Chamberlain, G. and M. Rothschild (1981), "A note on the probability of casting a decisive vote", Journal of Economic Theory, 25:152–162.

Champsaur, P. and G. Laroque (1981), "Fair allocations in large economies", Journal of Economic Theory, 25:269–282.

Chapman, B. (1983), "Rights as constraints: Nozick versus Sen", Theory and Decision, 15:1–10.

Chichilnisky, G. (1976), "Manifolds of preference and equilibria", Ph.D. Dissertation. Berkeley: University of California.

Chichilnisky, G. (1979), "On fixed point theorems and social choice paradoxes", Economic Letters.

Chichilnisky, G. (1980a), "Social choice and the topology of spaces of preferences", Advances in Mathematics, 37:165–176.

Chichilnisky, G. (1980b), "Continuous representation of preferences", Review of Economic Studies, 47:959–963.

Chichilnisky, G. (1980c), "Intensity of preference and von Neumann–Morgenstern utilities", Working Paper. Essex University.

Chichilnisky, G. (1981), "Structural instability of decisive majority rules", Journal of Mathematical Economics, 9:207–221.

Chichilnisky, G. (1982a), "Social aggregation rules and continuity", Quarterly Journal of Economics, 96:337–352.

Chichilnisky, G. (1982b), "Topological equivalence of the Pareto condition and the existence of a dictator", Journal of Mathematical Economics, 9:223–233.

Chichilnisky, G. (1983), "Social choice and game theory: Recent results with a topological approach", in: Pattanaik and Salles (1983).

Chichilnisky, G. and G. Heal (1979), "On strong manipulability", Working Paper. Essex University.

Chichilnisky, G. and G. Heal (1981), "Incentive compatibility and local simplicity", Working Paper. Essex University.

Chichilnisky, G. and G. Heal (1983), "Necessary and sufficient conditions for a resolution of the social choice paradox", Journal of Economic Theory, 31:68–87.

Chipman, J. S. (1974), "The welfare ranking of Pareto distributions", Journal of Economic Theory, 9:275–282.

Chipman, J. S. (1976), "The Paretian heritage", Revue Européenne des Science Sociales et Cahiers Vilfredo Pareto, 14:65–173.

Chipman, J. S., L. Hurwicz, M. K. Richter and H. F. Sonnenschein (1971), Preference, Utility and Demand. New York: Harcourt.

Cohen, L. and S. Matthews (1980), "Constrained Plott equilibria, directional equilibria and global cycling sets", Review of Economic Studies, 47:975–986.

Cohen, L. J. et al. (1981), Logic, Methology and the Philosophy of Science. Amsterdam: North-Holland.

Coleman, J. S. (1966), "The possibility of a social welfare function", American Economic Review, 56:1105–1122.

Coleman, J. (1973), The Mathematics of Collective Action. London: Heinemann.

Condorcet, M. de (1785), Essai sur l'Application de l'Analyse a la Probabilité des Decisions Rendues à le Pluralité des Voix. Paris.

Contini, B. (1966), "A note on Arrow's postulates for a social welfare function", Journal of Political Economy, 74:278–280.

Coughlin, P. J. (1981), "Necessary and sufficient conditions for δ-relative majority voting equilibria", Econometrica, 49:1223–1224.

Coughlin, P. J. (1983), "Libertarian concessions of the private Pareto rule", Economica, forthcoming.

Coughlin, P. J. and S. Nitzan (1981), "Electoral outcomes with probabilistic voting and Nash social welfare maxima", Journal of Public Economics, 15:113–121.

Coughlin, P. J. and S. Nitzan (1983), "Directional and local electoral equilibria with probabilistic voting", in: Pattanaik and Salles (1983).

Craven, J. (1971), "Majority voting and social choice", Review of Economic Studies, 38:265–267.

Crawford, V. P. (1977), "A game of fair division", Review of Economic Studies, 44:235–248.

Crawford, V. P. (1979), "A procedure for generating Pareto-efficient egalitarian–equivalent allocations", Econometrica, 47:49–60.

Crawford, V. P. and W. P. Heller (1979), "Fair division with indivisible commodies", Journal of Economic Theory, 21:10–27.

Daniel, T. E. (1975), "A revised concept of distributional equity", Journal of Economic Theory, 11:94–109.

Daniels, N., ed. (1975), Reading Rawls. Oxford: Blackwell.

Danielsson, S. (1974), "The group preference problem as a problem of distributive justice". Uppsala: Filosofiska Institutionen, Uppsala Universitet.

Dasgupta, P. (1974), "On alternative criteria for justice between generations", Journal of Public Economics, 3:405–423.

Dasgupta, P. (1982), The Control of Resources. Oxford: Blackwell.

Dasgupta, P., P. Hammond and E. Maskin (1979), "The implementation of social choice rules: Some general results on incentive compatibility", Review of Economic Studies, 46:185–216.

Dasgupta, P. and G. Heal (1979), Economic Theory and Exhaustible Resources. London: James Nisbet and Cambridge University Press.

d'Aspremont, C. and L. Gevers (1977), "Equity and informational basis of collective choice", Review of Economic Studies, 44:199–210.

Davis, O. A., M. H. De Groot and M. J. Hinich (1972), "Social preference orderings and majority rule", Econometrica, 40:147–157.

Day, R. H. (1971), "Rational choice and economic behaviour", Theory and Decision, 1:229–251.

Deb, R. (1975), "Rational choice and cyclical preferences", Ph.D. Dissertation. London University.
Deb, R. (1976), "On constructing generalized voting paradoxes", Review of Economic Studies, 43:347–352.
Deb, R. (1977), "On Schwartz's rule", Journal of Economic Theory, 16:103–110.
Debreu, G. (1959), Theory of Value: An Axiomatic Study of Economic Equilibrium. New York: Wiley.
Debreu, G. (1960), "Topological methods in cardinal utility theory", in: K. J. Arrow, S. Karlin and P. Suppes, eds., Mathematical Methods in Social Sciences, 1959. Stanford: Stanford University Press.
DeMeyer, F. and C. R. Plott (1970), "The probability of a cyclical majority", Econometrica, 38:345–354.
DeMeyer, F. and C. R. Plott (1971), "A welfare function using 'relative intensity' of preferences", Quarterly Journal of Economics, 85:179–186.
Denzau, A. T. and R. P. Parks (1975), "The continuity of majority rule equilibrium", Econometrica, 43:853–866.
Dechamps, R. and L. Gevers (1978), "Leximin and utilitarian rules: A joint characterisation", Journal of Economic Theory, 17:143–163.
Deschamps, R. and L. Gevers (1979), "Separability, risk-bearing and social welfare judgments", in: Laffont (1979).
Diamond, P. A. (1967), "Cardinal welfare, individualistic ethics, and interpersonal comparisons of utility: Comment", Journal of Political Economy, 75:765–766.
Dobb, M. (1955), On Economic Theory and Socialism. London: Routledge and Kegan Paul.
Downs, A. (1957), An Economic Theory of Democracy. New York: Harper.
Dummett, M. and R. Farquharson (1961), "Stability in voting", Econometrica 29:33–43.
Dutta, B. (1977), "Existence of stable situations, restricted preferences and strategic manipulation under democratic group decision rules", Journal of Economic Theory, 15:99–111.
Dutta, B. (1980a), "Strategic voting in a probabilistic framework", Econometrica, 58:447–456.
Dutta, B. (1980b), "On the possibility of consistent voting procedures", Review of Economic Studies, 47:603–616.
Dutta, B. (1983), "Further results on voting with veto", in: Pattanaik and Salles (1983).
Dutta, B. and P. K. Pattanaik (1978), "On nicely consistent voting systems", Econometrica, 46:163–170.
Dworkin, R. (1978), Taking Rights Seriously, 2nd ed. London: Duckworth.
Encarnacion, Jr., J. (1969), "On independence postulates concerning choice", International Economic Review, 10:134–140.
Farkas, D. and S. Nitzan (1979), "The Borda rule and Pareto stability: A comment", Econometrica, 47:1305–1306.
Farquharson, R. (1956), "Straightforwardness in voting paradoxes", Oxford Economic Papers, 8:80–89.
Farquharson, R. (1969), Theory of Voting. New Haven: Yale University Press.
Farrell, M. J. (1976), "Liberalism in the theory of social choice", Review of Economic Studies, 43:3–10.
Feldman, A. (1979), "Nonmanipulable multivalued social decision functions", Public Choice, 34:39–50.
Feldman, A. (1980a), Welfare Economics and Social Choice Theory. Boston: Martinus Nijhoff.
Feldman, A. (1980b), "Strongly nonmanipulable multivalued collective choice rules", Public Choice, 35:503–509.
Feldman, A. and A. Kirman (1974), "Fairness and envy", American Economic Review, 64:995–1005.
Feldman, A. and D. Weiman (1979), "Envy, wealth and class hierarchies", Journal of Public Economics, 11:81–91.
Ferejohn J. A. (1977), "Decisive coalitions in the theory of social choice", Journal of Economic Theory, 30:35–42.
Ferejohn J. A. (1978), "The distribution of rights in society", in: Gottinger and Leinfellner (1978) 119–132.
Ferejohn, J. A. and D. Grether (1974), "On a class of rational social decision procedures", Journal of Economic Theory, 8:471–482.
Ferejohn, J. A. and D. Grether (1977a), "Some new impossibility theorems", Public Choice, 30:35–42.
Ferejohn, J. A. and D. Grether (1977b), "Weak path independence", Journal of Economic Theory, 14:19–31.

Ferejohn, J. A., D. Grether, S. A. Matthews and E. W. Packel (1980), "Continuous-valued binary decision procedures", Review of Economic Studies, 47:787–796.

Fine, B. J. (1975a), "A note on 'interpersonal aggregation and partial comparability'", Econometrica, 43:173–174.

Fine, B. J. (1975b), "Individual liberalism in a Paretian society", Journal of Political Economy, 83:1277–1281.

Fine, B. J. and K. Fine (1974), "Social choice and individual ranking", Review of Economic Studies, 41:303–322, 459–475.

Fine, K. (1972), "Some necessary and sufficient conditions for representative decision on two alternatives", Econometrica, 40:1083–1090.

Fine, K. (1973), "Conditions for the existence of cycles under majority and non-minority rules", Econometrica, 41:889–899.

Fishburn, P. C. (1970a), "Suborders on commodity spaces", Journal of Economic Theory, 2:321–328.

Fishburn, P. C. (1970b), "Arrow's impossibility theorem: Concise proof and infinite voters", Journal of Economic Theory, 2:103–106.

Fishburn, P. C. (1970c), "The irrationality of transitivity in social choice", Behavioral Sciences, 15:119–123.

Fishburn, P. C. (1970d), "Comments on Hansson's 'group preferences'", Econometrica, 38:933–935.

Fishburn, P. C. (1971), "Should social choice be based on binary comparisons?", Journal of Mathematical Sociology, 1:133–142.

Fishburn, P. C. (1972a), "A location theorem for single-peaked preferences", Journal of Economic Theory, 4:94–97.

Fishburn, P. C. (1972b), "Even-chance lotteries in social choice theory", Theory and Decision, 3:18–40.

Fishburn, P. C. (1972c), "Lotteries and social choice", Journal of Economic Theory, 5:189–207.

Fishburn, P. C. (1973a), The Theory of Social Choice. Princeton: Princeton University Press.

Fishburn, P. C. (1973b), "Transitive binary social choices and interprofile conditions", Econometrica, 41:603–615.

Fishburn, P. C. (1973c), "Summation social choice functions", Econometrica, 41:1183–1196.

Fishburn, P. C. (1974a), "On collective rationality and a generalized impossibility theorem", Review of Economic Studies, 41:445–459.

Fishburn, P. C. (1974b), "Subset choice conditions and the computation of social choice sets", Quarterly Journal of Economics, 88:320–329.

Fishburn, P. C. (1974c), "Choice functions on finite sets", International Economic Review, 15:729–749.

Fishburn, P. C. (1974d), "Impossibility theorems without the social completeness axiom", Econometrica, 42:695–704.

Fishburn, P. C. (1974e), "Social choice functions", SIAM Review, 16:63–90.

Fishburn, P. C. (1975a), "Semiorders and choice functions", Econometrica, 43:975–977.

Fishburn, P. C. (1975b), "A probabilistic model of social choice: A comment", Review of Economic Studies, 42:297–301.

Fishburn, P. C. (1976a), "Dictators on blocks: Generalizations of social choice impossibility theorems", Journal of Combinatorial Theory, 20:153–170.

Fishburn, P. C. (1976b), "Representable choice functions", Econometrica, 44:1033–1043.

Fishburn, P. C. (1977), "Condorcet social choice functions", SIAM Journal of Applied Mathematics, 33:469–489.

Fishburn, P. C. and W. V. Gehrlein (1976), "Borda's rule, positional voting, and Condorcet's simple majority principle", Public Choice, 28:79–88.

Fishburn, P. C. and W. V. Gehrlein (1977a), "Collective rationality versus distribution of power for binary social choice functions", Journal of Economic Theory, 15:72–91.

Fishburn, P. C. and W. V. Gehrlein (1977b), "Toward a theory of elections with probabilistic voting", Econometrica, 45:1907–1923.

Fishburn, P. C., W. V. Gehrlein and E. Maskin (1979), "Condorcet proportions and Kelly's conjectures", Discrete Applied Mathematics, 229–252.

Foley, D. (1967), "Resource allocation in the public sector", Yale Economic Essays, 7:73–76.

Foster, J. (1984), "On economic poverty: A survey of aggregate measures", Advances in Econometrics, forthcoming.

Frey, B. S. (1978), Modern Political Economy. Oxford: Martin Robertson.

Frey, B. S. (1983), Democratic Economic Policy. Oxford: Martin Robertson.

Gaertner, W. (1983), "Equity- and inequity-type Borda rules", Mathematical Social Sciences, 4:137–154.

Gaertner, W. and A. Heinecke (1978), "Cyclically mixed preferences – A necessary and sufficient condition for transitivity of the social preference relation", in: Gottinger and Leinfellner (1978) 169–186.

Gaertner, W. and L. Krüger (1981), "Self-supporting preferences and individual rights: The possibility of Paretian libertarianism", Economica, 48:17–28.

Gaertner, W. and L. Krüger (1983), "Alternative libertarian claims and Sen's paradox", Theory and Decision, 15:211–230.

Gärdenfors, P. (1973), "Positional voting functions", Theory and Decision, 4:1–24.

Gärdenfors, P. (1975), "Fairness without interpersonal comparisons", Working Paper No. 15. Lund: Mattias Fremling Society.

Gärdenfors, P. (1976), "Manipulation of social choice functions", Journal of Economic Theory, 13:217–228.

Gärdenfors, P. (1979), "On definitions of manipulation of social choice functions", in: Laffont (1979).

Gärdenfors, P. (1981), "Rights, games and social choice", Nous, 15.

Gardner, R. (1977), "The Borda game", Public Choice, 30:43–50.

Gardner, R. (1980), "The strategic inconsistency of Paretian liberalism", Public Choice, 35:241–252.

Garman, M. and M. Kamien (1968), "The paradox of voting: Probability calculations", Behavioral Science, 13:306–316.

Gehrlein, W. V. and P. C. Fishburn (1976), "The probability of the paradox of voting: A computable solution", Journal of Economic Theory, 13:14–25.

Gehrlein, W. V. and P. C. Fishburn (1979), "Proportion of profiles with the majority candidate", Computation and Mathematics with Applications, 5:117–124.

Gevers, L. (1979), "On interpersonal comparability and social welfare orderings", Econometrica, 47:75–90.

Gibbard, A. (1973), "Manipulation of voting schemes: A general result", Econometrica, 41:587–601.

Gibbard, A. (1974), "A Pareto-consistent libertarian claim", Journal of Economic Theory, 7:338–410.

Gibbard, A. (1977), "Manipulation of schemes that mix voting with chance", Econometrica, 45:665–681.

Gibbard, A. (1978), "Social decision, strategic behavior, and best outcomes", in: Gottinger and Leinfellner (1978) 153–168.

Goldman, S. M. and C. Sussangkarn (1978), "On the concept of fairness", Journal of Economic Theory, 19:210–216.

Gottinger, H. W. and W. Leinfellner (1978), Decision Theory and Social Ethics: Issues in Social Choice. Dordrecht: Reidel.

Graaff, J. de V. (1946), "Fluctuations in income concentration", South African Journal of Economics, 14.

Graaff, J. de V. (1957), Theoretical Welfare Economics. Cambridge: Cambridge University Press; 2nd ed. with a foreword by P. A. Samuelson (1967).

Graaff, J. de V. (1965), "Unpublished notes on the probability of cyclical majorities". Cambridge: Churchill College.

Graaff, J. de V. (1977), "Equity and efficiency as components of the general welfare", South African Journal of Economics, 45:362–375.

Graaff, J. de V. (1983), "Normative measurement", mimeographed. Oxford: All Souls College.

Grandmont, J.-M. (1978), "Intermediate preferences and majority rule", Econometrica, 46:317–330.

Green, J. and J.-J. Laffont (1979), Incentives in Public Decision Making. Amsterdam: North-Holland.

Grether, D. M. and C. R. Plott (1982), "Nonbinary social choice: An impossibility theorem", Review of Economic Studies, 49:143–149.

Grout, P. (1977), "A Rawlsian intertemporal consumption rule", Review of Economic Studies, 44:337–346.

Groves, T. and J. Ledyard (1977), "Optimal allocation of public goods: A solution to the 'free rider' problem", Econometrica, 45:783–809.

Guha, A. S. (1972), "Neutrality, monotonicity and the right of veto", Econometrica, 40:821–826.
Guilbaud, G. Th. (1952), "Les théories de l'intérêt général et la problème logique de l'aggrégation", Economie Appliquée, 5:501–584; English translation in: P. F. Lazarsfeld and N. W. Henry, eds., Readings in Mathematical Social Sciences. Chicago: Science Research Associates, 1966.
Hahn, F. H. (1973), On the Notion of Equilibrium in Economics: An Inaugural Lecture. Cambridge: Cambridge University Press.
Hamada, K. (1973), "A simple majority rule on the distribution of income", Journal of Economic Theory, 6:243–264.
Hamada, K. and N. Takayama (1978), "Censored income distributions and the measurement of poverty", Bulletin of the International Statistical Institute, 47.
Hammond, P. J. (1975), "A note on extreme inequality aversion", Journal of Economic Theory, 11:465–467.
Hammond, P. J. (1976a), "Equity, Arrow's conditions and Rawls' difference principle", Econometrica, 44:793–804.
Hammond, P. J. (1976b), "Why ethical measures of inequality need interpersonal comparisons", Theory and Decision, 7:263–274.
Hammond, P. J. (1977a), "Dual interpersonal comparisons of utility and the welfare economics of income distribution", Journal of Public Economics, 6:51–71.
Hammond, P. J. (1977b), "Dynamic restrictions on metastatic choice", Economica, 44:337–380.
Hammond, P. J. (1978), "Economic welfare with rank order price weighting", Review of Economic Studies, 45:381–384.
Hammond, P. J. (1979a), "Straightforward individual incentive compatibility in large economies", Review of Economic Studies, 46:263–282.
Hammond, P. J. (1979b), "Equity in two person situations: Some consequences", Econometrica, 47:1127–1136.
Hammond, P. J. (1981), "Liberalism, independent rights, and the Pareto principle", in: Cohen (1981).
Hammond, P. J. (1982), "Utilitarianism, uncertainty and information", in: Sen and Williams (1982).
Hammond, P. J. (1983), "Ex-post optimality as a dynamically consistent objective for collective choice under uncertainty", in: Pattanaik and Salles (1983).
Hansson, B. (1968), "Choice structures and preference relations", Synthese, 18:443–458.
Hansson, B. (1969a), "Voting and group decision functions", Synthese, 20:526–537.
Hansson, B. (1969b), "Group preferences", Econometrica, 37:50–54.
Hansson, B. (1972), "The existence of group preferences", Working Paper No. 3. Lund: Mattias Fremling Society.
Hansson, B. (1973), "The independence condition in the theory of social choice", Theory and Decision, 4:25–49.
Hansson, B. (1976), "The existence of group preferences", Public Choice, 28:89–98.
Hansson, B. and H. Sahlquist (1976), "A proof technique for social choice with variable electorate", Journal of Economic Theory, 13:193–200.
Harsanyi, J. C. (1955), "Cardinal welfare, individualistic ethics, and interpersonal comparisons of utility", Journal of Political Economy, 63:309–321.
Harsanyi, J. C. (1958), "Ethics in terms of hypothetical imperatives", Mind, 69:305–316.
Harsanyi, J. C. (1975), "Non-linear social welfare functions, or do welfare economists have a special exemption from Bayesian rationality", Theory and Decision, 6:311–332.
Harsanyi, J. C. (1977a), "Non-linear social welfare functions: A rejoinder to Professor Sen", in: R. Butts and J. Hintikka, eds., Logic, Methodology and Philosophy of Science. Dordrecht: Reidel.
Harsanyi, J. C. (1977b), Rational Behaviour and Bargaining Equilibrium in Goods and Social Situations. Cambridge: Cambridge University Press.
Harsanyi, J. C. (1978), "Bayesian decision theory and utilitarian ethics", American Economic Review: Papers and Proceedings, 68:223–228.
Harsanyi, J. C. (1979), "Bayesian decision theory, rule utilitarianism and Arrow's impossibility theorem", Theory and Decision, 11:289–318.
Hayek, F. A. (1960), The constitution of liberty. London: Routledge.
Heal, G. (1973), The Theory of Economic Planning. Amsterdam: North-Holland.
Heal, G. (1983), "Contractability and public decision-making", in: Pattanaik and Salles (1983).

Heiner, R. A. and P. K. Pattanaik (1983), "The structure of general probabilistic group decision rules", in: Pattanaik and Salles (1983).
Herzberger, H. G. (1973), "Ordinal preference and rational choice", Econometrica, 41:187–237.
Hildreth, C. (1953), "Alternative conditions for social ordering", Econometrica, 21:81–89.
Hinich, M. (1977), "Equilibrium in spatial voting: The median voter result is an artifact", Journal of Economic Theory, 16:208–219.
Hinich, M., J. Ledyard and P. Ordeshook (1972), "Non-voting and the existence of equilibrium under majority rule", Journal of Economic Theory, 4:144–153.
Hoag, C. G. and G. H. Hallett (1926), Proportional Representation. New York: Macmillan.
Hooker, C. A., J. J. Leach and E. F. McClennen, eds. (1978), Foundations and Applications of Decision Theory. Dordrecht: Reidel.
Houthakker, H. S. (1950), "Revealed preference and utility function", Economica, 17:159–174.
Hurwicz, L. (1962), "On informationally decentralized systems", in: R. Radner and B. McGuire, eds., Decision and Organisation. Amsterdam: North-Holland.
Hurwicz, L. and D. Schmeidler (1978), "Construction of outcome functions guaranteeing Pareto optimality of Nash equilibria", Econometrica, 46:1447–1474.
Inada, K. (1955), "Alternative incompatible conditions for a social welfare function", Econometrica, 23:396–399.
Inada, K. (1964a), "On the economic welfare function", Econometrica, 32:316–338.
Inada, K. (1964b), "A note on the simple majority decision rule", Econometrica, 32:525–531.
Inada, K. (1969), "On the simple majority decision rule", Econometrica, 37:490–506.
Inada, K. (1970), "Majority rule and rationality", Journal of Economic Theory, 2:27–40.
Intriligator, M. (1973), "A probabilistic model of social choice", Review of Economic Studies, 40:553–560.
Intriligator, M. (1979), "Income redistribution: A probabilistic approach", American Economic Review, 69:97–105.
Jain, S. K. (1977a), "The stability of binary social decision rules", Discussion Paper. Delhi: Indian Statistical Institute.
Jain, S. K. (1977b), "Characterization of rationality conditions in terms of minimal decisive sets". Delhi: Indian Statistical Institute.
Jamison, D. T. and L. J. Lau (1973), "Semiorders and the theory of choice", Econometrica, 41:901–912; "A Correction", 43:975–977 (1975).
Jamison, D. T. and L. J. Lau (1977), "The nature of equilibrium with semiordered preferences", Econometrica, 45:1595–1605; "A Correction", 47:1047–1048 (1979).
Jamison, D. T. and E. Luce (1972), "Social homogeneity and the probability of intransitive majority rule", Journal of Economic Theory, 5:79–87.
Jeffrey, R. C. (1971), "On interpersonal utility theory", Journal of Philosophy, 68.
Johansen, L. (1970), "An examination of the relevance of Kenneth Arrow's general possibility theorem for economic planning", in: Optimation et Simulation de Macro-décisions, Économie Mathématique et Econométrie No. 3. Namur: Gembloux.
Johansen, L. (1976), "The theory of public goods: Misplaced emphasis", Oslo: Institute of Economics, University of Oslo.
Kahneman, D. and A. Tversky (1979), "Prospect theory: An analysis of decision under risk", Econometrica, 47.
Kakwani, N. (1980a), Income Inequality and Poverty. New York: Oxford University Press.
Kakwani, N. (1980b), "On a class of poverty measures", Econometrica, 48:437–446.
Kakwani, N. (1981), "Note on a new measure of poverty", Econometrica, 49:525–526.
Kalai, E. and N. Megiddo (1980), "Path independent choices", Econometrica, 48:781–784.
Kalai, E. and E. Muller (1977), "Characterization of domains admitting non-dictatorial social welfare functions and nonmanipulable voting procedures", Journal of Economic Theory, 16:457–469.
Kalai, E., E. Muller and M. A. Satterthwaite (1977), "Social welfare functions when preferences are convex and continuous: Impossibility results", Public Choice, forthcoming.
Kalai, E. and Z. Ritz (1980), "Characterization of private alternative domains admitting Arrow social welfare functions", Journal of Economic Theory, 22:23–36.
Kalai, E. and D. Schmeidler (1977), "Aggregation procedures for cardinal preferences: A formulation and proof of Samuelson's impossibility conjecture", Econometrica, 45.

Kalai, E. and M. Smordinsky (1975), "Other solutions to Nash's bargaining problem", Econometrica, 43:513–518.

Kaneko, M. (1975), "Necessary and sufficient conditions for transitivity in voting theory", Journal of Economic Theory, 11:385–393.

Kaneko, M. (1980), "An extension of the Nash bargaining problem and the Nash social welfare function", Theory and Decision, 12:135–148.

Kaneko, M. and K. Nakamura (1979), "The Nash social welfare function", Econometrica, 47:423–436.

Kanger, S. (1972), "Law and logic", Theoria, 38:105–132.

Kanger, S. (1975), "Choice based on preference", mimeographed. Uppsala: Department of Philosophy, Uppsala University.

Karni, E. (1978), "Collective rationality, unanimity and liberal ethics", Review of Economic Studies, 45:571–574.

Karni, E. and D. Schmeidler (1976), "Independence of non-feasible alternatives and independence of non-optimal alternatives", Journal of Economic Theory, 12:488–493.

Kats, A. (1972), "On the social welfare function and the parameters of income distribution", Journal of Economic Theory, 5:377–382.

Kats, A. and S. Nitzan (1976), "Global and local equilibrium in majority voting", Public Choice, 25:105–106.

Kelly, J. S. (1971), "The continuous representation of a social preference ordering", Econometrica, 39:593–597.

Kelly, J. S. (1974a), "Necessity conditions in voting theory", Journal of Economic Theory, 8:149–160.

Kelly, J. S. (1974b), "Voting anomalies, the number of voters and the number of alternatives", Econometrica, 42:239–251.

Kelly, J. S. (1976a), "The impossibility of a just liberal", Economica, 43:67–75.

Kelly, J. S. (1976b), "Rights-exercising and a Pareto consistent libertarian claim", Journal of Economic Theory, 13:138–153.

Kelly, J. S. (1976c), "Algebraic results of collective choice rules", Journal of Mathematical Economics, 3:285–293.

Kelly, J. S. (1977), "Strategy-proofness and social choice functions without single-valuedness", Econometrica, 45:439–446.

Kelly, J. S. (1978), Arrow Impossibility Theorems. New York: Academic Press.

Kelsey, D. (1982), "Topics in social choice", D.Phil. Thesis. Oxford University.

Kelsey, D. (1983a), "The structure of social decision functions", mimeographed.

Kelsey, D. (1983b), "Social choice without the Pareto principle", Review of Economic Studies, forthcoming.

Kemp, M. C. (1954), "Arrow's general possibility theorem", Review of Economic Studies, 21:240–243.

Kemp, M. C. and A. Asimakopulos (1952), "A note on social welfare functions and cardinal utility", Canadian Journal of Economics and Political Science, 18:195–200.

Kemp, M. C. and Y.-K. Ng (1976), "On the existence of social welfare functions, social orderings and social decision functions", Economica, 43:59–66.

Kemp, M. C. and Y.-K. Ng (1977), "More on social welfare functions: The incompatibility of individualism and ordinalism", Economica, 44:89–90.

Kern, L. (1978), "Comparative distributive ethics: An extension of Sen's examination of the pure distribution problem", in: Gottinger and Leinfellner (1978).

Kim, K. H. and F. W. Roush (1980a), Introduction to Mathematical Consensus Theory. New York: Marcel Dekker.

Kim, K. H. and F. W. Roush (1980b), "Binary social welfare functions", Journal of Economic Theory, 23:416–419.

Kirman, A. P. and D. Sondermann (1972), "Arrow's theorem, many agents and invisible dictators", Journal of Economic Theory, 5:267–277.

Kolm, S. Ch. (1969), "The optimum production of social justice", in: Margolis and Guitton (1969).

Kolm, S. Ch. (1972), Justice et Equité. Paris: Centre National de la Recherche Scientifique.

Kramer, G. H. (1973), "On a class of equilibrium conditions for majority rule", Econometrica, 41:285–297.

Kramer, G. H. (1977), "Theories of political processes", in: M. Intriligator, ed., Frontiers of Quantitative Economics III. Amsterdam: North-Holland.

Kramer, G. H. (1978), "A dynamic model of political equilibrium", Journal of Economic Theory, 16:310–334.
Kramer, G. H. and A. K. Klevorick (1974), "Existence of a 'local' cooperative equilibrium in a class of voting games", Review of Economic Studies, 41:539–548.
Krantz, D. H., R. D. Luce, P. Suppes and A. Tversky (1971), Foundations of Measurement. New York: Academic Press.
Kuga, K. and H. Nagatani (1974), "Voter antagonism and the paradox of voting", Econometrica, 42:1045–1067.
Kuhn, A. (1978), "On relating individual and social decisions", in: Hooker, Leach and McClennen (1978).
Kundu, A. and T. Smith (1983), "An impossibility theorem on poverty indices", International Economic Review, 24:423–434.
Laffont, J.-J., ed. (1979), Aggregation and Revelation of Preferences. Amsterdam: North-Holland.
Laffont, J.-J. and E. Maskin (1981), "The theory of incentives: An overview", Economic Theory Discussion Paper 41. Cambridge University.
Leinfellner, W. (1978), "Marx and the utility approach to the ethical foundation of microeconomics", in: Gottinger and Leinfellner (1978).
Leinfellner, W. and E. Köhler, eds. (1974), Developments in the Methodology of Social Sciences. Dordrecht: Reidel.
Levi, I. (1974), "On indeterminate probabilities", Journal of Philosophy, 71.
Levi, I. (1980), The Enterprise of Knowledge. Cambridge: MIT Press.
Levi, I. (1982), "Liberty and welfare", in: Sen and Williams (1982).
Lindahl, L. (1977), Position and Change: A Study in Law and Logic. Dordrecht: Reidel.
Lindbeck, A. (1983), "Interpreting income distribution in a welfare state: The case of Sweden", European Economic Review, 21:227–256.
Little, I. M. D. (1950), A Critique of Welfare Economics. Oxford: Clarendon Press; 2nd ed. (1957).
Little, I. M. D. (1952), "Social choice and individual values", Journal of Political Economy, 60:422–432.
Luce, R. D. (1956), "Semiorders and a theory of utility discrimination", Econometrica, 24:178–191.
Luce, R. D. and H. Raiffa (1957), Games and Decisions. New York: Wiley.
Machina, M. (1982), "Expected utility analysis without the independence axiom", Econometrica, 50:277–324.
Machina, M. and R. Parks (1981), "On path independent randomized choices", Econometrica, 49:1345–1347.
MacIntyre, I. and P. K. Pattanaik (1981), "Strategic voting under minimally binary group decision functions", Journal of Economic Theory, 25:338–352.
Mackay, A. F. (1980), Arrow's Theorem: The Paradox of Social Choice. New Haven: Yale University Press.
Majumdar, T. (1956), "Choice and revealed preference", Econometrica, 24:71–73.
Majumdar, T. (1961), The Measurement of Utility. London: Macmillan.
Majumdar, T. (1969a), "A note on Arrow's postulates for a social welfare function: A comment", Journal of Political Economy, 77:528–531.
Majumdar, T. (1969b), "Sen's theorem on transitivity of majority decisions: An alternative approach", in: T. Majumdar, ed., Choice and Growth. Calcutta: Oxford University Press.
Majumdar, T. (1973), "Amartya Sen's algebra of collective choice", Sankhya, 35:533–542.
Majumdar, T. (1983), Investment in Education and Social Choice. Cambridge: Cambridge University Press.
Malinvaud, E. and M. O. L. Bacharach (1967), Activity Analysis in the Theory of Growth and Planning. London: Macmillan.
Margolis, J. and H. Guitton, eds. (1969), Public Economics. London: Macmillan.
Marx, K. (1875), Critique of Gotha Programme; English translation in: K. Marx and F. Engels, Selected Works, Vol. II. Moscow: Foreign Language Publishing House.
Marx, K. (1887), Capital: A Critical Analysis of Capitalist Production, Vol. I. London: Sonnenschein.
Mas-Colell, A. and H. F. Sonnenschein (1972), "General possibility theorems for group decisions", Review of Economic Studies, 39:185–192.

Maskin, E. (1976a), "Social welfare functions on restricted domain", Review of Economic Studies, forthcoming.

Maskin, E. (1976b), "On strategyproofness and social welfare functions when preferences are restricted", mimeographed. Cambridge: Darwin College and Harvard University.

Maskin, E. (1978), "A theorem on utilitarianism", Review of Economic Studies, 45:93–96.

Maskin, E. (1979a), "Implementation and strong Nash equilibria", in: Laffont (1979).

Maskin, E. (1979b), "Decision-making under ignorance with implications for social choice", Theory and Decision, 11:319–337.

Matthews, S. (1977), "Undominated directions in simple dynamic games", Social Science Working Paper 169. Caltech.

Matthews, S. (1978), "Pairwise symmetry conditions for voting", Social Science Working Paper 209. Caltech.

Matthews S. (1979), "A simple directional model of electoral competition", Public Choice, 34:141–156.

Matsumoto, Y. (1982), "Choice functions: Preference, consistency and neutrality", D.Phil. Thesis. Oxford University.

May, K. O. (1952), "A set of independent, necessary and sufficient conditions for simple majority decision", Econometrica, 20:680–684.

May, K. O. (1953), "A note on complete independence of the conditions for simple majority decision", Econometrica, 21:172–173.

Mayston, D. J. (1974), The Idea of Social Choice. London: Macmillan.

Mayston, D. J. (1975), "Alternatives to irrelevant alternatives", Discussion Paper No. 61. Department of Economics, University of Essex.

Mayston, D. J. (1980), "Where did prescriptive welfare economics go wrong?", in: D. A. Currie and W. Peters, eds., Contemporary Economic Analysis, Vol. 2. London: Croom Helm.

Mayston, D. J. (1982), "The generation of a social welfare function under ordinal preferences", Mathematical Social Sciences, 3:109–129.

McClennen, E. F. (1978), "The minimax theory and expected utility reasoning", in: Hooker, Leach and McClennen (1978).

McGarvey, D. C. (1953), "A theorem in the construction of voting paradoxes", Econometrica, 21.

McKelvey, R. D. (1975), "Policy related voting and electoral equilibrium", Econometrica, 43:815–843.

McKelvey, R. D. (1976), "Intransitivities in multidimensional voting models and some implications for agenda control", Journal of Economic Theory, 12:472–482.

McKelvey, R. D. (1979), "General conditions for global intransitivities in formal voting models", Econometrica, 47:1085–1112.

McKelvey, R. D. and R. G. Niemi (1978), "A multistage game representation of sophisticated voting for binary procedures", Journal of Economic Theory, 18:1–22.

McLenan, A. (1980), "Randomized preference aggregation: Additivity of power and strategy proofness", Journal of Economic Theory, 22:1–11.

McManus, M. (1975), "Inter-tastes consistency in social welfare functions", in: M. Parkin and A. R. Nobay, eds., Current Economic Problems. Cambridge: Cambridge University Press.

McManus, M. (1976), "A numerical social welfare function", Discussion Paper A.185. University of New South Wales.

McManus, M. (1978), "Social welfare optimization with tastes as variables", Weltwirtschaftliches Archiv, 114:101–123.

McManus, M. (1982), "Some properties of topological social choice functions", Review of Economic Studies, 49:447–460.

McManus, M. (1983), "Positive association and its relatives", in: Pattanaik and Salles (1983).

Mill, J. S. (1859), On Liberty; reprinted in: J. S. Mill, Utilitarianism, On Liberty, Representative Government. London: Everyman's Library.

Miller, N. R. (1977), "'Social preference' and game theory: A comment on 'The dilemma of a Paretian liberal'", Public Choice, 30:23–28.

Mirkin, B. G. (1972), "Description of some relations on the set of real-line intervals", Journal of Mathematical Psychology, 9:243–252.

Mirkin, B. G. (1974), Russian text; English translation (1979), Group Choice, edited by P. C. Fishburn. Washington, DC: Winson.

Mirrlees, J. A. (1971), "An exploration in the theory of optimal income taxation", Review of Economic Studies, 175–208.
Mirrlees, J. A. (1982), "The economic uses of utilitarianism", in: Sen and Williams (1982).
Mishan, E. J. (1957), "An investigation into some alleged contradictions in welfare economics", Economic Journal, 67:445–454.
Mizutani, S. (1978), "Collective choice and extended orderings", M.Phil. Dissertation. London University.
Monjardet, B. (1967), "Remarques sur une classe de procédures de votes et les 'théorèmes de possibilité'", in: La Décision. Aix-en-Provence: Colloque du CNRS.
Monjardet, B. (1978), "Une autre preuve du théorème d'Arrow", RAIRO Recherche Operationnelle, 12:291–296.
Monjardet, B. (1979), "Duality in the theory of social choice", in: Laffont (1979).
Monjardet, B. (1983), "On the use of ultrafilters in social choice theory", in: Pattanaik and Salles (1983).
Moon, J. W. (1976), "A problem on rankings by committees", Econometrica, 44:241–246.
Moulin, H. (1979), "Dominance solvable voting schemes", Econometrica, 47:1337–1351.
Moulin, H. (1980), "On strategy-proofness and singlepeakedness", Public Choice, 35:437–455.
Moulin, H. (1981), "The proportional veto principle", Review of Economic Studies, 48:407–416.
Moulin, H. (1983), The Strategy of Social Choice. Amsterdam: North-Holland.
Moulin, H. and B. Peleg (1982), "Cores of effectivity functions and implementation theory", Journal of Mathematical Economics, 10:115–145.
Mueller, D. C. (1967), "The possibility of a social welfare function: Comment", American Economic Review, 57:1304–1311.
Mueller, D. C. (1973), "Constitutional democracy and social welfare", Quarterly Journal of Economics, 87:60–80.
Mueller, D. C. (1978), "Voting by veto", Journal of Public Economics, 10:57–76; reprinted in: Laffont (1979).
Mueller, D. C. (1979), Public Choice. Cambridge: Cambridge University Press.
Mukherji, A. (1977), "The existence of choice functions", Econometrica, 45:889–894.
Muller, E. (1982), "On graphs and social welfare functions", International Economic Review, 23:609–622.
Muller, E. and M. A. Satterthwaite (1977), "The equivalence of strong positive association and strategy-proofness", Journal of Economic Theory, 14:412–418.
Murakami, Y. (1961), "A note on the general possibility theorem of the social welfare function", Econometrica, 29:244–246.
Murakami, Y. (1968), Logic and Social Choice. New York: Dover.
Myerson, R. B. (1983), "Utilitarianism, egalitarianism, and the timing effect in social choice problems", Econometrica, 49:883–897.
Nader-Isfahani, R. (1979), "Cardinal utility", Ph.D. Dissertation. London University.
Nakamura, K. (1975), "The core of a simple game with ordinal preferences", International Journal of Game Theory, 4:95–104.
Nakamura, K. (1978), "Necessary and sufficient conditions on the existence of a class of social choice functions", Economic Studies Quarterly, 29:259–267.
Nakamura, K. (1979), "The vetoers in a simple game with ordinal preferences", International Journal of Game Theory, 8:55–61.
Nalebuff, B. (1981), Unpublished notes on domain conditions for resolving the conflict of the "Paretian liberal". Oxford: Nuffield College.
Nash, J. F. (1950), "The bargaining problem", Econometrica, 18:155–162.
Ng, Y.-K. (1971), "The possibility of a Paretian liberal: Impossibility theorems and cardinal utility", Journal of Political Economy, 79:1397–1402.
Ng, Y.-K. (1975), "Bentham or Bergson? Finite sensibility, utility functions and social welfare functions", Review of Economic Studies, 42:545–569.
Ng, Y.-K. (1979), Welfare Economics. London: Macmillan.
Ng, Y.-K. (1983), "Some broader issues of social choice", in: Pattanaik and Salles (1983).
Nicholson, M. B. (1965), "Conditions for the 'voting paradox' for committee decisions", Metroeconomica, 7:29–44.

Niemi, R. G. and H. Weisberg (1968), "A mathematical solution for the probability of the paradox of voting", Behavioral Sciences, 13:317–323.
Niemi, R. G. and H. Weisberg (1972), Probability Models of Collective Decision Making. Columbus: Merill.
Nitzan, S. (1975), "Social preference orderings in a probabilistic model", Public Choice, 24:93–100.
Nitzan, S. and A. Rubinstein (1981), "A further characterization of the Borda ranking methods", Public Choice, 36:153–158.
Nozick, R. (1973), "Distributive justice", Philosophy and Public Affairs, 3:45–126.
Nozick, R. (1974), Anarchy, State and Utopia. Oxford: Blackwell.
Olson, M. (1965), The Logic of Collective Action. Cambridge: Harvard University Press.
Ordeshook, P. C. (1980), Game Theory and Political Science. New York: N.Y.U. Press.
Osborne, D. K. (1976), "Irrelevant alternatives and social welfare", Econometrica, 44:1001–1015.
Packard, D. J. and R. A. Heiner (1977), "Ranking functions and the independence conditions", Journal of Economic Theory, 16:84–102.
Parks, R. P. (1971), "Rationalizations, extensions and social choice paths", mimeographed. St. Louis: Washington University.
Parks, R. P. (1976a), "An impossibility theorem for fixed preferences: A dictatorial Bergson–Samuelson social welfare function", Review of Economic Studies, 43:447–450.
Parks, R. P. (1976b), "Further results on path independence, quasitransitivity, and social choice", Public Choice, 26:75–87.
Pattanaik, P. K. (1968a), "A note on democratic decision and the existence of choice sets", Review of Economic Studies, 35:1–9.
Pattanaik, P. K. (1968b), "Risk, impersonality and the social welfare function", Journal of Political Economy, 76:1152–1169.
Pattanaik, P. K. (1971), Voting and Collective Choice. Cambridge: Cambridge University Press.
Pattanaik, P. K. (1973), "On the stability of sincere voting situations", Journal of Economic Theory, 6:558–574.
Pattanaik, P. K. (1974), "Stability of sincere voting under some classes of non-binary group decision procedures", Journal of Economic Theory, 8:206–224.
Pattanaik, P. K. (1975), "Strategic voting without collusion under binary and democratic group decision rules", Review of Economic Studies, 42:93–103.
Pattanaik, P. K. (1976a), "Collective rationality and strategy-proofness of group decision rules", Theory and Decision, 7:191–204.
Pattanaik, P. K. (1976b), "Counter-threats and strategic manipulation under voting schemes", Review of Economic Studies, 43:11–18.
Pattanaik, P. K. (1976c), "Threats, counter-threats and strategic voting", Econometrica, 44:91–104.
Pattanaik, P. K. (1978), Strategy and Group Choice. Amsterdam: North-Holland.
Pattanaik, P. K. and M. Salles, eds. (1983), Social Choice and Welfare. Amsterdam: North-Holland.
Pattanaik, P. K. and M. Sengupta (1974), "Conditions for transitive and quasi-transitive majority decision", Economica, 41:414–423.
Pattanaik, P. K. and M. Sengupta (1980), "Restricted preferences and strategy-proofness of a class of group decision functions", Review of Economic Studies, 47:965–973.
Pazner, E. A. (1977), "Pitfalls in the theory of fairness", Journal of Economic Theory, 14:458–466.
Pazner, E. A. and D. Schmeidler (1972), "Decentralization, income distribution and the role of money in socialist economies", Technical Report No. 8. Foerder Institute of Economic Research, Tel-Aviv University.
Pazner, E. A. and D. Schmeidler (1974), "A difficulty in the concept of fairness", Review of Economic Studies, 41:441–443.
Pazner, E. A. and D. Schmeidler (1978), "Egalitarian equivalent allocations: A new concept of economic equity", Quarterly Journal of Economics, 93.
Pazner, E. A. and E. Wesley (1977), "Stability of social choice in infinitely large societies", Journal of Economic Theory, 14:252–262.
Pazner, E. A. and E. Wesley (1978), "Cheatproofness properties of the plurality rule in large societies", Review of Economic Studies, 45:85–91.
Peleg, B. (1978), "Consistent voting systems", Econometrica, 46:153–162.

Peleg, B. (1979a), "A note on manipulability of large voting schemes", Theory and Decision, 11:401–412.
Peleg, B. (1979b), "Representation of simple games by social choice functions", International Journal of Game Theory, 7:81–94.
Peleg, B. (1983), "On simple games and social choice correspondences", in: Pattanaik and Salles (1983).
Peleg, B. (1984), Game Theoretic Analysis of Voting in Committees. Cambridge: Cambridge University Press.
Perelli-Minetti, C. R. (1977), "Nozick on Sen: A misunderstanding", Theory and Decision, 8:387–393.
Phelps, E. S., ed. (1973), Economic Justice. Harmondsworth: Penguin.
Phelps, E. S. (1977), "Recent developments in welfare economics: Justice et équité", in: M. D. Intriligator, ed., Frontiers of Quantitative Economics, Vol. 3. Amsterdam: North-Holland.
Pigou, A. C. (1920), The Economics of Welfare. London: Macmillan.
Plott, C. R. (1967), "A notion of equilibrium and its possbility under majority rule", American Economic Review, 57:788–806.
Plott, C. R. (1971), "Recent results in the theory of voting", in: M. D. Intriligator, ed., Frontiers of Quantitative Economics. Amsterdam: North-Holland.
Plott, C. R. (1972), "Ethics, social choice theory and the theory of economic policy", Journal of Mathematical Sociology, 2:181–208.
Plott, C. R. (1973), "Path independence, rationality and social choice", Econometrica, 41:1075–1091.
Plott, C. R. (1976), "Axiomatic social choice theory: An overview and interpretation", American Journal of Political Science, 20:511–596.
Plott, C. R. (1978), "Rawls' theory of justice: An impossibility result", in: Gottinger and Leinfellner (1978).
Plott, C. R. (1979), "Application of laboratory experimental methods to public choice", mimeographed.
Plott, C. R. and M. E. Levine, "A model of agenda influence on committee decisions", American Economic Review, 68:146–160.
Pollak, R. A. (1979), "Bergson–Samuelson social welfare functions and the theory of social choice", Quarterly Journal of Economics, 93:73–90.
Quine, M. V. (1940), Mathematical Logic. Cambridge: Harvard University Press.
Rae, D. (1969), "Decision rules and individual values in constitutional choice", American Political Science Review, 63:40–56.
Ramachandra, V. S. (1972), "Liberalism, non-binary choice and the Pareto principle", Theory and Decision, 3:49–54.
Ramsey, F. P. (1928), "A mathematical theory of savings", Economic Journal, 38:543–559.
Rawls, J. (1958), "Justice as fairness", Philosophical Review, 57:653–662.
Rawls, J. (1971), A Theory of Justice. Cambridge: Harvard University Press, and Oxford: Clarendon Press.
Rawls, J. (1982), "Social unity and primary goods", in: Sen and Williams (1982).
Ray, P. (1973), "Independence of irrelevant alternatives", Econometrica, 41:987–991.
Richelson, J. (1977), "Conditions on social choice functions", Public Choice, 31:79–110.
Richelson, J. (1978), "Some further results on consistency, rationality and social choice", Review of Economic Studies, 45:343–346.
Riker, W. H. (1961), "Voting and the summation of preferences", American Political Science Review, 60:900–911.
Riker, W. H. and P. C. Ordeshook (1968), "A theory of the calculus of voting", American Political Science Review, 62:25–42.
Riker, W. H. and P. C. Ordeshook (1973), An Introduction to Positive Political Theory. Englewood Cliffs: Prentice-Hall.
Riley, J. (1983), "Collective choice and individual liberty", D.Phil. Thesis. Oxford University.
Ritz, Z. (1981), "Restricted domains, Arrow social welfare functions and non-corruptable and non-manipulable social choice correspondences", mimeographed. Urbana-Champaign: University of Illinois.
Robbins, L. (1932), An Essay on the Nature and Significance of Economic Science. London: Allen & Unwin.

Roberts, K. W. S. (1977), "Welfare theoretic social choice", D.Phil. Dissertation. Oxford University.

Roberts, K. W. S. (1979), "The characterization of implementable choice rules", in: Laffont (1979).

Roberts, K. W. S. (1980a), "Possibility theorems with interpersonally comparable welfare levels", Review of Economic Studies, 47:409–420.

Roberts, K. W. S. (1980b), "Interpersonal comparability and social choice theory", Review of Economic Studies, 47:421–439.

Roberts, K. W. S. (1980c), "Social choice theory: The single and multiple-profile approaches", Review of Economic Studies, 47:441–450.

Roemer, J. (1982), A General Theory of Exploitation and Class. Cambridge: Harvard University Press.

Rosenthal, R. W. (1975), "Voting majority sizes", Econometrica, 43:293–299.

Ritz, Z. (1981), "Restricted domains, Arrow social welfare functions and non-corruptable and non-manipulable social choice correspondences", mimeographed. Urbana–Champaign: University of Illinois.

Rothenberg, J. (1961), The Measurement of Social Welfare. Englewood Cliffs: Prentice-Hall.

Rowley, C. K. and A. T. Peacock (1975), Welfare Economics: A Liberal Restatement. London: Martin Robertson.

Rubinstein, A. (1979), "A note about the 'nowhere denseness' of societies having an equilibrium under majority rule", Econometrica, 47:511–514.

Rubinstein, A. (1980a), "Ranking the participants in a tournament", SIAM Journal of Applied Mathematics, 38:108–111.

Rubinstein, A. (1980b), "Stability of decision systems under majority rule", Journal of Economic Theory, 23:150–159.

Rubinstein, A. (1981), "The single profile analogues to multiple profile theorems: Mathematical logic's approach", mimeographed. Murray Hill: Bell Laboratories.

Salles, M. (1975), "A general possibility theorem for group decision rules with Pareto-transitivity", Journal of Economic Theory, 11:110–118.

Salles, M. (1976), "Characterization of transitive individual preferences for quasi-transitive collective preference under simple games", International Economic Review, 17:308–318.

Salles, M. and R. E. Wendell (1978), "A further result on the core of voting games", International Journal of Game Theory, 6:35–40.

Samuelson, P. A. (1947), Foundations of Economic Analysis. Cambridge: Harvard University Press.

Samuelson, P. A. (1967a), "Arrow's Mathematical Politics", in: S. Hook, ed., Human Values and Economic Policy. New York: N.Y.U. Press.

Samuelson, P. A. (1967b), "Foreword", in: Graaff (1967).

Samuelson, P. A. (1977), "Reaffirming the existence of reasonable Bergson–Samuelson social welfare functions", Economica, 44:81–88.

Saposnik, R. (1974), "Power, the economic environment and social choice", Econometrica, 42:461–470.

Saposnik, R. (1975a), "On transitivity of the social preference relation under simple majority rule", Journal of Economic Theory, 10:1–7.

Saposnik, R. (1975b), "Social choice with continuous expression of individual preferences", Econometrica, 43:683–690.

Satterthwaite, M. A. (1975), "Strategy-proofness and Arrow's conditions: Existence and correspondence theorems for voting procedures and social welfare functions", Journal of Economic Theory, 10:187–217.

Satterthwaite, M. A. and H. Sonnenschein (1981), "Strategy-proof allocation mechanisms at differentiable points", Review of Economic Studies, 48:587–598.

Scanlon, T. M. (1975), "Preference and urgency", Journal of Philosophy, 72:665–669.

Schick, F. (1969), "Arrow's proof and the logic of preference", Journal of Philosophy, 36:127–144.

Schick, F. (1978), "Toward a theory of sociality", in: Hooker, Leach and McClennen (1978).

Schmeidler, D. and H. Sonnenschein (1978), "Two proofs of the Gibbard–Satterthwaite theorem on the possibility of a strategy-proof social choice function", in: Gottinger and Leinfellner (1978).

Schmeidler, D. and K. Vind (1972), "Fair net trades", Econometrica, 40:637–642.

Schmeidler, D. and M. Yaari (1970), "Fair allocations", unpublished manuscript.

Schmitz, N. (1977), "A further note on Arrow's impossibility theorem", Journal of Mathematical Economics, 4:189–196.

Schofield, N. (1977), "Transitivity of preferences on a smooth manifold of alternatives", Journal of Economic Theory, 14:149–172.

Schofield, N. (1978), "Instability of simple dynamic games", Review of Economic Studies, 40:575–594.

Schofield, N. (1980), "The theory of dynamic games", in: P. Ordeshook, ed., Game Theory and Political Science. New York: N.Y.U. Press.

Schofield, N. (1983a), "Equilibria in simple dynamic games", in: Pattanaik and Salles (1983).

Schofield, N. (1983b), "Generic instability of majority rule", Review of Economic Studies, 50:695–705.

Schwartz, T. (1970), "On the possibility of rational policy evaluation", Theory and Decision, 1:89–106.

Schwartz, T. (1972), "Rationality and the myth of the maximum", Nous, 6:97–117.

Schwartz, T. (1974), "Notes on the abstract theory of collective choice", in: Mathematics in the Social Sciences, 90–726. Pittsburgh: Carnegie-Mellon University.

Schwartz, T. (1976), "Choice functions, 'rationality' conditions, and variations on the weak axiom of revealed preferences", Journal of Economic Theory, 13:414–427.

Scott, D. and P. Suppes (1958), "Foundational aspects of theories of measurement", Journal of Symbolic Logic, 23:113–128.

Seidl, C. (1975), "On liberal values", Zeitschrift für Nationalökonomie, 35:257–292.

Sen, A. K. (1966), "A possibility theorem on majority decisions", Econometrica, 34:75–79.

Sen, A. K. (1969), "Quasi-transitivity, rational choice and collective decisions", Review of Economic Studies, 36:381–393.

Sen, A. K. (1970a), Collective Choice and Social Welfare. San Francisco: Holden-Day, and Edinburgh: Oliver & Boyd; distribution taken over by Amsterdam: North-Holland.

Sen, A. K. (1970b), "The impossibility of a Paretian liberal", Journal of Political Economy, 72:152–157.

Sen, A. K. (1970c), "Interpersonal aggregation and partial comparability", Econometrica, 38:393–409; "A correction", 40:959 (1972).

Sen, A. K. (1971), "Choice functions and revealed preference", Review of Economic Studies, 38:307–317.

Sen, A. K. (1973), On Economic Inequality. Oxford: Clarendon Press.

Sen, A. K. (1974), "Informational bases of alternative welfare approaches: Aggregation and income distribution", Journal of Public Economics, 3:387–403.

Sen, A. K. (1976a), "Liberty, unanimity and rights", Economica, 43:217–245.

Sen, A. K. (1976b), "Welfare inequalities and Rawlsian axiomatics", Theory and Decision, 7:243–262.

Sen, A. K. (1976c), "Poverty: An ordinal approach to measurement", Econometrica, 44:219–231.

Sen, A. K. (1976d), "Real national income", Review of Economic Studies, 43:19–39.

Sen, A. K. (1977a), "Social choice theory: A re-examination", Econometrica, 45:53–89.

Sen, A. K. (1977b), "On weights and measures: Informational constraints in social welfare analysis", Econometrica, 45:1539–1572.

Sen, A. K. (1977c), "Rational fools: A critique of the behavioral foundations of economic theory", Philosophy and Public Affairs, 6:317–344.

Sen, A. K. (1977d), "Non-linear social welfare functions: A reply to Professor Harsanyi", in: R. Butts and J. Hintikka, eds., Logic, Methodology and Philosophy of Science. Dordrecht: Reidel.

Sen, A. K. (1979a), "Interpersonal comparisons of welfare", in: M. Boskin, ed., A Festschrift for Tibor Scitovsky, Economics and Human Welfare. New York: Academic Press.

Sen, A. K. (1979b), "Personal utilities and public judgments: Or what's wrong with welfare economics?", Economic Journal, 89:537–558.

Sen, A. K. (1979c), "Strategies and revelations: Informational constraints in public decisions", in: Laffont (1979).

Sen, A. K. (1979d), "The welfare basis of real income comparisons", Journal of Economic Literature, 17:1–45.

Sen, A. K. (1982), Choice, Welfare and Measurement. Oxford: Blackwell, and Cambridge: MIT Press.

Sen, A. K. (1983), "Liberty and social choice", Journal of Philosophy, 80:5–28.

Sen, A. K. and P. K. Pattanaik (1969), "Necessary and sufficient condition for rational choice under majority decision", Journal of Economic Theory, 1:178–202.

Sen, A. K. and B. Williams (1982), Utilitarianism and Beyond. Cambridge: Cambridge University Press.

Sengupta, M. (1974), "On a concept of representative democracy", Theory and Decision, 5:249–262.
Sengupta, M. (1978a), "On a difficulty in the analysis of strategic voting", Econometrica, 46:331–344.
Sengupta, M. (1978b), "Strategy-proofness of a class of Borda rules", mimeographed. School of Economics, La Trobe University.
Sengupta, M. (1980a), "Monotonicity, independence of irrelevant alternatives and strategy-proofness of group decision functions", Review of Economic Studies, 47:393–408.
Sengupta, M. (1980b), "The knowledge assumption in the theory of strategic voting", Econometrica, 49:1301–1304.
Sengupta, M. and B. Dutta (1979), "A condition for Nash stability under binary and democratic group decision functions", Theory and Decision, 10:293–310.
Sertel, M. and A. van der Bellen (1979), "Synopses in the theory of choice", Econometrica, 47:1367–1390.
Sertel, M. and A. van der Bellen (1980), "Routes and paths of comparison and choice", Public Choice, 35:205–218.
Shepsle, K. A. (1970), "A note on Zeckhauser's 'majority rule with lotteries on alternatives': The case of the paradox of voting", Quarterly Journal of Economics, 84:705–709.
Sjöberg, M. (1975), "Remarks on semiorders and the theory of choice", mimeographed. Department of Philosophy, Uppsala University.
Skala, H. J. (1978), "Arrow's impossibility theorem: Some new aspects", in: Gottinger and Leinfellner (1978).
Sloss, J. (1973), "Stable outcomes in majority voting games", Public Choice, 15:19–48.
Slutsky, S. (1975), "Abstentions and majority equilibrium", Journal of Economic Theory, 11:292–304.
Slutsky, S. (1977), "A characterization of societies with consistent majority decision", Review of Economic Studies, 44:211–226.
Slutsky, S. (1979), "Equilibrium under α-majority voting", Econometrica, 47:1113–1125.
Smith, J. H. (1973), "Aggregation of preferences and variable electorate", Econometrica, 41:1027–1041.
Smith, T. E. (1974), "On the existence of most-preferred alternatives", International Economic Review, 15:184–194.
Smith, T. E. (1979), "Rationality of indecisive choice functions on triadic choice domains", Theory and Decision, 10:113–310.
Smith, V. L. (1979), "An experimental comparison of three public good decision mechanisms", Scandinavian Journal of Economics, 81:198–215.
Sobel, J. (1979), "Fair allocations of a renewable resource", Journal of Economic Theory, 21:235–248.
Starr, R. M. (1973), "Optimal production and allocation under uncertainty", Quarterly Journal of Economics, 87:81–95.
Stevens, D. N. and J. E. Foster (1978), "The possibility of democratic pluralism", Economica, 45:391–400.
Strasnick, S. (1976), "Social choice theory and the derivation of Rawls' difference principle", Journal of Philosophy, 73:85–99.
Strasnick, S. (1978), "Extended sympathy comparisons and the basis of social choice", Theory and Decision, 10:311–328.
Sugden, R. (1981), The Political Economy of Public Choice. Oxford: Martin Robertson.
Suppes, P. (1966), "Some formal models of grading principles", Synthese, 6:284–306; reprinted in: Suppes (1969).
Suppes, P. (1969), Studies in the Methodology and Foundations of Science. Dordrecht: Reidel.
Suppes, P. (1978), "The distributive justice of income inequality", in: Gottinger and Leinfellner (1978).
Suzumura, K. (1973), "Acyclic preference and the existence of choice functions", Discussion Paper No. 068. Kyoto Institute of Economic Research, Kyoto University.
Suzumura, K. (1976a), "Rational choice and revealed preference", Review of Economic Studies, 43:149–158.
Suzumura, K. (1976b), "Remarks on the theory of collective choice", Economica, 43:381–390.
Suzumura, K. (1978), "On the consistency of libertarian claims", Review of Economic Studies, 45:329–342; "A correction", 46:743 (1979).
Suzumura, K. (1980), "Liberal paradox and the voluntary exchange of rights-exercising", Journal of Economic Theory, 22:407–422.

Suzumura, K. (1983a), Rational Choice, Collective Decisions and Social Welfare. Cambridge: Cambridge University Press.
Suzumura, K. (1983b), "Resolving conflicting views of justice in social choice", in: Pattanaik and Salles (1983).
Svensson, L.-G. (1977), Social Justice and Fair Distributions, Lund Economic Studies 15.
Svensson, L.-G. (1980), "Equity among generations", Econometrica, 48:1251–1256.
Szpilrajn, E. (1930), "Sur l'extension de l'ordre partiel", Fundamenta Mathematicae, 16:386–389.
Takayama, N. (1979), "Poverty, income inequality and their measures: Professor Sen's approach reconsidered", Econometrica, 47:747–760.
Tarjan, T. G. (1975), "Transitive majority decisions", mimeographed. Budapest: Institute of Economics, Hungarian Academy of Sciences.
Taylor, M. (1969), "Proof of a theorem on majority rule", Behavioral Sciences, 14:228–231.
Taylor, M. (1970), "The problem of salience in the theory of collective decision-making", Behavioral Sciences, 15:415–430.
Taylor, M. (1971), "Mathematical political theory", British Journal of Political Science, 1:339–382.
Thon, D. (1979), "On measuring poverty", Review of Income and Wealth, 25.
Tideman, T. and G. Tullock (1976), "A new superior principle for collective choice", Journal of Political Economy, 84:1145–1159.
Tullock, G. (1964), "The irrationality of intransitivity", Oxford Economic Papers, 16:401–406.
Tullock, G. (1967), "The general irrelevance of the general possibility theorem", Quarterly Journal of Economics, 81:256–270.
Tullock, G. (1968), Toward a Mathematics of Politics. Ann Arbor: University of Michigan Press.
Ulph, D. (1978), "Extended sympathy and social choice: A note", mimeographed. London: Department of Political Economy, University College.
Usher, D. (1981), The Economic Prerequisite to Democracy. Oxford: Blackwell.
Van Herwaarden, F., A. Kapteyn and B. M. S. Van Praag (1977), "Twelve thousand individual welfare functions: A comparison of six samples in Belgium and the Netherlands", European Economic Review, 9:283–300.
Van Praag, B. M. S. (1971), "The welfare function of income in Belgium: An empirical investigation", European Economic Review, 2:337–369.
Varian, H. (1974), "Equity, envy and efficiency", Journal of Economic Theory, 9:63–91.
Varian, H. (1975), "Distributive justice, welfare economics and the theory of fairness", Philosophy and Public Affairs, 4:223–247.
Varian, H. (1976a), "On the history of the concepts of fairness", Journal of Economic Theory, 13:486–487.
Varian, H. (1976b), "Two problems in the theory of fairness", Journal of Public Economics, 5:249–260.
Vickrey, W. (1945), "Measuring marginal utility by reactions to risk," Econometrica, 13:319–333.
Vickrey, W. (1960), "Utility, strategy and social decision rules", Quarterly Journal of Economics, 75:507–535.
Vickrey, W. (1977), "Economic rationality and social choice", Social Research, 44:691–707.
Von Neumann, J. and O. Morgenstern (1947), Theory of Games and Economic Behaviour. Princeton: Princeton University Press.
Wagstaff, P. (1976), "Proof of a conjecture in social choice", International Economic Review, 17:491–497.
Waldner, I. (1972), "The empirical meaningfulness of interpersonal utility comparisons", Journal of Philosophy, 69.
Walker, M. (1977), "On the existence of maximal elements", Journal of Economic Theory, 16:470–474.
Ward, B. (1965), "Majority voting and alternative forms of public enterprise", in: J. Margolis, ed., The Public Economy of Urban Communities. Baltimore: Johns Hopkins Press.
Weddepohl, H. N. (1970), "A vector representation of majority voting", Research Memorandum EIT 16. Tilburg Institute of Economics.
Weymark, J. A. (1983), "Arrow's theorem with social quasi-orderings", Public Choice, 42:235–246.
Williams, B. A. O. (1973), "A critique of utilitarianism", in: J. J. C. Smart and B. Williams, eds., Utilitarianism: For and Against. Cambridge: Cambridge University Press.

Williamson, O. E. and T. J. Sargent (1967), "Social choice: A probabilistic approach", Economic Journal, 77:797–813.

Wilson, R. B. (1969), "An axiomatic model of logrolling", American Economic Review, 59:331–341.

Wilson, R. B. (1970), "The finer structure of revealed preference", Journal of Economic Theory, 2:348–353.

Wilson, R. B. (1971), "A game-theoretic analysis of social choice", in: B. Leiberman, ed., Social Choice. New York: Gordon and Breach.

Wilson, R. B. (1972a), "The game-theoretic structure of Arrow's general possibility theorem", Journal of Economic Theory, 5:14–20.

Wilson, R. B. (1972b), "Social choice theory without the Pareto principle", Journal of Economic Theory, 5:478–486.

Wilson, R. B. (1975), "On the theory of aggregation", Journal of Economic Theory, 10:89–99.

Wriglesworth, J. (1982a), "The possibility of democratic pluralism: A comment", Economica, 49: 43–48.

Wriglesworth, J. (1982b), "Using justice principles to resolve the 'impossibility of a Paretian liberal'", Economic Letters, 10:217–221.

Wriglesworth, J. (1983a), "Libertarian conflicts in social choice", D.Phil. Thesis. Oxford University; to be published by Cambridge University Press.

Wriglesworth, J. (1983b), "Respecting individual rights in social choice", mimeographed. London: Queen Mary College.

Yaari, M. H. (1978), "Rawls, Edgeworth, Shapley, Nash: Theories of distributive justice re-examined", Journal of Economic Theory, 24:1–30.

Young, H. P. (1974a), "An axiomatization of Borda's rule", Journal of Economic Theory, 9:43–52.

Young, H. P. (1974b), "A note on preference aggregation", Econometrica, 42:1129–1131.

Young, H. P. (1975), "Social choice scoring functions", SIAM Journal of Applied Mathematics, 28:824–838.

Young, J. P. (1977), "Extending Condorcet's rule", Journal of Economic Theory, 16:335–353.

Zeckhauser, R. (1969), "Majority rule with lotteries on alternatives", Review of Economic Studies, 42:696–703.

Zeckhauser, R. (1973), "Voting systems, honest preferences and Pareto optimality", American Political Science Review, 67:934–946.

Chapter 23

AGENCY AND THE MARKET*

KENNETH J. ARROW

Stanford University

A very widespread economic situation is that of the relation between a principal and an agent. Even in ordinary and in legal discourse, the principal–agent relation would be significant in scope and economic magnitude. But economic theory in recent years has recognized that analogous interactions are almost universal in the economy, at least as one significant component of almost all transactions.

The common element is the presence of two individuals. One (the agent) is to choose an action among a number of alternative possibilities. The action affects the welfare of the other, the principal, as well as that of the agent's self. The principal, at least in the simplest cases, has the additional function of prescribing payoff rules, that is, of determining in advance of the choice of action, a rule which obliges him or her what fee to pay as a function of his or her observations on the results of the action. The problem acquires interest only when there is uncertainty at some point, and, in particular, when the information available to the two participants is unequal. The main but not only case in the literature is that where the agent's action is not directly observable by the principal and where in addition the outcome is affected but not completely determined by the agent's action. (If the latter were not ture, the principal could in effect infer the agent's action by observing the outcome.) In technical language, the outcome is a random variable whose distribution depends on the action taken.

More generally, there may be many agents for a single principal. Each takes an action, and the output of the system is a random function of all the actions. The principal cannot observe the actions themselves but may make some observations, for example, of the output and possibly others. Again the principal sets in

*Reprinted, with permission of the Harvard Business School Press, from "The Economics of Agency" in *Agency: The Structure of Business*, edited by John W. Pratt and Richard Zeckhauser. Copyright 1985 by the President and Fellows of Harvard College. The author wishes to express his gratitude to John W. Pratt, John G. Riley, and Richard Zeckhauser, whose comments have materially improved the exposition of this paper.

Handbook of Mathematical Economics, vol. III, edited by K.J. Arrow and M.D. Intriligator
© 1986, Elsevier Science Publishers B.V. (North-Holland)

advance a schedule stating the fees to be paid to the individual agents as a function of the observations made by the principal.

A similar but not identical principal–agent relation occurs when the agent makes an observation not shared with the principal and bases his/her action on that observation. The action itself may be observable, but the principal does not know whether or not it is the most appropriate.

The principal–agent theory is in the standard economic tradition. Both principal and agent are assumed to be making their decisions optimally in view of their constraints. Intended transactions are realized. The function of this theory has the dual aspect usual in economic theory; it can be interpreted both normatively and descriptively. It can be interpreted as advice in the construction of contracts to guide and influence principal–agent relations in the real world, in short, as a foundation for social engineering. It can also be interpreted as an attempt to explain observed phenomena in the empirical economic world, particularly exchange relations which are observed but not explained by more standard economic theory.

Before specifying the model more completely, it is useful to give a few examples of each of the two kinds of principal–agent problems. As will be seen, many situations that are not classified under that heading in ordinary discourse can be considered as such. I will call the two types of principal–agent problems *hidden action* and *hidden information*, respectively. In the literature, they are frequently referred to as *moral hazard* and *adverse selection*. These terms have been borrowed from the practice and theory of insurance and are really applicable only to special cases.

The most typical *hidden action* is the effort of the agent. Effort is a disutility to the agent, but it is at the same time a value to the principal in the sense that it increases the random outcome (technically, the distribution of the outcome to a higher effort stochastically dominates that to a lower effort, i.e. the probability of achieving any given level outcome, or better is higher with higher effort). The physician–patient relation is a notorious case. Here, the patient is the principal, and the physician is the agent. The very basis of the relation is the superior knowledge of the physician. Hence, the patient cannot check to see if the actions of physician are as diligent as they could be.

A second non-obvious example that of torts. One individual takes an action which results in damage to another, for example, one automobile hitting another. The care which the first driver takes cannot easily be observed, but the outcome is very visible indeed. Although it may seem an odd use of language, one has to consider the damager as the agent and the one damaged as the principal. Again, in pollution control, society may be regarded as the principal, and the polluter, whose actions cannot be fully monitored, as the agent.

An example of very special economic importance is the relation between stockholders and management. The stockholders are principals, who certainly cannot observe in detail whether or not the management, their agent, is making appropriate decisions. A very similar relation formally, though in a different context, is that of sharecropping; the landlord, the principal here, prefers a relation which supplies incentives for better production as against a straight wage payment, since the landlord cannot directly observe the tenant's diligence; on the other hand, the tenant, too poor to bear excessive risks, wants to avoid a fixed rent, which would maximize incentives but would expose him or her to all the risks of weather and price. Fire insurance dulls incentives for care and even creates incentives for arson; this is the origin of the term, "moral hazard". Health insurance creates similar problems, though with less moral overtones; payment of medical fees by the insurer reduces risks to the insured but creates an incentive to excessive medical care, more than the patient would have if he or she had to pay the entire price. The employment relation, in general, is one in which effort and ability acquired through training and self-improvement are hard to observe. This has led to a theory which explains the existence of firms as a device for measuring effort.

These have been examples of the hidden-action type of principal–agent relation. There is another class, sometimes discussed under different headings, the hidden-knowledge type. Here the agents differ from the principal in having made some observation which the principal has not made. The agents use (and should use) this observation in making their decisions; however, the principal cannot check whether or not the agents have used their information wisely from the principal's viewpoint. A case much studied from different points of view in the economic literature is that of a decentralized socialist economy. The knowledge of productivity cannot be centralized. Hence, the individual productive units have information about the possibilities of production not available to the central planning unit. The question arises, how this information can be tapped. The productive units may well have incentives not to reveal their full potentiality, because it will be easier to operate with less taxing requirements. A similar problem occurs in decentralization within a firm. This branch of the literature has acquired the name of "incentive compatibility".

The original problem of "adverse selection" is drawn from insurance of several kinds, of which life insurance is typical. The population being insured is heterogeneous from the viewpoint of probability of risk, say of death. In some cases, at least, the insured have better knowledge of this probability than the insurance company which is unable to differentiate. If the same premium is charged to everyone, then the high-risk individuals will purchase more insurance and the low-risk ones less. This will lead to an inefficient allocation of risk-bearing [Rothschild and Stiglitz (1975)]. Public utilities, such as telephones, also face heterogeneous populations, though again, as in insurance, the utility provider cannot know to which class the purchaser belongs. Nevertheless, as has been pointed out in recent literature, some differentiation can be made by offering alternative rate schedules and letting the customers choose which to follow. In these cases, the insurance company or the public utility is the principal, the customer, with more knowledge not available to the principal, is the agent [Spence (1977), Roberts (1979), and Maskin and Riley (1983)].

To illustrate the theoretical issues for the hidden-knowledge model, consider a monopolistic public utility facing two types of customers, labelled H and L for *high* and *low* demanders, respectively. Assume the absence of income effects. Let $U_t(x)$ be the money equivalent of amount x of the public utility for type t ($t = \mathrm{H, L}$) so that $U_t(0) = 0$, and characterize high and low demand by the condition that $U'_{\mathrm{H}}(x) > U'_{\mathrm{L}}(x)$ for all x. It is assumed that the characteristics of the product preclude resale.

The public utility knows the proportion of high demanders but not the identity of these individuals. It offers a total payment schedule, $T(x)$, a function of the amount purchased. Assuming a constant marginal cost of production, c, the monopolists' markup for x units is

$$M(x) = T(x) - cx.$$

For convenience, let $V_t(x) = U_t(x) - cx$, the consumer's surplus over social cost. Since $V'_{\mathrm{H}}(x) > V'_{\mathrm{L}}(x)$, all x, there is a difference in willingness to pay which the monopolist can exploit.

Since individuals are free to refrain from purchase, no offer by the company can yield a negative consumer's surplus. The monopolist can try to extract all consumer's surplus by all-or-none offers. Let $\bar{x}_t$ maximize $V_t(x)$. If the monopolist can identify the types of the consumers, it will offer buyers of type t $\bar{x}_t$ units and charge a markup of $\bar{M}_t = V_t(\bar{x}_t)$.

In the absence of identification, this scheme breaks down. If the monopolist offers the consumer a choice of these two offers, the high demanders will always choose $(\bar{x}_{\mathrm{L}}, \bar{M}_{\mathrm{L}})$. Since $V'_{\mathrm{H}}(x) > V'_{\mathrm{L}}(x)$, it follows that $V_{\mathrm{H}}(\bar{x}_{\mathrm{L}}) > V_{\mathrm{L}}(\bar{x}_{\mathrm{L}}) = \bar{M}_{\mathrm{L}}$, so that type H individuals get a positive consumer's surplus by choosing the offer appropriate to type L individuals and only zero by the alternative choice. To induce type H individuals to buy $\bar{x}_{\mathrm{H}}$, the markup demanded must be reduced so that they are no worse off than they would be choosing $(\bar{x}_{\mathrm{L}}, \bar{M}_{\mathrm{L}})$, i.e. the markup demanded must satisfy the condition

$$V_{\mathrm{H}}(\bar{x}_{\mathrm{H}}) - M^{\circ}_{\mathrm{H}} = V_{\mathrm{H}}(\bar{x}_{\mathrm{L}}) - \bar{M}_{\mathrm{L}}. \qquad \text{(C)}$$

This can be accomplished without identification by choosing $M(x) = \bar{M}_{\mathrm{L}}$ for $x \leqq \bar{x}_{\mathrm{L}}$, and $M(x) = M^{\circ}_{\mathrm{H}}$ for $x > \bar{x}_{\mathrm{L}}$.

This allocation is Pareto efficient, since all consumers are paying marginal cost. The monopolist is extracting all surplus from the low demanders but not from the high demanders. However, the allocation does not yield maximum profits to the monopolist. To do so requires creation of inefficiency. The amount to be bought by the low demanders will be reduced by a small amount. This will reduce the surplus to be extracted from them. On the other hand, the constraint imposed on extraction of surplus from the high demanders to prevent them from switching to the offer intended for the low demanders will become less binding. It turns out that the loss is second-order in the reduction in purchase amount while the gain is first-order. In symbols, let the amount to be purchased by type L consumers be reduced from $\bar{x}_L$ to $\bar{x}_L - dx$. This is enforced by locating the discontinuous increase in markup at that point. The markup must be reduced correspondingly; choose $M_L^* = V_L(\bar{x}_L - dx)$. Since V_L is maximized at $\bar{x}_L$, it must be that the difference $M_L^* - \overline{M}_L$ is of the second order in dx.

To induce the type H consumers to choose $\bar{x}_H$ rather than $(\bar{x}_L - dx, M_L^*)$, the markup to them must be set so that

$$V_H(\bar{x}_H) - M_H^* = V_H(\bar{x}_L - dx) - M_L^*.$$

By comparison with (C), it is seen that

$$M_H^* - M_H^\circ = (M_L^* - \overline{M}_L) + [V_H(\bar{x}_L) - V_H(\bar{x}_L - dx)].$$

The first term on the right is, as stated, of the second order in dx. But, since $V'(\bar{x}_L) > 0$, the second term is positive and of the first order. Hence, for dx sufficiently small, the loss in markup from the type L consumers is of the second order, the gain in markup from the type H consumers is of the first order, and there is a net gain. This is true no matter what the proportions of the two types of consumers are, though of course the optimal policy of the monopoly depends on them. The optimal monopoly policy can be enforced without identification of the types of consumers by letting $M(x) = M_L^*$ for $x \leqq \bar{x}_L - dx$, and $M(x) = M_H^*$ for $x > \bar{x}_L - dx$.

Constraints such as (C), which ensure that the different types are induced to accept the allocations allotted to them, are referred to as *self-selection* constraints. The example illustrates a very general principle in hidden-knowledge models; the optimal incentive schedule typically requires distortions (deviations from first-best Pareto-optimal) at all but one point.

Two further illustrations of hidden knowledge in economic decision-making are as diverse as auctions with private information [Vickrey (1961), Maskin and Riley (1984), and Milgrom and Weber (1982)] and optimal income taxation [Mirrlees (1971)]. Consider bidding for oil leasing when the bidders are each permitted to engage in exploratory drilling and other geophysical studies. Each then has an observation unknown to the others and to the seller, most usually the government in the United States today. The problem is to design auction rules to achieve some objective. Much of the current literature is devoted to maximizing the seller's revenues, rather than social welfare in some broader sense. The problem of optimal income taxation is that any income tax creates a distortion of the choice between labor and leisure. This deficiency could in principle be overcome completely if the social price of leisure (i.e. the productivity or wage rate of the individual) were observable. But in general, this information is available to the taxpayer but not to the government. In one case the geophysical estimates of oil field size, in the other case individual wage rates are private information and therefore hidden knowledge to the principal.

The above discussion of hidden-knowledge principal–agent problems has concentrated on the case of a single principal. Further complications arise when principals compete for agents [Spence (1973), Rothschild and Stiglitz (1975), and Riley (1975)]. To take the opposite extreme, suppose there are a large number of potential principals who will enter the market to exploit any profitable alternative. Consider, for example, an insurance market with a large number of competing insurance companies, each of which, because of risk pooling, is approximately risk-neutral. As argued earlier, any premium per dollar of coverage will be more attractive to those with higher loss probabilities; insurance companies will then have an incentive to sort risk classes by offering lower premiums per dollar coverage to those willing to accept higher deductibles. However, in contrast to the monopoly case, each insurance company must now take into account the effect of other available alternatives on the type of individuals attracted to its own offerings. To use Spence's terminology, it is not enough that low risk classes are able to "signal" their differences by accepting larger deductibles; such signals must also be competitively viable.

The issue of what kind of signalling survives competitive pressures turns out to be a delicate one. In general, there does not exist a Walrasian (or Nash) equilibrium with the property that no principal has an incentive to introduce new profitable alternatives. However, recent work by Wilson (1977) and Riley (1979) has argued that equilibrium can be sustained if principals rationally anticipate certain responses to their behavior.

Let me now turn to a simple formulation of the hidden-action model. The agent (for the moment, assume there is only one) chooses an action a. The result of his/her choice is an outcome x, which is a random variable whose distribution depends on a. The principal has chosen beforehand a *fee function* $s(x)$ to be paid to the agent. For the simplest case, assume that the outcome x is income, i.e. a transferable and measurable quantity. Then the net receipts of the principal will be $x - s(x)$. The principal and agent are both, in general, risk averters. Hence, each values whatever income he or she receives by a utility function with diminishing marginal utility. Let U be the utility function of the principal, V that of the agent. Further, let $W(a)$ be the disutility the agent attaches to action a. It will be assumed separable from the utility of income, i.e. the marginal utility of income is independent of the action taken (the amount of effort). Note that the action is taken before the realization of the uncertainty and is therefore not uncertain to the agent, though it is unknown to the principal.

Since, even for a given action, the outcome x is uncertain, both principal and agent are motivated to maximize the expected value of utility. Given the principal's choice of fee function $s(x)$, the agent wishes to maximize the expected value of $V[s(x)]-W(a)$. In effect, therefore, the principal can predict the action taken for any given fee schedule. The choice of fee schedules is, however, restricted by competition for agents. The agent has alternative uses for his or her time. Hence, the utility achievable by the agent with the principal under consideration must be at least equal to that achievable in other activities. The fee schedule chosen by the principal must then satisfy this constraint. (The literature has usually referred to this condition as that of "individual rationality", a term first used by J. von Neumann and O. Morgenstern, but this name is easily misinterpreted. The term, *participation constraint*, has come into use recently and seems more appropriate.)

It is interesting to note that the principal–agent relation defined as here by a fee function is a significant departure from the usual arm's length fixed-price relation among economic agents postulated in economic theory. The principal does not buy the agent's services at a fixed price set by the competitive market nor does the principal simply buy output from the agent. The relation is not even describable by a contingent contract, in which payments and services rendered are agreed-on functions of an exogenous random variable; the principal observes the outcome but cannot analyze it into its two components, the agent's action and the exogenous uncertainty. Even though the underlying principles are impeccably neoclassical, in that each party is acting in its self-interest and is subject to the influence of the market, the variable to be determined is not a price but a complicated functional relationship.

The principal–agent problem combines two inextricable elements. One is simple risk-sharing; even if there were no problem of differential information, there would be some sharing of the outcome if both parties are risk-averse. Indeed, if the agent were risk-neutral, the principal–agent problem would admit of a trivial solution; the agent would bear all the risks, and then the differential information would not matter. That is, the principal would retain a fixed amount for him/herself and pay all the remainder to the agent, who therefore has no dilution of incentives [Shavell (1979)]. In the terminology used above, the fee function would equal the outcome less a fixed amount, $s(x)=x-c$, where the constant c is determined by the participation constraint. Thus a landlord renting land to a tenant farmer would simply charge a fixed rent independent of output, which in general depends on both the tenant's effort, unobservable to the landlord, and the vagaries of the weather. However, this solution ceases to be optimal as soon as the agent is risk-averse. Since all individuals are averse to sufficiently large risks, the simple solution of preserving incentives by assigning all risks to the agent fails as soon as the risks are large compared with the agent's wealth. The president of a large corporation can hardly be held responsible for its income fluctuations.

In the general case of a risk-averse agent, the fee will be a function of the outcome, in order to supply incentives, but the risk will be shared. If the ability of the agent to affect outcomes approaches either zero or infinity, then the efficiency level which could be achieved under full information to the principal can be approached with an optimally chosen fee function. More generally, there is a trade-off between incentives and efficiency of the system considering both principal and agent [Shavell (1979)].

For an application, consider the case of insurance with moral hazard. There will be some insurance written, but it will not be complete. In the terminology of the insurance industry, there will be *coinsurance*, that is, the insured will bear some of the losses against which the insurance is written. Coinsurance is customary in health insurance policies, where the insured has considerable control over the amount of health expenditures. Similarly, in a system of legal liability for torts, in the absence of insurance, the payment should increase with the amount of damages inflicted, to provide incentive for avoiding the inflicting of damages, but by an amount less than the increase in damages, so that there is a sharing of the unavoidable risks.

More recent literature has stressed the possibility of monitoring. By this is meant that the principal has information in addition to the outcome, an observation y. If y conveys any information about the unobserved action a, beyond that revealed by x (technically, if x is not a sufficient statistic for the pair x, y with respect to a), then one can always improve by making the fee depend upon y as well as x. In the case of torts, the information used in a negligence standard represents additional knowledge beyond the outcome, though the last is all that is required for a strict liability standard. It turns out that if the liable party (the agent in this interpretation) is risk-neutral, then strict liability is optimal. But otherwise an appropriate negligence standard is an improvement [Shavell (1979) and Holström (1979)]. Harris and Raviv (1978, 1979) have argued that the custom of paying lawyers (in most circumstances) by time as well as by a contingent fee illustrates monitoring. If this idea were applied to health insurance, it would suggest that an improvement could be achieved by making insurance payments depend on some measure of the amount of medical services, such as frequency of visits.

It has been shown that if the monitoring information is essentially an imperfect measure of the action taken, i.e. $y = a + u$, where u is a random variable with mean zero, then an optimal fee policy takes the form of paying a very low figure, independent of outcome, if the measured action is sufficiently low, and paying according to a more complicated schedule otherwise.

The whole discussion, to this point, has concentrated on a single agent and a single time period. New possibilities for incentives arise when there are many agents for a single principal or repeated relations between agent and principal. The many-agent case offers new opportunities for inference of hidden actions (or of hidden information) if the uncertainty of the relation between the action (or the agent's observation) is the same for all the agents. In that situation, an estimate of the uncertainty can be obtained by comparison of the performances of the different agents, and therefore the individual actions can be approximately identified. One can meaningfully compare the performance of each agent with the average, for example, or use the ordinal ranking of the agents' outcomes as a basis for fees [Holmström (1982)].

A different and as yet only slightly explored problem can arise in the case of many agents with a single principal. Suppose the principal cannot observe the outcome of each individual but only the output of the group of agents as a whole. This is obviously an important case in production carried out jointly, with many complementary workers. Even in the case of certainty in the relation between actions and collective outcomes, there are difficulties. Holmström (1982) has considered the problem of a team, whose output depends on the unobservable actions of all members. Each team member has a disutility for his or her action. Assume for simplicity that utility is linear in the output. Then one can speak of a social optimum, that vector of actions which maximizes total output minus the sum of disutilities for actions. The question is, can the team devise some incentive scheme which will induce the members to perform the socially optimal actions. This will necessarily be a game, since the reward to each is a function of the output and therefore of the actions of all. When there is no uncertainty, an incentive scheme can be devised with the desired outcome in mind. Let a_i be the action to be chosen by individual i, $x(a_1,\dots, a_n)$ the production function which gives the output of the team as a function of the actions of all members, and $W_i(a_i)$ the disutility of individual i as a function of his or her action. Then the socially optimal set of actions is that which maximizes $x(a_1,\dots, a_n)-\sum W_i(a_i)$. Call the actions so defined, $a_1^*,\dots, a_n^*$, and let $x^* = x(a_1^*,\dots, a_n^*)$ be the output at this optimum. Choose any set of lump-sum rewards, $b_1,\dots, b_n$, which add up to x^*, subject to the condition that $b_i > W_i(a_i^*)$ for each i. Then set up the following game: Individual i chooses a_i. If the result of all these actions is to produce an output which is less than optimal, no one receives anything. If the total output, $x(a_1,\dots, a_n)$ is greater than or equal to x^*, then individual i receives b_i. It is easy to see that a Nash equilibrium of the game is for each individual to choose the appropriate action, a_i^*; that is, for each individual i, choosing a_i^* is optimal given the payoffs, providing each other individual j chooses a_j^*. But the proposed game is hardly satisfactory. It involves in effect collective punishment. More analytically, there are many Nash equilibria, of which $(a_1^*,\dots, a_n^*)$ is only one. If some individuals shirk a little, it pays the others to work somewhat harder to achieve the same output. Hence, the scheme does not enforce the optimal outcome, though it permits it.

When there are repeated relations between a principal and an agent, there are new opportunities for incentives. Experience rating in insurance illustrates the situation; the premium rate charged today depends on past outcomes. In effect, the information on which the fee function is based is greatly enriched. Radner (1981) has demonstrated the possibilities for achieving almost fully optimal outcomes in hidden-action situations. Suppose the principal has a desired level of action, a^*, that the agent is to implement. In any one trial, the action is hidden, in that the outcome differs from the action by a random variable, i.e. $x_t = a_t + u_t$, where the random variables u are identically and independently distributed, with mean zero. If the agent is in fact performing the desired action a^*, then the distribution of the x's is known. Hence, if enough are observed, the principal should be able to detect statistically whether or not the agent is performing actions below the desired level. Specifically, the principal can keep track of the cumulative sum of the outcomes. If it ever falls below a known function of time, then the principal can assume that the performance of the agent is below that desired. More exactly, the principal imposes a very severe penalty if there is some time T such that

$$\sum_{t=1}^{T} x_t < Ta^* - k \log\log T.$$

For properly chosen k, the probability of imposing a penalty when the agent is in fact carrying out the desired action can be made very low, while the probability of eventually imposing the penalty if the agent is shirking is one.

I have sketched some of the leading ideas in the rapidly-burgeoning literature on the economic theory of the principal–agent relation. We may step back a bit from the pure theory and ask in a general way to what extent our understanding of economic processes has been enhanced. On the positive side, there is little question that a good many economic relations inexplicable in previously standard analysis can now be understood. Contractual relations are frequently a good deal more complicated than the simple models of exchange of commodities and services at fixed prices would suggest. Sharecropping, incentive compensation to executives and other employees, the role of dismissal as an incentive, coinsurance, and other aspects of insurance all find a place in this literature not found in standard economic analysis.

But it is perhaps more useful to consider the extent to which the principal–agent relation in actuality differs from that in the models developed to date. Most importantly, the theory tends to lead to very complex fee functions. It turns out to be difficult to establish even what would appear to be common-sense properties of monotonicity and the like. We do not find such complex relations in reality. Principal–agent theory gives a good reason for the existence of sharecrop contracts, but it is a very poor guide to their actual content. Indeed, as John Stuart Mill pointed out long ago, the terms tend to be regulated by custom. They are remarkably uniform from farm to farm and from region to region. Principal–agent theory would suggest that the way the produce is divided between landlord and tenant would depend on the probability distribution of weather and other exogenous uncertainties and on the relation between effort and output, both of which certainly vary from one region to another; the latter has varied over time as well. Similarly, the coinsurance provisions in health insurance policies are much simpler than could possibly be accounted for by principal–agent theory.

In some cases where principal–agent theory seems clearly applicable, there is very little trace in reality. In many respects, the physician–patient relation exemplifies the principal–agent relation almost perfectly. The principal (the patient) is certainly unable to monitor the efforts of the agent (the physician). The relation between effort and outcome is random, but presumably there is some connection. Yet the fee schedule is in no way related to outcome. (It is true that liability for malpractice serves in a way as a modification of the fee schedule in the direction indicated by principal–agent theory; but it is not applicable to what might be termed run-of-the-mill shirking, and it requires very special kinds of evidence.) In general, indeed, compensation of professionals has only mild traces of the complex fee schedules implied by theory.

Even when there are compensation systems that seem closer in form to the theoretical, there are significant differences. Consider the incentive compensation schemes for corporate executives. They invariably have a large discretionary component. What is the purpose of this? Why should the incentive payment not be based entirely on observable magnitudes, profits, rates of return, and the like?

These difficulties can be explained within the terms of the principal–agent logic but in a way that points beyond the usual bounds of economic analysis. One basic problem is the cost of specifying complex relations. There is a large, though not easily defined, cost to a contract which specifies payments which depend on many variables. There is a cost to the very statement of the contract, a cost to understanding it and its implications, and a cost to verifying which terms apply in a given situation. Hence, there is a pressure for simple contracts, the more so since in fact any of our models are much too simple to capture all the aspects of a relation which those in it would deem relevant.

A second aspect of reality is the variety of means of monitoring and the difficulty of defining exactly what they are. The world is full of performance evaluations based on some kind of direct observations. These evaluations may not always be objective, reproducible observations of the kind used in our theories (perhaps the only kind about which it is possible to construct a theory). Executives are judged by their superiors and students by professors on criteria which could not have been stated in advance. Outcomes and even supplementary objective measures simply do not exhaust the information available upon which to base rewards.

A third limitation of the present models is the restricted reward or penalty system used. It is always stated in terms of monetary payments. Actually, the present literature has already begun to go beyond this limit by considering the possibility of dismissal. Still further extensions are needed to capture some aspects of reality. Clearly, there is a whole world of rewards and penalties in social rather than monetary form. Professional responsibility is clearly enforced in good measure by systems of ethics, internalized during the education process and enforced in some measure by formal punishments and more broadly by reputations. Ultimately, of course, these social systems have economic consequences, but they are not the immediate ones of current principal–agent models.

All three of these limiting elements, cost of communication, variety and vagueness of monitoring, and socially mediated rewards, go beyond the usual boundaries of economic analysis. It may be ultimately one of the greatest accomplishments of the principal–agent literature to provide some structure for the much-sought goal of integrating these elements with the impressive structure of economic analysis.

References

Gjesdal, F. (1982), "Information and incentives: The agency information problem", Review of Economic Studies, 49:373–390.

Grossman, S. and O. D. Hart (1983), "An analysis of the principal–agent problem", Econometrica, 51:7–46.

Harris, M. and A. Raviv (1978), "Some results of incentive contracts with applications to education and employment, health insurance and law enforcement", American Economic Review, 68:20–30.

Harris, M. and A. Raviv (1979), "Optimal incentive contracts with imperfect information", Journal of Economic Theory, 20:231–259.

Holmström, B. (1979), "Moral hazard and observability", Bell Journal of Economics, 74–91.

Holmström, B. (1982), "Moral hazard in teams", Bell Journal of Economics, 13:324–340.

Marcus, A. J. (1982), "Risk sharing and the theory of the firm", Bell Journal of Economics, 13:369–378.

Maskin, E. S. and J. G. Riley (1983), "Monopoly with incomplete information", Discussion Paper. Los Angeles: University of California.

Maskin, E. S. and J. G. Riley (1984), "Optimal auctions with risk averse buyers", Econometrica, 52, forthcoming.

Milgrom, P. R. and R. J. Weber (1982), "A theory of auctions and competitive bidding", Econometrica, 50:1089–1122.

Mirrlees, J. (1971), "An exploration in the theory of optimum income taxation", Review of Economic Studies, 38:175–208.

Radner, R. (1981), "Monitoring cooperative agreements in a repeated principal–agent relationship", Econometrica, 49:1127–1148.

Riley, J. G. (1975), "Competitive signalling", Journal of Economic Theory, 10:174–186.

Riley, J. G. (1979), "Informational equilibrium", Econometrica, 47:331–360.

Roberts, K. (1979), "Welfare considerations of nonlinear pricing", Economic Journal, 89:66–83.

Rothschild, M. and J. E. Stiglitz (1976), "Equilibrium in competitive insurance markets", Quarterly Journal of Economics, 80:629–649.

Shavell, S. (1979), "Risk sharing and incentives in the principal and agent relationship", Bell Journal of Economics, 10:55–73.

Spence, A. M. (1973), Market Signalling: Information Transfer in Hiring and Related Processes. Cambridge: Harvard University Press.

Spence, A. M. (1977), "Nonlinear prices and welfare", Journal of Public Economics, 8:1–18.

Vickrey, W. (1961), "Counterspeculation, auctions, and competitive sealed tenders", Journal of Finance, 16:8–37.

Wilson, C. A. (1977), "A model of insurance markets with incomplete information", Journal of Economic Theory, 16:167–207.

Chapter 24

THE THEORY OF OPTIMAL TAXATION

J. A. MIRRLEES

Nuffield College, Oxford

1. Economic theory and public policy

A good way of governing is to agree upon objectives, discover what is possible, and to optimize. At any rate, this approach is the subject of optimal tax theory. From this point of view "optimal tax theory" is an unduly narrow term to describe the subject, but it is neater than "theory of optimal public policy". In any case, I shall not be discussing the optimization of macroeconomic models, which are used to treat several aspects of public policy. Much – though not all – of what has so far been done in optimal tax theory uses the standard model of competitive equilibrium, with rational consumers and profit-maximizing, price-taking firms. In this way one avoids debate about the dubious relationships of disequilibrium macroeconomics or oligopoly theory, and concentrates on essentials.

The central element in the theory is information. Public policies apply to individuals only on the basis of what can be publicly known about them. There is little difficulty about paying the same subsidy to every individual in the economy: there is not much more difficulty in making the subsidy depend on age. Uniform positive taxes may be a little more difficult. Taxes and subsidies proportional to trade in specified goods or services may also be difficult to administer with perfect accuracy. But, subject to some minor imperfections, we can take it that most such taxes use information that is cheaply and publicly available. Not all conceivable public policies have this convenient property. One of the basic theorems of welfare economics asserts that, where a number of convexity and continuity assumptions are satisfied, an optimum is a competitive equilibrium once initial endowments have been suitably distributed. To make distribution requires, in general, complete information about individual consumers, for the transfers must be lump-sum in character, that is, independent of the individual's behaviour. It is generally agreed by economists that the lump-sum transfers necessary to achieve an optimum are scarcely ever feasible.[1] There is no way of obtaining the

[1] Hahn (1973) asserts that lump-sum taxation has in fact been used. This is true, though his examples are bad ones; but it is beside the point. The question is whether *optimal* lump-sum transfers are possible.

Handbook of Mathematical Economics, vol. III, edited by K.J. Arrow and M.D. Intriligator
© 1986, Elsevier Science Publishers B.V. (North-Holland)

information about individuals that is required except in a society of individuals who are truthful regardless of selfish considerations. A theorem supporting this view is given in Section 3 below.

Widespread agreement among economists that optimal lump-sum taxation is impossible in practice came long before analysis of optimal non-lump-sum taxation. This is surprising. Possibly too many economic theorists were chiefly interested in the supposed merits of the undistorted competitive price system; but socialist economists did not fill the gap. Perhaps distaste for the welfare function was a more effective barrier to progress. It is true that Bergson and Samuelson used welfare functions in their work on the fundamentals of welfare economics. But those more closely concerned with policy issues would not have thought the welfare function, embodying interpersonal comparisons of welfare, a practical tool of analysis. In this century, economists have usually preferred to analyse empirical propositions of doubtful validity rather than analyse the consequences of value judgements, even when these might have been expected to command more widespread agreement.

There are, it seems to me, only two promising approaches to making well-based recommendations about public policy. One is to use a welfare function of some form and develop the theory of optimal policy. The other is to model the existing state of affairs in some manageable way, and on that basis to display the likely effects of changes in government policy, these effects being displayed in sufficient detail to make rational choice among alternative policies possible. If a welfare function were used to evaluate the changes predicted, the second approach would come fairly close to the first, and in fact there is then a close theoretical relationship. But the second method could concern itself with presentation of effects rather than their evaluation. For example, the effects of policy changes on income distribution can be presented graphically. This approach is open to many objections as it is practised, and it is not easy to see how these faults could be avoided. In the first place, the particular way of presenting effects is not the outcome of systematic analysis, but is chosen quite informally. Secondly, the presentation is liable to divert attention completely from matters that could be important. In the income-distribution example, people presented with income-distribution pictures are unlikely to consider how these judgements should be affected by differences in relative prices. Thirdly, summary variables may be used which no plausible welfare judgements would validate. The use of Gini coefficients in the presentation of income-distribution effects is, I think a case in point.[2] The user of such figures is all too likely to regard bigger as better. The fact that the summary variable is precisely intended not to be a welfare function, or argument in a welfare function, is no help in avoiding misuse.

There are then some practical arguments in favour of using welfare functions to analyse public policy. But unless there are stronger cases for some welfare

[2] Sen (1973, pp. 29–34) makes a moderate case for this measure.

functions than for others, the formal derivation of properties of welfare-maximizing policies is a pointless exercise. It turns out that some of these properties are independent of the welfare function; but optimal policies are not. For much of the theory, one must bear in mind what kind of welfare function is likely to be satisfactory. Furthermore, some of the most interesting results obtainable in this area are numerical calculations for specific welfare functions. For this reason, too, optimal tax theory is a field where econometric work is of considerable interest to the theorist, and the needs of theory a guide to the econometrician.

The models to be discussed are firmly based on a distinction between public and private information. The government deals with an economy of consumers, producers, and possibly other corporate institutions, such as charitable bodies. These private individuals and institutions may know things the government does not know, such as a specific person's income-earning potential. The simplest assumption is that, in respect of such individual characteristics, the government either observes and knows the precise truth, or knows nothing to distinguish the individual from anyone else. Thus we usually exclude the realistic possibility that the government could at a cost improve its information; or that the government has information about individuals that is not completely reliable. But the theory can be expected to throw light on the magnitude of the gain from additional information of this kind. Something will be said about the use of imperfect lump-sum taxation, based on individual characteristics observed with errors, in Section 3.

Another aspect of public policy omitted from the basic models is the evasion and enforcement of government policies. From one point of view, the problem of enforcement is one of getting information. A firm reports its profits and pays tax accordingly: the profit tax is a policy tool that relates tax payment to reported profits. Actual profits may or may not be equal to reported profits; so there are other rules relating tax payments – this time known as fines and imprisonment – jointly to reported profits and a more accurate measurement of actual profits made, at a cost, by government agents. Again, in certain countries, what the government servants report actual profits to be may be influenced by bribes. This brings in another set of considerations, where transactions are necessarily personal, unlike transactions in the standard competitive model. Since, in the basic optimal tax models, states of information are fixed, personal transactions, whose terms are specific to the individuals involved, need not be considered. But transactions of this kind – which are common in the real world, particularly in the capital market – would be an important subject of study in a complete theory of the administration, enforcement and evasion of the tax system.[3]

[3]A more straightforward treatment of administrative costs has been initiated by Heller and Shell (1974).

The range of public policy contemplated in optimal tax theory is quite wide. Besides taxes and subsidies themselves, which may be related to any transactions between individuals, firms, other corporate bodies, foreign countries and individuals, and government and its agencies, the theory should also be prepared to encompass the use of quantity controls and restrictions, and the control of information flows, for example in training programmes or public advice. Also the government and its agencies can make expenditures or set up productive activities itself. Public expenditures may be undertaken to meet international obligations, or to benefit individuals, corporate bodies, or groups of these. A first requirement of the theory is that one finds a convenient, simple notation that will encompass all such policy variables without unduly complicating the analysis. In fact, despite the range of possible policies, the basic relationships are usually quite simple and similar. It is good to cultivate the art of seeing specific policy instruments as instances of the general possibilities of policy whose modelling we are to discuss.

It will be noticed that the list of policy instruments in the previous paragraph does not include certain policies which rely for their operation on disequilibrium states of the economy. Deficit finance, price control, wage and income policies are instances of non-equilibrium policies. It should be possible to apply the methods of optimal tax theory also to models allowing disequilibrium.[4] This seems to be an interesting area for further research.

In the next section, the common mathematical form of optimal tax problems will be explained, and certain basic features and issues discussed. In subsequent sections, we shall look at a variety of cases. After dealing with lump-sum taxation in Section 3, we examine linear taxation in Sections 4 and 5. Sections 6 and 7 are devoted to the theory of income taxation and non-linear taxation generally. The discussion is concluded largely in terms of taxes and subsidies. Models with individual uncertainty about the effect of policies are discussed briefly in Section 8. Some remarks and results about computation and approximation are collected in Section 9. After some concluding remarks, constituting Section 10, Section 11 provides some brief notes on the literature.

This paper does not contain a thorough survey of the literature on optimal tax theory. Neither the time nor the facilities for such a survey were available. It is rather an account of what seem to me the fundamental parts of the theory, with emphasis on the mathematical problems. Much of the published literature deals with economies in which all individuals are identical. Since this case does not seem to me especially interesting or useful, it will not be given much attention. Interesting and important areas which are neglected are the analysis of an international economy, where the impossibility of lump-sum transfers should have many interesting consequences; and the study of variable population.

[4] Dixit (1976) has looked at some issues in a temporary equilibrium model.

2. Optimization subject to maximization constraints

Problems in optimal tax theory have a characteristic form. To bring this out, consider three typical models.

In the first, the government sets commodity taxes $t=(t_1,\dots,t_n)$ proportional to trade in the n commodities. Producers face prices $p=(p_1,\dots,p_n)$, and their production activity is uniquely determined by these prices. Writing y for the aggregate net production vector, and x^h for the net demand vector of consumer h (there being H consumers), market clearing requires

$$\sum_{h=1}^{H} x^h = y(p). \tag{2.1}$$

At the same time, consumers maximize utility, and we have for $h=1,2,\dots,H$,

$$\left.\begin{aligned} &x^h \text{ maximizes } u^h(x)\\ &\text{subject to}\quad (p+t)\cdot x \le b^h(p)\\ &\text{and}\quad x\in X_h \end{aligned}\right\}, \tag{2.2}$$

where b_h is the profit income of the consumer, which in the absence of profit taxation is simply a function of p; and X_h is the consumption set of consumer h, u^h his utility function.

A rather general form for the welfare function that government seeks to maximize is

$$W(x^1,x^2,\dots,x^H).$$

In the problem outlined, W is to be maximized subject to the constraints (2.1) and (2.2). The first of these constraints is of familiar type. The second group of constraints looks quite unlike those encountered in elementary constrained maximization problems, for it involves maximization itself with respect to some of the variables, in this case the x^h, while other variables p and t are parametric.

It will be noticed that, when the u^h are strictly concave and the X_h convex, the apparently complicated form of (2.2) is of no great consequence, because we can write

$$x^h = x^h(p+t, b^h(p)),$$

just as the supply functions $y(p)$ may be derived from profit maximization. This feature is specific to problems with linear taxation.

The second problem makes a common, but generally unsatisfactory, assumption that all consumers react in the same way to the government's policy variables. The government provides a facility, such as education, to some homogeneous groups of consumers. The supply of the facility is measured by a real number z which happens to be equal to its cost. The cost is met from the taxes paid by the beneficiaries, and taxes T_0 obtained from the rest of the community. The tax paid by the beneficiaries is a function $T_1(y)$, of their labour supply y. This function is to be taken as given. The welfare function has as arguments the utility $u(y, z)$ of beneficiaries, and the tax T_0 paid by the rest of the community. Thus the problem is

$$\underset{y, z, T_0}{\text{maximize}}\ W(u(y,z),T_0), \tag{2.3}$$

$$\text{subject to} \quad z = T_0 + T_1(y), \tag{2.4}$$

$$\text{and } y \text{ maximizes } u(y,z). \tag{2.5}$$

This problem is a rather special and artificial one, but shows how naturally a maximization constraint arises. In this case there would be no reasonable presumption that u be strictly concave in y for all z, and therefore no reason to suppose that we can replace (5) by writing $y = y(z)$.

The third problem is that of optimal income taxation, where there are two commodities, a consumption good and labour, and the population is an infinite one where individuals are characterised by a continuous parameter h, distributed with density function f. The income tax takes a net amount $t(wy)$ of consumption good from a consumer who supplies labour y, the wage rate being w. Consumer h has utility $u(x, y, h)$, x being his consumption, and $x = wy - t(wy)$. The welfare function is

$$W = \int u(wy - t(wy), y, h) f(h)\,\mathrm{d}h. \tag{2.6}$$

This is to be maximized subject to the constraints that

$$y(h) \text{ maximizes } u(wy - t(wy), y), \tag{2.7}$$

for all h; and the production constraint

$$\int [wy - t(wy)]\, f(h)\,\mathrm{d}h \le G\left(\int yf(h)\,\mathrm{d}h\right). \tag{2.8}$$

Furthermore, the wage is the marginal product of labour,

$$w = G'\left(\int yf(h)\,\mathrm{d}h\right). \tag{2.9}$$

In this formulation it has been assumed that all profits go to the government: otherwise the consumer's budget constraint would have to be modified.

Each of these problems can be written in the form

$$\left.\begin{array}{l} \text{maximize } W(x,z) \\ \text{subject to} \quad (x,z)\in A \\ \text{and } x \text{ maximizes } U(x',z) \\ \text{subject to} \quad x'\in X(z) \end{array}\right\}. \tag{2.10}$$

Generally, the set A represents technological feasibility, and the relationship between production and prices. The maximization constraint represents consumer and producer behaviour. The set $X(z)$ is the intersection of the set of definition of the function $U(.,z)$ and other constraints imposed by government.

There could be many maximization constraints, but in each of the above problems they can be written as one. For instance, in the third example, the function $y(h)$ is chosen to maximize

$$\int u(wy-t(wy),y)f(h)\,\mathrm{d}h,$$

and this single maximization encompasses the behaviour of all consumers. This is possible because of the absence of consumption externalities. It will also be noticed that in this case the constraint imposed on consumers by taxation is incorporated into the utility function, and the set $X(z)$ is simply the set of $y(h)$ that are consistent with non-negative consumption and labour, i.e. that satisfy $0\le y(h)$, $t(wy(h))\le wy(h)$. It usually seems best so to transform a problem that the sets $X(z)$ reflect only consumption feasibility, and can often be understood implicitly from finiteness of the function U. It will be seen below that transformations of problems into convenient form play an important part in the theory. The first and third problems, as set out above, are not in a good form for mathematical analysis: in fact they are much simpler than they look when the economics is first set up mathematically.

In some cases the control variables z and the behavioural variables x are numbers or vectors in finite-dimensional vector space. In other cases, such as our third problem, they are functions. (z might even be a subset of finite- or infinite-dimensional space, but I know of no problem that has been analysed directly in this form.)

Granted that (2.10) is the form of problems in optimal tax theory, we have to deal with two issues. The first is that concavity of W and convexity of A are not usually implied by the natural assumptions of the problem. Therefore theorems of concave programming are not applicable; and first-order conditions for optimality are unlikely to be sufficient conditions. These issues will be taken up as they appear in the various models: the optimal tax theorist must always bear them in mind, and look for ways of circumventing them.

The second issue is the nature and treatment of the constraint that [leaving $X(z)$ to be understood]

$$x \text{ maximizes } U(x', z). \tag{2.11}$$

If U is differentiable and strictly concave, (2.11) is equivalent to

$$U_x = 0, \tag{2.12}$$

which can be handled as a normal set of constraints, although it is unlikely to define a convex set. But that takes us back to the first issue. In many interesting cases, U is not concave in x, at least not for all z: this is so for the second and third examples above.

There are two ways of handling (2.11). We could replace (2.11) by the rather large set of constraints, with new variables,

$$U(x, z) \geq (x', z), \quad \text{all } x'. \tag{2.13}$$

In almost all interesting cases, this is an uncountable infinity of inequalities, which may therefore be delicate to handle: but the reduction to (2.13) can be useful. The alternative method is to examine directly the set of (x, z) defined by (2.11).[5]

It may be helpful to do this first for a special case (which has no economic significance).

Example 1

x and z are scalars.

$$\text{Find } z \text{ to maximize } -(x-1)^2-(z-2)^2$$

$$\text{subject to} \quad x \text{ maximizes} \quad U(x, z) = z\mathrm{e}^{-(x+1)^2} + \mathrm{e}^{-(x-1)^2}.$$

We begin by describing the constraint set. The first-order condition for maximization with respect to x is

$$z(x+1)\mathrm{e}^{-(x+1)^2} + (x-1)\mathrm{e}^{-(x-1)^2} = 0,$$

[5] The recently developed branch of differential topology known as catastrophe theory (based on work of R. Thom) studies the set of (x, z) such that $U_x = 0$, and particularly the set of z for which the local behaviour of x satisfying $U_x = 0$ is especially noteworthy. The study of the set of (x, z) such that x maximizes U is in some ways closely related. Bröcker (1975, p. 145) refers to the *Maxwell convention* as describing this kind of problem. But the features of these sets that are of interest in optimization are, by and large, quite different from those that are of interest in the dynamic analysis of systems, which has so far been the main motivation of catastrophe theory. In particular, the *catastrophe points* z at which $\det U_{xx} = 0$ are rather unimportant optimizations.

i.e.

$$z=\frac{1-x}{1+x}\mathrm{e}^{4x}. \tag{2.14}$$

For x between 0.344 and 2.903 there are three values of x satisfying (2.14), and it still remains to discover which of them actually does the maximizing.

To settle this, we observe that

$$\begin{aligned} U(z,x)-U(z,-x) &= (z-1)(\mathrm{e}^{-(x+1)^2}-\mathrm{e}^{-(x-1)^2}) \\ &= -(z-1)(\mathrm{e}^{4x}-1)\mathrm{e}^{-(x+1)^2}, \end{aligned}$$

so that for fixed $z>1$, U is less for positive x than for negative; while if $z<1$, U is greater for positive x than for negative. Therefore the maximum of U occurs for positive x when $z<1$, for negative x when $z>1$. In either case (as is readily verified) this identifies the desired solution of (2.14) uniquely. The points of the locus (2.14) for which x maximizes U form two closed connected subsets of the locus. When $z=1$, U is maximized by $x=\pm 0.957$.

It is clear, by sketching contours $(x-1)^2+(z-2)^2=$ constant in a diagram that the solution of the maximization problem is

$$x=0.957, \qquad z=1.$$

This solution is not obtained if one treats the problem as a conventional constrained maximization problem with the first-order condition (2.14) as constraint. The Lagrangian is then

$$-(x-1)^2-(z-2)^2+\lambda\left(z-\frac{1-x}{1+x}\mathrm{e}^{4x}\right),$$

whose derivatives are zero when

$$\begin{aligned} &2(z-2)=\lambda, \\ &2(x-1)=\frac{4x^2\cdot 2}{(1+x)^2}\mathrm{e}^{4x}\lambda, \\ &z=\frac{1-x}{1+x}\mathrm{e}^{4x}, \end{aligned}$$

i.e. when

$$\begin{aligned} &z=\frac{1-x}{1+x}\mathrm{e}^{4x}, \\ &2z(2-z)=\frac{(1-x^2)^2}{(1+x)(2x^2-1)}. \end{aligned}$$

There are three solutions:

(I) $x = 0.895, \quad z = 1.99;$

(II) $x = 0.420, \quad z = 2.19;$

(III) $x = -0.980, \quad z = 1.98.$

The first clearly gives the largest value for the maximand, $-(x-1)^2-(z-2)^2$, but our previous analysis shows that x does not maximize $U(x, z)$ for this value of z. As a matter of fact, x is a *local* maximum, but not a *global* maximum. The second solution is ineligible on all possible grounds: x is a local minimum of $U(x, z)$. The third solution, on the other hand, has the property that x is a global maximum of $U(x, z)$, so that it does satisfy the constraint of the original problem. But it is not the solution of that problem, and indeed gives a much lower value of the maximand than is actually possible.

This example shows that it is not legitimate to attempt to solve the problem by substituting first-order conditions for the maximization constraint. Furthermore, and this deserves emphasis, the example, though complicated, is in no sense special. Any moderate variation of the functions involved yields a problem with the same properties.

In order to understand the form of the set[6]

$$M = \{(x, z): x \text{ maximizes } U(x', z)\},$$

in general, we should take U to be a smooth (C^∞) function on $(m+n)$-dimensional Euclidean space. We do not want to examine M for all possible smooth U, but for "almost all" U, excluding pathological or special cases. In general, for each z, U has a finite number of distinct maxima, $x_i(z)$ $(i=1,\dots,r)$. Provided that the matrix U_{xx} of second derivatives is of full rank m at each of these maxima, the x_i are smooth mappings of z. Then

$$U(x_i(z), z) - U(x_1(z), z) = 0, \quad i = 2,\dots,r, \tag{2.15}$$

and also

$$U_x(x_i(z), z) = 0, \quad i = 1,\dots,r. \tag{2.16}$$

Regarding (2.15) and (2.16) as equations for $z, x_1,\dots, x_r$, we have $r-1+rm$ equations and $n+rm$ unknowns. That is to say, the set of $(z, x_1(z),\dots, x_r(z))$ is contained in the inverse image of $(0,\dots,0)$ by the mapping

$$(z, x_1,\dots,x_r) \to (U(x_2, z) - U(x_1, z),\dots, U(x_r, z) - U(x_1, z),\dots,$$
$$U_x(x_1, z),\dots, U_x(x_r, z)),$$

[6]The discussion of M owes a great deal to discussions with Kevin Roberts, who formulated the theorem about the essential maximum to the number of maxima recorded below.

from E^{n+rm} to E^{r-1+rm}. For almost all functions U, $(0,\ldots,0)$ should be a regular value of this mapping when $x_1,\ldots,x_r$ are distinct. Provided that is the case, there will be a $(n-r+1)$-dimensional neighbourhood of $(z, x_1,\ldots, x_r)$ that also maps into $(0,\ldots,0)$. In other words the set of z for which U has r distinct maxima is of dimension $n-r+1$; and the corresponding subset of M has the same dimension. In particular, there are no z with $r > n+1$, i.e. more than $n+1$ maxima, for general U.

In Mirrlees and Roberts (1980), written after the present chapter, the following theorem was proved:

For almost all[7] C^{∞} functions U, the number of distinct maxima is less than or equal to $n+2$ for all z, and the dimension of the set of points of M corresponding to z with r distinct maxima is less than or equal to $n+1-r$.

It should not be supposed that, since dimension falls with the number of distinct maxima, points with a single maximum are almost certain to give the answer in actual optimization problems. Points (x, z) corresponding to r maxima essentially form the boundary to the set of (x, z) corresponding to $r-1$ maxima. Thus, broadly speaking, the solution to an optimizing problem is just as likely to be a value of z with many maxima as with few, subject to the overall bound $n+2$.

The economic significance of this is that an optimum may well leave consumers indifferent among several options, only one of which the government would like to see chosen. Also the optimum can easily be something of a corner solution. To bring this out consider how one would have to solve a general problem of the form

$$\left.\begin{array}{l}\text{maximize } W(x,z)\\ \text{subject to} \quad G(x,z)=0\\ \text{and } x \text{ maximizes } U(x,z)\end{array}\right\}. \tag{2.17}$$

Using the theorem stated above, we can express this in a more convenient form for almost all U. Not only do we know that when x maximizes $U(x, z)$, $U_x(x, z) = 0$, but also there are only a finite number of x' that maximize U, and these also satisfy $U_x = 0$. Thus x maximizes U if and only if $U_x = 0$ and

$$\left.\begin{array}{l}U(x,z) \geq U(x',z), \quad \text{all } x'\\ \text{such that} \quad U_x(x',z)=0\end{array}\right\}, \tag{2.18}$$

and we can have equality in the constraints (2.18) for at most $n+2$ values of x'.

[7] The set of such functions contains a countable intersection of open dense sets in the Whitney or strong topology.

In this way we can replace the constraint "x maximizes U" by a finite number of equations and inequalities. The problem can therefore be treated as a standard Kuhn–Tucker problem.

Provided that certain regularity conditions are satisfied, it is necessary for (x, z) to be an optimum, that there exist a scalar λ, an m-vector μ, and scalars ν', one for each x' satisfying $U_x = 0$, such that

$$L(x, z) = w + \lambda G + U_x \cdot \mu + \sum \nu' \{U(x, z) - U(x', z)\} \tag{2.19}$$

have zero derivatives with respect to x and z. The summation is over all x' satisfying $U_x = 0$, and each $\nu' \geq 0$ with strict inequality only if $U(x, z) = U(x', z)$. Differentiation of L yields

$$W_x + \lambda G_x + U_{xx} \cdot \mu = 0, \tag{2.20}$$

$$W_z + \lambda G_z + U_{xz} \cdot \mu + \sum \nu' \{U_z(x, z) - U_z(x', z)\} = 0. \tag{2.21}$$

(2.20) has simplified because the last terms drop out, as $U_x(x, z)$ and $U_x(x', z)$ both vanish.

In principle, the equations we have found are enough to determine a finite number of solutions, one of which is the optimum. The chief difficulty is that the set M and its structure must be known before (2.21) can be found explicitly. To use the Lagrangian method, we would need to try successively z for which the maximim is unique (when the last terms drop out), then z with two maxima, and so on until all possibilities have been tried. Unfortunately, the determination of the set M of maxima for every z must usually be difficult and require much computation.

Nevertheless, certain lessons can be drawn. Granted the difficulties in handling the general problem, it is important to find conditions under which it simplifies, particularly under which one can be sure that the optimum occurs where there is a unique maximum for U. It is also important not to be lulled into believing that solutions in these cases have a character that is universally applicable.

One of the most striking features of these problems is that, as the number of control variables (the dimension of z) increases, the possible extent of consumer indifference in the optimum increases. This suggests that when the government policies are functions, i.e. infinite-dimensional, it can be optimal for consumers to have continuous ranges of indifference. It can even be the case that this idea simplifies the task of solution, because indifference over a range determines the form of optimal policy over the range; just as knowledge that the optimum is at an $(n+2)$-maximum virtually determines optimal z in the class of problems we have been discussing. On occasion it is possible to discover quite easily conditions sufficient to imply that the optimum has this form.

3. Lump-sum transfers

In this and the following sections, a common model will be used. It will be useful to establish notation.

x^h = net demand vector (i.e. consumption net of endowment) of consumer,

u^h, X^h = utility function and consumption set of consumer h.

Either there are a finite number of consumers H, or h is continuously distributed and non-negative with density function f.

y = aggregate net supply vector of private producers,

Y = aggregate production set,

y^j = net supply vector of producer j,

Y^J = production set of producer j,

z = net supply vector of government, being the public production vector minus the public consumption vector,

Z = set of feasible z,

q = prices faced by consumers,

p = prices faced by private producers.

It will be assumed that u^h is differentiable and concave and X^h convex. This is rather stronger than assuming convex preferences, but convenient. Private production sets are convex unless otherwise stated. Each u^h is a strictly increasing function of its arguments.

Let the welfare function be individualistic, i.e.

$$W = \Omega(u^1, \ldots, u^H) \tag{3.1}$$

in the case of a finite population. W is smooth and an increasing function of all u^h. It is interesting first to analyse the problem when all possible policies are available to government, partly because we can introduce some techniques that prove useful later. If all policies are possible, the government can impose on each consumer separately a budget set $B^h(p)$, and have each producer maximize profits. (There is no interest here in considering more general forms of production control.) Then the constraints in the optimization are

$$\left.\begin{array}{l} x^h \text{ maximizes } u^h(x) \\ \text{subject to} \quad x \in B^h(p) \cap X^h \end{array}\right\}, \tag{3.2}$$

$$\left.\begin{array}{l} y^j \text{ maximizes } p \cdot y \\ \text{subject to} \quad y \in Y^j \end{array}\right\}, \tag{3.3}$$

$$\sum_h x^h = \sum_j y^j + z, \tag{3.4}$$

$$z \in Z. \tag{3.5}$$

By the fundamental theorem of welfare economics, it is known that the solution to this complicated looking problem takes the simple form (when every consumer is in the interior of his consumption set at the optimum)

$$B^h(p) = \{x : p \cdot x - b^h\}, \tag{3.6}$$

where the scalars b^h satisfy

$$\sum_h b^h = \sum_j p \cdot y^{j*} + p \cdot z^*, \tag{3.7}$$

y^{j*} and z^* being the optimal values of y^j and z.

Using the indirect utility function, we state a rule for the optimal lump-sum transfer b^h. Let

$$v^h(q, b^h) = \max\{u^h(x) : q \cdot x \leq b^h,\ x \in X^h\}, \tag{3.8}$$

and

$$V(q, b^1, \ldots, b^h) = \Omega(v^1(z, b^1), \ldots, v^H(q, b^H)). \tag{3.9}$$

Then optimal $b^* = (b^{1*}, \ldots, b^{H*})$ maximizes

$$V(p, b) \text{ subject to the constraint (3.7).} \tag{3.10}$$

The first-order condition for this is that

$$\partial V / \partial b^h = \lambda, \qquad h = 1, \ldots, H, \tag{3.11}$$

for some scalar λ. This familiar condition may also be expressed by using the expenditure function

$$E^h(q, u^h) = \max\{q \cdot x : u^h(x) \geq u^h\}. \tag{3.12}$$

With this notation we can say that optimal utility levels $u^* = (u^{1*}, \dots, u^{H*})$ maximize

$$\Omega(u) \quad \text{subject to} \quad \sum_h E^h(p, u^h) \leq \sum_j p \cdot y^{j*} + p \cdot z^*. \tag{3.13}$$

The assumption that $u^h(x)$ is a concave function implies that E^h is a convex function of u^h: $E^h_{uu} \geq 0$. The first-order conditions for (3.13) are

$$\partial \Omega / \partial u^h = \lambda E^h_u. \tag{3.14}$$

The objections to assuming it possible to make b^h a function of h are, first, that consumers may not choose to give the government correct information about their utility functions; and, second, that, even if consumers were willing to tell the truth, it would be costly to obtain the information. These objections will each be formalised.

To capture the first objection we need a formulation of welfare with more content. The most powerful welfare functions are those based on the idea that individuals are basically the same, but vary in endowment, abilities, and sensibilities. These differences can be taken to be differences in the significance of trade for utility. A simple formulation (ignoring differences in material endowment) is

$$u^h(x) = u(h_1 x_1, \dots, h_n x_n), \tag{3.15}$$

with the consumer described by n parameters $h_1, \dots, h_n$. If, for example, commodity n is labour, and labour has disutility, larger h_n means labour is harder, or equivalently, the ability (or inclination) to provide labour is less. Similarly h_1 can represent the ability to appreciate wine. If individuals are identical, welfare ought to be a symmetrical function of utilities. For concreteness and convenience, take an additive function

$$\Omega(u) = \int u^h f(h_1, \dots, h_n) \, \mathrm{d}h_1, \dots, \mathrm{d}h_n. \tag{3.16}$$

If the government must rely completely on individual report for its knowledge of an individual's h, and individuals are truthful only when they do not lose by it, either B^h must be independent of h (so that the government does not use observations of h), or u^h must be independent of h. We shall see below that, under some plausible assumptions, the latter is the better alternative. It may be more interesting to suppose that the government can obtain information about h by some form of testing. The leading examples are abilities, where an individual can easily pretend to less ability than he truly has, but would find it difficult to prove he has more. (Uncertainties of observation will be mentioned later.) Supposing then that individuals can misreport h_i only by claiming it is greater than in fact it is, an h_i-dependent policy can be administered only if, in the outcome,

$$v(h) = u^h(x^h) \tag{3.17}$$

is a non-increasing function of h_i. The following result is then of interest.

Theorem 3.1

Let the utility function be (3.15), and the welfare function additive. In the first-best optimum, v is an increasing function of h_i if commodity i is always a normal (i.e. not an inferior) commodity.

Proof (illustrating the convenience of the expenditure function in these problems)

We have seen that, at the optimum,

$$E_u^h(p, v(h)) = \lambda. \tag{3.18}$$

With the utility function (3.15), E^h takes the form

$$E^h(p,u) = E\left(\frac{p_1}{h_1},\ldots,\frac{p_n}{h_n},u\right). \tag{3.19}$$

Therefore differentiation of (3.18) with respect to h_i yields

$$E_{uu}\frac{\partial v}{\partial h_i} - E_{ui}\frac{p_i}{h_i^2} = 0,$$

Thus

$$E_{uu}\frac{\partial v}{\partial h_i} = \frac{p_i}{h_i}\frac{\partial}{\partial u}x_i^c, \tag{3.20}$$

where $x_i^c = (\partial/\partial p_i)E^h$ is the compensated demand function for commodity i. Normality means that $(\partial/\partial u)x_i^c > 0$; and concavity of u implies $E_{uu} > 0$. Therefore (3.20) implies $\partial v/\partial h_i > 0$ as claimed. □

This theorem shows how unlikely it is that optimal lump-sum taxation is feasible. But the constraint that v be a non-increasing function of the h_i still implies that all taxation should be lump-sum in character. It does not, in all circumstances, imply that utility should be the same for everyone. The next result includes one of the cases where equal utility is optimal. We go back to a more general form for u^h.

Theorem 3.2

Let welfare be individualistic, and consumers be characterized by m parameters $h_1,\ldots, h_m$. If it is required that utility be a non-increasing function of the h_i, the optimal budget sets have the form

$$B^h = \{x : p\cdot x \le b^h\}.$$

If $m=1$, and the marginal utility of income at constant prices is a non-decreasing function of h_1, all consumers have the same utility at the optimum.

Proof (illustrating the use of indifference surfaces; and of convexity inequalities)

Let $\xi(x_2,\ldots, x_n, u, h_1,\ldots, h_n)$ be the amount of commodity one required to provide utility u when $x_2,\ldots, x_n$ are trade levels in the other commodities. $\xi=\infty$ if u is unattainable. $\xi=0$ if $x_2,\ldots, x_n$ are already enough to provide more that u. With this notation, the constraints in the optimization problem take the form

$$v(h)=v(h_1,\ldots,h_m) \quad \text{is non-increasing in all arguments,} \tag{3.21}$$

$$x = y + z, \tag{3.22}$$

$$y \text{ maximizes } p\cdot Y, \tag{3.23}$$

$$x_1=\int\xi(x_2(h),\ldots,x_n(h),v(h),h)f(h)\,\mathrm{d}h_1,\ldots,\mathrm{d}h_m, \tag{3.24}$$

$$x_i=\int x_i(h)f(h)\,\mathrm{d}h_1,\ldots,\mathrm{d}h_n, \qquad i=1,\ldots,n. \tag{3.25}$$

We may as well assume that Y admits free disposal, since extra production can be used to increase utility, and therefore welfare, without breaking constraint (3.21). Fixing v and z at their optimal levels, consider x, defined by (3.24) and (3.25), as the functions $x_2(.),\ldots, x_n(h)$ vary. We shall never obtain a point $x-z$ in the interior of Y, because if we did it would be possible to change v in such a way as to increase welfare. Therefore the interior of Y does not intersect the set of points $x-z$ with $x_1 \ge \int \xi f\,\mathrm{d}^n h$, $x_i \ge \int x_i(h) f\,\mathrm{d}^n h$. This latter set is convex, since preferences are convex. Therefore we can separate by a hyperplane yielding prices p.

These prices satisfy (3.23), and we also have

$$p_1\xi(x_2(h),\dots,x_n(h),v(h),h)+\sum_{i=2}^{n}p_ix_i(h)=\max_x, \tag{3.26}$$

for almost all h at the optimum. We may as well satisfy it for all h. (3.26) implies that $\partial\xi/\partial x_i=-p_i/p_1$ $(i=2,\dots,n)$, i.e. that consumers maximize utility subject to budget constraints of the form stated in the theorem.

To prove the second part of the theorem, introduce the expenditure functions $E(p,v(h),h)$, with a single parameter h. Let u^* be the maximum constant utility level consistent with the optimum output levels, and let $v(.)$ be a non-increasing function, which is also consistent with these output levels, consumers always facing prices p. Then

$$\int E(p,u^*,h)f\mathrm{d}^nh=\int E(p,v,h)f\mathrm{d}^nh. \tag{3.27}$$

Since E is a convex function of v,

$$E(p,u^*,h)-E(p,v(h),h)\le E_u(p,u^*,h)(u^*-v(h)). \tag{3.28}$$

Let h_0 be the largest value such that $v(h)\ge u^*$. (If there is none such, u^* yields more welfare than v.) Then, since, by assumption, E_u is a non-increasing function of h,

$$E_u(p,u^*,h)(u^*-v(h))\le E_u(p,u^*,h_0)(u^*-v(h)). \tag{3.29}$$

Combining (3.28) and (3.29), and integrating over h,

$$\int E(p,u^*,h)f\mathrm{d}^nh-\int E(p,v,h)f\mathrm{d}^nh$$

$$\le E_u(p,u^*,h)\left\{\int u^*f\mathrm{d}^nh-\int vf\mathrm{d}^nh\right\}.$$

Since, by (3.27), the left-hand side is zero, $\int u^*f\mathrm{d}^nh\ge\int vf\mathrm{d}^nh$. Therefore, as claimed, the optimum has constant utility. □

This last argument fails with more than one parameter because there may be no h_0 for which (3.29) holds. Conditions can be found that imply constant utility, but it looks as if there are cases where it is not optimal. When it is, the theorem implies that it is better to have constant utility than any budget set that is the same for all h. I do not know whether this is always true.

It is obviously unreal to suppose that a government can get perfect information about individual characteristics even when individuals have nothing to lose by reporting it. We can consider a model in which these characteristics are imperfectly observed by government.[8] For simplicity, suppose the population characterized by a simple parameter, h. An individual seems to government to have characteristic k, but knows he has characteristic h. The distribution of h and k, which is not degenerate, is described by a joint density function $f(h, k)$. With an additive welfare function, and indirect utility function $v(p, b, h)$, welfare in a competitive equilibrium is

$$W = \int\int v(p, b(k), h) f(h, k) \mathrm{d}h \mathrm{d}k. \tag{3.30}$$

Aggregate demand is

$$\int\int x(p, b(k), h) f(h, k) \mathrm{d}h \mathrm{d}k. \tag{3.31}$$

Theorem 3.3

Let v_b be a strictly monotonic function of h (for each p and b); and let there be a commodity, say $i = 1$, for which x_1 is a strictly monotonic function of h. If the frontier of Y is smooth, then no competitive equilibrium with only lump-sum taxation is optimal.

Proof

We first determine optimal lump-sum transfers given that there are no other taxes. The derivative of welfare with respect to $b(k)$ is

$$W_k = \int v_b(p, b(k), h) f(h, k) \mathrm{d}h.$$

The derivative of $p \cdot y$ with respect to $b(k)$ is

$$p \cdot \int x_b(p, b(k), h) f(h, k) \mathrm{d}h = \int f(h, k) \mathrm{d}h.$$

Since Y is smooth at y, transfers b are optimal if and only if W_k is proportional to $\int f(h, k) \mathrm{d}h$. Thus, for some λ,

$$\int v_b(p \cdot b(k), h) f(h, k) \mathrm{d}h = \lambda \int f(h, k) \mathrm{d}h. \tag{3.32}$$

[8] The material on imperfect lump-sum taxation is joint work with Peter Diamond.

With optimal transfers, λ is the change in W made possible (by changing b) if $p \cdot y$ is changed by a unit.

It will now be shown that a change in p_1 (corresponding to commodity taxation of the first commodity), along with appropriate changes in b, can increase welfare. The derivative of W with respect to p_1 is

$$\int\int \frac{\partial v}{\partial p_1} f \mathrm{d}h \,\mathrm{d}k = -\int\int v_b x_1 f \mathrm{d}h \,\mathrm{d}k.$$

The derivative of aggregate demand is $\int\int(\partial/\partial p_1)xf\mathrm{d}h\,\mathrm{d}k$, whose value at prices p is

$$p \cdot \int\int \frac{\partial}{\partial p_1} xf \mathrm{d}h \,\mathrm{d}k = -\int\int x_1 f \mathrm{d}h \,\mathrm{d}k.$$

It will be shown that

$$\int\int v_b x_1 f \mathrm{d}h \,\mathrm{d}k \neq \lambda \int x_1 f \mathrm{d}h \,\mathrm{d}k. \tag{3.33}$$

It follows that it cannot be optimal for p_1 to be the consumer price for commodity one. This will prove the theorem.

To demonstrate (3.33), we use (3.32) to obtain

$$\begin{aligned}&\int\int (v_b - \lambda) x_1 f \mathrm{d}h \,\mathrm{d}k \\ &= \int\int (v_b - \lambda)\{x_1(p, b(k), h) - x_1(p, b(k), h_k)\} f \mathrm{d}h \,\mathrm{d}k.\end{aligned} \tag{3.34}$$

where we can define h_k by $v_b(p, b(k), h_k) = \lambda$. Since $v_b - \lambda$ is strictly monotonic, and so is x_1, for each k, the right-hand side of (3.34) is not zero. This proves (3.33) and completes the proof of the theorem. □

The assumption that the private production set has a smooth frontier merely excludes pathological cases. The general lesson is that imperfect information normally implies that non-lump-sum taxation ought to be used. In the model here, it would usually be desirable to use lump-sum transfers as well. There is one problem with lump-sum taxation based on inaccurate information which is of great practical importance and is hidden by the model, or at least the way it has been handled. Suppose, to fix ideas, that consumer prices are p. One would anticipate that for certain values of h and k there will be no feasible consumption plan satisfying $p \cdot x \leq b(k)$. Men of high ability should pay large taxes: what should be done about men of apparently high ability who are unable to earn much, and how can those be distinguished who simply do not feel like it?

Throughout this section, and throughout subsequent sections, it is assumed that the government is well informed about the population, as a statistical aggregate. The government may be unable to use information about an individual as a basis for applying policy to him, but the construction of policies is based on knowledge of his characteristics. This dichotomy between individual and statistical information cannot be strictly justified. In a small population, any information an individual gives affects his own fate. This leads to the theory of preference revelation,[9] which is however of no value to the student of public policy, since it uses only the uselessly weak criterion of Pareto efficiency. A welfare-theoretic treatment of the issues, using a Bayesian formulation, would be of interest. But for large populations, it seems reasonable to use a model in which there is fixed prior information about the distribution of characteristics in the population. It is unlikely that for most policy issues this will give misleading results.

It will now be assumed that there is no information basis for lump-sum taxation, because we thereby concentrate attention on the central difficulties. Lump-sum taxation is easily introduced into the theory. Something will be said about this later.

4. Producers and efficiency

In the standard general model of competitive equilibrium, consumers are related to producers in two ways, as traders, and as owners receiving pure profits. If there are constant returns to scale in private production, equilibrium profits are zero. We shall make this assumption for the present and return to it below. In the absence of profits, consumers are completely described by their utility functions, consumption sets, and budget constraints. If government has no information allowing it to discriminate among individuals, the budget set B, consisting of those demand vectors that are available to the consumer, is the same for all individuals. For example, if there are commodity taxes proportional to trades and a uniform lump-sum tax (often called a poll tax or subsidy), the budget set is

$$B = \{x : q \cdot x \leq b\}, \tag{4.1}$$

where $q = p + t$. Notice the important point that we can regard q and b as the control variables rather than t and b. In general, B can be taken to be the control variable rather than B as a function of p.

In Sections 6 and 7 we shall analyse cases where the government is not further constrained in its choice of B, which may be defined by linear inequalities as in (4.1), or some more general set. In most of optimal tax theory, B has been

[9] See Groves and Ledyard (1977).

assumed subject to constraint, for example that it be linear, or even more severely constrained, with some commodities untaxed. In the present section, the choice of B is not the focus of interest, but the control of private producers and the choice of government expenditures and production plans. The rules that should govern these choices depend on the extent to which the government is constrained in the control it can apply to consumers. One of the lessons of optimal tax theory that matters most in practice is that optimal production rules are not as much affected by the existence of constraints on consumer taxation, and in particular on lump-sum taxation, as might once have been thought.

Theorem 4.1 (Efficiency Theorem for Linear Taxation)

Let the welfare function be individualistic. If the government is constrained to use linear taxation, i.e. to choose a budget set of the form (4.1), then at the optimum, $y+z$ is in the frontier of the aggregate net production set $Y+Z$. This result is true even if it is possible to subject producers to differential commodity taxation.

Proof (simple topology)
Suppose first that all production is under government control, so that the optimization problem is

$$\left.\begin{array}{l} \text{maximize } W \\ \text{subject to } \sum_h x^h \in Y+Z \\ x^h \text{ maximizes } u^h(x) \text{ for } x \in X^h \\ \text{and } \quad q \cdot x \leq b \end{array}\right\}. \tag{4.2}$$

Under our concavity assumption, the maximizing x^h is a continuous function of q and b. If the solution to the problem is q^*, b^*, no welfare-increasing variation of q and b yields feasible aggregate demands. In particular if $b > b^*$ and q^* remains fixed,

$$\sum x^h(q^*, b) \notin Y+Z.$$

Since $\sum x^h$ is continuous in b, it follows that

$$y^* + z^* = \sum x^h(q^*, b^*) \in \text{ frontier of } Y+Z.$$

This implies that y^* is in the frontier of Y, and, by convexity, that there exists p such that y^* maximizes $p \cdot Y$. Therefore the optimum for problem (4.2) is also the optimum for the more constrained optimizations where production is private and competitive with or without differential taxation. This proves the theorem. □

The proof of the theorem is pretty trivial. The result obviously holds whenever the range of budget sets that can be imposed on consumers by government is sufficiently wide that arbitrarily small expansions of any budget set are possible. In particular, the addition of new tax and control possibilities leaves the conclusion unaffected. The importance of the result is that it implies simple rules for shadow prices. There are shadow prices s for z^* in the frontier of Z if Z is convex and s are support prices at z^*, i.e. z^* maximizes $s \cdot Z$; or if the frontier of Z is smooth at z^*, and s defines a tangent hyperplane at z^*. In either case we have:

Corollary 4.2

Under the assumptions of Theorem 4.1, optimal public net production z^* is in the frontier of Z, and if shadow prices exist, there are shadow prices which are equal to producer prices at the optimum.

The theorem and its corollary imply that, when the assumptions of constant returns, competitive conditions for private production, unconstrained linear taxation, and individualistic welfare, are satisfied, there should be no taxation of intermediate goods, i.e. of trade between producers, and that public and private discount rates for production decisions should be the same.

It is interesting to enquire what happens to the efficiency result when the assumptions of the theorem are relaxed. Individualistic welfare is not an issue: it would be hard to devise interesting welfare assumptions for which the result did not hold. I shall comment on non-constant returns, non-competitive conditions, and tax constraints, in turn.

If private producers do not have constant returns, we can restore constant returns by defining new dummy commodities, a fixed factor for each producer, owned by consumers in the same proportions as they have shares in the firm.[10] In other words the firm is itself regarded as a commodity. Since these fixed factors do not affect utility, utility functions are not strictly concave in terms of all commodities, but supplies are continuous functions, provided we make the usual assumption that consumers are prepared to supply even when the price is zero.[11] Then the theorem remains valid. This means that efficiency holds if the fixed

[10]Avinash Dixit has encouraged me to take this approach.

[11]If a firm that could exist does not, it may be hard for the government to take advantage of its potential existence in setting taxes and subsidies. If it cannot, it is possible to construct examples in which the optimum is inefficient. There are even examples where no optimum exists. See Mirrlees (1972). In that paper, I also discuss briefly the case of what are there called managerial inputs. In the terminology used above, it is assumed impossible to distinguish between the managerial input and the fixed input for tax purposes. In this case efficiency is generally undesirable. Hahn's (1973, p. 104) argument to the contrary is fallacious because it ignores the effect of price changes on the marginal profitability of managerial effort.

factors can be taxed independently, or, equivalently, profit taxes are levied at possibly different rates on different firms.

If all profits have to be taxed at the same proportional rate, the relative value of different shares to the consumer is the same as the relative values of the firms, measured in producer prices. Thus the budget sets that can be imposed on consumers are constrained by the producer prices ruling. (Taxation on transactions between firms can restore the effect of firm-specific profit taxation, but this also violates the uniform treatment of firms.) A similar point might be made about the difficulty of taxing labour income derived from different firms at different rates, although labour for different firms should often be treated as different commodities. The fixed-factor aspect of the issue is really beside the point. In any case, profits can be interpreted as the return to the initial entrepreneur or inventor who set up the firm (and perhaps took his gains by floating the firm as a corporation). Then they are returns to a variable factor, and not particularly different from prices in any other market.

What comes out of this discussion is the importance of the assumption that consumer prices (or equivalently tax rates) can be chosen independently of producer prices. Governments do not act as though this were true. Then the efficiency theorem is not valid – though it may be a good approximation.

Non-competitive behaviour by firms does not change the efficiency theorem, but rather its interpretation, provided that any profits can be taxed as desired. In this case Y should be interpreted not as the production set of private producers but as the set of net supply vectors that can be elicited as producer taxation and other government controls vary. Then shadow prices for government production decisions can be obtained as the tangent hyperplane to the new set Y, and will not in general be simply related to producer prices.

Constraints on the tax powers of government have been much analysed in the literature.[12] We have seen that they may be implied by uniform tax treatment of producers. Many of the constraints dealt with in the literature are introduced without any compelling reason. The non-taxability of certain commodities and the imposition of profit constraints on public producers may be instanced. By and large these constraints are a way of capturing administrative considerations rather than limitations imposed by lack of information. Ideally, a theory of administration and implementation would be developed before considering what are the most relevant and interesting constraints on taxation to model.

Another reason why tax constraints are important is that governments are often prepared to seek advice on public production and expenditure decisions when they are not prepared, in the medium run, to change a tax system whose form they believe to be constrained by its political image, and perceived effect on

[12] Dasgupta and Stiglitz (1971). Guesnerie (1975) deals with non-competitive producer behaviour. The shadow price theorem (Theorem 4.3) comes from Diamond and Mirrlees (1976).

particular groups. The last result of this section gives some information about shadow prices under circumstances where the efficiency theorem does not apply. It takes as premise the optimality of efficiency within the public sector, which is probably valid under very general circumstances, since some policy change would almost always increase welfare if the resources were available, though no theorem on this point seems to be available.

Theorem 4.3

Let policy possibilities be constrained only by producer prices (not quantities). Suppose that for any optimum, z^* is in the frontier of Z. If y° is the production vector for a competitive, constant returns producer in the optimum, there exist shadow prices s for z^* such that

$$s \cdot y^\circ = 0. \tag{4.3}$$

Proof

Let θ be a real number such that $|\theta| < 1$. If the producer who has been singled out produced θy° and the public sector produced $z^* + (1-\theta)y^\circ$, there would be no change in policies and no effective change in equilibrium. Then welfare is unchanged. The producer in question is perfectly willing to produce θy° instead of y°. Thus $z^* + (1-\theta)y^\circ$ would be another optimum for public production if it were feasible. It follows that

$$z^* + (1-\theta)y^\circ \in \text{frontier of } Z, \qquad |\theta| < 1.$$

Therefore there exists a tangent hyperplane at z^* containing all vectors $z^* + (1-\theta)y^\circ$. Let the shadow prices defined by this hyperplane be s. Then $s \cdot y^\circ = 0$, as was claimed. □

This result is of use wherever there are a number of sectors which can be adequately modelled as constant returns competitive sectors. It implies in particular that shadow prices of commodities traded at fixed prices in world markets are proportional to border prices, a result useful in benefit–cost analysis. It must be emphasized that (4.3) is not applicable if in the optimum the constant returns firm should close down: it is not always valid to use y° derived from input–output tables for an existing economy.

5. Linear taxation

As we have seen, there is no loss of generality in assuming that private-sector producers have constant returns to scale. With this assumption, the efficiency theorem (Theorem 3.1) means that the optimal choice of linear taxation is achieved by finding q^* and b^* that maximize $V(q, b)$ subject to $x(q, b) \in Y + Z$, where $x(q, b)$ is the aggregate net demand function of consumers. It must be emphasized that $q \geq 0$ in this optimization. If production sets had smooth frontiers, there would be a unique shadow price vector s associated with $x^* = x(q^*, b^*)$. Since in that case the aggregate production frontier is approximately given by $s \cdot y = s \cdot x^*$ in the neighbourhood of the optimum, we would expect that the derivatives V_q and V_b should be proportional to $s \cdot x_q$ and $s \cdot x_b$ at the optimum, provided the optimum is not on the boundary in price space, i.e. q^* is strictly positive.

To obtain a general theorem yielding these conditions, we need certain regularity conditions. A fairly simple one will be used here: we introduce the following assumption, which says, in a rather strong way, that inefficiency is feasible in the neighbourhood of the optimum:

(I) There exists y° in the relative interior of Y and continuously differentiable functions $q^\circ(\theta)$, $b^\circ(\theta)$ defined for $0 \leq \theta \leq 1$ such that $q^\circ(\theta) \geq 0$ and

$$x(q^\circ(\theta), b^\circ(\theta)) = (1-\theta)x^* + \theta y^\circ. \tag{5.1}$$

Notice that $q^\circ(0) = q^*$, $b^\circ(0) = b^*$. When $q^* \gg 0$ (i.e. $q_i^* > 0$ for all i), (I) is implied simply by the assumption:

(J) The matrix $(x_q(q^*, b^*), x_b(q^*, b^*))$ is of full rank.

(J) implies that all x in a neighbourhood of x^* correspond to some (q, b) with $q \geq 0$; and (I) is therefore trivially satisfied, provided that Y consists of more than a single point. This assumption (J) is a fairly acceptable one, which would be satisfied in almost all cases,[13] but it is insufficient when q^* has zero components. Assumption (I) is by no means the weakest assumption that would work in the following theorem, but it yields a fairly simple proof, and problems not satisfying it are unlikely to arise in practice.

Theorem 5.1

Let V and x be continuously differentiable functions of q and b for $q \geq 0$, and Y a convex set. If q^*, b^* maximize V subject to $x \in Y$, and assumption (I) is satisfied, there exists a non-zero vector s and a scalar λ such that

$$x^* \text{ maximizes } s \cdot Y \tag{5.2}$$

$$V_q(q^*, b^*) \leq \lambda s \cdot x_q(q^*, b^*), \tag{5.3}$$

$$V_b(q^*, b^*) = \lambda s \cdot x_b(q^*, b^*). \tag{5.4}$$

[13] (J) is not satisfied when there are fixed factors, but (I) generally is.

Since V and x are homogeneous of degree zero in q and b,

$$[V_q - s\cdot x_q]\cdot q + [V_b - s\cdot x_b]b = 0.$$

Therefore q^* being non-negative, (5.3) and (5.4) imply that

$$\frac{\partial V}{\partial q_i}(q^*, b^*) = s\cdot\frac{\partial x}{\partial q_i}(q^*, b^*) \quad \text{when} \quad q_i^* > 0. \tag{5.6}$$

Proof

We work in the smallest linear manifold L containing Y. Let C be the cone of non-zero vectors s in L such that x^* maximizes $s\cdot Y$. Since y° is in the interior of Y in L, $s\cdot y^\circ < s\cdot x^*$ for all s in C. Now (5.1) implies, differentiating with respect to θ and setting $\theta = 0$, that

$$x_q(q^*, b^*)q^{\circ\prime}(0) + x_b(q^*, b^*)b^{\circ\prime}(0) = y^\circ - x^*. \tag{5.7}$$

$q_i^\circ(0) \geq 0$ for any i such that $q_i^\circ(0) = q_i^* = 0$. By multiplying $q(\theta), b(\theta)$ by a positive scalar if necessary [which does not change $x(q(\theta), b(\theta))$], we can ensure that $q_i^{\circ\prime}(0) \geq 0$ for all i. Thus (5.7) implies that there exists $a^\circ \geq 0$ and α° such that

$$x_q^*\cdot a^\circ + x_b^*\alpha^\circ = y^\circ - x^*.$$

Since $s\cdot y^\circ < s\cdot x^*$ for all s in C, this implies that

$$s\cdot x_q^*\cdot a^\circ + s\cdot x_b^*\alpha^\circ < 0, \qquad s\in C. \tag{5.8}$$

This inequality will prove to be of crucial importance in the proof.

Consider smooth functions $q(\theta), b(\theta)$ $(0 \leq \theta \leq 1)$ such that $q(0) = q^*, b(0) = b^*, a = q'(0) \geq 0, \alpha = b'(0)$. If

$$V_q(q^*, b^*)\cdot a + V_b(q^*, b^*)\cdot\alpha > 0, \tag{5.9}$$

$V(q(\theta), b(\theta)) > V(q^*, b^*)$ for all small θ. Consequently

$$x(q(\theta), b(\theta)) \notin Y.$$

It follows that, for some $s\in C$,

$$s\cdot x_q(q^*, b^*)\cdot a + s\cdot x_b(q^*, b^*)\alpha \geq 0. \tag{5.10}$$

Thus (5.9) implies (5.10) for some $s \in C$. Equivalently,

$$s \cdot x_q^* \cdot a + s \cdot x_b^* \alpha < 0, \quad \text{all} \quad s \in C, \quad \text{and} \quad a \geq 0, \tag{5.11}$$

implies

$$V_q^* \cdot a + V_b^* \alpha \leq 0. \tag{5.12}$$

Suppose it were only true that

$$s \cdot x_q^* \cdot a + s \cdot x_b^* \alpha \leq 0, \quad \text{all} \quad s \in C, \quad \text{and} \quad a \geq 0. \tag{5.13}$$

Then for any positive number γ, (5.11) is satisfied by $a' = a + \gamma a^{\circ}$ and $\alpha' = \alpha + \gamma\alpha^{\circ}$. This follows from (5.8). Then (5.12) holds for a' and α'. Letting $\gamma \to 0$, we see that (5.12) also holds for a and α.

Since (5.13) implies (5.12), we can apply the duality theorem for convex cones to deduce that the vector (V_q^*, V_b^*) is in the closure of the cone

$$D = \left\{ \left(s \cdot x_q^* - d, s \cdot x_b^* \right) : s \in C, d \geq 0 \right\}.$$

In other words, there exists a scalar λ and $s \in C$ such that

$$V_q^* \leq \lambda s \cdot x_q^*, \qquad V_b^* = \lambda s \cdot x_b^*.$$

The scalar λ must be inserted to allow for the (exceptional) possibility that $\lambda = 0$. □

Most of the literature on optimal commodity taxation is concerned with manipulating and interpreting the first-order conditions of this theorem. Many papers have been written on the case of identical consumers (with identical endowments) with $b = 0$. Since it is hard to see why b must be zero, this case seems to be of little practical interest. In the case of identical consumers, the conditions obtained by using the direct utility function and constraining maximization by the first-order conditions for consumer choice, are of some interest, particularly for additively separable utility,[14] but the indirect utility approach seems to be much more useful for the many-consumer economy.

The chief manipulations used in interpreting (5.3) and (5.4) are the following. If welfare is individualistic,

$$V(q,b) = \Omega\left(v^1(q,b), \ldots, v^H(q,b)\right),$$

and, writing Ω for $\partial\Omega/\partial v^h$,

$$V_q = \sum \Omega_h v_q^h = -\sum \Omega_h v_b^h x^h = -\sum \beta_h x^h, \tag{5.14}$$

[14]Atkinson and Stiglitz (1972).

where

$$\beta_h = \Omega_h v_b^h$$

is often called the "welfare weight," or "marginal social utility of income". (5.14) says that $-V_q$ is a weighted sum of demands. One also finds that

$$V_b = \sum \beta_h. \tag{5.15}$$

Thus $-V_q/V_b$ is a weighted average of demands, and this interpretation encourages one to divide (5.3) by (5.4).

The right-hand sides of (5.3) and (5.4) can be written, interpreting $q-s=t$ as tax rates

$$\begin{aligned} s\cdot x_q &= -(q-s)\cdot x_q - x \\ &= -\frac{\partial}{\partial t}[t\cdot x(s+t,b)-b], \end{aligned} \tag{5.16}$$

$$\begin{aligned} s\cdot x_b &= -(q-s)\cdot x_b + 1 \\ &= -\frac{\partial}{\partial b}[t\cdot x(s+t,b)-b], \end{aligned} \tag{5.17}$$

Writing

$$T(t,b,s) = t\cdot x(s+t,b)-b, \tag{5.18}$$

for the net revenues of government, (5.16) and (5.17) can be written

$$s\cdot x_q = -T_t, \qquad s\cdot x_b = -T_b,$$

and the first-order conditions (5.3) and (5.4) become

$$\sum \beta_h x^h \geq \lambda T_t, \tag{5.19}$$

$$\sum \beta_h = -\lambda T_b. \tag{5.20}$$

Assuming $q \gg 0$, $\lambda > 0$ for emphasis, and dividing (5.19) by (5.20),

$$\sum \beta_h x^h \Big/ \sum \beta_h = (\partial b/\partial t)_{T\,\text{constant}}. \tag{5.21}$$

In words, the welfare-weighted average of demands should be equal to the constant-revenue effect of tax-rate changes on the general subsidy b.

Another manipulation should be mentioned, though it may have been overrated. Writing x^{ch} for the compensated demand functions, we have

$$\begin{aligned} s\cdot x_q^h &= -(q-s)\cdot x_q^h - x^h \\ &= -t\cdot x_q^{ch} + t\cdot x_b^h x^h - x^h \\ &= -x_q^{ch}\cdot t - (1 - t\cdot x_b^h)x^h, \end{aligned} \tag{5.22}$$

by Slutsky symmetry. Now $x_q^{ch}\cdot t$ is, to a first-order approximation, the changes in demands brought about by the introduction of taxes, provided income effects are ignored. One can also interpret $x_q^{ch}\cdot t = [(\partial/\partial\theta)x^{ch}(s+\theta t, b)]_{\theta=1}$ as showing the effects on compensated demand of intensification of the tax system. Thus (5.3) implies that

$$\sum_h \{\beta_h - \lambda(1 - t\cdot x_b^h)\}x^h \geq \sum_h x_q^{ch}\cdot t. \tag{5.23}$$

The welfare weights on demands are here modified to take account of the revenue effects of changes in the consumer's lump-sum income. (5.4) implies that

$$\sum_h \{\beta_h - \lambda(1 - t\cdot x_b^h)\} = 0. \tag{5.24}$$

It follows from (5.24) that the left-hand side of (5.23) is the covariance of x^h, and the adjusted weights (called the *social marginal utility of income* by Diamond)

$$\gamma_h = \beta_h - \lambda(1 - t\cdot x_b^h). \tag{5.25}$$

Among the problems in this area that seem to be of theoretical interest, mention should be made of separability questions, as to the conditions under which some commodities should be untaxed, or groups of commodities taxed at the same rates. In this connection, it is important to notice that in the model there are always many equivalent tax systems. If q^*, b^* and s are optimal consumer prices and subsidy, and shadow prices, it is optimal to set producer prices

$$p = \mu s,$$

and tax rates

$$t = \nu q^* - \mu s,$$

while paying a general uniform subsidy

$$b = \nu b^*.$$

This tax system is optimal for any positive μ and ν. In general, any commodity can be made an untaxed commodity by suitable choice of μ and ν. If the natural interpretation of a problem, e.g. untaxed fixed factors representing the absence of profit taxation, imposes part of the normalization, the tax system can no longer be chosen so freely. This point has sometimes led to confusion and error.

It is also interesting to enquire how the optimal tax rules are altered when there are constraints on the choice of linear tax systems, for example when certain goods cannot be taxed. In such problems, the private producers may, and usually should, face prices that are not proportional to shadow prices s, and it is useful to speak of consumer taxes $q - s$ and producer taxes $p - s$, although the constraint may take the form of requiring that they be equal for certain commodities.[15]

In the model discussed, there has been no dependence of consumer utilities on public expenditures, that is, no role for what are called public goods. If such expenditures are the sole responsibility of government, and their provision is not associated with new controls on consumers, they are easily accommodated in the model. We simply write $V(q, b, g), x(q, b, g)$, where g is public consumption expenditure. The same methods as were used to establish the first-order conditions for optimal taxation prove that it is necessary for optimality that

$$V_g = \lambda(s \cdot x_g + s). \tag{5.26}$$

If welfare is individualistic, this can be rewritten as before,

$$\sum \beta_h m^h = \lambda(-t \cdot x_g + s),$$

where $m^h = -(\partial b^h / \partial g)_{u^h \text{ constant}}$ is the marginal value of the public expenditures at constant q. Thus at the optimum

$$s = \frac{1}{\lambda} \sum \beta_h m^h + t \cdot x_g. \tag{5.27}$$

The revenue effect could in practice by very important. A revenue gain arising from provision of the good strengthens the case for it.

[15] Dasgupta and Stiglitz (1971). This work is clarified, and to some extent corrected, by Munk (1977).

6. Nonlinear taxation in a one-dimensional population

So long as the government is constrained to choose linear tax systems, consumers, provided they have convex preferences, have well-defined consumption choices, so that the maximization constraint defines a nice set. If there is no constraint on the tax system, other than independence of individual information, it may be desirable to impose a budget set which leaves some consumers indifferent among widely different consumption plans. For a finite population we intuitively expect that this will be optimal. The most able consumer need be no better off than if he did the same as the next most able consumer, but in general the government would want him to do something different, i.e. choose a different point on the same indifference surface.

The case of a large finite population seems unlikely to be of much interest, because computation would be extremely demanding. Accordingly, we go to the continuum case, where under some circumstances it is to be expected that the optimum budget set can be defined by nice functions. The population is described by a non-negative scalar parameter h with density function f. The allocations that can be brought about by government policy are given by

$$x(h) \text{ maximizes } u(x,h) \quad \text{for} \quad x \in X^h \cap B, \tag{6.1}$$

for some set B. The first task is to find more manageable control variables than the set B. One way of doing this would be to single out a numeraire good and express B by the inequality

$$x_1 \leq c(x_2,\dots,x_n). \tag{6.2}$$

This approach turns out to be extremely complicated, and an alternative must be devised. The difficulty with using the function c in (6.2) as the control variable seems to be that variations in it can have complicated effects on the variables of the problem.

An approach that is manageable is to define the function

$$v(h) = \max\{u(x,h): x \in X^h \cap B\}, \tag{6.3}$$

and use an "envelope theorem" for it. If the maximizing x is a differentiable function of h, and $x(h)$ is always in the interior of X^h,

$$v(h_1) \geq u(x(h_2),h_1), \tag{6.4}$$

at least for h_2 near h_1. This is because, B being independent of h, $x(h_2)$ is available to a consumer of type h_1 if he wants it. (6.4) implies that, as h_1 varies, $v(h_1) - u(x(h_2),h_1)$ attains a local minimum (which happens to be zero) when

$h_1 = h_2$. It follows that

$$v'(h) = u_h(x(h), h). \tag{6.5}$$

If (6.5) were equivalent to (6.1) for some B, we should have reduced our maximization constraint to a simple differential equation, which ought not to be too difficult to handle; and is in any case the kind of constraint met with in control theory.

The argument leading to (6.5) leaned heavily on the unwarranted assumption that $x(h)$, and consequently $v(h)$, is a differentiable function of h. There were also some loose ends about the consumption sets. A precise lemma is needed. Before stating it, some standing assumptions about utility functions and consumption sets are introduced. These lay down some standard properties, and insist that as h increases the consumption set expands in a very regular way.

(C_1) u is a continuously differentiable function of x and h, concave in x.

(C_2) X^h is a convex set; and for all $h, k,\ k > h$, the closure of X^h is contained in X^k.

(C_3) For all x in X^h there exists $\varepsilon > 0$ such that $x \in X^k$ when $|k - h| < \varepsilon$.

(B) $X^h \cap \{x : u(x,k) \le u(x^\circ, k)\}$ is bounded if $h < k$, and $x^\circ \in X^k$.

The first assumption requires no comment, nor does the first part of (C_2). The second part says that X^h is a non-decreasing function of h and actually increases along any "open" part of its boundary. (C_3) requires that X^h vary continuously with h and that "closed" parts of the boundary remain fixed. The last assumption is a little weaker than the requirement that indifference hypersurfaces be bounded. It allows the possibility that the indifference hypersurface $u(x,k) = u(x^\circ, k)$ is asymptotic to an "open" part of the consumption frontier, but only if that part of the frontier is moving outwards, even at infinity.

The assumptions are satisfied, for example, by a function u satisfying (C_1) with $X^h = \{-h < x_1 \le 0,\ x_i \ge 0,\ i = 2,\dots,n\}$, and all indifference surfaces cutting the co-ordinate planes $x_i = 0$ when $i = 2,\dots,n$. (Think of commodity 1 as labour.) In effect, bigger h is now taken to mean greater ability, unlike the special cases in Section 3 where it was convenient to use the opposite convention.

Assumption (B) is unduly strong, but it is hard to see how to prove the result we want without something like it.

Lemma 6.1

Let the above assumptions hold. If there exists B such that for all $h, x(h)$ maximizes $u(x,h)$ for $x \in X^h \cap B$, and $v(h) = u(x(h), h)$,

$$v(h) - v(0) = \int_0^h u_h(x(k), k)\,\mathrm{d}k. \tag{6.6}$$

Proof

Let $\eta > 0$. It will first be shown that the set

$$A = \{(x(k), k') : 0 \leq k, k' \leq h + \eta\}$$

is bounded. Let $h_1 > h + \eta$. Since for all k, $x(k) \in B$, and $x(k) \in X^{h_1}$ for $k \leq h + \eta$,

$$u(x(k), h_1) \leq u(x(h_1), h_1), \qquad k \leq h + \eta.$$

Therefore

$$x(k) \in X^{h+\eta} \cap \{x : u(x, h_1) \leq u(x(h_1), h_1)\},$$

and is bounded, by assumption (B). Thus the set A is bounded. It follows that the partial derivative $u_h(x(k), k')$ is bounded in A, and thence, by the mean value theorem, that

$$\alpha_\varepsilon(k) = \tfrac{1}{3}\{u(x(k), k + \varepsilon) - u(x(k), k)\}$$

is bounded for $0 \leq k \leq h$, $|\varepsilon| < \eta$, $k + \varepsilon \geq 0$.

Since $\alpha_\varepsilon(k) \to u_h(x(k), k)$ as $\varepsilon \to 0$, Lebesgue's theorem on bounded convergence implies that

$$\lim_{\varepsilon \to 0} \int_\eta^h \alpha_\varepsilon(k)\,\mathrm{d}k = \int_\eta^h u_h(x(k), k)\,\mathrm{d}k. \tag{6.7}$$

Now

$$\begin{aligned}
\varepsilon \int_\eta^h \alpha_\varepsilon(k)\,\mathrm{d}k &= \int_\eta^h \{u(x(k), k + \varepsilon) - u(x(k), k)\}\,\mathrm{d}k \\
&\leq \int_\eta^h \{v(k + \varepsilon) - v(k)\}\,\mathrm{d}k \\
&= \int_0^\varepsilon \{v(h + x) - v(\eta + x)\}\,\mathrm{d}x.
\end{aligned}$$

Therefore

$$\begin{aligned}
&\lim_{\varepsilon \to 0+} \frac{1}{\varepsilon} \int_0^\varepsilon \{v(h + x) - v(\eta + x)\}\,\mathrm{d}x \\
&\geq \lim_{\varepsilon \to 0} \int_\eta^h \alpha_\varepsilon(k)\,\mathrm{d}k \\
&\geq \lim_{\varepsilon \to 0-} \frac{1}{\varepsilon} \int_0^\varepsilon \{v(h + x) - v(\eta + x)\}\,\mathrm{d}x.
\end{aligned}$$

The left-hand and right-hand limits exist and are both equal to $v(h)-v(\eta)$, since v is a continuous function. Therefore, from (6.7), we have

$$v(h)-v(\eta)=\int_{\eta}^{h}u_h(x(k),k)\,\mathrm{d}k.$$

Finally we let $\eta\to 0$, and the lemma is proved. □

The strategy that will now be followed is to use the lemma to prove that certain conditions are *sufficient* for optimality. Naturally this can be proved only under rather strong assumptions on the utility function; but, since sufficiency theorems are of the first value in doing computations, the restrictions are worth their cost. To motivate the sufficiency conditions, I shall first derive them in a rather heuristic way.

We saw in Section 3 that, under plausible assumptions, the first-best optimum requires that utility decrease with ability. This suggests that the constraint (6.6) which (partially) expresses the constraint that B be uniform works as an inequality preventing $v(h)$ from being too low in relation to $v(0)$,

$$v(h)-v(0)-\int_{0}^{h}u_h(x(k),k)\,\mathrm{d}k\geq 0. \tag{6.8}$$

In this form it is a linear constraint in v. If we are to apply the ideas of programming theory to obtain sufficient conditions, the left-hand side of the inequality should be a concave function of the control variables. This suggests that we treat $v(.)$ as one of the control variables, and eliminate one of the commodities. Specifically, let us treat commodity one as numeraire, denoting it by ξ, and write x' for the vector of commodities 2 to n. Then ξ is defined as a function of x', v and h by

$$v=u(\xi,x',h). \tag{6.9}$$

It is readily shown that (C_1) implies that ξ is a *convex* function of x' and v, and a differentiable function of all the variables.

With this transformation, v and x' are to be regarded as the control variables. The assumption that will let the sufficiency theorem go through is

(CON′) $u_h(\xi(x',v,h),x',h)$ is a convex function of x and v.

As it stands this is not in satisfactory form. It is equivalent to:

(CON) For any vector a, $(\partial/\partial h)(a\cdot u_{xx}(x,h)\cdot a/u_{\xi}(x,h))\geq 0$.

In words, this states that the degree of concavity of u (which is measured by $-a\cdot u_{xx}\cdot a$) does not increase, relative to the marginal utility of numeraire, when

h increases. The condition is numeraire dependent. To have the best chance of applying the sufficiency theorem successfully, one should choose as numeraire a commodity such that $u_{\xi h}/u_{\xi}$ is as large as possible, i.e. the commodity for which $\partial(u_{x'}/u_{\xi})/\partial h \leq 0$.

To prove that (CON′) and (CON) are equivalent, one makes a routine change of variables. Writing $w=(v, x')$, $x=(\xi, x')$ and $\psi(w,h)=u_h(x,h)$, we have $u_{hxx}=w_x \cdot \psi_{ww} \cdot w_x + \psi_w \cdot w_{xx}$ (subscripts denoting differentiation). It is easily seen that $\psi_w \cdot w_{xx}=(u_{h\xi}/u_{\xi})u_{xx}$. Thus ψ_{ww} is positive semi-definite if and only if

$$u_{hxx}-(u_{h\xi}/u_{\xi})u_{xx}=u_{\xi}\frac{\partial}{\partial h}(u_{xx}/u_{\xi})$$

is positive semi-definite. The equivalence of (CON) and (CON′) follows at once.

Assume an additive welfare function $\int vf\,\mathrm{d}h$, and consider the problem

$$\left.\begin{array}{l}\text{maximize } \int vf\,\mathrm{d}h \\ \text{subject to (6.8) and } \left(\int \xi(x',v,h)f\,\mathrm{d}h, \int x'f\,\mathrm{d}h\right) \in Y\end{array}\right\}. \tag{6.10}$$

Following our work on the linear problem, it should be legitimate to replace the production constraint (6.10) by

$$\int\{\xi(x',v,h)+s'\cdot x'\}f\,\mathrm{d}h \leq \alpha, \tag{6.11}$$

where the shadow price of numeraire has been set at unity, and s' are the shadow prices of the other commodities.

If Lagrange's method of undetermined multipliers is applicable, we can find conditions for optimality by setting equal to zero the derivatives of the Lagrangian

$$L=\int vf\,\mathrm{d}h-\lambda\int\{\xi+s'\cdot x'\}f\,\mathrm{d}h+\int\mu(h)\left\{v(h)-v(0)-\int_0^h u_h\,\mathrm{d}k\right\}\mathrm{d}h,$$

where λ should be positive. The sign of $\mu(h)$ will be considered later. If we reverse the order of integration in the double integral, we obtain

$$L=\int_0^{\infty}\left\{(v-\lambda\xi-\lambda s'\cdot x')f+\mu v-\mu v(0)-\int_h^{\infty}\mu(k)\,\mathrm{d}k\cdot u_h\right\}\mathrm{d}h. \tag{6.12}$$

On differentiating with respect to $x'(h)$ we have

$$\lambda(\xi_x+s')f+\int_h^{\infty}\mu\,\mathrm{d}k(u_{h\xi}\xi_{x'}+u_{hx'})=0, \tag{6.13}$$

provided that $\xi(h), x'(h)$ is in the interior of X^h. If it is on the boundary we have an inequality (e.g. for people who do not choose to work). Differentiation with respect to $v(h)$ yields

$$(1-\lambda\xi_v)f+\mu-\int_h^{\infty}\mu\,\mathrm{d}k\cdot u_{h\xi}\xi_v=0, \tag{6.14}$$

and differentiation with respect to $v(0)$,

$$\int_0^{\infty}\mu\,\mathrm{d}h=0. \tag{6.15}$$

Consider the sign of μ. In the light of (6.15), we cannot want to have $\mu \geq 0$. But we see from (6.12) that L is a concave function of the control variables provided that

$$M(h)=\int_h^{\infty}\mu\,\mathrm{d}k \geq 0, \tag{6.16}$$

for all h. This completes the heuristic derivation of first-order conditions, except for some suggestive simplifications. We note that

$$\xi_v = 1/u_{\xi}, \tag{6.17}$$

$$\xi_{x'} = -u_{x'}/u_{\xi}, \tag{6.18}$$

which suggests we define the marginal rates of substitution, or marginal consumer prices as

$$q=q(\xi,x',h)=-\xi_{x'}=u_{x'}/u_{\xi}. \tag{6.19}$$

Also

$$u_{h\xi}\xi_{x'}+u_{hx'}=u_{\xi}\frac{\partial}{\partial h}(u_{x'}/u_{\xi})=u_{\xi}q_h. \tag{6.20}$$

These formulas are used to obtain the conditions in the sufficiency theorem.

Theorem 6.2

Assume (C_1), (C_2), (C_3), (B), and (CON). Let the allocation $\xi^*(.)$, $x'^*(.)$, and s', ν and $\mu(.)$ satisfy the following conditions:

$$(\xi^*(h), x'^*(h)) \in X^h, \quad \text{for all } h. \tag{6.21}$$

For all h, k, such that $(\xi^*(k), x'^*(k)) \in X^k$,

$$u(\xi^*(k), x'^*(k), h) \leq u(\xi^*(h), x'^*(h), h), \tag{6.22}$$

$$\left(\int \xi^* f \mathrm{d}h, \int x'^* f \mathrm{d}h\right) \text{ maximizes } (1, s')\cdot Y, \tag{6.23}$$

$$\{q(\xi^*(h), x'^*(h), h) - s'\} f(h) = u_\xi^* q_h^* \int_h^\infty \mu \,\mathrm{d}k, \tag{6.24}$$

(for consumers in the interior of X^h, and an appropriate boundary condition in other cases),

$$\mu(h) - \frac{u_{h\xi}^*}{u_\xi^*}\int_h^\infty \mu \,\mathrm{d}k = \left(\frac{1}{u_\xi^*} - \nu\right) f, \tag{6.25}$$

$$\nu > 0, \quad \int_h^\infty \mu \,\mathrm{d}k \geq 0, \quad \text{all } h, \tag{6.26}$$

$$\int_0^\infty \mu \,\mathrm{d}k = 0. \tag{6.27}$$

Then the given allocation is an optimum.

In this statement, ν is $1/\lambda$ and μ replaces μ/λ in (6.13)–(6.16).

Proof

The argument is a routine calculation based on the assumed concavity properties. We consider an alternative allocation satisfying the constraints of the problem, i.e.

$$\xi(h), x'(h) \text{ maximize } u(\xi, x', h),$$

$$\text{subject to} \quad (\xi, x') \in X^h \cap B, \tag{6.28}$$

$$\left(\int \xi f \mathrm{d}h, \int x' f \mathrm{d}h\right) \in Y, \tag{6.29}$$

and show that ξ^*, x'^* provides utility at least as great. [It is a feasible allocation by (6.21), (6.22), and (6.25).]

Lemma 6.1 implies that

$$v(h) - v(0) - \int_0^h u_h(\xi(k), x'(k), k)\,\mathrm{d}k = 0, \tag{6.30}$$

$$v^*(h) - v^*(0) - \int_0^h u_h(\xi^*(k), x'^*(k), k)\,\mathrm{d}k = 0. \tag{6.31}$$

(6.30) follows from (6.28), and (6.31) from (6.22) [where the set B^* consists simply of all $\xi^*(h), x'^*(h)$]. Subtracting (6.31) from (6.30), multiplying by $\mu(h)$ and integrating from 0 to ∞, we get

$$\int_0^\infty \left\{\mu(v-v^*)-\mu\int_0^h (u_h-u_h^*)\,\mathrm{d}k\right\}\mathrm{d}h=\{v(0)-v^*(0)\}\int\mu\,\mathrm{d}k$$
$$=0 \quad \text{by (111)}.$$

Reversing the order of integration, we deduce that

$$\int\mu(v-v^*)\,\mathrm{d}h=\int_0^\infty\int_h^\infty \mu\,\mathrm{d}k(u_h-u_h^*)\,\mathrm{d}h$$
$$\geq\int_0^\infty\int_h^\infty \mu\,\mathrm{d}k\left\{\frac{u_{h\xi}^*}{u_\xi^*}(v-v^*)+u_\xi^* q_h^*(x'-x'^*)\right\}\mathrm{d}h, \tag{6.32}$$

by using (6.26) and (CON), and using our earlier calculations for the partial derivatives of u_h with respect to v and x'.

Combining (6.32) with conditions (6.24) and (6.25), we obtain

$$\int\left(\frac{1}{u_\xi^*}-\nu\right)(v-v^*)f\,\mathrm{d}h\geq\int(q^*-s')\cdot(x'-x'^*)f\,\mathrm{d}h$$
$$\geq\int q^*\cdot(x'-x'^*)f\,\mathrm{d}h+\int(\xi-\xi^*)f\,\mathrm{d}h, \tag{6.33}$$

by (6.23). Since ξ is a convex function of v and x',

$$\xi-\xi^*\geq\xi_v^*(v-v^*)+\xi_{x'}^*\cdot(x'-x'^*)$$
$$=\frac{1}{u_\xi^*}(v-v^*)-q^*\cdot(x'-x'^*).$$

Combining this with (6.33), we have finally

$$-\nu\int(v-v^*)f\,\mathrm{d}h\geq 0. \tag{6.34}$$

Since $\nu>0$, this implies that $\int v^*f\,\mathrm{d}h\geq\int vf\,\mathrm{d}h$. □

The two problems with this theorem are, first, that (CON), expressing decreasing concavity of u, is a little obscure though not implausible; and, second, that even when (CON) is satisfied, there may not exist any allocation satisfying the conditions of the theorem. As to the first problem, it is useful to note certain special cases where (CON) holds. If u has the form

$$u = u_1(x', h) + u_2(\xi),$$

convexity of u_{1h} with respect to x' is equivalent to (CON), and it is readily checked whether or not this holds. If u has the form

$$u = u_1(x') + u_2(\xi, h),$$

it is sufficient for (CON) that u_{2h} be an increasing convex function of ξ (since ξ is itself convex in x' and v).

In this context it is also interesting to note that the theorem can be generalized by assuming a welfare function

$$W = \int G(v) f \mathrm{d}h,$$

with G concave, increasing; i.e. by taking a monotone transform of utility before using Lemma 6.1. The only change in the theorem is that ν is replaced by $\nu G'(v^*(h))$. By this transformation to a new utility function u_h may sometimes be made convex when it would not otherwise have been.

The second problem, that it may be impossible to satisfy the conditions of the theorem, arises because there are allocations satisfying (6.6) that are not utility-maximizing allocations. One would expect to be able to satisfy the conditions if (6.6) replaced the stronger condition (6.21), but that may not be what one wants.

To check whether or not a particular allocation $x(h)$ as h varies maximizes utility for some constant budget set B, the following partial converse to Lemma 6.1 is useful:

Lemma 6.3

Suppose that for all h,

$$x(h) \in X^h,$$

$$v(h) = u(x(h), h),$$

$$v(h) - v(0) = \int_0^h u_h(x(k), k) \mathrm{d}k,$$

$$u_h(x(k), h) \text{ is a non-decreasing function of } k, \tag{6.35}$$

for k such that $x(k) \in X^h$. Then there exists B such that, for all h,

$$x(h) \text{ maximizes } u(x,h) \quad \text{for} \quad x \in B \cap X^h.$$

Proof

It is sufficient to show that for all h, h_0 such that $x(h_0) \in X^h$, $u(x(h),h) \geq u(x(h_0),h)$. Since $u_h(x(k),h)$ is non-decreasing in k, we have

$$\begin{aligned} u(x(h),h) - v(h_0) &= \int_{h_0}^{h} u_h(x(k),k)\,\mathrm{d}k \geq \int_{h_0}^{h} u_h(x(h_0),k)\,\mathrm{d}k \\ &= u(x(h_0),h) - u(x(h_0),h_0), \end{aligned}$$

proving the lemma. □

When x is differentiable, a routine calculation shows that (6.35) is equivalent to

$$q_h'(x(k),h) \cdot \frac{\mathrm{d}}{\mathrm{d}k} x'(k) \geq 0.$$

It is interesting to compare this with a form of the second-order necessary condition for maximization (also easily proved),

$$q_h'(x(h),h) \cdot \frac{\mathrm{d}}{\mathrm{d}h} x'(h) \geq 0.$$

In the two-commodity case, and particularly in the simple optimal income-tax problem, x' is a scalar. Suppose that h can be measured in such a way that $\partial(u_{x'}/u_{\xi})/\partial h < 0$. Then both necessary and sufficient supplements to the envelope condition (6.6) have the simple form that x' be a non-increasing function of h, and, equivalently, that ξ be a non-decreasing function of h. In general the class of allocations consistent with the maximization constraint cannot be so easily identified.

Suppose now that an attempt to apply the sufficiency theorem fails because we cannot find a solution satisfying (6.22). Then it must be realized that we should not have neglected the other constraints on maximizing allocations [besides the condition (6.6) of Lemma 6.1]. It must also be the case that $v(h)$ does not become smaller than $u(x(h_1),h)$ as h increases from h_1. It might be optimal to have $v(h)$ just remaining equal to $u(x(h_1),h)$ over some interval $[h_1,h_2]$. Then we must allow for the additional constraint $v(h) \geq u(x(h_1),h)$ in our maximization problem. This introduces a new term $\int_{h_1}^{h_2} \rho(h)\{v(h) - (\xi(h_1), x'(h_1), h)\}\,\mathrm{d}h$ into the Lagrangian, with $\rho(h) \geq 0$. If $[h_1,h_2]$ is the whole interval on which the additional constraint binds, we see at once that, since $\xi(h_1)$, $x'(h_1)$ occurs predominantly in the new term of the Lagrangian,

$$\int_{h_1}^{h_2} \rho(h) u_\xi(x(h_1), h)\{q(x(h_1), h_1) - q(x(h_1), h)\}\,\mathrm{d}h = 0. \tag{6.36}$$

It is further found that condition (6.24) is unchanged, while condition (6.25) becomes

$$\rho(h) + \mu(h) - \frac{u^*_{h\xi}}{u^*_\xi} \int_h^\infty \mu\,\mathrm{d}k = \left(\frac{1}{u^*_\xi} - \nu\right) f. \tag{6.37}$$

One can prove in the same way as before that, if the conditions of the theorem hold on intervals where $v(h) > u(x(h_1), h)$ and the modified conditions [(6.37) replacing (6.25), and (6.36) added] hold on intervals where $v(h) = u(x(h_1), h)$, an optimum has been found.

In the two-commodity case, $v(h) = u(x(h_1), h)$ and $v'(h) = u_h(x(h), h)$ generally imply that $x(h)$ is constant. Thus these awkward intervals correspond to bunching of consumers, many of whom choose the same demands. In the many commodity case, this need no longer be so.

Returning to the conditions of the theorem, we see that (6.24) strongly suggests that $x(h)$ is a continuous function of h. This seems to be correct under assumption (CON). It appears that discontinuities occur only when assumption (CON) is violated. When it is violated, we can no longer hope to use sufficient conditions for an optimum, but must make do with necessary conditions. For these we can rely on Pontrjagin's Maximum Principle, suitably generalized to take account of possible discontinuities.[16] The conditions given in the theorem are then necessary conditions for an optimum.

The conditions for optimal non-linear taxation are interesting in a number of ways. Condition (6.24) is the most striking, for it not only shows that the effective marginal tax rates on consumer h have the signs of $\partial q/\partial h$, but also gives a simple formula relating different marginal tax rates,

$$\frac{q_i - s_i'}{q_j - s_j'} = \frac{q_{ih}}{q_{jh}}. \tag{6.38}$$

The general principle is that the proportional marginal tax rate $(q_i - s_i')/q_i$, or equivalently $(q_i - s_i')/s_i'$, should be higher for commodity i than for commodity j if and only if $(\partial/\partial h)(u_i/u_j) > 0$, i.e. when the marginal rate of substitution

[16] Relevant results and methods can be found in Swinnerton–Dyer (1959).

would be increased by an increase in h. This suggests a theorem of Atkinson and Stiglitz, whose formal proof is omitted here.

Theorem 6.4

If utility takes the form

$$u(x,h)=U(u_1(x'),u_2(\xi,h)),$$

the optimal allocation can be obtained by imposing a budget constraint of the form

$$p\cdot x'\leq c(\xi). \tag{6.39}$$

It is interesting to note that the analysis of this section can also be done in a fully dual way,[17] treating marginal prices q and utility as control variables. We can think of offering consumers a set of linear budget constraints C instead of a set of demand vectors. Writing $E(q,u,h)$ for the expenditure function, and $v(q,b,h)$ for the indirect utility function, we can set up the problem as maximization of $\int u(h)f(h)\mathrm{d}h$ subject to

$$\left.\begin{array}{l}\int E_q(q(h),u(h),h)f(h)\mathrm{d}h\in Y\\ q(h),\ E(q(h),u(h),h)\text{ maximizes } v(q,b,h)\\ \text{subject to}\quad (q,b)\in C\cap Q^h\end{array}\right\}, \tag{6.40}$$

where Q^h is the set of linear budget constraints that are consumption-feasible.

The entire previous analysis can be applied to (6.40), and we obtain as first-order conditions

$$\mu(h)-\frac{v^*_{hb}}{v^*_b}\int_h^{\infty}\mu\,\mathrm{d}k=\left(\frac{1}{v^*_b}-\nu\right)f, \tag{6.41}$$

$$-s\cdot x^{\mathrm{c}}_q f=v^*_b x_h\int_h^{\infty}\mu\,\mathrm{d}k, \tag{6.42}$$

where x_h is the derivative of demands holding q and b constant. In fact (6.41) is exactly the same equation as (6.25). (6.42) does not look the same as (6.24), but the two can easily be shown to be equivalent, by using the equation

$$x^{\mathrm{c}}_q\cdot q_h=-x_h,$$

[17] This approach is due to Kevin Roberts.

which can be obtained by differentiating the equations $x^{c}(q(x,h),u(x,h),h)=x$ and $x^{c}(q,v(q,b,h),h)=x(q,b,h)$ with respect to h.

Equation (6.42) has an interesting similarity to the first-order conditions for optimal linear taxation, for they can be expressed [cf. (5.22)] in the form

$$-\int s\cdot x_q^{c} f\,\mathrm{d}h=\frac{1}{\lambda}\int (v_b-s\cdot x_b)xf\,\mathrm{d}h. \tag{6.43}$$

If we write (6.42) in the form, obtained by using $q\cdot x_q^{c}=0$ and Slutsky symmetry,

$$x_q^{c}\cdot(q-s)=\left(v_b^{*}\int_h^{\infty}\mu\,\mathrm{d}k\right)x_h/f(h), \tag{6.44}$$

it says that the approximate compensated effect on consumer h's demands of imposing the optimum tax system is proportional to the derivative of demands with respect to the population percentile.

It is worth emphasizing that this dual approach to the problem provides a technique that allows us to apply non-linear taxation to some groups of commodities while applying only proportional taxes to others, for it is very easy to insist that some q_i be independent of h.[18]

Finally, to mention the obvious, (6.25) would in practice be treated as a differential equation in $\int_h^{\infty}\mu\,\mathrm{d}k$. It is written in the form above to allow for the possibility that μ is discontinuous, and that happens only where x is discontinuous.

7. *m*-dimensional populations

Although the one-dimensional population is an extremely useful model for computations and examination of particular issues, it is not, in that respect, an accurate representation of reality. Theorem 6.2 can be generalized to the m-dimensional case, with m parameters $h_1,\dots,h_m$ ranging over the non-negative orthant. The function $\mu(h)$ becomes an m-dimensional vector field, and the main equations of the theorem become

$$(q-s')f=u_{\xi}s_h'\cdot M=u_{\xi}\sum_1^m\frac{\partial s'}{\partial j_j}M_j,$$

where

$$M_j=\int_{h_j}^{\infty}\mu_j(h_1,\dots,h_{j-1},k,h_{j+1},\dots,h_n)\,\mathrm{d}k, \tag{7.1}$$

[18] The problem was solved by a different method in Mirrlees (1976).

and

$$\sum_{1}^{m}\mu_j - \frac{1}{u_\xi}u_{\xi h}\cdot M = \left(\frac{1}{u_\xi} - \nu\right)f, \tag{7.2}$$

(ignoring corner solutions). In the case where μ (and x) vary continuously, (7.2) can be written

$$\Delta\cdot M - \frac{1}{u_\xi}u_{\xi h}\cdot M = \left(\frac{1}{u_\xi} - \nu\right)f,$$

where $\Delta\cdot M = \sum\partial M_j/\partial h_j$ is the divergence of M. The boundary conditions in terms of M (which should be non-negative for the sufficiency theorem to go through) are

$$M_j = 0 \quad \text{when} \quad h_j = 0,$$
$$M_j \to 0 \quad \text{as} \quad h_j \to \infty.$$

The above equations will not be derived here. Lemma 6.1 is easily generalized to the m-dimensional case, carrying with it the important fact that u_h is an integrable vector field. Then the equation $v(h) - v(0) = \int_0^h u_h\cdot \mathrm{d}k$ is brought in as a constraint in m different ways, following m different rectangular paths of integration, to enforce integrability on the solution to the optimization problem. The m Lagrangian functions μ_i correspond to these m constraints.

To find an optimum, we would look for a solution to the system of equations, in the functions $v, x'; M_1,\dots, M_m$ of $h = (h_1,\dots, h_m)$:

$$u_\xi s_h'\cdot M = (q - s')f, \tag{7.3}$$

$$\Delta\cdot M - \frac{1}{u_\xi}u_{\xi h}\cdot M = \left(\frac{1}{u_\xi} - \nu\right)f, \tag{7.4}$$

$$\Delta v = u_h, \tag{7.5}$$

$$M_j = 0 \quad \text{when} \quad h_j = 0, \infty. \tag{7.6}$$

M occurs in these equations only where it is shown explicitly. [(7.5) is the generalized envelope theorem.] When $m < n$, i.e. the number of characteristics is less than the number of commodities, (7.3)–(7.6) can be reduced to a second-order partial differential equation for v with mixed boundary conditions specifying the values of functions of v and Δv where $h_j = 0, \infty$. To do this, we would first solve (7.3) for x' as a function $x'(M, v, h)$ of M, v and h. In general this is a mapping of full rank from M to x', provided $m<n$; and so is the mapping from x', M, h, v to Δv given by (7.5). Consequently the mapping from M to Δv obtained from (7.5) and $x'(M, v, h)$ can be inverted, giving M as a function of Δv, v and h. Substitution in (7.4) gives the promised equation for v.

This procedure breaks down when $m \geq n$. In that case one can eliminate v and x' from (7.3) and (7.4) to obtain v as a function of $\Delta \cdot M$, m, and h. Substitution in (7.5) yields a second-order system of m partial differential equations for the m functions M_j, with boundary conditions specified in (7.6). Even when $m=n=2$, these look hard to handle. But there are many aspects of the solution one would like to know about. Since the budget set frontier is $(n-1)$-dimensional and the population m-dimensional, any point (ξ, x') is chosen by an $(m-n+1)$-dimensional set of people. One would like to know what these sets are like. Since (7.3) no longer gives any information about marginal tax rates if nothing is known about M, it is now a much deeper question, how to characterize the commodities that should be most heavily taxed. It would be interesting to enquire what special structure of the utility function, as a function of h particularly, would simplify the equations and yield information about the solution. One would like to use that to indicate what should guide us in setting up one-parameter models for practical work.

In the model with large m, the boundary conditions (7.6) seem to play a very important part in determining the solution. This means that the economist's instinct to rely on differential first-order conditions to derive properties of the solution is no help in these cases. I think this is the root difficulty in making the m-dimensional model produce any results.

8. Consumer uncertainty

In all the models considered, consumers have been perfectly informed about themselves and the possibilities open to them. There has been no uncertainty about taxes or prices, or about the circumstances in which these taxes and prices will apply. There is a large range of unexplored problems here. The only case for which much is known is that in which individuals all make their decisions in advance of knowing the states of nature that distinguish them. This is the case of pure moral hazard. Denoting the initially unknown state of the consumer – his future ability, health, or luck – by θ, and the observed outcome on which government policy can be based – wage, retirement date, or prize – by y, we assume a functional relationship

$$y=g(\theta, x), \tag{8.1}$$

where x is the consumer's choice variable. On the basis of y, government delivers

$$z = \zeta(y) \tag{8.2}$$

to the consumer, who chooses x to maximize

$$\mathrm{E}u(\zeta(g(\theta, x)), x). \tag{8.3}$$

There is a resource (or revenue) constraint, which for a large identical population with independent and identically distributed states θ can be written in the form

$$\mathrm{E}h(y, \zeta(y), x) = 0, \tag{8.4}$$

where y is given by (8.1). The leading case is that in which expected utility is also the government's maximand, though other welfare functions are also of interest.

I shall not go into the methods of analysing this kind of problem. It is a problem in which the general issues raised in Section 2 loom large; and one further issue arises which I have not discussed earlier. Since the set defined by the maximization constraint can be complicated and difficult to work with, it is best to look for cases in which certain kinds of fairly simple solutions exist. There are three (if the maximization constraint is not inessential).

(1) Problems where the solution is a function ζ for which the expected utility function is known to be a concave function of x. In such a case the maximization constraint is equivalent to its first-order condition, and the optimization can then be treated by Kuhn–Tucker methods. It is usually quite easy to see for what functions ζ, $\mathrm{E}u$ is concave in x: to get an adequately large class one may have to specialise the utility function. The difficulty is to find conditions under which one can be sure in advance that the optimum ζ falls into this class. This requires a direct argument that any other ζ can be improved upon.

(2) Following the discussion in Section 2 it is a real possibility that expected utility should have up to $m+1$ global maxima, where m is the dimensionality of the available set of policy functions ζ. If all functions (analytic, integrable or whatever) are available, a continuum of global maxima is a possible optimum, not only in exceptional cases. It is therefore a good idea to look for cases in which

$$\mathrm{E}u(\zeta(g(\theta, x)), x) = u_0 \tag{8.5}$$

is a constant at the optimum. It may not be very difficult to find under what conditions such a policy is optimal, and when it is, both computation and further analysis are relatively easy.[19]

[19]An example of special economic interest is treated by Diamond and Mirrlees (1977).

(3) In cases where utility is unbounded, which may be useful approximations to reality, it is possible that no optimum exists, because government can always increase expected utility by reducing the level of reward to some low probability set of possible outcomes. It is obvious that minimum reward is an optimal policy when effort is then increased so that events with minimum reward do not occur. In most of the interesting cases effort can never ensure that disastrous outcomes will never happen. Yet it can be (nearly) optimal to impose extremely severe "punishment" when these events do occur. Solutions of this form can occur in perfectly reasonable models, which contrast sharply in this respect with models where there is no consumer uncertainty. The possibility of providing incentives, usually sticks rather than carrots, through the consequences of rare events is of considerable interest, and should be examined in all cases.[20]

These three possibilities seem to exhaust the manageable solutions to problems with consumer uncertainty; but they do not by any means exhaust the possible solutions. It may be that some of the most interesting results in this area will come from identifying the borderlines between the different classes of optimum rather than by attacking the optimization problem directly.

9. Computation and approximation

A major aim of optimal tax theory is to obtain numerical information about optimal policies. In most of these problems, non-concavity is an important intrinsic property, and first-order conditions may not determine the optimum uniquely, even when the more intractable problems of non-connected constraint sets explained in Section 2 do not occur. For example, in the simple linear income-tax problem, where there are two tax parameters in a simple two-commodity world, we have essentially a one-variable maximization, but it must be carried out by explicit search over the possible range, not by hill-climbing, or solving first-order conditions.[21] As soon as additional parameters are introduced, computational problems begin to be severe. Even the standard, and empirically oversimple, linear expenditure system, when labour is included, leads to non-concave problems. It would seem that optimal tax theory can contribute to the computation of optimal commodity taxes chiefly by narrowing down the range of tax rates it is sensible to try.

A major advantage of the non-linear theory is, therefore, that there is a sufficiency theorem, such that solution of certain differential equations is sufficient to give an optimum, provided a basic condition is satisfied. When the population

[20] See Mirrlees (1974) for an example.

[21] Stern (1976) discusses and carries out computations for the optimal linear income tax.

is more than one-dimensional, the computational problems again become severe. However, one-dimensional models would seem to be a promising tool for computing optimal commodity taxes for many-commodity models, provided an empirically acceptable model can be devised. Of course if the model of Theorem 6.4 is applicable, and on present knowledge it seems as good as any other, optimal tax calculations are reduced essentially to a two-commodity income-tax problem, which poses no insurmountable computational difficulties.

Since, in general, computation and simulation are not particularly easy (and this is even more true of the models mentioned in Section 8), other techniques of numerical exploration can be useful. It seems to be illuminating to set up a number of questions as approximation problems, asking for properties of the optimum when certain parameters are small. I have been able to obtain approximate formulae for optimal commodity taxes when the distribution of characteristics in the economy has a low variance; and when the degree of inaccuracy in observations used for lump-sum taxation is small; but there is not space to develop these calculations here.

In a similar vein, it is interesting to analyse the asymptotic form of non-linear optimal tax policies for very high (or very low) values of the characteristics. But this often gives inaccurate, or even seriously misleading, information.[22] Indeed it is a general principle of work on approximations that one should try to discover something about the accuracy of the approximations. It would be valuable to show that certain classes of approximations are tolerably accurate by carrying out complete calculations in a few representative cases. This may even be the best line to follow for calculating optimal commodity taxes.

10. In conclusion

Computational and empirical issues seem likely to loom large in optimal taxation in future. It is not always easy to devise simple models that are simple enough to be manageable theoretically and rich enough to be empirically relevant. Like growth theory and planning theory, to instance only two examples in the recent history of economics, optimal tax theory has fairly quickly reached a stage where good theorems may be hard to come by, while the theory contains many suggestions or possibilities for practical implementation.

Yet there is still much theoretical work to be done, and the best theorems may be still to come. The whole area of consumer uncertainty where consumers are not identical remains to be explored. Little has been done on variations in population size. Aspects of the real world, like overtime rates, discrete labour choices, misperceptions, and, above all, disequilibria, could be incorporated in manage-

[22] The optimal income-tax problem analysed by Mirrlees (1971) provides examples of this.

able models. International issues, such as tax agreements and treaties, or incentives acting upon countries (e.g. aid agreements) could be examined. Problems of tax evasion and administration have only begun to be looked at.

This account of optimal tax theory has by no means covered all the theoretical work that has been done. On the contrary, it has concentrated on certain fundamental models, and the methods for solving them, and has said rather little about properties of the solutions. I conclude with a few notes on the literature, to guide the reader to what has been said about the topics taken up here.

11. Notes on the literature

Optimal tax theory began with Ramsey (1927), who solved the problem of raising revenue by commodity taxes from a single consumer. Pigou (1947) discussed Ramsey's solution, but the next contributions published were those of Boiteux (1956), Corlett and Hague (1953–54) and Meade (1955). Boiteux still assumed lump-sum taxation, as it happens quite unnecessarily, and looked at optimal pricing by public enterprises subject to a budget constraint. This is essentially equivalent to Ramsey, but Boiteux introduced the use of indirect utility functions. Corlett and Hague considered a special case of the problem of improving matters by introducing taxes where none were before, and Meade solved the corresponding optimization problem. Work on discount rates for public investment during the sixties often implicitly assumed imperfections, such as absence of lump-sum taxation, but general models of optimal taxation seem not to have appeared before 1970. Several contributions appeared at the beginning of the seventies: Baumol and Bradford (1970), Diamond and Mirrlees (1971), Feldstein (1972), and Kolm (1970) may be mentioned among many. Diamond and Mirrlees introduced the many-consumer economy without lump-sum taxes, stated and proved the efficiency theorem, provide a discussion of existence, and give a case where the optimum can be obtained explicitly. An application of this work to the measurement of national income is presented in Mirrlees (1969).

This work and later contributions are discussed in a brief survey by Sandmo (1976), which includes a useful bibliography. Sandmo's paper forms part of a symposium in the July–August number of the *Journal of Public Economics*, which contains several useful papers. Much of the recent work in optimal tax theory has appeared in that journal.

I conclude with a few selected references for the individual sections. The references provided are by no means complete, even for the period to 1977 when the chapter was written. Two valuable books containing extensive accounts of optimal tax theory have appeared, Atkinson and Stiglitz (1980), and Tresch (1981).

Section 2

The material presented here has not previously appeared in print. Problems of the type discussed were classified in Spence and Zeckhauser (1971). Some of the difficulties arising from the maximization constraint were noticed in Helpman and Laffont (1975). On evasion see Srinivasan (1973).

Section 3

The treatment of lump-sum taxation as based on individual information is related to the work on 'signalling' and 'screening': Spence (1973) and Stiglitz (1976). Some aspects were mentioned in Mirrlees (1974).

Section 4

Efficiency and other shadow-price results are important for cost-benefit analysis: Diamond (1968), Little and Mirrlees (1974), and Dasgupta and Stiglitz (1974) may be consulted. Efficiency when there are positive profits was first discussed in Dasgupta and Stiglitz (1972). See also Mirrlees (1972).

Section 5

In addition to the works referred to above, the following should be mentioned: Dasgupta and Stiglitz (1971), Atkinson and Stiglitz (1972), Diamond (1975), and Atkinson and Stiglitz (1976). Dixit (1975) and Guesnerie (1977) discuss the welfare effects of commodity tax changes.

Section 6

A special case of nonlinear taxation, with extensive results and numerical calculations, is given in Mirrlees (1971). Theorem 6.1 above generalizes the result that underpinned that paper, a result that never seemed worth publishing for a special case. The more general model is discussed under differentiability assumptions in Mirrlees (1976). See also Atkinson (1973), Phelps (1973), Atkinson and Stiglitz (1976), Sadka (1976), and Seade (1977). An interesting approach, not concerned with optimality, but with bargaining, is Aumann and Kurz (1977).

Section 7

The optimality conditions were given in Mirrlees (1976). Mirrlees and Spence have work in progress on special cases of optimization with many characteristics.

Section 8

The papers by Mirrlees (1974), where the inadequacy of treating first-order conditions as constraints is not adequately appreciated, and Helpman and Laffont (1975), referred to above, are relevant here. Diamond and Mirrlees (1977) give a fairly full explicit analysis of an interesting special case, where the consumer chooses retirement, and the government the social insurance system.

References

Atkinson, A. B. (1973), "How progressive should income tax be?", in: M. Parkin, ed., Essays in Modern Economics, pp. 90–109. London: Longmans.
Atkinson, A. B. and J. E. Stiglitz (1972), "The structure of indirect taxation and economic efficiency", Journal of Public Economics, 1:97–119.
Atkinson, A. B. and J. E. Stiglitz (1976), "The design of tax structure: Direct versus indirect taxation", Journal of Public Economics, 6:55–75.
Atkinson, A. B. and J. E. Stiglitz (1980), Lectures on Public Economics. New York: McGraw-Hill.
Aumann, R. J. and M. Kurz (1977), "Power and taxes", Econometrica, 45:1137–1162.
Baumol, W. J. and D. F. Brandford (1970), "Optimal departures from marginal cost pricing", American Economic Review, 60:265–283.
Bröcker, T. H. (1975), Differential Germs and Catastrophes. Cambridge: Cambridge University Press.
Boiteux, M. (1956), "Sur la gestion des monopoles publics astreint à l'équilibre budgétaire", Econometrica, 24:22–40.
Corlett, W. J. and D. C. Hague (1953–54), "Complementarity and the excess burden of taxation", Review of Economic Studies, 21:21–30.
Dasgupta, P. and J. E. Stiglitz (1971), "Differential taxation, public goods and economic efficiency", Review of Economic Studies, 38:151–174.
Dasgupta, P. and J. E. Stiglitz (1972), "On optimal taxation and public production", Review of Economic Studies, 39:87–104.
Dasgupta, P. and J. E. Stiglitz (1974), "Benefit–cost analysis and trade policies", Journal of Political Economy, 82:1–33.
Diamond, P. A. (1968), "The opportunity costs of public investment: A comment", Quarterly Journal of Economics, 82:682–686.
Diamond, P. A. (1975), "A many-person Ramsey tax rule", Journal of Public Economics, 4:335–342.
Diamond, P. A. and J. A. Mirrlees (1971), "Optimal taxation and public production I-II", American Economic Review, 61:8–27, 261–278.
Diamond, P. A. and J. A. Mirrlees (1976), "Private constant returns to scale and public shadow prices", Review of Economic Studies, 43:41–48.
Diamond, P. A. and J. A. Mirrlees (1977), "A model of social insurance with retirement", Journal of Public Economics, 10:295–336.
Dixit, A. K. (1975), "Welfare effects of tax and price changes", Journal of Public Economics, 4:103–124.
Dixit, A. K. (1976), "Public finance in a Keynesian temporary equilibrium", Journal of Economic Theory, 12.
Feldstein, M. S. (1972), "Distributional equity and the optimal structure of public prices", American Economic Review, 62:32–36.
Groves, T. and J. Ledyard (1977), "Optimal allocation of public goods: A solution to the 'free-rider' problem", Econometrica, 45:783–810.
Guesnerie R. (1975), "Production of the public sector and taxation in a simple second best model", Journal of Economic Theory, 10:127–156.
Guesnerie, R. (1977), "On the direction of tax reform", Journal of Public Economics, 7:179–202.
Hahn, F. H. (1973), "On optimum taxation", Journal of Economic Theory, 6:96–106.
Heller, W. P. and K. Shell (1974), "On optimal taxation with costly administration", American Economic Review, Papers and Proceedings, 64:338–345.
Helpman, E. and J.-J. Laffont (1975), "On moral hazard in general equilibrium theory", Journal of Economic Theory, 10:8–23.
Kolm, S. Ch. (1970), La Theorie des Contraintes de Valeur et Ses Applications. Paris: Dunod.
Little, I. M. D. and J. A. Mirrlees (1974), Project appraisal and planning for developing countries. London: Heinemann, and New York: Basic Books.

Meade, J. E. (1955), Trade and welfare. London: Oxford University Press.
Mirrlees, J. A. (1969), "The evaluation of national income in an imperfect economy", Pakistan Development Review, 9:1–13.
Mirrlees, J. A. (1971), "An exploration in the theory of optimum income taxation", Review of Economic Studies, 38:175–208.
Mirrlees, J. A. (1972), "On producer taxation", Review of Economic Studies, 39:105–111.
Mirrlees, J. A. (1974), "Notes on welfare economics, information, and uncertainty", in: M. Balch, D. McFadden and S. Wu, eds., Essays in Equilibrium Behaviour Under Uncertainty. Amsterdam: North-Holland.
Mirrlees, J. A. (1976), "Optimal tax theory: A synthesis", Journal of Public Economics, 6:327–358.
Mirrlees, J. A. and K. W. S. Roberts (1980), "Functions with multiple maxima", mimeographed. Oxford: Nuffield College.
Munk, K. J. (1978), "Differential taxation and economic efficiency reconsidered", Review of Economic Studies.
Phelps, E. S. (1973), "Wage taxation for economic justice", Quarterly Journal of Economics, 87: 331–354.
Pigou, A. C. (1947), A Study in Public Finance, 3rd ed. London: Macmillan; 1st ed. (1928).
Ramsey, F. P. (1927), "A contribution to the theory of taxation", Economic Journal, 37:47–61.
Sadka, E. (1976), "On income distribution, incentive effects and optimal income taxation", Review of Economic Studies, 43:261–267.
Seade, J. K. (1977), "On the shape of optimal tax schedules", Journal of Public Economics, 7:203–235.
Sen, A. K. (1973), On Economic Inequality. Oxford: Clarendon Press.
Spence, M. (1973), "Job market signalling", Quarterly Journal of Economics, 87:355–379.
Spence, M. and R. Zeckhauser (1971), "Insurance, information, and individual action", American Economic Review, 61:380–387.
Srinivasan, T. N. (1973), "Tax evasion: A model", Journal of Public Economics 2:339–346.
Stern, N. H. (1976), "On the specification of models of optimum income taxation", Journal of Public Economics, 6:123–162.
Stiglitz, J. E. (1975), "Information and economic analysis", in: M. Parkin and A. R. Nobay, eds., Current Economic Problems. Cambridge: Cambridge University Press.
Swinnerton-Dyer, H. P. F. (1959), "On an extremal problem", Proceedings of the London Mathematical Society.
Tresch, R. W. (1981), Public Finance: A Normative Theory. Plano, TX: Business Publications.

Chapter 25

POSITIVE SECOND-BEST THEORY*

A Brief Survey of the Theory of Ramsey Pricing

KARE P. HAGEN**

Norwegian School of Economics and Business Administration, Bergen

and

EYTAN SHESHINSKI***

The Hebrew University, Jerusalem

"One appoints inspectors of weights and measures but not prices." *Babylonian Talmud* (Baba-Bathra, 89A)
"The role of inspectors would be to assure that nobody sells at too high a price. The logic of the situation indicates that this is unnecessary. If one merchant wants to charge a high price, another one who needs money will give the merchandize cheaply and the customers will go to him. And so the first one will have to sell cheaply as well." *Rabbi Shmuel Ben-Meier* (1080–1160 A.D.) *op. cit.*

1. Introduction

While understanding the favorable role of competition, the commentators of the Talmud could hardly be expected to dwell on possible exceptions and the consequent desirability of government intervention. In current theory, the most familiar case of "market failure" is that of *natural monopoly*.

This paper surveys some issues in positive second-best theory, specifically the theory of the optimal pricing of goods (private and public) produced by public firms, that is, firms whose objective is the maximization of social welfare. It is assumed that these firms, characteristically, display increasing returns to scale. In these situations, first-best optima may require lump-sum taxes and subsidies. In view of the size of the public sector in most industrialized countries; it is difficult to imagine that public activities can be financed without distortionary effects elsewhere in the economy.

It may be claimed, however, that the theory should be sufficiently general to explain why first-best optima are infeasible rather than merely stating this as an

* Due to an error in the manuscript, the name of Kare P. Hagen as co-author has been omitted in an earlier printing.

** Hagen's contribution is based on his article, "Optimal pricing in public firms in an imperfect market economy", *Scandinavian Journal of Economics*, 81: 475–493 (1979).

*** I wish to thank Roger Guesnerie for helpful discussions and Robert Aumann for the quotation below.

Handbook of Mathematical Economics, vol. III, edited by K.J. Arrow and M.D. Intriligator
© 1986, Elsevier Science Publishers B.V. (North-Holland)

exogenous fact. That would be a rather ambitious task which is beyond the scope of this survey. Here we ask a much more modest question. Given certain structural facts which prevent prices from being set equal to marginal costs everywhere in the economy, what would then be the optimal rules to follow for public production and prices under public control? These price distortions may be due to the infeasibility of lump-sum taxation, i.e., price distortion created by the government, or they may arise due to monopolistic pricing in the private sector which may have to be accepted for political reasons. Clearly this does not imply that monopoly in some industry *ought* to be dealt with by pricing policy by public firms. The analysis below merely shows what factors would be relevant to take into consideration if price under public control were to be set on the basis of economic efficiency, if the inviolability of monopoly has to be accepted for some reason.

It is crucial for the second-best argument that the public firm considers its economic environment as given, such that optimal pricing rules only cover aspects under its control. On the other hand, it may be claimed that if these rules are to be of any practical interest, they must not require the public firm to have knowledge which, in practice, would be nearly impossible or extremely costly for it to ascertain. In this respect we have to admit that the formulas for the second-best pricing rules obtained here may be rather demanding from an informational point of view. Indeed, the most interesting possibility for development in the theory of "second-best" pricing rules seems to be the explicit introduction of uncertainty into the "Ramsey pricing models".

Throughout this chapter we have assumed perfect possibilities for lump-sum income transfers in order to focus on the efficiency aspect of optimal pricing. If lump-sum redistribution is impossible, deviations from marginal costs for prices under public control may be motivated by distributional considerations; that is, the government may want to use its excise tax power to improve the income distribution.

The results reported here are not new and can be found scattered in the economic literature (see bibliography). For the sake of unity, we thus start with a general formulation that forms a basis for special cases which are analyzed in more detail subsequently.

Section 2 is a general formulation of the pricing rule adopted by a multi-product firm whose objective is the maximization of social welfare ("Ramsey prices"). The firm's technology is characterized by increasing returns to scale and thus a financial constraint on profits (losses) may preclude attainment of the first-best allocation. The "public firm" is viewed as a "Stackelberg leader", competing in some markets with private, profit-maximizing firms. Section 3 specializes the analysis to the case where the private firms behave competitively. Section 4 discusses the questions of "cross-subsidization" and the conditions under which undercutting the competition by the public firm is socially desirable. Section 5 analyzes the optimal pricing rule in imperfectly competitive markets. Section 6 relates "Ramsey prices" to the issue of price sustainability in perfectly contestable markets [Baumol et al. (1982)]. Section 7 looks at the possibility of more general, credible decision rules for a dominant public firm, when private firms do not necessarily take prices as given. As special cases, the marginal cost pricing rule is favorably compared with the fixed output rule. Section 8 discusses the joint decision for the optimal supply of public goods and the pricing problem analyzed previously. Finally, Section 9 analyzes some issues in predation and Ramsey pricing in a dynamic context.

2. Ramsey pricing

We consider an economy consisting of l individuals labelled h, m private firms indexed j, and a public sector.[1] For simplicity, the public sector is assumed to consist of one public firm. There are $n+1$ private goods indexed $i=0,1,\dots,n$, consumed or supplied by the individuals, the private firms and the public sector.

We use the following notations:

$x^h \equiv (x_0^h, x_1^h,\dots, x_n^h) = (n+1)$-dimensional commodity vector representing consumer h's consumption plan,

$y^j \equiv (y_0^j, y_1^j,\dots, y_n^j) = (n+1)$-dimensional commodity vector representing private firm j's production plan,

$z \equiv (z_0, z_1,\dots, z_n) = (n+1)$-dimensional commodity vector representing the net production plan for the public sector, i.e., total public supply minus public consumption.

We apply the sign convention that negative components in the consumption plans represent the net supply of services, while net demand is measured in positive quantities. As for production plans, output is measured in positive quantities, while input is measured in negative quantities. In the subsequent analysis the commodity with index zero will be used as a *numéraire* good.[2]

The index set E represents the set of goods whose prices are subject to public regulation. Goods of which the public sector is the sole supplier or consumer clearly belong to this set, although in order to control the price of a commodity, it is not necessary for the government to have complete control of its supply or demand.

Production in the public sector takes place using labor and other private goods supplied by individuals and private firms as inputs. The public sector supplies

[1] The model generalizes Boiteux (1971) by allowing for private firms competing with the public firm.

[2] Note that we normalize both consumer and producer prices. This may be motivated by an assumption of perfect competition in the market for the *numéraire*. In any case, this normalization will not matter since we assume perfectly redistributable incomes.

consumer goods and intermediate goods to private firms. We assume that technically efficient production plans for the public sector are defined by the implicit production function

$$g(z)=0.$$

Moreover, we assume that for structural reasons which are exogenous to this model, the activity in the public sector is subject to a budget constraint given by

$$b-\sum_{i} p_i z_i = 0,$$

where $p \equiv (p_0, p_1, \dots, p_n)$ is a $(n+1)$-dimensional price vector with $p_0 = 1$.

A binding budget constraint for the public sector can of course be motivated by increasing returns to scale in the public sector such that the amount of lump-sum financing (b) is insufficient to cover the public deficit at a first-best optimum. If b is equal to zero, we are imposing a zero profit constraint on the public firm operating in the markets for private goods. This would mean that the public sector had to be financed entirely through distortionary commodity taxation.

It may also be noted that we treat the public sector as an aggregate. In a model with a disaggregated specification of the public sector, we would have many public firms with different production technologies. However, in assuming one overall budget constraint for the public sector, we must clearly have production efficiency in the public sector at a second-best optimum. Hence, the optimal pricing and production rules must be the same for all public firms so that the method of treating the public sector in an aggregate fashion entails no loss of generality.

For convenience of analysis and to focus on the efficiency aspect of optimal pricing, personal incomes are assumed to be perfectly redistributable through lump-sum transfers. By means of this assumption, we do not have to specify how the public budget b is financed (if $b<0$), and we do not have to specify the distribution of profits in the private sector. Also, compensated demand functions have some nice properties which will be utilized repeatedly in the subsequent analysis.

The decision variables under public control are the prices under public control p_e, $e \in E$, the net production plan for the public firm denoted z, and the income distribution $\{r^h\}$, where r^h denotes the non-labor income of individual h.

Assuming that individual consumption plans are ranked according to the strictly increasing and strictly quasi-concave utility functions $U^h(x^h)$ and that only equilibrium values are relevant for social welfare, efficient rules for pricing and production in the public sector and an optimal income distribution are obtained through solving the following constrained maximization problem:

$$\max_{(p_e,\{r^h\},z)} \sum_h \lambda^h U^h(x^h), \qquad h \geq 0, \quad \forall h,$$

subject to (dual variables)

$$\sum_h x_i^h - z_i - \sum_j y_i^j = 0, \qquad (\alpha_i,\ i = 0,1,\dots,n),$$

$$g(z) = 0, \qquad (\beta),$$

$$b - \sum_i p_i z_i = 0, \qquad (\gamma).$$

The necessary maximum conditions are

$$\sum_h \sum_i \lambda^h U_i^h \frac{\partial x_i^h}{\partial p_e} - \sum_i \alpha_i \left(\sum_h \frac{\partial x_i^h}{\partial p_e} - \sum_j \frac{\partial y_i^j}{\partial p_e} \right) + \gamma z_e = 0, \qquad e \in E, \tag{2.1}$$

$$\lambda^h \sum_i U_i^h \frac{\partial x_i^h}{\partial r^h} - \sum_i \alpha_i \frac{\partial x_i^h}{\partial r^h} = 0, \qquad h = 1,\dots,l, \tag{2.2}$$

$$\alpha_i = \beta g_i - \gamma p_i = 0, \qquad i = 0,1,\dots,n, \tag{2.3}$$

where U_i^h and g_i denote partial derivatives of the functions $U^h(\cdot)$ and $g(\cdot)$ with respect to the ith argument.

Under the assumption that b exceeds the unconstrained profits (possibly negative as in the case of increasing returns to scale in the public firm), $\gamma > 0$. Write $V(-b)$ for the maximum value of $\sum_h \lambda^h U^h(x^h)$. Then we have

$$\gamma = \lim_{s \to 0} [V(-b+s) - V(-b)]/s.$$

Hence γ measures the value of marginally relaxing the constraint.

Multiplying both sides of (2.2) by $1/U_0^h$, observing that according to the first-order conditions for consumer optima $U_i^h/U_0^h = p_i$, and using the property of individual demand that $\sum_i p_i(\partial x_i^h/\partial r^h) = 1$, the necessary condition (2.2) for an optimal income distribution simplifies to

$$\lambda^h U_0^h = \sum_i \alpha_i \frac{\partial x_i^h}{\partial r^h}, \qquad h = 1,\dots,l. \tag{2.4}$$

Substituting (2.4) into (2.1) and using the fact that for each h, $\sum_i p_i(\partial x_i^h/\partial p_e) =$

x_e^h, the necessary conditions for an optimal price structure p_e simplify to

$$\sum_i \alpha_i \left[\sum_h \left(\frac{\partial x_i^h}{\partial p_e} + x_e^h \frac{\partial x_i^h}{\partial r^h} \right) - \sum_j \frac{\partial y_i^j}{\partial p_e} \right] = \gamma z_e, \qquad e \in E. \tag{2.5}$$

We define

$$\frac{\partial \hat{x}_i}{\partial p_e} \equiv \sum_h \left(\frac{\partial x_i^h}{\partial p_e} + x_e^h \frac{\partial x_i^h}{\partial r^h} \right),$$

representing compensated price derivatives of the private consumption demand for commodity i. Similarly,

$$\frac{\partial \hat{z}_i}{\partial p_e} \equiv \frac{\partial \hat{x}_i}{\partial p_e} - \sum_j \frac{\partial y_i^j}{\partial p_e},$$

representing compensated price derivatives of the net market demand for commodity i.

We eliminate the dual variable β through the normalizations $\delta_i \equiv \alpha_i / \beta g_0$ and $\mu \equiv \gamma / \beta g_0$. Then we define $c_i^0 \equiv g_i / g_0$. Hence, c_i^0 denotes the marginal cost of producing commodity i in the public sector if $z_i > 0$, or the marginal technical rate of substitution between input i and the *numéraire* if $z_i < 0$.

With these definitions, conditions (2.5) and (2.3) can be rewritten as

$$\sum_i \delta_i \frac{\partial \hat{z}_i}{\partial p_e} = \mu z_e, \qquad e \in E, \tag{2.6}$$

$$\delta_i = c_i^0 - \mu p_i, \qquad i = 0, 1, \ldots, n. \tag{2.7}$$

We note from (2.7), $\delta_0 + \mu = 1$, and since $\delta_0 > 0$ from strict monotonicity of the utility functions, we must have that $0 < \mu < 1$.

Substituting (2.7) into (2.4) and (2.6), we get

$$\frac{\lambda^h U_0^h}{\lambda^l U_0^l} = \frac{\sum_i c_i^0 (\partial x_i^h / \partial r^h) - \mu}{\sum_i c_i^0 (\partial x_i^l / \partial r^l) - \mu}, \qquad h = 2, \ldots, l, \tag{2.8}$$

$$\sum_i (c_i^0 - \mu p_i) \frac{\partial \hat{z}_i}{\partial p_e} = \mu z_e, \qquad e \in E. \tag{2.9}$$

According to condition (2.8), an optimal income distribution is obtained by equating, for any pair of consumers labelled l and h, the ratios of social weights attributed to marginal income transfers in favor of l and h to the ratios of marginal social costs associated with these transfers. If prices deviate from marginal costs in the public sector and Engel elasticities differ among consumers, the marginal social cost of income transfers will be different for different consumers, which calls for setting $\lambda^h U_0^h \neq \lambda^l U_0^l$ for $h \neq l$.

We define $c_i^j \equiv -\partial y_0^j / \partial y_i^j$ as the marginal cost of producing commodity i in firm j if $y_i^j > 0$, and if $y_i^j < 0$, c_i^j denotes the marginal rate of substitution between input i and the *numéraire* in firm j.

Subtracting $(1-\mu)\sum_i p_i(\partial \hat{z}_i / \partial p_e)$ on both sides of (2.9) yields

$$\sum_i \left(c_i^0 - p_i\right) \frac{\partial \hat{z}_i}{\partial p_e} = \mu z_e - (1-\mu) \sum_i p_i \frac{\partial \hat{z}_i}{\partial p_e}. \tag{2.10}$$

Letting t_i^j denote the mark-up on the price of commodity i in firm j (which may be negative), that is $t_i^j \equiv p_i - c_i^j$, and writing $x_i \equiv \sum_h x_i^h$, from well-known properties of compensated market demand functions, we have

$$\begin{aligned} \sum_i p_i \frac{\partial \hat{z}_i}{\partial p_e} &= \sum_i p_i \frac{\partial \hat{x}_i}{\partial p_e} - \sum_j \sum_i t_i^j \frac{\partial y_i^j}{\partial p_e} - \sum_j \sum_i c_i^j \frac{\partial y_i^j}{\partial p_e} \\ &= -\sum_j \sum_i t_i^j \frac{\partial y_i^j}{\partial p_e} = -\sum_j \sum_i \left(p_i - c_i^j\right) \frac{\partial y_i^j}{\partial p_e}, \end{aligned}$$

and substituting into (2.10) we obtain our central conditions for optimal pricing of private goods in the public sector

$$\sum_i \left(c_i^0 - p_i\right) \frac{\partial \hat{z}_i}{\partial p_e} = \mu z_e + (1-\mu) \sum_j \sum_i \left(p_i - c_i^j\right) \frac{\partial y_i^j}{\partial p_e}, \qquad e \in E. \tag{2.11}$$

3. Competitive fringe

We see immediately from (2.11) that if the budget constraint for the public sector is not binding and if the private sector is perfectly competitive (i.e., marginal costs are set equal to prices throughout the private sector), efficiency will require marginal cost pricing in the public sector. It should also be clear that if the budget constraint is binding and/or there are price distortions in the private sector, we then have a second-best situation, in which case it will generally be optimal for the public sector to deviate from marginal cost pricing.

In order to discuss one complication at a time, we begin by assuming that the budget constraint for the public sector is binding and that the private sector is perfectly competitive. In this case the optimal pricing rule (2.11) simplifies to

$$\sum_i \left(c_i^0 - p_i\right) \frac{\partial \hat{z}_i}{\partial p_e} = \mu z_e, \qquad e \in E. \tag{3.1}$$

Except for the dual variable μ and the compensated nature of the demand functions, conditions (3.1) are the same as the necessary conditions for profit-maximizing prices for a *monopolist* endowed with a technology given by $g(z)=0$ and with control over the prices p_e, $e \in E$.

Hence, we have the well-known theorem: *For all commodities $e \in E$, the public firm should, in the market for outputs, set prices as if it were a monopolist with all the compensated demand elasticities inflated by a factor $1/\mu$. In the markets for inputs, the public firm should behave as a monopsonist with all the supply elasticities inflated by the same factor $1/\mu$.* This implies that the public firm should use price discrimination in the markets for both inputs and outputs whenever possible. The inflating factor $1/\mu > 1$ is determined such that the budget constraint for the public sector holds as an equality.

The intuitive reason for this result is clearly that as long as the budget constraint for the public sector is binding, the shadow price of the *numéraire* will be higher in the public sector than in the private sector.

Although the efficiency conditions (3.1) for the case where $\mu = 1$ are formally equivalent to the necessary conditions for profit-maximizing monopoly mark-ups, it seems that, except in some rather special cases, no general conclusions can be drawn with respect to the relationships between prices and mark-ups in the efficiency case and in the monopoly case, i.e., the case where all goods were produced and marketed by one large monopoly. As one such special case, assume that all prices are under public control and that there are linear compensated net market demand functions of the form

$$z_j = \sum_i A_{ji}\left(p_{0i} - p_i\right), \qquad j = 1, \ldots, n, \tag{3.2}$$

where the constants $A_{ji} = A_{ij}$ from the symmetry conditions on net market demand, and p_{0i} are the prices that simultaneously choke off compensated net demand in all markets. Moreover, we assume constant marginal costs so that the deviation of prices from marginal costs must be justified by the presence of a fixed or overhead cost.

Inserting into (3.1), using the symmetry conditions and rearranging terms, we get

$$\sum_i \left[(1+\mu)p_i - c_i^0 - \mu p_{0i}\right] A_{ij} = 0, \qquad j=1,\ldots,n. \tag{3.3}$$

From non-singularity of the Jacobian matrix $\|A_{ij}\|$ the only solution to (3.3) is given by $(1+\mu)p_i - c_i^0 - \mu p_{0i} = 0$, and hence the efficiency prices are given by $\hat{p}_i = (c_i^0 + \mu p_{0i})/(1+\mu)$ and the corresponding monopoly prices by $\bar{p}_i = (c_i^0 + p_{0i})/2$. Clearly, $p_{0i} > c_i^0$, $\forall i$, and hence it is easily seen that $\bar{p}_i > \hat{p}_i$, $\forall i$, when $\mu < 1$. The efficiency mark-ups are given by $\hat{p}_i - c_i^0 = \mu(p_{0i} - c_i^0)/(1+\mu)$ and are hence proportional to the corresponding monopoly mark-ups so that, while efficiency prices are always below the corresponding monopoly prices in the case of linear compensated net market demand and constant marginal costs, mark-up ratios between any pair of commodities (except the *numéraire*) will be the same at a second-best Pareto optimum and a monopoly profit maximum.[3] In the general case, the solution of (3.1) would involve a fixed point problem and there is no reason to expect any proportionality between efficiency and the corresponding monopoly mark-ups.

Going back to the general optimality conditions (3.1) and making the simplifying assumption that $\partial \hat{z}_i / \partial p_e = 0$, $\forall i \neq 0$, e (in which case commodity e and the *numéraire* must be net substitutes), (3.1) simplifies to

$$((p_e - c_e^0)/p_e)\hat{\eta}_e = \mu, \qquad e \in E, \tag{3.4}$$

where

$$\hat{\eta}_e = \frac{1}{\pi_e}\left[\frac{p_e}{\hat{x}_e}\frac{\partial \hat{x}_e}{\partial p_e} - (1-\pi_e)\sum_j \left(\frac{\hat{y}_e^j}{\sum_j \hat{y}_e^j}\right)\left(\frac{p_e}{\hat{y}_e^j}\frac{\partial \hat{y}_e^j}{\partial p_e}\right)\right], \tag{3.5}$$

and $\pi_e = \hat{z}_e/\hat{x}_e$, $0 \leq \pi_e \leq 1$, the share of the public sector in market e. The left-hand side of (3.4) are sometimes called "*Ramsey Numbers*". In the absence of private firms in market e, $\hat{y}_e^j = 0$, $\forall j$ (i.e., $\pi_e = 1$), $\hat{\eta}_e$ is the own (absolute value of the) compensated price elasticity of demand for commodity e. If there are constant own compensated price elasticities of demand in the above example, we then get the well-known result in optimal taxation theory that the optimal relative mark-up, or *ad valorem* tax, will be the same for all taxable commodities and in

[3]As would be the case when utility functions are linear in the *numéraire* and homogeneously separable in the taxed goods.

particular, with unitary own compensated price elasticities of demand, it will be equal to the normalized shadow price of the public budget constraint.

Willig and Bailey (1979) have calculated "Ramsey Numbers" in different markets, testing the hypothesis that they are significantly less than unity (distinguishing a "socially desirable" Ramsey firm from a "socially undesirable" profit-maximizing monopolist). Unfortunately, their tests take $\hat{\eta}_e$ to be the elasticity of market demand, thus disregarding relevant cases of competitive fringe, as suggested in (3.4).

4. Market division and welfare aspects of competition

As a special case of the above discussion, consider an economy with two goods 1 and 2, and identical private firms which offer commodity 1 at a fixed price, $\bar{p}_1$, i.e., $\{\hat{y}_1(p_1)|\hat{y}_1=0 \text{ if } p_1<\bar{p}_1,\ 0\leq\hat{y}_1\leq\infty \text{ if } p_1=\bar{p}_1, \text{ and } \hat{y}_1=\infty \text{ if } p_1>\bar{p}_1\}$. Private firms do not offer commodity 2. Thus, one may call market 1 the *competitive market* and market 2 the (public firm's) *monopolized market*. Under what conditions is it socially optimal for the public firm to undercut the competition in the competitive market?

This question, discussed by Arrow (1983), is related to the argument that public firms tend to engage in *cross-subsidization*.[4] Specifically, these firms tend to undercut competition, making up any losses on the competitive commodity by increasing prices in monopolized markets so as to generate the required revenue. Note that if the public firm would be an *unregulated* monopoly, there is no room for this cross-subsidization because the monopolized markets already yield the maximum possible profit. The argument about cross-subsidization for profit-constrained regulated public firms is based on the fact that, under regulation, there are unexploited opportunities for monopoly profit. Therefore a loss in the competitive market will lead to recoupment in the monopolized market.

We assume that costs of the public firm are *subadditive* [see Baumol, Panzar and Willig (1982)], which is a natural interpretation of economies to scale in the multi-product case. It is then never optimal for the public firm to share the competitive market with the other firms. The comparison is thus between the case when the public firm undercuts the competition and when it abandons the competitive market.

[4] If "cross-subsidization" is to be forbidden, it should be welfare-diminishing. The natural definition, as noted by Arrow (1983), is that there is cross-subsidization from a monopolized to a competitive sector if prices are higher in the former than they would be if the competitive market were abandoned by the public firm. The reason is that from the social welfare point of view, all that matters are the prices charged to consumers. So, if the price in the monopolized market is lower – as against the situation if the competitive market were abandoned by the public firm – then the public firm should stay in the competitive market.

The profit constraint on the public firm may be written

$$P(z_1, z_2) = \hat{p}_1 z_1 + \hat{p}_2 z_2 - c(z_1, z_2) = b, \tag{4.1}$$

where $\hat{p}_i(z_i)$, $i = 1,2$, is the (inverse) demand for commodity i (i.e., demands are assumed independent). If the public firm abandons market 1, the output of commodity 2 would be determined by (4.1) with $z_1 = 0$,

$$P(0_1, z_2^*) = b. \tag{4.2}$$

Here, z_2^* is the "*stand-alone*" output in market 2 (assumed to be unique). Under what conditions can meeting the competition and satisfying (4.1) yield a lower price for commodity 1 than a "stand-alone" policy? A lower price is equivalent to a higher output. Let $\bar{z}_1$ be the quantity demanded at the competitive price $\bar{p}_1$, i.e., $\bar{p}_1 = \hat{p}_1(\bar{z}_1)$. The question is thus: z_1', z_2' satisfying (4.1) with $z_1' \geq \bar{z}_1$ and $z_2' > z_2^*$?

Suppose $P(z_1', z_2^*) > b$. Then, if z_2 is increased, profits will eventually fall below b, and therefore $z_2' > z_2^*$ for which $P(z_1', z_2') = b$. Conversely, if $P(z_1', z_2^*) < b$, it is reasonable to suppose that profits will remain below b for larger values of z_2.[5] Hence, the condition that undercutting competition in market 1 yields a lower priced commodity 2 is equivalent to the condition

$$P(z_1', z_2^*) = \hat{p}_1(z_1') z_1' + \hat{p}_2(z_2^*) z_2^* - c(z_1', z_2^*) > b. \tag{4.3}$$

Now, since (4.2) can be written $\hat{p}_2(z_2^*) z_2^* - c(0, z_2^*) = b$, substitution in (4.3) yields

$$\hat{p}_1(z_1') z_1' - IC_1(z_1', z_2^*) > 0, \tag{4.4}$$

where $IC_1(z_1, z_2) = c(z_1, z_2) - c(0, z_2)$, $\forall z_1, z_2$, are the "Incremental Costs" of commodity 1.

The proposition is that *a sufficient condition for undercutting competition to be socially desirable is that revenues in the competitive market cover incremental costs when output in the monopolized market is at the "stand-alone" level.*

Similar propositions (i.e., necessary and sufficient conditions for undercutting or abandonment of a competitive market) can be proved for a *profit-maximizing* monopoly, instead of a Ramsey firm, and these can be used to test whether a regulated firm diverges from the (constrained) social optimum or engages in predatory pricing.

[5] There may conceivably be alternative regimes where profits exceed and fall below average costs. We do not consider this possibility.

Another question can be raised in this context: in "second-best" situations of the kind discussed above, does "increased efficiency of competition" enhance or decrease welfare? Specifically, suppose that private firms' supply (demand) functions, y_i^j, depend on an efficiency parameter θ, as well as prices, $y_i^j = y_i^j(p, \theta)$. Let an increase in θ raise efficiency in some private firms and not be detrimental to any, i.e., $\partial y_i^j / \partial \theta \geq 0$ if $y_i^j > 0$ ($\partial y_i^j / \partial \theta < 0$ if $y_i^j < 0$). The optimum level of welfare, V, now clearly depends on θ: $V = V(-b, \theta)$.

It has been shown [Sheshinski (1983)] that *increased efficiency in some active private firms* (i.e., for a given j, $y_i^j > 0$ or $y_i^j < 0$ for some i) *may yield* $(\partial V / \partial \theta) < 0$. This result is not confined to the case where private firms are active in the *constrained optimum* but are inactive in the *first-best* social optimum (i.e., the solution to the maximization problem in Section 2 *without* the profit constraint on the public firm). A characterization of the conditions under which welfare decreases when the efficiency of a competitive fringe increases is as yet unavailable.

5. Imperfectly competitive fringe

We next consider the problem of optimal pricing in the public sector in what is perhaps a more interesting case where prices are not set equal to marginal costs everywhere in the private sector. These price distortions may be due to a given set of commodity taxes and subsidies which, for political or institutional reasons, may be considered exogenous to the problem at hand, or they may be due to monopoly pricing in the private sector. In fact, the actual reasons for price distortions in the private sector are unimportant for the problem of optimal pricing in the public firm, as long as these distortions are given for that firm.

It is seen from condition (2.7) that an optimal production plan for the public sector is determined by equating marginal cost c_i^0 in the public firm to the normalized shadow price δ_i for commodity i, plus the normalized shadow price of the public budget times the market price for commodity i. With price distortions in the private sector, shadow prices will deviate from market prices. Hence, regardless of whether or not the public budget constraint is binding, it will generally be the case that efficiency requires prices and marginal costs in the public sector to differ. We note from (2.7) and (2.11) that this will be the case regardless of whether the public firm controls the price or is a price-taker and adjusts quantities at given prices.

To take a simple stylized example, assume that the public firm produces electricity, to be indexed e, and uses gas (indexed g) as one of its inputs (not necessarily in the production of electricity). We assume that electricity is a substitute for gas in the private sector and a complement to appliances which are indexed a. All its other cross-elasticities are assumed to be negligibly small. Appliances are produced and used entirely in the private sector. For simplicity, we assume that a change in p_e does not change net market demand for gas so that $\partial \hat{z}_g/\partial p_e = 0$. Hence, by assumption, a change in the price of electricity will only change the composition of net market demand for gas in the private sector.

The public firm controls only the price of electricity and from (2.11) the optimal price is given by

$$p_e = c_e^0 - \frac{\bar{\gamma} z_e}{\partial \hat{z}_e/\partial p_e} - (1-\mu)\sum_j \sum_{i=a,g} (p_i - c_i^j)\frac{\partial y_i^j/\partial p_e}{\partial \hat{z}_e/\partial p_e}. \tag{5.1}$$

We see from (5.1) that the public firm should set the price under its control different from marginal cost for two reasons. First, a compensated price change will have an effect on the public budget and the second term on the right-hand side of (5.1) measures the social value of this budget effect per unit change in the production of electricity. From the assumption of strict convexity of individual preferences and private production sets, we have $\partial \hat{z}_e/\partial p_e < 0$. Thus, as $z_e > 0$ (output), the budget effect unambiguously calls for setting the price higher than marginal cost. Second, we have a re-allocation effect as given by the last term on the right-hand side of (5.1) which measures the social cost or gain resulting from the re-allocation of resources in the private sector caused by a marginal compensated price change for commodity e. As the consumption side is assumed to be perfectly competitive, only the effect of a change in the prices under public control on demand and supply of private firms will matter for this re-allocation effect.

We note that $p_i > c_i^j$ with monopolistic pricing ($y_i^j > 0$) and $p_i < c_i^j$ for a monopsony ($y_i^j < 0$). This is also the case with commodity taxes and subsidies, although for subsidies the inequality sign must be reversed. Moreover, if commodities i and e are substitutes (complements) then $\partial y_i^j/\partial p_e > (<)\ 0$ for outputs and $\partial y_i^j/\partial p_e < (>)\ 0$ for inputs. Hence, the re-allocation effect alone would call for setting the price of electricity above (below) marginal cost if it is generally the case that substitutes (complements) to electricity are subject to monopolistic pricing or commodity taxation and complements (substitutes) are subsidized.

This result is quite instructive since, in the present model, the allocation of resources is governed by the price mechanism. Thus, if a re-allocation of resources to a monopolized sector increases social welfare, the government can do so by raising the price of substitutes and lowering the price of complementary goods, provided that these prices are under public control whereas the monopoly price is not.

Returning to the specific example above, we may assume that the price of gas is subject to a monopolistic mark-up. The government controls the price of electricity but has no control over the price of gas. If we assume that the industry supply of appliances is perfectly competitive, the re-allocation effect would unambiguously call for setting the price of electricity above marginal cost. If electrical appliances were subsidized, this would strengthen the re-allocation argument for setting p_e above marginal cost. However, if appliances were also subject to mark-ups, the re-allocation effect in this particular market would pull the optimal price for electricity in the opposite direction and the overall re-allocation effect on the optimal price for electricity would be indeterminate.

The matter is complicated even further if a change in the price of electricity changes the net demand (or supply) of gas in the private sector. In this case we would have to add the term $-(p_g - c_g^0)(\partial \hat{z}_g / \partial p_e)/(\partial z_e / \partial p_e)$ on the right-hand side of (5.1), in which case the optimal price for electricity would also depend on the relevant shadow price for gas in the public sector. On the other hand, if the government in this example controls the prices of electricity, gas and appliances, we would be left with the conventional taxation problem where the optimal solution is given by (3.1).

We now assume that the public budget constraint is not binding and that we still have price distortions in the economy. In this case we cannot appeal to the theory of optimal taxation to justify the existence of such distortions, as marginal cost pricing would yield a first-best optimum if it were feasible. The persistence of price distortions created either through monopolistic pricing in the private sector or through commodity taxation must therefore be regarded as political restrictions determined outside the model.

We assume that the government has partial control over prices and we examine how various feasible price changes will affect social welfare. This may be viewed as a problem of comparative statics – or positive economics – in the sense that we consider the change in the equilibrium allocations corresponding to various feasible price – or tax – reforms and then find the resulting change in social welfare.

We set $t_i = p_i - c_i^0 = p_i - c_i^j$ for all j so that we impose the same absolute price distortion on goods produced in both the private sector and the public firm. This may be motivated by assuming that the public firm chooses its production plan such as to obtain aggregate production efficiency. If all price distortions are due to commodity taxation, this formulation implies that taxes on intermediate goods within the private sector and between the private and the public sector cancel out so that in effect, only private and government sales to and purchases from consumers are subject to commodity taxation. Of course, $t_0 = 0$ by definition.

As the public budget constraint is assumed to be not binding, $\bar{\gamma} = 0$, and setting prices under public control different from marginal costs must therefore be motivated by the re-allocation effect in the markets with exogenously given price distortions. Assume that we undertake a (small) finite change in the commodity prices by $\mathrm{d}p_i$, $i = 1, \ldots, n$. According to conditions (2.1), (2.5) and (2.11), such a change in the price structure will increase social welfare if

$$\sum_{k=1}^{n}\sum_{i=1}^{n} t_i \frac{\partial \hat{z}_i}{\partial p_k} \mathrm{d}p_k + \sum_{k=1}^{n}\sum_{i=1}^{n} t_i \sum_j \frac{\partial y_i^j}{\partial p_k} \mathrm{d}p_k \geq 0, \tag{5.2}$$

and since the change in net public production must equal the change in net private demand, the sufficient condition for an increase in social welfare simplifies to

$$\sum_{k=1}^{n}\sum_{i=1}^{n} t_i \frac{\partial \hat{x}_i}{\partial p_k} \mathrm{d}p_k \geq 0. \tag{5.3}$$

We first assume that all prices are changed in proportion to the distortions, that is, $\mathrm{d}p_k = t_k \mathrm{d}\lambda$, $\forall k$. This means a uniform price movement towards or away from marginal costs according to whether $\mathrm{d}\lambda$ is negative or positive.

In this case the sufficiency condition reduces to

$$\sum_{k=1}^{n}\sum_{i=1}^{n} t_i \frac{\partial \hat{x}_i}{\partial p_k} t_k \mathrm{d}\lambda \geq 0, \tag{5.4}$$

and since the Slutsky matrix $\partial \hat{x}_i / \partial p_k$ is negative definite, (5.4) is satisfied if and only if $\mathrm{d}\lambda < 0$ when $t_i \neq 0$ for at least one i. Hence, if *all* prices move towards marginal costs in proportion to the prevailing distortions, this will increase social welfare. Similar results have been obtained in the context of international trade theory by Foster and Sonnenschein (1970).

Looking at partial price changes, a change in the price of commodity e will increase social welfare if

$$\sum_{i=1}^{n} t_i \frac{\partial \hat{x}_i}{\partial p_e} \mathrm{d}p_e \geq 0. \tag{5.5}$$

We define $\theta_i \equiv t_i / p_i$, that is, the relative price distortion or the *ad valorem* tax on commodity i. With given distortions on commodities other than e, it follows from (5.5) that the optimal *ad valorem* tax θ_e^* is given by

$$\theta_e^* = \frac{\sum_{i \neq e} \theta_i p_i (\partial \hat{x}_i / \partial p_e)}{\sum_{i \neq e} p_i (\partial \hat{x}_i / \partial p_e)}, \tag{5.6}$$

that is, a weighted sum of the relative price distortions on commodities other than

e. If commodity e is a net substitute for all commodities, then it follows immediately from (5.6) that $\min_{i \neq e}\theta_i < \theta_e^* < \max_{i=e}\theta_i$.

From the homogeneity of compensated demand functions,

$$\theta_e \sum_{i=0}^{n} p_i \frac{\partial \hat{x}_i}{\partial p_e} \mathrm{d} p_e = 0,$$

and deducting from (5.5) and rearranging, we have that a partial price change $\mathrm{d} p_e$ will increase welfare if

$$\sum_{i=0}^{n} (\theta_i - \theta_e) p_i \frac{\partial \hat{x}_i}{\partial p_e} \mathrm{d} p_e \geq 0, \tag{5.7}$$

where of course $\theta_0 = 0$. Hence, assuming that $\theta_i > 0$, $\forall i \neq 0$, we have from (5.7) that a sufficient condition for a decrease in p_e to increase social welfare is that commodity e is a net complement to all commodities with a greater relative distortion than that on commodity e and a net substitute to commodities with a lower relative price distortion. In particular, assume that commodities are ordered according to the relative price distortion so that $\theta_n > \theta_{n-1} > \cdots > \theta_1$. Then, if all commodities (including the *numéraire*) are net substitutes, a partial reduction in the price of commodity n will increase social welfare until $\theta_n = \theta_{n-1}$ and then social welfare could be improved by reducing p_n and p_{n-1} until $\theta_n = \theta_{n-1} = \theta_{n-2}$ and so on, until $\theta_i = 0$, $\forall i \neq 0$.

As a special case of (5.7) we consider the case where the relative price distortions are the same for all commodities (θ may be a common *ad valorem* tax). Then (5.7) reduces to

$$-\theta \frac{\partial \hat{x}_0}{\partial p_e} \mathrm{d} p_e \geq 0. \tag{5.8}$$

Hence, if the price is initially above marginal cost by the same factor for all commodities (except the *numéraire*), then it will be optimal to raise the price of commodities which are complementary (in consumption) and reduce the price of commodities which are substitutes for the *numéraire* good. Thus, even though prices are a partial price movement away from marginal costs for all commodities, we have here an example where a partial price movement away from marginal cost will increase social welfare. However, the economic rationale for this result is quite simple. With an optimal redistribution of income, θ_i is the marginal social value of an additional unit of the *numéraire* allocated to the purchase of commodity i. Since, by assumption, θ_i is the same for all $i \neq 0$, a re-allocation of resources away from the production of the *numéraire* to *any* distorted sector will, in this particular case, improve social welfare. This can be achieved by taxing complements and subsidizing substitutes for the *numéraire* good, which is precisely what condition (5.8) says.

In the case where all price distortions in the private sector are caused by commodity taxation, all private producers face the same prices and under competitive behavior, there will, in equilibrium, be production efficiency in the private sector. Moreover, under the assumption that the absolute price distortion shall be the same for goods produced in both the private and public sector, there will also be aggregate production efficiency. However, if there are no such constraints on the production plan of the public firm, then the optimality condition (2.11) does not imply aggregate production efficiency. This follows from the fact that an optimal production plan in the public firm is obtained through equating marginal production costs to shadow prices and with price distortions in the private sector, shadow prices will depend on whether commodities are taken out of consumption or private production. To see this, assume an economy with only two goods, indexed 0 and e, and that there is a commodity tax levied on private producers of commodity e which the public firm takes as given. In this case condition (2.11) implies (under the assumption that $\mu = 0$)

$$c_e^0 = p_e \frac{\partial \hat{x}_e / \partial p_e}{\partial \hat{z}_e / \partial p_e} + c_e \frac{(-\partial y_e / \partial p_e)}{\partial \hat{z}_e / \partial p_e}.$$

Hence, in this particular case, the public firm should equate marginal cost to a weighted sum of producer and consumer prices with non-negative weights adding up to unity.

The above example is perhaps somewhat artificial since the public firm takes the price distortion in the private sector as given, even when the public budget constraint is not binding. With a binding budget constraint and if all goods are taxable in the sense that all consumer and producer prices can be changed independently, it has been shown by Diamond and Mirrlees (1971) that aggregate production efficiency will always be desirable, even in the presence of price distortion – provided that profits, if any, in the private sector can be transferred to the government.

If we assume that the government controls all consumer and producer prices, then private producer prices would be the relevant shadow prices to use in the public sector, provided that the tax structure were optimal. Hence, in this particular case, the public firm should aim at aggregate production efficiency, so that it should produce until marginal costs in the private and public sector are equal. Private marginal costs will of course deviate from consumer prices at a second-best tax optimum, which implies that an optimal production plan in the public firm is characterized by marginal costs that differ from consumer prices.

6. Contestable markets: Relation between sustainable and Ramsey prices

We have analyzed the optimal pricing and production of a multi-product public firm, characterized by increasing returns to scale and facing a competitive fringe. The technology available to private firms has not been specified except in assuming (implicitly) that, at the optimum, it provided non-negative profits to the active private firms. Baumol, Panzar and Willig (1982) have taken a different approach. In their analysis, the "public firm" is a profit-maximizing monopoly ("incumbent"), disciplined by potential entrants who have access to the *same* technology, with entry and exit being frictionless (no "sunk" costs).[6] They argue forcefully that under certain conditions, the market equilibrium is (second-best) socially optimal even though there may be only one active firm. Specifically, the monopoly will charge "Ramsey prices" which enable the firm to cover costs. This result they call the "*Weak Invisible Hand Theorem*" [Baumol, Bailey and Willig (1977)].

Baumol et al. consider a monopoly using a technology expressed by a *subadditive* cost function, i.e., $c(z^1+z^2)\leq c(z^1)+c(z^2)$, $\forall z^1, z^2 \geq 0$, where $z^j = (z_1^j, z_2^j,\dots, z_n^j)$ is an n-dimensional output vector.[7] Cost subadditivity implies that one firm can produce more cheaply than two (or any number of) firms producing the same total outputs and hence is offered as a unifying definition of natural monopoly. The monopoly faces a vector of demands $x(p)=(x_1(p), x_2(p), \dots, x_n(p))$ where $p=(p_1, p_2,\dots, p_n)$ are consumer prices. Denote by $N=\{1,2,\dots, n\}$ the set of all goods and let $S\subseteq N$ be a subset of N. Thus, z^s, x^s and p^s are the projections of z, x and p, respectively, on E_+^s. The convention that $c(z^s)=c(z^S,0^{N/S})$, where N/S denotes the complement of S with respect to N, will be used.

Consider a potential entrant having access to the same cost function $c(\cdot)$ and incurring zero entry and exit costs. The entrant may produce any vector of quantities $\hat{z}^s$ at prices $\hat{p}^s$. Then, a price vector $\bar{p}$ is sustainable[8] if every triple $(S, \hat{z}^s, \hat{p}^s)$ satisfying

$$\text{(I)}\quad \hat{p}^s\leq \bar{p}^S, \qquad \text{(II)}\quad \hat{z}^s\leq x^s(\hat{p}^s, \bar{p}^{N/S}), \tag{6.1}$$

[6] The market for air travel serves as a canonical example. Such a market may support only one airline, but the active airline must price at cost to prevent a price-cutting rival airline from flying in and skimming off customers.

[7] We focus on the production of outputs, $z\geq 0$. The analysis, however, can be applied to inputs too.

[8] This is called sustainability against *partial* entry, because the entrant may produce any quantities up to those determined by market demands. Sustainability against full entry would require that entrants supply market demands.

also satisfies

$$\hat{p}^s z^S - c(\hat{z}^s) \leq 0. \tag{6.2}$$

When condition (6.2) is not satisfied then we say that markets *N/S cross-subsidize* the commodities in *S*.

Two further conditions on cost functions are assumed. First, *decreasing ray average cost*, i.e.,

$$c(\gamma z) < \gamma c(z) \quad \text{for} \quad \forall \gamma > 1. \tag{6.3}$$

Second, *transray convexity*, i.e.,

$$c(\lambda z^1 + (1-\lambda) z^2) \leq \lambda c(z^1) + (1-\lambda) c(z^2), \qquad 0 < \lambda < 1. \tag{6.4}$$

The "*Weak Invisible Hand Theorem*" [Baumol, Bailey and Willig (1977)] states that *under conditions* (6.1)–(6.4) *Ramsey prices are sustainable*. Thus, under these conditions, a monopoly which uses prices to deter entry leads to an efficient allocation without governmental regulation.

A number of comments on this result are now in order. Assumptions (6.3) and (6.4) are contradictory when they are assumed globally and when the cost function has no fixed costs, $c(0) = 0$.[9] Baumol et al. (1977) have noticed this problem and thus required that (6.4) hold only on the hyperplane which is tangent to the zero-profit curve at the Ramsey prices [see Baumol, Bailey and Willig (1977, p. 356)]. Clearly, these assumptions crucially depend on the location of the Ramsey optimum allocation.

Furthermore, in the separable cost case, i.e.,

$$c(z) = \sum_{j=1}^{n} c_j(z_j),$$

$$c_j\colon E^1_+ \to E^1_+ \quad \text{and} \quad c(0) = 0,$$

these conditions are contradictory. In fact, it has been shown by Mirman, Tauman and Zang (1982) that in this case the *only* sustainable prices are average cost prices, which may obviously differ from "Ramsey Prices".

Finally, Brock and Scheinkman (1983) advanced the notion of *quantity sustainability*. While in Baumol et al. (1977) entrants expect that the incumbent's *prices* remain fixed [hence, definition (6.1)–(6.2) above may be called *price*

[9] In (6.4), let $z^2 \to 0$. Then $c(\lambda z^1) \leq \lambda c(z^1)$, $0 \leq \lambda \leq 1$, which is equivalent to $c(\gamma z^1) \geq \lambda c(z^1)$, $\gamma \geq 1$.

sustainability], they assume the polar case where entrants expect that the incumbents' quantities remain fixed. This definition, they point out, is more relevant in the case where the incumbent's costs are all sunk, because then it is in its interest to maintain larger output levels after entry than in the case where costs are escapable. They show that under a well-behaved demand function, price sustainability implies quantity sustainability. Moreover, it is possible that a quantity sustainable price vector yields positive profits to the monopoly. Hence, even in the separable cost case, quantity sustainable prices are not in general average cost prices. Thus, it is more likely for a weak invisible hand result to hold when the notion of price sustainability is replaced by the notion of quantity sustainability.

Baumol et al. (1982) claim that their notion of contestability avoids the *ex post* oligopolistic interactions based on entrants' conjectures, typical of current game-theoretic industrial-organization models [for example Kreps and Wilson (1982) and Milgrom and Roberts (1981)]. Attempts to lay out a framework specified as a game between incumbents and challengers [e.g. Maskin and Tirole (1982) and Mirman et al. (1982)] have shown that perfect contestability emerges as an equilibrium outcome in *some*, but not in all, dynamic games, depending on the assumptions and rules of the game (asymmetric) information (who moves first, etc.).

7. On the public firms' decision rules

We have assumed that private firms respond, competitively or non-competitively, to *price changes* induced by the public firm. In the competitive case this is a natural assumption. In a non-competitive environment, it is perhaps more natural to regard the public firm as a *dominant agent*, i.e., as one capable of imposing its decision rules, such that the other firms have to adapt to them. This view has been taken by Harris and Wiens (1980), followed by Beato and Mas-Colell (1983).

To simplify the discussion, assume that there is only one commodity (in addition to the *numéraire*) and one private firm. Consumption is x, the quantity produced by the public firm is z and by the private firm y: $x = z + y$.

A decision rule for the public firm is, in general form, an arbitrary function ψ: $E^1_+ \to E^1_+$, which assigns a public production $\psi(y)$ to every private production y and which belongs to some admissible set of functions Λ. Particular examples are the marginal cost pricing rule, to be discussed below, or a rule which assigns a constant level, say $\bar{z}$, to any level of y. Given the rule $\psi(y)$, private firms' profit maximization with respect to y determines the equilibrium configuration $(y, \psi(y))$. Optimization of social welfare consists of finding the best admissible decision rule given the information available on demand and the cost function of the private firm.

As pointed out by Beato and Mas-Colell (1983), in this general form, it is not clear what the admissible set Λ should be. In fact, if no restrictions are imposed and information is perfect then the first-best can be attained by the decision rule: $\psi(y) = \max[\bar{x} - y, 0]$, where $\bar{x}$ is the first-best production. This rule fixes the price at its optimum level with the effect that the private firm, robbed of its monopoly power, will produce the socially optimal output. This result, however, crucially depends on the public firm's decision rule being perfectly *credible*, irrespective, for example, of possible losses incurred by the public firm. A natural, but ad-hoc, restriction on ψ would thus be a *no-loss* condition. Furthermore, under uncertainty about private costs, the problem of finding the best ψ may not be degenerate. The optimum will depend, of course, on the characteristics of the uncertainty. However, no results are yet available on this question.

Rather than pursuing the full optimality approach, Beato and Mas-Colell (1983) have taken a "bounded rationality" line, comparing the performance of two "simple" decision rules: constant public output and marginal cost pricing (MCP).

A simple diagram may exhibit their analysis. Assume that there is one consumer, with an additively separable utility, U, linear in the *numéraire* x_0, $U = u(x) + x_0 = u(z+y) + x_0$, where u is strictly concave. Costs (in terms of the *numéraire*) of the public and private firms are $c(z)$ and $\hat{c}(y)$, respectively. For simplicity, both cost functions are assumed to be convex and hence no zero-profit constraint need be imposed on the public firm. For a given amount of the *numéraire*, welfare, or net utility, W, is given by

$$W(z, y) = u(z+y), - c(z) - \hat{c}(y). \tag{7.1}$$

Let us examine first the "constant output" decision rule. Assuming that z is given, the private firm maximizes profits, P, with respect to y,

$$\max_{y} P(z, y) = p(z+y)y - \hat{c}(y), \tag{7.2}$$

where

$$p(z+y) = u'(z+y).$$

Denote the graph of the solution to (7.2) by g: $y = g(z)$. The assumptions imply that g is continuous and decreasing, as in Figure 7.1. The social optimum is now obtained by maximizing (7.1) where $y = g(z)$. This is point L (quantity z_0), where a social indifference curve is tangent to g.

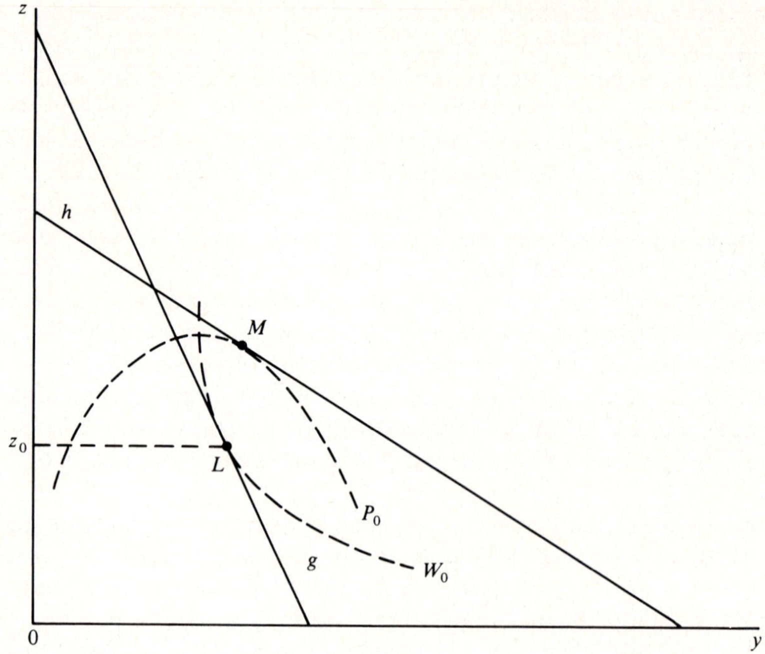

Figure 7.1

Now consider the MCP rule

$$p(z+y)-c'(z)=0, \tag{7.3}$$

whose graph is denoted by h: $z=h(y)$. The assumptions imply that h is also continuous and intersects g as described in Figure 7.1. The private firm maximizes its profits with respect to y assuming that $z=h(y)$,

$$\max_y P(h(y)+y)=p(h(y)+y)y-\hat{c}(y). \tag{7.4}$$

The solution is at M, where an iso-profit curve, P_0, is tangent to h.

It is quite clear that, in terms of W, the relation between M and L is ambiguous. In particular, the MCP rule is not dominated by the constant output rule. In fact, Beato and Mas-Colell (1983) show that with linear demands and constant marginal costs in the public firm, the MCP rule is superior to a constant output rule.

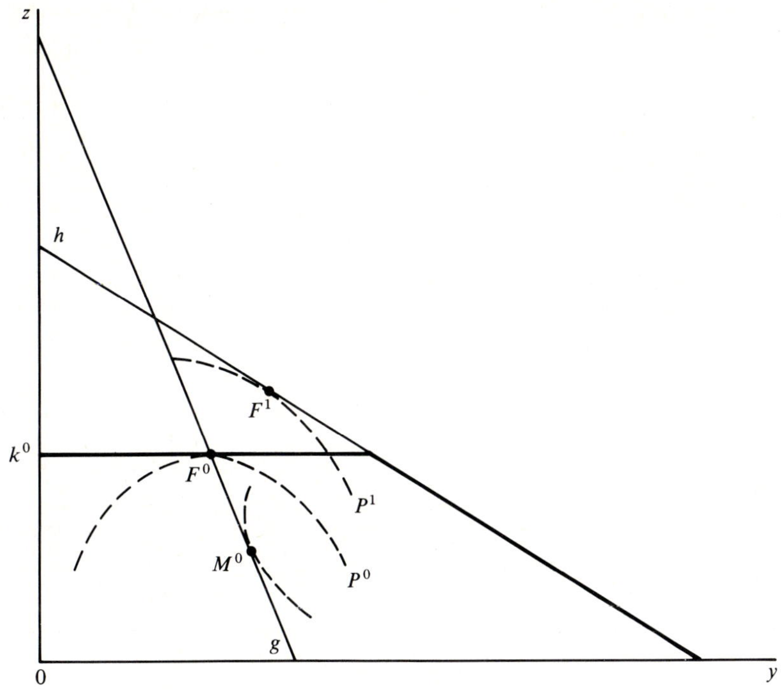

Figure 7.2

The ambiguity regarding the ranking of decision rules can best be seen by considering capacity constraints in the above example. Thus, suppose that the public firm has a capacity constraint k, i.e., $z \leq k$. Let the optimum solution with a *constant output* decision rule be denoted by (y^M, z^M) (point M^0 in Figure 7.2) and with the *MCP* rule by (y^F, z^F) (point F^0). Then it is easy to construct an example where for "low" capacity, k^0, $U(y^M, z^M) > U(y^F, z^F)$, while for "high" or no capacity constraint, the welfare ordering is reversed (point F^1 preferred to M^0).[10] As the example indicates, it may be desirable for the public firm to build capacity and to switch, as capacity expands, from one rule to the other. As the Beato and Mas-Colell (1983) calculations indicate, the region (in terms of parameters) where the MCP rule fares worse than constant output is when (the public firm's) capacity is low and marginal costs are high.

[10] The numerical example: $p = 10 - y - z$. The private firm's cost function $c(y) = \frac{1}{2}y^2$ [so $g(z) = (10 - z)/4$ for $z \leq 10$], the public firm's cost $c(z) = 3z$ and $k = 2.5$. Then $(y^M, z^M) = (3,1)$ and $(y^F, z^F) = (2.5, 2.5)$. Since $U = 10(y + z) - \frac{1}{2}(y + z)^2 - \frac{1}{2}y^2 - 3z$, clearly $U(3,1) > U(2.5, 2.5)$.

Their calculations also show that there is no obvious relation between the size of profits (for the public and for the private firms) and welfare levels. In fact, if a profit constraint is imposed, one of these decision rules may not be feasible.

8. Optimal supply of public goods

The introduction of public goods into the present model would not, in principle, present any further difficulty, as we could think of public goods as of government production which is given away rather than sold. Yet optimal pricing and production rules for public firms which produce both private and public goods will be discussed in more detail.

We let the vector v represent the supply of public goods, and we focus on the polar case where exclusion is not feasible and all consumers have to consume the same amount of each public good. Hence, individual preferences will be represented by the utility functions $U^h(x^h, v)$ and we set $\partial U^h/\partial v_k \equiv U_k^h > 0)$, $\forall k$.

The efficiency frontier of the public production technology is now given by the implicit production function $g(z, v) = 0$.

Necessary conditions for efficient supply of public goods are then given by conditions analogous to (2.1),

$$\sum_h \lambda^h \left(U_k^h + \sum_i U_i^h \frac{\partial x_i^h}{\partial v_k} \right) - \sum_i \alpha_i \left(\sum_h \frac{\partial x_i^h}{\partial v_k} - \sum_j \frac{\partial y_i^j}{\partial v_k} \right) - \beta g_k = 0, \quad \forall k, \tag{8.1}$$

where

$$g_k \equiv \partial g / \partial v_k.$$

We write $\pi_k^h \equiv U_k^h / U_0^h \equiv$ consumer h's marginal rate of substitution between public good k and the *numéraire*, or consumer h's (individualized) price for public good k in terms of the *numéraire*. Observing that $\sum_i p_i (\partial x_i^h / \partial v_k) = 0$, $\forall h$, and using (2.4), (8.1) can be rewritten as

$$\sum_h \sum_i \delta_i \frac{\partial x_i^h}{\partial r^h} \pi_k^h = c_k^0 + \sum_i \delta_i \left(\sum_h \frac{\partial x_i^h}{\partial v_k} - \sum_j \frac{\partial y_i^j}{\partial v_k} \right), \tag{8.2}$$

where c_k^0 is the marginal production cost of public good k. Substituting from (2.7) and observing that $\sum_i p_i (\partial x_i^h / \partial r^h) = 1$, $\forall h$, and $\sum_i c_i^j (\partial y_i^j / \partial v_k) = 0$, $\forall j$, after some manipulations, (8.2) can be rewritten as

$$\sum_h \left(1-\mu-\sum_i (p_i - c_i^0)\frac{\partial x_i^h}{\partial r^h}\right)\pi_k^h = c_k^0 - \sum_h\sum_i (p_i - c_i^0)\frac{\partial x_i^h}{\partial v_k} + \sum_i\sum_j (p_i - c_i^0)\frac{\partial y_i^j}{\partial v_k} - (1-\mu)\sum_i\sum_j (p_i - c_i^j)\frac{\partial y_i^j}{\partial v_k}. \tag{8.3}$$

The compensated demand derivatives of private goods with respect to public goods supply are defined by

$$\frac{\partial \hat{x}_i}{\partial v_k} \equiv \sum_h \left(\frac{\partial x_i^h}{\partial v_k} - \frac{\partial x_i^h}{\partial r^h}\pi_k^h\right),$$

and, accordingly,

$$\frac{\partial \hat{z}_i}{\partial v_k} \equiv \frac{\partial \hat{x}_i}{\partial v_k} - \sum_j \frac{\partial y_i^j}{\partial v_k}$$

is the partial derivative of the compensated net market demand for private good i with respect to the supply of public good k. Using these definitions, condition (8.3) can be rewritten as

$$\sum_h \pi_k^h = \frac{1}{1-\mu}\left[c_k^0 - \sum_i (p_i - c_i^0)\frac{\partial \hat{z}_i}{\partial v_k}\right] - \sum_i\sum_j (p_i - c_i^j)\frac{\partial y_i^j}{\partial v_k}. \tag{8.4}$$

With marginal cost pricing of private goods everywhere in the economy and with no constraints on public spending, (8.4) simplifies to the familiar Samuelsonian conditions for optimal public goods supply [see Samuelson (1954)]. With perfect competition in the private sector, (8.4) is identical to the conditions for optimal supply of public goods derived by Drèze and Marchand (1976) and Lau, Sheshinski and Stiglitz (1978).

The left-hand side of (8.4) is of course the marginal social value measured in terms of the *numéraire* of increasing the supply of public good k. The term in brackets on the right-hand side is the net resource requirement in the public sector needed for a marginal increase in the supply of public good k. This net resource requirement may be greater or smaller than marginal production cost, depending on whether the public sector prices private goods above or below marginal costs and on whether these private goods are complementary to or substitutes for public good k. The net resource requirement for a marginal increase in the supply of public good k is inflated by a factor $1/(1-\mu) \geq 1$, which reflects the fact that the shadow price of the *numéraire* good is higher in the public sector than in the private sector in the case where the public budget constraint is binding. The last term on the right-hand side of (8.4) is due to price distortions in the private sector and expresses the social cost or gain resulting from the re-allocation of resources in the private sector caused by a marginal (compensated) increase in the supply of public good k.

An example may perhaps be in order to illustrate the optimality condition (8.4). Suppose that a public TV station is contemplating an additional television channel. Assume further that this requires highly specialized program personnel of which the TV station is the sole employer. Let this input be indexed e and assume for simplicity that $\partial \hat{z}_i / \partial v_k = 0$, $\forall i \neq e$. We can also assume that the optimal pricing rule derivable from the first-order conditions (2.11) tells the TV station to behave monopsonistically in the market for commodity e, setting $p_e < -\partial z_0 / \partial z_e$. Hence, in this case, the net resource requirement of providing an additional TV channel will be smaller than marginal cost c_k^0. Moreover, an additional TV channel may lead to increased production of TV sets in the private sector and, to the extent that TV sets are subject to mark-up pricing, this will entail a social gain which should be added on the benefit side.

We now turn to the case where all price distortions are caused by commodity taxes and subsidies and we assume that the government controls all production, so that $y_i = 0$ and $x_i = z_i$, $\forall i$. In this case, we can show that the familiar Samuelsonian conditions have to be modified to take into account what may be called an allocative effect and a distortive effect caused by an increase in public goods supply.

Under the above assumption, condition (8.4) simplifies to

$$\sum_h \pi_k^h = \frac{1}{1-\mu}\left(c_k^0 - \sum_i t_i \frac{\partial \hat{x}_i}{\partial v_k}\right), \tag{8.5}$$

which is similar to the results of Diamond and Mirrlees (1971) and Lau, Sheshinski and Stiglitz (1978). In the case where all production of private goods takes place in the public sector, condition (2.11) for optimal pricing can be rewritten as

$$\bar{\gamma} = -\sum_i \frac{t_i}{x_e} \frac{\partial \hat{x}_i}{\partial p_e}, \quad \forall e,$$

and substituting into (8.5) we get

$$\sigma \sum_h \pi_k^h + \sum_i t_i \frac{\partial \hat{x}_i}{\partial v_k} = c_k^0, \tag{8.6}$$

where

$$\sigma \equiv 1 + \sum_i \frac{t_i}{x_e} \frac{\partial \hat{x}_i}{\partial p_e}.$$

Condition (8.6) for optimal supply of public goods says that marginal production costs for public goods should be equated to a factor σ times the sum of individual consumer prices, plus the social gain (or loss) resulting from the changes in the demand for private goods caused by the complementarity and substitutability of private and public goods. The latter effect, which is given by the change in net tax payments by households, represents the allocative effect of increased public goods supply.

As public expenditures are financed through commodity taxation, an increased supply of public goods will lead to increased taxation. This will have a distortive effect which is given by the factor σ. From the negative definiteness of the Slutsky matrix it can be shown that under the condition of positive public expenditure ($\sum_i t_i x_i > 0$),

$$\sum_i \frac{t_i}{x_e} \frac{\partial \hat{x}_i}{\partial p_e} \leq 0;$$

hence $\sigma \leq 1$ and the distortive effect will always be non-positive.

9. Some intertemporal issues

Interpreting $x_1, x_2, \ldots$ as the same commodities at *different dates*, the previous discussion can be interpreted as applying to the problem of optimum public investment, i.e., the socially optimum shadow discount rates for public investment. This assumes, however, that the public firm is regulated so that the *present value* of its profits achieve a given target level (b). This seems a sensible formulation of the dynamic Ramsey problem since it recognizes the existence of capital markets for borrowing and lending. However, a time inconsistency problem may arise when explicit dynamics are introduced [Brock (1982)].

Suppose that production in the public firm is subject to a "learning" element, which can be simply modelled by modifying the public firm's technology to

$g(z, \varepsilon) = 0$, where ε is a productivity parameter and the change in ε over time is given by a standard distributed lags function of outputs,

$$\frac{\mathrm{d}\varepsilon}{\mathrm{d}t} = \sum_i a_i z_i - \theta\varepsilon, \tag{9.1}$$

where a_i ($a_i \geq 0$) is the "contribution" of a unit of output i to efficiency and θ ($\theta \geq 0$) a fixed depreciation rate. Clearly, *all* variables have now a time dimension. Assume first that the profit constraint applies at *each* point in time,

$$b - \sum p_i z_i = 0, \quad \forall t. \tag{9.2}$$

The objective is to maximize the infinite-horizon present value of the maximand in Section 2, with a given positive social discount rate, r.

Elementary control theory yields that equation (2.11), which applies to the case of a competitive fringe, then becomes

$$\sum_i (c_i^0 - \xi_i - p_i)\frac{\partial \hat{z}_i}{\partial p_e} = \mu \hat{z}_e, \quad \forall t, \tag{9.3}$$

where

$$\xi_i(t) = a_i \int_t^{\infty} \mathrm{e}^{-(r+\theta)(s-t)}\left(-\frac{\partial c_i^0}{\partial \varepsilon}\right)\mathrm{d}s, \qquad t \geq 0. \tag{9.4}$$

The "shadow price" ξ_i represents the prospective contribution of z_i to future cost reduction along the optimum path. We may interpret $c_i^0 - \xi_i$ as the "*true*" marginal cost of output i. Notice that in (9.3), μ measures the value of relaxing the constraint (9.2) and hence it may vary in possibly complicated ways over time. Clearly, operationalizing (9.3) is a complex issue.

The profit constraint in present value terms is

$$\int_0^{\infty} \mathrm{e}^{-rt}\left(\sum_i p_i z_i\right)\mathrm{d}t = b/r, \tag{9.5}$$

and similar first-order conditions can be deduced. However, as Brock and Dechert (1982) have argued, the solution may be *time-inconsistent*. That is, starting at date $t > 0$, with initial conditions being the optimum values obtained for t starting at 0, the optimum values for any $s > t$ are different from those obtained when starting at 0. Clearly, there is little point in laying out an optimal plan if it is not optimal to follow the rest of the plan at subsequent dates.

References

Areeda, P. and D. F. Turner (1975), "Predatory pricing and related practices under Section 2 of the Sherman Act", Harvard Law Review, 88:637–733.

Arrow, K. J. (1983), Testimony in U.S. Government vs. AT&T.

Atkinson, A. B. and J. E. Stiglitz (1980), Lectures on Public Economics. New York: McGraw-Hill.

Averch, H. and L. L. Johnson (1962), "Behavior of the firm under regulatory constraint", American Economic Review, 52:1053–1069.

Bailey, E. E. and J. C. Panzar (1981), "The contestability of airline markets during the transition to deregulation", Law and Contemporary Problems, 44:125–145.

Baumol, W. J., E. E. Bailey and R. D. Willig (1977), "Weak invisible hand theorems and the sustainability of multiproduct natural monopoly", American Economic Review, 67:350–365.

Baumol, W. J., J. C. Panzar and R. D. Willig (1982), Contestable Markets and the Theory of Industry Structure. New York: Harcourt–Brace–Jovanovich.

Beato, P. and A. Mas-Collel (1983), "The marginal cost pricing rule as a regulation mechanism in mixed markets", HIER Discussion Paper No. 957.

Billera, L. J., D. C. Heath and J. Raanan (1978), "Internal telephone billing rates – A novel application of non-atomic game theory", Operations Research, 26:956–965.

Boiteux, M. (1960), "Peak-load pricing", The Journal of Business, 33:157–179.

Boiteux, M. (1971), "On the management of public monopolies subject to budgetary constraints", Journal of Economic Theory, 3:219–240.

Bös, D. (1981), Economic Theory of Public Enterprise. New York: Springer Verlag.

Brock, W. A. and J. A. Scheinkman (1983), "Free entry and the sustainability of natural monopoly: Bertrand revisited by Cournot", in: Evans (1983).

Brown, D. J. and G. Heal (1980a), "Two-part tariffs, marginal cost pricing and increasing returns in a general equilibrium model", Journal of Public Economics, 13:25–49.

Brown, D. J. and G. Heal (1980b), "Marginal cost pricing revisited", Mimeographed. Econometric Society World Congress at Aix-en-Provence, Aug.

Christensen, L. R. and W. H. Green (1976), "Economies of scale in U.S. electric power generation", Journal of Political Economy, 84:655–676.

Demsetz, H. (1968), "Why regulate utilities?", Journal of Law and Economics, 11:55–65.

Diamond, P. A. and J. A. Mirrlees (1971a), "Optimal taxation and public production I: Production efficiency", American Economic Review, 61:8–27.

Diamond, P. A. and J. A. Mirrlees (1971b), "Optimal taxation and public production II: Tax rules", American Economic Review, 61:261–278.

Drèze, J. H. (1964), "Some postwar contributions of French economists to theory and public policy", American Economic Review, 54, Part 2 (suppl.): 1–64.

Drèze, J. H. (1980), "Public goods with exclusion", Journal of Public Economics, 13:5–24.

Drèze, J. H. and M. Marchand (1976), "Pricing, spending, and gambling rules for non-profit organizations", in: R. E. Grieson, ed., Public and Urban Economics: Essays in Honor of William S. Vickrey. Lexington: Lexington, 59–89.

Evans, D. S., ed. (1983), Breaking up Bell: Essays on Industrial Organization and Regulation. Amsterdam: North-Holland.

Faulhaber, G. R. (1975), "Cross-subsidization: Pricing in public enterprises", American Economic Review, 65:966–977.

Foster, E. and H. Sonnenschein (1970), "Price distortions and economic welfare", Econometrica, 38:281–297.

Green, H. A. J. (1961), "The social optimum in the presence of monopoly and taxation", Review of Economic Studies, 28:66–78.

Grossman, S. J. (1981), "Nash equilibrium and the industrial organization of markets with large fixed costs", Econometrica, 49:1149–1172.

Guesnerie, R. (1980), "Second-best pricing rules in the Boiteux tradition: Derivation, review and discussion", Journal of Public Economics, 13:51–80.

Hagen, K. P. (1979), "Optimal pricing in public firms in an imperfect market economy", Scandinavian Journal of Economics, 81:475–493.

Harris, R. and E. Wiens (1980), "Government enterprise: An instrument for the internal regulation of industry", Canadian Journal of Economics, 13:125–132.

Hotelling, H. (1938), "The general welfare in relation to problems of taxation and of railway and utility rates", Econometrica, 6:242–269.

Joskow, P. L. (1976), "Contributions to the theory of marginal cost pricing", Bell Journal of Economics, 7:197–206.

Kahn, A. E. (1970–71), The Economics of Regulation: Principles and Institutions, 2 vols. New York: Wiley.

Kreps, D. M. and R. Wilson (1982), "Reputation and imperfect information", Journal of Economic Theory, 27:253–279.

Lau, L., E. Sheshinski and J. Stiglitz (1978), "Efficiency in the optimum supply of public goods", Econometrica, 46:269–284.

Mas-Collel, A. (1980), "Noncooperative approaches to the theory of perfect competition: Presentation", Journal of Economic Theory, 22:121–135.

Maskin, E. S. and J. Tirole (1982), "A theory of dynamic oligopoly, I: Overview and quantity competition with large fixed costs", Manuscript. Cambridge: M.I.T., Department of Economics.

Milgrom, P. and J. Roberts (1982), "Predation, reputation, and entry deterrence", Journal of Economic Theory, 27:280–312.

Mirman, L., Y. Tauman and I. Zang (1982), "Issues in the theory of contestable markets", Manuscript. Champaign: University of Illinois, Department of Economics.

Mirman, L., Y. Tauman and I. Zang (1983), "Ramsey prices, average cost prices and price sustainability", Discussion Paper No. 561. Evanston: Center for Mathematical Studies in Economics and Management Science, Northwestern University.

Ng, Y. K. and M. Weisser (1974), "Optimal pricing with a budget constraint: The case of the two-part tariff", Review of Economic Studies, 41:337–345.

Panzar, J. C. and R. D. Willig (1977), "Free entry and the sustainability of natural monopoly", Bell Journal of Economics, 8:1–22.

Pigou, A. C. (1928), A Study of Public Finance. London: Macmillan.

Posner, R. (1976), Antitrust Law: An Economic Perspective. Chicago: University of Chicago Press.

Ramsey, F. (1927), "A contribution to the theory of taxation", Economic Journal, 37:47–61.

Samet, D. and Y. Tauman (1980), "A characterization of price mechanisms and the determination of marginal-cost prices under a set of axioms", Discussion Paper No. 8008. Louvain: C.O.R.E.

Schmalense, R. (1981), "Utility regulation and economic efficiency", in: The Control of Natural Monopolies. Lexington: Lexington.

Sheshinski, E. (1983), "Issues in Ramsey pricing", Discussion Paper No. 10/83. Jerusalem: Hebrew University, Department of Economics.

Spence, M. (1977), "Nonlinear prices and welfare", Journal of Public Economics, 8:1–18.

Stiglitz, J. E. (1981), "Potential competition may lower welfare", American Economic Review, 71:184–189.

Willig, R. D. and E. Bailey (1979), "The economic gradient method", American Economic Review, 69:96–101.

Zajac, E. E. (1978), Fairness or Efficiency: An Introduction to Public Utility Pricing. Cambridge: Ballinger.

Chapter 26

OPTIMAL ECONOMIC GROWTH, TURNPIKE THEOREMS AND COMPARATIVE DYNAMICS*

LIONEL W. McKENZIE*

University of Rochester

I. Optimal Paths and Duality

1. Introduction

We will be concerned with the long-term tendencies of paths of capital accumulation that maximize, in some sense, a utility sum for society over an unbounded time span. However, the structure of the problem is characteristic of all economizing over time whether on the social scale, or the scale of the individual or the firm. The mathematical methods that will be used are closely allied to the old mathematical discipline, calculus of variations. However, our problem is made simpler by substituting discrete for continuous time so that the Euler differential equation is replaced by a difference equation. On the other hand, the problem is complicated by the use of an infinite horizon and the adoption as a primary objective the characterization of the asymptotic behavior of optimal paths. We are particularly interested in the tendency of optimal paths which start from different initial positions to converge to the same limit path as time goes to infinity. We will go beyond the traditional approach in another direction to consider paths of capital stocks which meet the boundaries of the regions within which they must lie given the conditions of the problem, in particular, the requirement that the capital stocks be non-negative. Of course, this is one of the principal modern innovations in the theory of maximization from the work of writers such as Kuhn and Tucker (1951), Bellman (1957), and Pontryagin (1962).

*I have received assistance from many readers in preparing this chapter. Above all, I am grateful to my students, Swapan Dasgupta and Makoto Yano who have made numerous contributions to the text and corrected many errors. I have also received valuable aid on specific points from Jose Scheinkman, Peter Hammond, Teh M. Huo, Ali Khan, and Tapan Mitra. Finally, I am especially indebted to William Brock, David Cass, David Gale, and Roy Radner for my understanding of optimal growth theory.

Handbook of Mathematical Economics, vol. III, edited by K.J. Arrow and M.D. Intriligator
© 1986, Elsevier Science Publishers B.V. (North-Holland)

A crucial condition for the maximum to be achieved, whether as a necessary condition or as one of the sufficient conditions, has been concavity of the maximand, at least locally at the maximal path. This is to be expected from the conditions for a maximum of a function of a finite number of variables. In the calculus of variations the concavity that is needed is provided by the conditions of Weierstrass and Lagrange [see Bliss (1925) for a classical reference or Hestenes (1966) for a modern reference]. Moreover, when global results are sought, the concavity condition is assumed throughout a relevant region. This is also to be expected from the theory with a finite number of variables. In our theory concavity of the utility function will always be assumed, even uniformly over a relevant region for the global maximum and over time. The utility is defined directly on the capital stocks at the beginning and the end of a standard period of time, and the concavity is with respect to these variables. It should be mentioned, however, that some results have been achieved in one-sector models in which concavity is not assumed everywhere [see Skiba (1978), Majumdar and Mitra (1982), and Dechert and Nishimura (1983)].

In most of the discussion the utility function will be allowed to depend on time, as in the standard theory of the calculus of variations. Also the function to be maximized will be the sum of utility functions for each period over the future. This is described as a separable utility function over the sequence of future capital stocks and corresponds to the integral of calculus of variations. Since the consumption of one period does influence the utility of later consumption, the separability assumption is not exact. However, the error is no doubt reduced by lengthening the period, though this may not be much help in an application of the theory. Again there are results in the literature where the separability assumption is relaxed [see Samuelson (1971) and Iwai (1972)]. The treatment of utility in a period as dependent on initial and terminal stocks is not a restriction since the usual assumptions that make utility depend on consumption and consumption on production and terminal stocks will imply that an equivalent utility depending on capital stocks exists.

The theory that I will present will cover both discounted and undiscounted utility. We will seek to determine the asymptotic behavior of maximal paths, which display a tendency to cluster in the sufficiently distant future from whatever capital stocks they start. Other types of turnpike behavior that have been studied are clustering in early periods for finite optimal paths that start from the same initial stocks, but have different terminal stocks, and clustering in the middle parts of paths that may start and end with different stocks [see McKenzie (1976), McKenzie and Yano (1980), and Hieber (1981)]. In models with stationary utility functions, perhaps subject to discounting, the clustering has been seen as convergence to a stationary path along which capital stocks are constant. This view is reinforced by the fact that in stationary models the existence of stationary optimal paths, which are, moveover, supported by prices, is easy to prove by special means which are not useful for other optimal paths. Then this path and its prices can be used to establish the asymptotic convergence of other paths to it, with great ease in the undiscounted case. However, methods are now available from the work of Weitzman (1973) to derive the prices for other optimal paths directly so that the balanced path does not have a distinguished role in the asymptotic theory if existence is assumed or can be proved. Also methods are available which exploit concavity directly without introducing prices.

Our consideration will be confined to the deterministic model although using methods developed in this model analogous results have also been proved for the stochastic model in which the future is uncertain [see Evstigneev (1974), Brock and Mirman (1976), and Brock and Majumdar (1978)]. Also most of the argument will assume concavity of the relevant functions without requiring differentiability or interior solutions. However, some consideration will be given to differentiable cases where optimal paths are assumed to lie in the interior of the region of definition of the utility function. These stronger assumptions are analogous to the assumptions used in the comparative statics of general equilibrium models of competitive economies. Here they will permit some comparative dynamics to be done. The assumptions are in some ways even stronger than those usual in classical calculus of variations. However, the methods that become available are very powerful in the discrete model and, so far as I know, have not been extended to models where continuous time is the independent variable.

The original context for the optimal growth model was the problem of the level of saving that would maximize a utility sum over future time for a population. This problem was solved by the Cambridge mathematician Frank Ramsey (1928) for a one-good model, which may be thought of as an aggregated economy over an infinite future. The method used by him to handle the infinities involved is still useful today. However, the emphasis on asymptotic behavior for optimal paths appeared later in the multi-sector von Neumann model analyzed by Dorfman, Samuelson and Solow (1958). They dealt with finite paths where the objective was to maximize terminal stocks and their model contained two sectors. Since the model was stationary they could concentrate on the convergence of all optimal paths to a stationary optimal path. Later authors [Radner (1961), Morishima (1961), and McKenzie (1963)] extended the von Neumann model and the convergence theorems to many sectors. On the other hand, a Ramsey-style utility function on the consumption stream was introduced as the objective rather than terminal stocks. Also the horizon was extended to infinity. Asymptotic theorems for the one-sector Ramsey model were proved by Cass (1966), Koopmans (1965), and Samuelson (1965). Von Weizsäcker (1965) generalized the objective function somewhat by defining the overtaking criterion in which attention is turned to partial sums and optimality is assigned to a path whose partial utility sums eventually dominate when it is compared with an alternative path from the same initial stocks. He also dealt with a model in which utility and production functions change over time, but he aggregated the economy to a single sector. We will deal essentially with the Von Weizsäcker model in a disaggregated form, which is natural when the analysis is directed to asymptotic behavior of paths. The existence of infinite optimal paths in the stationary disaggregated model was proved by Gale (1967). Asymptotic theorems in this model were proved by Atsumi (1965), Gale (1967), and McKenzie (1968). The existence theorem was extended to models with discounted utility by Sutherland (1970), and the asymptotic theorems were extended to these models by Scheinkman (1976) and Cass and Shell (1976).

Although the primary sources of the optimal growth model are aggregate savings programs and capital accumulation programs for an economy, the theorems and methods of the subject find applications in other areas with increasing frequency. For example, applications are made to capital accumulation by the firm with adjustment costs by Brock and Scheinkman (1978) and Scheinkman (1978), and to competitive markets with perfect foresight by Brock (1974), or rational expectations by Brock (1980). In these models the social utility function is replaced by individuals' utility functions or by the profit functions of firms. Thus there is a movement toward a general theory of economic dynamics in which asymptotic theorems and comparative dynamic theorems form the bulk of the results and where the analysis is largely derived from the optimal growth literature. Excellent examples from the theory of competitive equilibrium are the recent works of Becker (1980), Bewley (1982), and Yano (1981), where the turnpike results from optimal growth theory are used to prove that competitive equilibria approach stationary states over time. It has been suggested that our subject is best described as the study of economizing over time [see Intriligator (1971)].

2. The basic model

We will use a reduced form of the objective function in which utility is expressed as a function of the initial and terminal stocks of a period. The utility function is written $u_t(x, y)$, where x is the vector of capital stocks at time $t-1$ and y is the vector of capital stocks at time t. Then u_t is the utility derived from activities during the time period from times $t-1$ to t, which we call the tth period. The reduced model is equivalent to the traditional extensive model in which utility is expressed as a function $u_t(c)$ of the consumption vector in the tth period. The extensive model introduces a production correspondence $f_t(x)$ which expresses output, not just capital goods, as depending on initial capital stocks. However, so long as the utility functions of different periods are independent, it is a necessary condition for an optimal program that c be chosen from $f_t(x) - y$, where y

represents terminal stocks, to maximize u_t. Thus the models are not significantly different. It should be noted that the full commodity space in which $f_t(x)$ lies may include labor services and perishable goods dated by their times of use within the period.

We may allow the utility function u_t, as well as the space E_t of capital stock vectors at time t, to depend on t. Then u_t maps a set D_t contained in the non-negative orthant of $E_{t-1} \times E_t$ into the real line, where E_{t-1} and E_t are Euclidean spaces of dimensions n_{t-1} and n_t, respectively. Let $|\cdot|$, for a vector argument, denote the Euclidean norm. We assume

(I) The utility functions $u_t(x, y)$ are concave and closed for all t. The sets D_t are convex.

(II) If $(x, y) \in D_t$ and $|x| < \xi < \infty$, there is $\zeta < \infty$ such that $|y| < \zeta$.

Assumption (I) provides the concavity and convexity that are recurrent features of calculus of variations and other theories of maximization. By u_t is *closed* is meant that $(x, y) \in$ boundary D_t implies $u_t(x, y) = \limsup(u_t(z, w)$ as $(z,w) \to (x, y)$ if $(x, y) \in D_t$ and $u_t(z, w) \to -\infty$ otherwise. Since we are seeking global results, the assumptions on concavity and convexity are global. The boundedness assumption (II) is made to avoid trivial cases. Note that (I) and (II) imply that $u_t(x, y)$ is bounded above for $|x| < \xi$.

A sequence of capital stocks $\{k_t\}$, $t \in N$, is a path of accumulation if N is a set of consecutive integers and $(k_{t-1}, k_t) \in D_t$ when $t-1$ and t are in N. The set N may be a finite or an infinite set.

We may note that the capital stocks are the state variables in the language of optimal control and there is no need to confine them to physical goods or things that can be appropriated as private or public property. For example, features of the environment, skills of workers, and mineral deposits may also be included. These offer ways in which future utility possibilities may be influenced by present choices. In addition, the dependence of the utility functions on time may take account of trends in technology, tastes, and environment in so far as they are independent of the choices made. Of course, the interpretation of the state variables will depend on the particular problem at hand. Our descriptions have been appropriate to the interpretation of u_t as a social utility function that is the objective of planning by the state.

In classical economics the concavity of the production correspondence which is part of the ground for assumption (I) is often explained in terms of the independence and linearity of basic productive activities, at least to an approximation. However, when external effects are present so that different activities influence one another, this ground of concavity is jeopardized [Starrett (1972)]. Also polluting substances in the environment are not allocated between activities the way capital goods are, so they do not fit into the paradigm of an allocation of stocks among independent, linear activities. These are important qualifications to the generality of the model.

3. The objective function

The objective function for a finite program from $t=0$ to $t=T$ is $\sum_{t=1}^{T} u_t(x_{t-1}, y_t)$. If the sum exists, the objective function for an infinite program $\{k_t\}$ beginning at $t=0$ is similarly $\sum_{t=1}^{\infty} u_t(k_{t-1}, k_t)$. However, the infinite sum may not exist and one of Ramsey's achievements was to show that this difficulty may be overcome in certain models with stationary utility functions by subtracting a constant from each term of the series to be summed. A more general method was introduced more recently by Von Weizsäcker (1965) and Atsumi (1965) and refined by Gale (1967) and Brock (1970). In this approach the infinite sum is replaced by a comparison of finite partial sums. The new criterion is called the overtaking criterion.

Two definitions are made. The stronger definition characterizes an optimal path. We will say that a path $\{k_t\}$ *catches up to* a path $\{k'_t\}$ starting at the same time, if for any $\varepsilon > 0$ there is $T(\varepsilon)$ such that $\sum_1^T (u_t(k'_{t-1}, k'_t) - u_t(k_{t-1}, k_t)) < \varepsilon$ for all $T > T(\varepsilon)$. Then a path $\{k_t\}$ is *optimal* if it catches up to every alternative path from the same initial stocks. In other words, an optimal path is asymptotically as good as any other path from the same starting point when they are compared by means of their initial segments.

We will say that a path $\{k'_t\}$ *overtakes* a path $\{k_t\}$ starting at the same time, if there is $\varepsilon > 0$ and $T(\varepsilon)$ such that $\sum_1^T (u_t(k'_{t-1}, k'_t) - u_t(k_{t-1}, k_t)) > \varepsilon$ for all $T > T(\varepsilon)$. Then a path $\{k_t\}$ is *maximal* if there is no path from the same initial stocks that overtakes it. This says that a maximal path does not become permanently worse than some alternative path when they are compared by means of their initial segments.

4. Support prices

We wish to allow for maximal paths that do not remain interior to the sets D_t at all times, or perhaps at any time. In these cases derivatives will not always exist for the utility functions along the path. For this reason it is convenient to introduce dual variables, which we call prices, as generalizations of derivatives. Then it is also possible to dispense with assumptions of differentiability in the interior of D_t as well. The existence of the appropriate prices for our purposes was proved by Weitzman (1973) when utility is summable. However, his method can be adapted to the overtaking criterion [McKenzie (1976) and Hieber (1981)]. A theorem corresponding to that of Weitzman has been proved for the continuous time model by Benveniste and Scheinkman (1982). This was extended to a continuous time model with the overtaking criterion by Takekuma (1982).

Consider a maximal path $\{k_t\}$, $t \in N$, where N is the set of non-negative integers. First, we normalize the utility function choosing the zeros of utility so that $u_t(k_{t-1}, k_t) = 0$ in every period. This is harmless since the choice of the zero level of utility in each period has no effect on the comparison of paths. Next we define a value function $V_t(x)$ which values a capital stock at time t by the utility sums that can be got from it in the future. Following the example of Peleg and Zilcha (1977) in the stationary model, we set

$$V_t(x) = \sup\left(\liminf_{T \to \infty} \sum_{t+1}^{T} u_\tau(h_{\tau-1}, h_\tau)\right), \tag{4.1}$$

over all paths $\{h_\tau\}$ with $h_t = x$. $V_t(x)$ is well defined when the right-hand side of (4.1) exists as a finite number or positive infinity. A little computation will show that the concavity of u_t and the convexity of D_t imply that $V_t(x)$ is concave and well defined on a convex set K_t. Since $V_t(k_t) = 0$ for all t, K_t is not empty. We may also note that $V_t(x)$ is well defined for any x for which there is a path $\{k'_\tau\}$ with $k'_t = x$ and $k'_{t+n} = k_{t+n}$.

Let P_t be the set of capital stocks y such that there is x with $(x, y) \in D_t$. P_t is the set of capital stocks that can be produced from some capital stocks held at time $t-1$. S is a *flat* in the Euclidean space E if there are vectors $y_i \in E$, $i \in I$, where I is a finite set, such that $z \in S$ is equivalent to $z = \sum_{i \in I} \alpha_i y_i$ for some numbers α_i such that $\sum_{i \in I} \alpha_i = 1$. Let S_0 be the smallest flat in E_0 that contains K_0, and for $t \geqq 1$, let S_t be the smallest flat in E_t that contains P_t and K_t. It is crucial to the derivation of support prices for $\{k_t\}$ to assume:

(III) Interior $(P_t \cap K_t) \neq \phi$ relative to S_t, for all $t \geqq 1$. Also $k_0 \in$ interior K_0 relative to S_0.

It is important to notice that assumption (III) is not independent of the maximal path $\{k_t\}$, since the sets K_t depend on $V_t(x)$ which is defined after normalizing utility on $\{k_t\}$.

Since k_0 lies in the relative interior of K_0, given any $x \in K_0$, there is x' such that $k_0 = \alpha x + (1-\alpha)x'$ with $0 < \alpha \leq 1$ and $x' \in K_0$. Then, from the concavity of u_t and $V_0(k_0) = 0$, it follows that $V_0(x) < \infty$. But $V_0(x) < \infty$ and $(x, y) \in D_1$ implies $V_1(y) < \infty$. Since by assumption (III), y may be chosen in the interior of $P_1 \cap K_1$ relative to S_1, $V_1(x) < \infty$ for all $x \in K_1$. This argument can be continued to any $t > 0$, so $V_t(x) < \infty$ for $x \in K_t$ for all t. In interpreting the model it should be recalled that any goods not held at $t = 0$ may be omitted from E_0 and any goods that cannot be produced from k_0 after t periods may be omitted from E_t.

From the definition (4.1) of $V_t(x)$ it is clear that the principle of optimality holds and we may also write

$$V_t(x) = \sup(u_{t+1}(x, y) + V_{t+1}(y)), \tag{4.2}$$

over all y such that $(x, y) \in D_{t+1}$ and $y \in K_{t+1}$. Make the induction assumption that there exists $p_t \in E_t$ (p_t may be 0) where $t \geqq 0$, such that

$$V_t(k_t) - p_t k_t \geqq V_t(x) - p_t x, \tag{4.3}$$

over all $x \in K_t$. Let $x = k_t$ in (4.2). Then the sup is attained at $y = k_{t+1}$ by the maximality of $\{k_\tau\}$. The substitution of (4.2) in (4.3) gives

$$u_{t+1}(k_t, k_{t+1}) + V_{t+1}(k_{t+1}) - p_t k_t \geqq u_{t+1}(x, y) + V_{t+1}(y) - p_t x, \tag{4.4}$$

for all $(x, y) \in D_{t+1}$ with $y \in K_{t+1}$. Denote the left-hand side of (4.4), a given number, by v_{t+1}. Then

$$v_{t+1} - u_{t+1}(x, y) + p_t x \geqq V_{t+1}(y). \tag{4.5}$$

We define two sets for each $t \geqq 0$,

$$A = \{(w, y) | y \in P_{t+1} \quad \text{and} \quad w > v_{t+1} - u_{t+1}(x, y) + p_t x$$
$$\text{for some } x \text{ with} \quad (x, y) \in D_{t+1}\},$$

and

$$B = \{(w, y) | y \in K_{t+1} \quad \text{and} \quad w \leqq V_{t+1}(y)\}.$$

By the existence of the maximal path $\{k_t\}$, $P_{t+1} \cap K_{t+1} \neq \phi$. Thus A and B are not empty. A and B are disjoint by the inequality (4.5). They are also convex. Thus by a separation theorem for convex sets [Berge (1963, p. 163)] A and B may be separated by a hyperplane contained in $R \times E_{t+1}$, where R is the real line. The separating hyperplane may be defined by a vector $(\pi, -p_{t+1}) \neq 0$, where p_{t+1} lies in the linear subspace parallel to S_{t+1} (that is, $q \in S_{t+1}$ implies $p_{t+1} + q \in S_{t+1}$). Then $\pi w - p_{t+1} y \geqq w' - p_{t+1} y'$ for all $(w, y) \in A$ and $(w', y') \in B$. This situation is illustrated in Figure 4.1.

Using the definitions of w, w', and v_{t+1} and relation (4.4), the separation of A and B implies

$$\pi\{u_{t+1}(k_t, k_{t+1}) + V_{t+1}(k_{t+1}) - p_t k_t - u_{t+1}(x, y) + p_t x\} - p_{t+1} y$$
$$\geqq \pi V_{t+1}(y') - p_{t+1} y', \tag{4.6}$$

for any (x, y) such that $(x, y) \in D_t$ and any $y' \in K_{t+1}$. If $\pi = 0$, (4.6) implies that $p_{t+1} \cdot (y' - y) \geqq 0$ for all $y' \in K_{t+1}$ and $y \in P_{t+1}$. However, $P_{t+1} \cap K_{t+1}$ has an interior in S_{t+1} by assumption (III), and p_{t+1} is parallel to S_{t+1}. Therefore, $p_{t+1} = 0$ as well, contradicting the requirement that $(\pi, p_{t+1}) \neq 0$. Thus $\pi \neq 0$ and we may set $\pi = 1$. Put $x = k_t$, $y = k_{t+1}$ and (4.6) becomes

$$V_{t+1}(k_{t+1}) - p_{t+1}k_{t+1} \geqq V_{t+1}(y') - p_{t+1}y', \tag{4.7}$$

for all $y' \in K_{t+1}$. Put $y' = k_{t+1}$ and we obtain

$$\{u_{t+1}(k_t, k_{t+1}) - p_t k_t\} + p_{t+1}k_{t+1} \geqq \{u_{t+1}(x, y) - p_t x\} + p_{t+1}y, \tag{4.8}$$

for all $(x, y) \in D_{t+1}$.

The induction is begun by supporting the value function $V_0(y)$ at $k_0 \in K_0$ in $R \times E_0$. The concavity of $V_0(y)$ implies there is $(\pi, p_0) \neq 0$ such that $p_0 \in E_0$

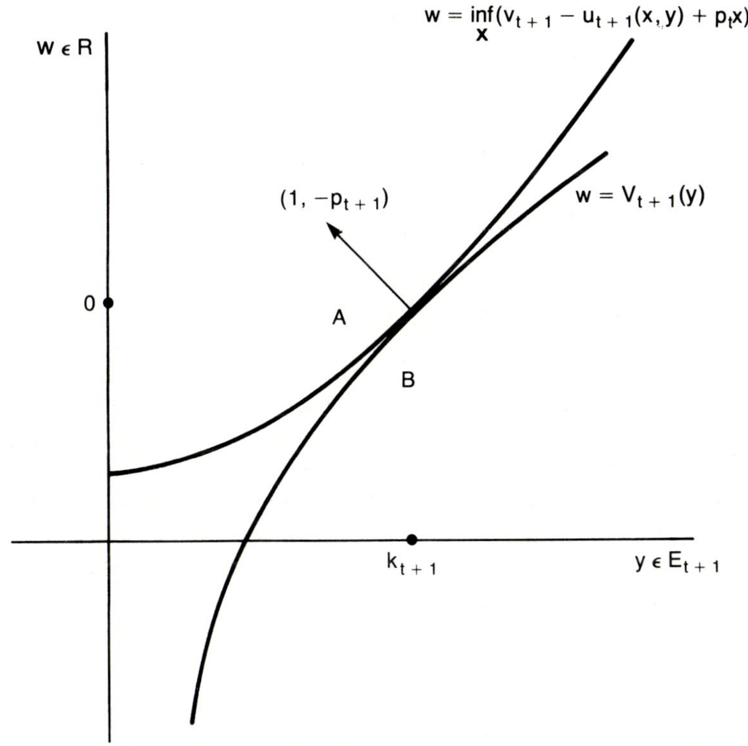

Figure 4.1

and

$$\pi V_0(k_0) - p_0 k_0 \geqq \pi V_0(x) - p_0 x, \tag{4.9}$$

for all $x \in K_0$. Choose p_0 in the linear subspace parallel to S_0 where S_0 is the smallest flat containing K_0. If $\pi = 0$, $p_0 \neq 0$ and $p_0(k_0 - x) \leqq 0$ for all $x \in K_0$. Since $k_0 \in$ relative interior K_0, the inequality (4.9) is impossible, and $\pi \neq 0$. We may choose $\pi = 1$, as before.

We have proved that prices exist supporting maximal paths in the following sense:

Lemma 4.1

Let $\{k_t\}$, $t = 0,1,\ldots,$ be a maximal path of accumulation. If assumptions (I), (II), and (III) are met, there exists a normalization of utility and a sequence of price vectors $p_t \in E_t$, $t = 0,1,\ldots,$ which satisfy

$$V_t(k_t) - p_t k_t \geqq V_t(y) - p_t y, \tag{4.10}$$

for all $y \in K_t$, and $V_t(k_t)$ is finite, and

$$u_{t+1}(k_t, k_{t+1}) + p_{t+1}k_{t+1} - p_t k_t \geqq u_{t+1}(x, y) + p_{t+1}y - p_t x, \tag{4.11}$$

for all $(x, y) \in D_{t+1}$.

By (4.10) the prices support the value function. By (4.11) they support the utility function. These properties of the prices play crucial roles in the arguments leading to turnpike theorems for maximal paths when assumptions of differentiability of the utility function and interiority of paths are not made. The fact that the Weitzman prices support the value function implies that they are Malinvaud prices (1953), that is, k_t has minimal value at p_t over the set of capital stocks from which the subsequent utility stream can be obtained. This is obvious from (4.10), since $V_t(y) = V_t(k_t)$ implies $p_t k_t \leqq p_t y$. Of course, Malinvaud prices are defined for efficient paths rather than maximal paths and, in particular, summable utility is not needed. A path $\{k_t\}$, $t = 1,2,\ldots,$ is said to be *efficient* if there is no path $\{k_t'\}$ with $k_0' = k_0$ such that $u_t(k_{t-1}', k_t') \geqq u_t(k_{t-1}, k_t)$ for all t with strict inequality for some t. It is clear that maximal paths must be efficient, but the contrary need not hold.

The converse of Lemma 4.1 is not true. However, a slight relaxation of the maximality conditions allows a converse result to be proved. The argument for Lemma 4.1 only requires that consecutive stocks along the path realize the (finite) supremum in (4.2), that is, for all t, it should be true that

$$V_t(k_t) = u_{t+1}(k_t, k_{t+1}) + V_{t+1}(k_{t+1}), \tag{4.12}$$

or equivalently that

$$V_0(k_0)=\sum_1^T u_t(k_{t-1},k_t)+V_T(k_T),\quad \text{all } T\geqq 1. \tag{4.13}$$

Then under assumptions (I), (II), and (III), the proof proceeds just as given. A path satisfying (4.13), where utility is normalized so that $V_0(k_0)$ is finite, may be called *potentially maximal*, since any "loss" from using an initial segment can be made arbitrarily small [a related idea for finite horizons may be found in Hammond and Mirrlees (1973) and Hammond (1975)]. At any time T, given an arbitrary $\varepsilon>0$, the initial segment of the potentially maximal path may be completed with a new choice of capital stocks beyond T, so that no path from the beginning can overtake the revised path by more than ε. As earlier, one path overtakes a second by ε if its finite sums eventually exceed those of the second path by ε at all subsequent times. If we call the revised path ε-maximal, the potentially maximal path can at any time be converted into an ε-maximal path where ε may be chosen arbitrarily small.

We will show that price supports imply that a path is potentially maximal. Suppose a price sequence $\{p_t\}$, $t=0,1,\ldots,$ $p_t\in E_t$, exists such that (4.10) and (4.11) are satisfied for $\{k_t\}$. Assume that $\{k_t\}$ is not potentially maximal. Then for some T there is $\varepsilon>0$ such that

$$V_0(k_0)\geqq\sum_1^T u_t(k_{t-1},k_t)+V_T(k_T)+\varepsilon. \tag{4.14}$$

But the definition of V_t implies there is some path $\{k_t'\}$, $t=0,1,\ldots,$ for which $k_0'=k_0$ and

$$V_0(k_0')\leqq\sum_1^T u_t(k_{t-1}',k_t')+V_T(k_T')+\varepsilon/2. \tag{4.15}$$

Comparing (4.14) and (4.15) we derive

$$\sum_1^T u_t(k_{t-1},k_t)+V_T(k_T)<\sum_1^T u_t(k_{t-1}',k_t')+V_T(k_T'). \tag{4.16}$$

However, from (4.11) we have

$$\sum_1^T\left(u_t(k_{t-1},k_t)-u_t(k_{t-1}',k_t')\right)\geqq p_T(k_T'-k_T)+p_0(k_0-k_0'), \tag{4.17}$$

and from (4.10)

$$V_T(k_T)-V_T(k'_T) \geqq p_T(k_T-k'_T). \tag{4.18}$$

Summing (4.17) and (4.18) and using $k_0=k'_0$ gives

$$V_T(k_T)+\sum_1^T u_t(k_{t-1},k_t) \geqq V_T(k'_T)+\sum_1^T u_t(k'_{t-1},k'_t).$$

This contradicts (4.16), so $\{k_t\}$ must be potentially maximal. Thus we have

Theorem 4.1

Under assumptions (I), (II), and (III) a path is potentially maximal if and only if it can be price supported in the sense of Lemma 1.

Notice that the assumptions are not needed to prove that a price supported path is potentially maximal.

The cake-eating example [Gale (1967, p. 4)] is the classic example of a path that is potentially maximal but not maximal. The set D contains the pairs of numbers (x, y) such that $y \geqq 0$, $x \geqq 0$, and $y \leqq x$. Utility $u(x, y)=v(z)$ where $z=x-y \geqq 0$ and $v(z)=\log(1+z)$. The path $k_t=k_0$, all t, is potentially maximal but not maximal. Indeed, no maximal path exists from positive initial stocks. The path $k_t=k_0$ is supported by the prices $p_t=1$, all t. The utility function is concave, but not strictly concave since $u(x, y)=u(x+z, y+z)$ for any $z \geqq -y$. Also $V_t(y)=y$, all t. However, u is strictly concave in terminal stocks separately. We may prove:

Theorem 4.2

If u is strictly concave in terminal stocks, a potentially maximal path is unique.

If a path is potentially maximal it satisfies (4.13). If there are two such paths $\{k_t\}$ and $\{k'_t\}$, let T be the first time that $k_t \neq k'_t$. Then by concavity of V_T and strict concavity of u_T in the terminal stocks, the average of the right-hand side of (4.13) for k_t and k'_t is less than the value of the right-hand side of (4.13) for the average of the two paths, k''_t, which is also feasible. That is,

$$V_0(k_0) < \sum_1^{T-1} u_t(k_{t-1},k_t)+u_T(k_{T-1},k''_T)+V_T(k''_T)-\varepsilon, \tag{4.19}$$

for some $\varepsilon>0$. By definition of $V_T(k''_T)$ there is a path $\{\tilde{k}_t\}$ from $t=T$ such that $\sum_{T+1}^{\infty} u(\tilde{k}_{t-1},\tilde{k}_t) > V_T(k''_T)-\varepsilon/2$, where $\tilde{k}_T=k''_T$. Let $\tilde{k}_t=k_t$ for $t<T$. Then from (4.19)

$$V(k_0) < \sum_1^{\infty} u_t(\tilde{k}_{t-1},\tilde{k}_t), \tag{4.20}$$

in contradiction to the definition of $V(k_0)$. Thus there can be only one potentially maximal path. Under conditions to be explored in Section 5 this path will be optimal. Theorem 4.2 was suggested by Peter Hammond.

It is sometimes valuable to know that capital values $p_t k_t$ are bounded as $t \to \infty$. Normalize utility on the potentially maximal path. A simple condition that guarantees boundedness of capital values is that $V_t(\alpha k_t)$ be bounded as $t \to \infty$, for any α sufficiently near 1. Consider

$$V_t(k_t) - p_t k_t \geqq V_t(\alpha k_t) - p_t(\alpha k_t),$$

or

$$(1-\alpha) p_t k_t \leqq V_t(k_t) - V_t(\alpha k_t).$$

Thus $p_t k_t$ is bounded above if $V_t(\alpha k_t)$ is bounded below for some $\alpha < 1$, since the normalization implies that $V_t(k_t) = 0$. Similarly, $\alpha > 1$ establishes a lower bound.

5. Optimal paths

A useful basis for establishing the existence of optimal paths depends on having price supports for the utility function in the sense of (4.11) such that capital values are bounded along the path. In the case of certain stationary optimal paths stationary supports can be found by special arguments. Since capital values are then necessarily bounded, the stationary paths are optimal. Then value loss type arguments may be applied to prove that optimal paths originate from all capital stocks whose value functions are well defined relative to the stationary optimal path.

For the sake of the existence theorems we make three special assumptions, suggested by the methods of Von Weizsäcker:

(W1) There is an infinite path $\{k_t\}$, $t = 0, \ldots$, whose utility functions are supported by a price sequence $\{p_t\}$ in the sense of (4.11).

(W2) $\operatorname{Lim\,sup} p_t k_t = M < \infty$, and if $\{k'_t\}$ is an infinite path with $k'_0 = k_0$, $\liminf p_t k'_t > M' > -\infty$.

Let the *value loss* $\delta_t(x, y) = u_t(k_{t-1}, k_t) + p_t k_t - p_{t-1} k_{t-1} - (u_t(x, y) + p_t y - p_{t-1} x)$, for any $(x, y) \in D_t$. By (4.11) $\delta_t(x, y) \geqq 0$.

(W3) For any $\varepsilon > 0$, there is $\delta > 0$, such that $|p_t(x - k_t)| > \varepsilon$ implies $\delta_{t+1}(x, y) > \delta$ for any $(x, y) \in D_{t+1}$.

Assumptions similar to these were used by Von Weizsäcker (1965) to prove existence for a one-sector model that is time-dependent.

(W2) places weak bounds on the limiting values of the capital stocks as $t \to \infty$, along feasible paths and along the path given by (W1). (W3) provides for a value loss for the input–output combination in period t when the value of input differs from the value of input on the given path. It is implied by uniform strict concavity of u along $\{k_t\}$, but it is weaker than that condition.

With these assumptions we may prove that $\{k_t\}$ is an optimal path. Consider any path $\{k'_t\}$ with $k'_0 = k_0$. Let $\delta_t = \delta_t(k'_{t-1}, k'_t)$, $u_t = u_t(k_{t-1}, k_t)$, $u'_t = u_t(k'_{t-1}, k'_t)$. Then $u'_t - u_t = p_t(k_t - k'_t) - p_{t-1}(k_{t-1} - k'_{t-1}) - \delta_t$. Summing, we obtain

$$\sum_1^T (u'_t - u_t) = p_0(k'_0 - k_0) + p_T(k_T - k'_T) - \sum_1^T \delta_t. \tag{5.1}$$

Since $k'_0 = k_0$, using (W2) gives

$$\limsup \sum_1^T (u'_t - u_t) \leqq M - M' - \lim \sum_1^T \delta_t. \tag{5.2}$$

Either $\{k_t\}$ catches up to $\{k'_t\}$ or $\limsup \sum_1^T (u'_t - u_t) > 0$. In the latter case (5.2) implies $\delta_t \to 0$. Then (W3) implies $p_T(k_T - k'_T) \to 0$, and (5.1) implies $\sum_1^T (u'_t - u_t) \leqq 0$ for large T, with $<$ unless $k'_t \equiv k_t$. This means $\{k_t\}$ catches up to $\{k'_t\}$. Since $\{k'_t\}$ is an arbitrary path with $k'_0 = k_0$, $\{k_t\}$ catches up to every path from k_0 and $\{k_t\}$ is optimal. We have proved [McKenzie (1974)].

Theorem 5.1

Under assumptions (W1), (W2), and (W3), the path $\{k_t\}$ is optimal.

Once an optimal path $\{k_t\}$ has been shown to exist from the initial stock k_0, optimal paths may be derived from all initial stocks in the set K_0, that is, the set of stocks for which the value function is well defined after normalization of utility by $u_t(k_{t-1}, k_t) = 0$, all t. The value function is well defined from a capital stock x if there exists a path $\{k'_t\}$ with $k'_0 = x$ such that $\liminf \sum_1^T u_t(k'_{t-1}, k'_t) > -\infty$, as $T \to \infty$. Consideration of (5.1) with (W2) and (W3) will show that this condition is met if and only if the value loss $\sum_1^T \delta_t$ is bounded as $T \to \infty$. The value loss is the shortfall of the utility sum less a part due to the first differential of u, when u is differentiable, or an analog defined by the support function in the general concave case. The value loss method works because the first-order effects on the utility sums depend only on the differences in value of the initial and terminal stocks, as (5.1) shows, and (W2) and (W3) place certain bounds on the limiting values of the terminal stocks.

Let K_0 be the set of capital stocks x with well defined values $V_0(x)$ when utility is normalized on the optimal path $\{k_t\}$. We prove [McKenzie (1974)]:

Theorem 5.2

If there is an optimal path $\{k_t\}$ from k_0, satisfying assumptions (I), (II), and (III), and if (W2) and (W3) are satisfied for one of its supporting price sequences $\{p_t\}$, there is an optimal path from every capital stock in the set K_0, defined relative to $\{k_t\}$.

By Lemma 4.1 the hypothesis of Theorem 5.2 implies (W1). Also from the discussion above, the set K_0 may equally well be defined as the set of stocks from which there exist paths with finite value loss. Let

$$L_0(x) = \inf\left(\lim \sum_1^T \delta_t(k'_{t-1}, k'_t), \quad T \to \infty\right),$$

over paths $\{k'_t\}$ such that $k'_0 = x$. $L_0(x)$ is well defined if and only if $V_0(x)$ is well defined, given (W2) and (W3). However, $\sum_1^T \delta_t$ has the advantage over $\sum_1^T u_t$ that its terms are non-negative, so the finite sums converge if they are bounded above. This fact underlies the original Ramsey (1928) arguments for one-sector models and was adapted to the multi-sector case by Atsumi (1965). However, its full implications for the existence problem were first drawn by Brock (1970).

The essential step in proving Theorem 5.2 is to show that the infimum in the definition of $L_0(x)$ is assumed by a well defined path from x, if $x \in K_0$. This path will also realize the supremum in the definition of $V_0(x)$. Let s index a sequence of paths from x and let $L_0^s(x)$ be the value loss on the sth path. We may assume that the sequence is chosen so that $L_0^s(x) \to L_0(x)$. Let $\{k_t^s\}$ be the sth path. By assumption (II), k_t^s, $s = 1, 2, \ldots$, is bounded for each t. Thus we may use the Cantor diagonal process to choose a subsequence such that (retain notation) $k_t^s \to \bar{k}_t$ for each t. By assumption (I), $(\bar{k}_{t-1}, \bar{k}_t) \in D_t$. Otherwise $u_t(k_{t-1}^s, k_t^s) \to -\infty$ and using the definition of value loss $\delta_t(k_{t-1}^s, k_t^s) \to \infty$ so that $\delta_t \geqq 0$ implies $L_0^s(x) \to \infty$ and $x \notin K_0$. Then $\{\bar{k}_t\}$ is a path of accumulation from x.

Let $\bar{L}_0$ be the value loss associated with $\{\bar{k}_t\}$. Then $\bar{L}_0 \geqq L_0(x)$. Suppose $\bar{L}_0 > L_0(x)$. Then for all large s, $\bar{L}_0 - L_0^s > \varepsilon$ for some $\varepsilon > 0$. Choose T so large that

$$\bar{L}_0 - \sum_1^T \delta_t(\bar{k}_{t-1}, \bar{k}_t) < \varepsilon/4. \tag{5.3}$$

Choose S so large that

$$\sum_1^T \delta_t(\bar{k}_{t-1}, \bar{k}_t) - \sum_1^T \delta_t(k_{t-1}^s, k_t^s) < \varepsilon/4, \qquad s > S. \tag{5.4}$$

Then, adding (5.3) and (5.4), we have

$$\bar{L}_0 - \sum_1^T \delta_t(k_{t-1}^s, k_t^s) < \varepsilon/2, \qquad s > S. \tag{5.5}$$

But $L_0^s \geqq \sum_1^T \delta(k_{t-1}^s, k_t^s)$, so $\bar{L}_0 - L_0^s < \varepsilon/2$ for $s > S$ which contradicts $\bar{L}_0 - L_0^s > \varepsilon$ for all large s. Therefore, $\bar{L}_0 = L_0(x)$, or the limit path realizes the minimal value loss over all infinite paths from x.

To prove that $\{\bar{k}_t\}$ is optimal we must show that it catches up to every path from x. Suppose $\{k_t'\}$ is an arbitrary path from x. Let $u_t' = u_t(k_{t-1}', k_t')$, $\bar{u}_t = u_t(\bar{k}_{t-1}, \bar{k}_t)$. By normalization we may put $u_t(k_{t-1}, k_t) = 0$, all t. Also $L_0(x)$ finite implies by (W3) that $p_T(k_T - \bar{k}_T) \to 0$, so by (5.1),

$$\sum_1^T \bar{u}_t \to p_0(x - k_0) - L_0(x) > -\infty.$$

If $\sum_1^T \delta_t(k_{t-1}', k_t') \to \infty$ with T, (5.2) implies

$$\limsup \sum_1^T u_t' = -\infty \quad \text{as} \quad T \to \infty.$$

Then

$$\limsup \sum_1^T (u_t' - \bar{u}_t) = \limsup \sum_1^T u_t' - \lim \sum_1^T \bar{u}_t = -\infty,$$

and $\{\bar{k}_t\}$ overtakes $\{k_t'\}$. On the other hand, if $\sum_1^T \delta_t(k_{t-1}', k_t')$ is bounded as $T \to \infty$, (W3) again implies $p_T(k_T - k_T') \to 0$, so by (5.1),

$$\sum_1^T u_t' \to p_0(x - k_0) - \sum_1^\infty \delta_t(k_{t-1}', k_t').$$

Since $\{\bar{k}_t\}$ minimizes value loss from x,

$$\sum_1^T (u_t' - \bar{u}_t) \to L_0(x) - \sum_1^\infty \delta_t(k_{t-1}', k_t') \leqq 0 \quad \text{as} \quad T \to \infty, \tag{5.6}$$

and $\{\bar{k}_t\}$ catches up to $\{k_t'\}$. Since $\{k_t'\}$ is an arbitrary path from x, $\{\bar{k}_t\}$ is optimal, and Theorem 5.2 is proved.

We observe that $L_0(x)$ is finite and $V_0(x)$ is well defined relative to the optimal path $\{k_t\}$ if there exists a path $\{k_t''\}$ with $k_0'' = x$ and $k_\tau'' = k_\tau$ for some $\tau \geqq 0$. Then $\{k_t\}$ is said to be *reachable* from x. In stationary models this is often provided for relative to the stationary optimal path.

It is clear from (5.1) that

$$\limsup \sum_1^T (u_t' - u_t) \leqq \limsup p_T(k_T - k_T'),$$

if $k_0' = k_0$. Thus $\{k_t\}$ is optimal if $\limsup p_T(k_T - k_T') \leqq 0$ holds for all paths $\{k_t'\}$ with $k_0' = k_0$. In particular, if $\lim p_t = 0$ and k_t is bounded over t, $\{k_t\}$ is optimal. These conditions are likely to be met in models where $u_t = \rho^t u$ for $0 < \rho < 1$, which are called *quasi-stationary*. We may state the assumption:

(W2′) k_t is bounded over t and $\lim p_t = 0$.

Then we have:

Theorem 5.3

Under assumptions (I), (II), (W1), and (W2′), the path $\{k_t\}$ is optimal.

Assumption (W2′) was introduced in the efficiency context by Malinvaud (1953) in the form $p_t k_t \to 0$, as $t \to \infty$.

II. Stationary Models and Turnpike Theory

6. The stationary model

A particular model to which Theorems 5.1 and 5.2 may be applied is that of Gale (1967) and McKenzie (1968). In this model the utility function is stationary, that is, $D_t = D$ and $u_t = u$ for all t. Stationarity may be introduced in a model with steadily growing population by use of per-capita quantities for capital stocks and per-capita utility in the objective. It may be shown that a constant path that gives maximum sustainable utility [that is, $k_t = k$, all t, and $u(k, k) \geqq u(x, y)$ for $(x, y) \in D$ and $y \geqq x$] is supported by prices in the sense of (4.11), so that it satisfies (W1). Since the prices may also be chosen to be constant and any path is bounded in this model, (W2) follows directly. (W3) also follows if $u(x, y)$ is strictly concave at (k, k). Then Theorems 4.1 and 4.2 apply to show that $\{k_t\}$, $t = 0, 1, \ldots,$ where $k_t = k$, all t, is optimal, and there is an optimal path from every $x \in K_t = K$, where K_t is defined relative to $\{k_t\}$ as in Section 4. On the assumptions often adopted in the stationary model K includes all positive stocks and all stocks from which positive stocks may be reached. Free disposal of surplus stocks, the expansibility of certain stocks, and $0 \in D$ are used to imply the wide scope of K.

In order to have a set of assumptions that imply (W1), (W2), and (W3), and are somewhat more specific than those conditions, we will describe the stationary model. The assumptions (I) and (II) of the basic model are retained and in addition we assume

(G1) $D_t = D$, $u_t = u$, for all t (*stationarity*).

(G2) There is $\zeta > 0$ such that $|x| > \zeta$ implies for any $(x, y) \in D$ that $|y| < \gamma|x|$ for $\gamma < 1$ (*bounded paths*).

(G3) If $(x, y) \in D$, then $(z, w) \in D$ for all $z \geqq x$, $0 \leqq w \leqq y$, and $u(z, w) \geqq u(x, y)$ holds (*free disposal*).

(G4) There is $(\bar{x}, \bar{y}) \in D$ for which $\bar{y} > \bar{x}$ (*existence of an expansible stock*).

Before stating the last assumption we must show that a constant path exists with constant prices satisfying (W1). Define the set $V = \{v | v = y - x$, where $(x, y) \in D\}$. Since $E_t = E^n$, an n-dimensional Euclidean space, all t, $V \subset E^n$. By free disposal, (G3), and the existence of an expansible stock $\bar{x}$, (G4), $0 \in$ interior V. Indeed, $y' - x' = v' \in V$ if $(x, y) \in D$ and $x < x' < y$ and $x < y' < y$. We will show that $y - x \geqq v \in V$ implies (x, y) is bounded. By (G4) there is $\bar{v} = \bar{y} - \bar{x} > 0$. Suppose there is $v \in V$ such that $D_v = \{(x, y) | y - x \geqq v\}$ is not bounded. Choose α to give $v' = \alpha\bar{v} + (1 - \alpha)v \geqq 0$, where $0 < \alpha < 1$. Let $(x', y') = \alpha(\bar{x}, \bar{y}) + (1 - \alpha)(x, y)$ for $(x, y) \in D_v$. Then $v' = y' - x' \geqq 0$ but (x', y') can be made arbitrarily large by choosing $(x, y) \in D_v$ arbitrarily large, contradicting either assumption (II) or (G2). Thus D_v is bounded for any $v \in V$.

Define $f(v) = \sup u(x, y)$ for $(x, y) \in D_v$. Since u is concave and closed by assumption (I) and D_v is bounded, the sup is attained for any $v \in V$. Let $W = \{(u, v) | u \leqq f(v)$ and $v \in V\}$. W is convex since f is concave and, putting $\bar{u} = f(0)$, $(\bar{u}, 0)$ is a boundary point of W. Thus there is $(\pi, p) \in E^{n+1}$ and $(\pi, p) \neq 0$, such that $\pi u + pv \leqq \pi\bar{u}$ for all $(u, v) \in W$. Since V is unbounded below by (G3), $p \geqq 0$. Suppose $\pi = 0$. Then $pv \leqq 0$ for all $v \in V$, or since 0 is interior to V, $p = 0$. Thus $\pi \neq 0$, and we may choose (π, p) so that $\pi = 1$. Then $u + pv \leqq \bar{u}$ for all $(u, v) \in W$. This implies

Lemma 6.1

There is $p \geqq 0$ such that $u(x, y) + py - px \leqq \bar{u}$ for all $(x, y) \in D$, where $\bar{u} = \max u(x, x)$ for $(x, x) \in D$.

Let $u(k,k)=\bar{u}$. Then the path $\{k_t\}$, $t=0,1,\dots$, with $k_t=k$ for all t is an infinite path supported, in the sense of (4.11), by the price sequence $\{p_t\}$, where $p_t=p$ for all t. We now assume:

(G5) The utility function u is strictly concave near the point (k,k), where $u(k,k)\geqq u(x,x)$ for all $(x,x)\in D$.

It follows from (G5) that $u(x,x)=\bar{u}$ implies $(x,x)=(k,k)$.

The value loss relative to the constant path $k_t=k$ is $\delta(x,y)=\bar{u}-u(x,y)-py+px$. Then $\delta(x,y)\geqq 0$. Since u is a concave function, δ is a convex function. We may use (G5) to prove a value loss lemma [Atsumi (1965) and Radner (1961)].

Lemma 6.2

For any $\varepsilon>0$, there is $\delta>0$, such that $|x-k|>\varepsilon$ implies $\delta(x,y)>\delta$ for any $(x,y)\in D$, where $u(k,k)=\bar{u}$ and p and $\bar{u}$ are given by Lemma 6.1.

Suppose Lemma 6.2 is not true. Then there exists a sequence (x^s,y^s), $s=1,2,\dots$, such that $|x^s-k|>\varepsilon$ and $\delta(x^s,y^s)\to 0$. Since $\delta(k,k)=0$ and δ is a convex function, $\delta(x^s,y^s)$ does not increase as (x^s,y^s) approaches (k,k) along a line segment. Thus we may put $|x^s-k|=\varepsilon$ for all s. Then the sequence (x^s,y^s), which is bounded by assumption (II), has a point of accumulation $(\bar{x},\bar{y})$ where $\delta(\bar{x},\bar{y})=0$. Also $(\bar{x},\bar{y})\in D$ by concavity and closedness of u on D. Then strict convexity of δ at (k,k) implies $\delta(x,y)<0$ for (x,y) between $(\bar{x},\bar{y})$ and (k,k) in contradiction to Lemma 6.1.

Lemma 6.2 implies (W3) for $k_t=k$, $p_t=p$. Lemma 6.1 implies (W1), and (W2) follows directly from (G2) and $p\geqq 0$. Thus Theorem 5.1 implies that $k_t=k$, $t=0,1,\dots$, is an optimal path. Also Theorem 5.2 implies that an optimal path exists from any $x\in K$, that is, from any x for which $V(x)>-\infty$ or equivalently $L(x)<\infty$, where these functions are defined relative to the stationary optimal path, $k_t=k$.

On the basis of Lemma 6.2 we may show that the prices derived in Lemma 6.1 are full Weitzman prices, that is:

Corollary

Given assumption (G5) the prices (p,p) of Lemma 6.1 and (k,k), where $u(k,k)=\bar{u}$, satisfy both (4.10) and (4.11), when (p_t,p_{t+1}) is set equal to (p,p) and (k_t,k_{t+1}) to (k,k).

That (p,p) and (k,k) satisfy (4.11) is the content of Lemma 6.1. Let k_0 lie in K and consider

$$u(k,k)=u(k_{t-1},k_t)+pk_t-pk_{t-1}+\delta(k_{t-1},k_t), \tag{6.1}$$

where $\{k_t\}$ is any path from k_0. Summing (6.1) gives

$$Tu=\sum_1^T u(k_{t-1},k_t)+pk_T-pk_0+\sum_1^T \delta(k_{t-1},k_t). \tag{6.2}$$

Let $u(k,k)=0$. Suppose $\liminf \sum_1^T u_t > -\infty$ as $T\to\infty$. Then by Lemma 6.2, $k_T\to k$. Therefore, taking the supremum of the right side of (6.2) over paths $\{k_t\}$ from k_0, we obtain

$$V(k)-pk \geqq V(k_0)-pk_0, \tag{6.3}$$

where $V(k)=0$. However, (6.3) is (4.10) for the present case.

We say that a stock x is *expansible* if there is $(x, y)\in D$ with $y>x$. We can prove [Gale (1967)]:

Lemma 6.3

If x is expansible, then $x\in K$.

Consider $\alpha^t(x, y)+(1-\alpha^t)(k,k)=(k_t,k'_{t+1})$, where $y>x$, $0<\alpha<1$, $t=0,1,2,\dots$. For $t=0$, $(k_t,k'_{t+1})=(x,y)$, and as $t\to\infty$, $(k_t,k'_{t+1})\to(k,k)$. But $k'_{t+1}=k-\alpha^t(k-y)$ and $k_{t+1}=k-\alpha^{t+1}(k-x)$. Then $k'_{t+1}>k_{t+1}$ if $y-\alpha x>(k-\alpha k)$. This holds for α near 1 since $y>x$. Thus by free disposal we may replace (k_t,k'_{t+1}) by (k_t,k_{t+1}) and $\{k_t\}$ is an infinite path approaching k.

Also, by concavity of u,

$$u(k_t,k'_{t+1})\geqq(1-\alpha^t)u(k,k)+\alpha^t u(x,y)=\alpha^t u(x,y),$$

and, using free disposal,

$$\sum_0^\infty u(k_t,k_{t+1})\geqq\frac{1}{1-\alpha}u(x,y),$$

proving that $x\in K$.

Summarizing the above results we may state:

Theorem 6.1

If in addition to assumptions (I) and (II) we accept assumptions (G1)–(G5), there is a stationary optimal path, supported by price vectors $p_t=p$ in the sense of (4.10) and (4.11), and there is an optimal path from any expansible stock.

Without the assumption of strict concavity at the stationary path that maximizes stationary utility, we cannot show that expansible stocks give rise to optimal paths. However, on the weaker assumption that this path is unique, the analogous result can be proved for maximal paths. See Brock (1970), where the terminology "weakly maximal" is used. The appropriate assumption is

(G5′) There is a point $(k,k)\in D$ such that $u(y,y)\geqq u(x,x)$ for all $(x,x)\in D$ implies $(y,y)=(k,k)$.

This assumption is only slightly weaker than requiring u to be strictly concave at (k,k) in the directions that lie in the diagonal. The possibility that u has a flat contour in other directions means that other paths originating at k may exist which oscillate about k without suffering value losses. See McKenzie (1968).

Make assumptions (G1)–(G4) and (G5′). Let p be the price vector of Lemma 6.1, where $u(k,k)=\bar{u}$. As before, define the set K relative to the path $k_t=k$, $t=0,1,\dots,$ where $K=\{x|V(x)>-\infty\}$. Equivalently $K=\{x|L(x)<\infty\}$, where $L(x)=L_0(x)$ is defined relative to $k_t=k$ and $p_t=p$. As in the proof of Theorem 5.2, for $x\in K$ there is a path $\{k'_t\}$, $t=0,1,\dots,$ that realizes minimum value loss $L(x)$ when $k'_0=x$. It is implied by (G2) that $\{k'_t\}$ is bounded. Thus $(1/T)\Sigma(k'_{t-1},k'_t)=(\bar{k}_{T-1},\bar{k}_T)$ has a limit point $(\bar{k},\bar{k})$. By closedness of u, $(\bar{k},\bar{k})\in D$. Let $u(k,k)=0$. Then

$$\sum_1^T u(k'_{t-1},k'_t)=p(k'_0-k'_T)-\sum_1^T \delta(k'_{t-1},k'_t), \tag{6.4}$$

from (5.1). Since $\Sigma_1^T\delta'_t\to L(x)$, (6.4) implies $(1/T)\Sigma_1^T u'_t\to 0$. On the other hand, by concavity of u, $(1/T)\Sigma_1^T u'_t\leqq u(\bar{k}_{T-1},\bar{k}_T)$. Thus $u(\bar{k},\bar{k})=0$ and $\bar{k}=k$ by assumption (G5′).

Suppose $\{k''_t\}$ is any other path from x. As in the proof of Theorem 5.2 it is enough to consider paths with finite value loss. Then by the above argument $(1/T)\Sigma k''_t$ also converges to k. However, from (5.1) we derive

$$\begin{aligned}&\sum_1^T\left(u(k''_{t-1},k''_t)-u(k'_{t-1},k'_t)\right)\\&\quad=p\cdot(k'_T-k''_T)+\sum_1^T\delta_t(k'_{t-1},k'_t)-\sum_1^T\delta_t(k''_{t-1},k''_t).\end{aligned} \tag{6.5}$$

Suppose $\liminf\Sigma_1^T(u''_t-u'_t)=\gamma>0$. Since $\lim\Sigma_1^T\delta'_t$, as $T\to\infty$, is minimal, $\lim(\Sigma_1^T\delta'_t-\Sigma_1^T\delta''_t)\leqq 0$. Thus (6.5) implies $\liminf p\cdot(k'_T-k''_T)\geqq\gamma$ must hold. But $k'_T\to k$ and $k''_T\to k$, which is a contradiction. Thus $\liminf\Sigma_1^T(u''_t-u'_t)\leqq 0$ and $\{k'_t\}$ is maximal. This establishes:

Theorem 6.2

If in addition to (I) and (II) we accept assumptions (G1)–(G4) and (G5′), there is a maximal path from any expansible stock.

7. The quasi-stationary model

The quasi-stationary model differs from the stationary model by the presence of a discount factor $0<\rho<1$ for utility, that is, $u_t(x, y)=\rho^t u(x, y)$ for $t \geqq 0$. We will first prove that a quasi-stationary model has a stationary optimal path that is supported by proportional price vectors [Sutherland (1970)]. As for the stationary model, from the stationary optimal path we may derive the existence of other optimal paths.

For the quasi-stationary model, we assume, in addition to (I) and (II) of the basic model:

(S1) $D_t = D \subset E^{2n}$, $u_t=\rho^t u$, for all t, where $0<\rho<1$ (*quasi-stationarity*).

(S2) Identical with (G2) (*bounded paths*).

(S3) Identical with (G3) (*free disposal*).

(S4) There is $(\bar{x}, \bar{y}) \in D$ for which $\rho\bar{y} > \bar{x}$ (*existence of a stock expansible by a factor exceeding* ρ^{-1}).

These assumptions are small modifications of those for the stationary model of Section 6, (G1)–(G4), to allow for the presence of ρ. Indeed, if ρ is put equal to 1, they are the same.

We will show that an optimal stationary path exists in the quasi-stationary model. This extends a theorem due to Peleg and Ryder (1974) to a general reduced form model. For ζ given by (S2), let Δ be the set $\{(x, x) | x \geqq 0$ and $|x| \leqq \zeta\}$. Δ is a compact convex subset of the diagonal of $E^n \times E^n$. For any $(x, x) \in \Delta$ define $f(x, x)=\{(z, w) | \rho w - z \geqq (\rho-1)x$ for $(z, w) \in D\}$. Since it contains the point $(\bar{x}, \bar{y})$ by assumption (S4), $f(x, x)$ is not empty. We will show that $f(x, x)$ is bounded. If $(z, w) \in f(x, x)$ the definition of f implies that

$$|z| \leqq \rho|w| + (1-\rho)|x|, \tag{7.1}$$

where $0<\rho<1$. Suppose $|z| \geqq \zeta$. Then by assumption (S2), $|w|<|z|$. Substituting in (7.1), $|z|<\rho|z|+(1-\rho)|x|$, or $|z|<|x|$. Since $|x| \leqq \zeta$ by (S2), this gives a contradiction. Thus the set $f(x, x)$ is bounded.

For $U \subset D$, let $g(U)=\{(z, w) \in U | u(z, w) \geqq u(z', w')$ for all $(z', w') \in U\}$. Consider $U=f(x, x)$. Since $u(x, y)$ is concave and closed by assumption (I) and $f(x, x)$ is bounded, the set $\{(z, w) \in U | u(z, w) \geqq u(x, x)\}$ is compact. Therefore, $g(U)$ is compact and not empty. Also by concavity, $g(U)$ is convex. Let $h(W)$, for $W \subset D$, be the set $\{(z, z) | (z, w) \in W\}$, which lies in Δ. Thus h is a projection on Δ along the first factor of the Cartesian product $E^n \times E^n$. Finally, define the correspondence $F=h \circ g \circ f$. F maps Δ into the set of non-empty, convex, compact subsets of Δ. See Figure 7.1.

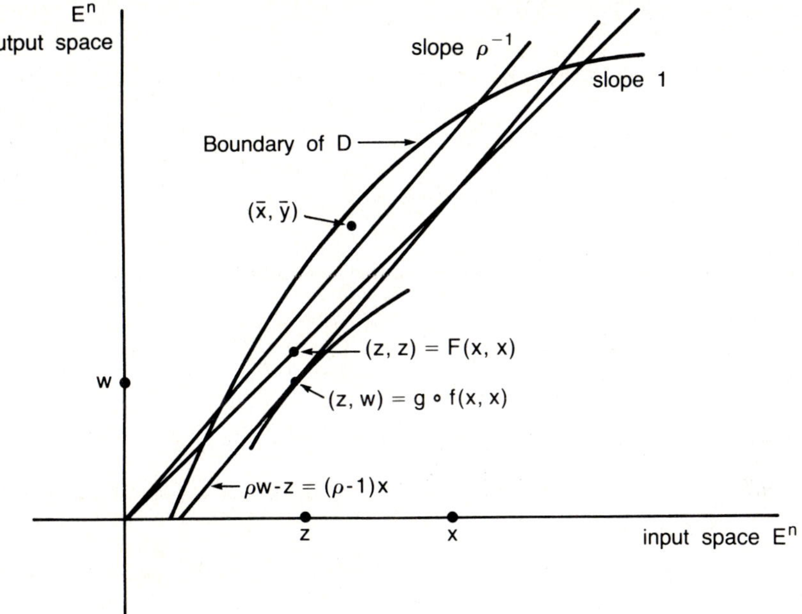

Figure 7.1

We will need:

Lemma 7.1

The correspondence F is upper semi-continuous.

Since h is a continuous correspondence, and both $g \circ f$ and h have compact range, it is sufficient to prove that $g \circ f$ is upper semi-continuous. Let $(z,w) \in g \circ f(x,x)$. Suppose $(z^s,w^s) \to (z',w')$ and $x^s \to x$, $s=1,2,\dots,$ where $(z^s,w^s) \in g \circ f(x^s,x^s)$ for all s. Suppose (z',w') is not in $g \circ f(x,x)$. Although f is not upper semi-continuous, $(z',w') \in f(x,x)$ holds, since $u(z',w') > u(\bar{x},\bar{y})$ and u closed implies $(z',w') \in D$. Then there exists $\varepsilon > 0$ such that $u(z,w) > u(z',w') + \varepsilon$, and there is s_1 such that $s \geqq s_1$ implies $u(x^s,y^s) \leqq u(z',w') + \varepsilon/3$, from closedness of u. Thus $s \geqq s_1$ implies

$$u(z,w) \geqq u(z^s,w^s) + 2\varepsilon/3. \tag{7.2}$$

Choose $x^0 \leqq x$, $x_i^0 < x_i$ if $x_i > 0$, $x_i^0 = x_i$ if $x_i = 0$. Since $f(x^0,x^0)$ is not empty, we may choose $(z^0,w^0) \in f(x^0,x^0)$. Finally, we may choose λ with $0 < \lambda < 1$ such that $(1-\lambda)(u(z^0,w^0) - u(z,w)) \geqq -\varepsilon/3$. Since $x^s \to x$, there is s_2

such that $s \geqq s_2$ implies

$$x^s \geqq (\lambda x + (1-\lambda)x^0) = x''. \tag{7.3}$$

Moreover, from the definition of f it is clear that

$$\lambda(z,w)+(1-\lambda)(z^0,w^0) \in f(x'',x''). \tag{7.4}$$

From (7.3) and (7.4) it follows that

$$\rho(\lambda w + (1-\lambda)w^0) - (\lambda z + (1-\lambda)z^0 \geqq (\rho-1)x^s,$$

for $s \geqq s_2$. In other words, $\lambda(z,w)+(1-\lambda)(z^0,w^0) \in f(x^s,x^s)$. But (z^s,w^s) maximal in $f(x^s,x^s)$ implies $u(z^s,w^s) \geqq u(\lambda(z,w)+(1-\lambda)(z^0,w^0))$ or, by concavity of u, $u(z^s,w^s) \geqq \lambda u(z,w)+(1-\lambda)u(z^0,w^0) = u(z,w)+(1-\lambda)(u(z^0,w^0)-u(z,w))$. Thus we have by choice of λ, for $s \geqq s_2$,

$$u(z^s,w^s) \geqq u(z,w) - \varepsilon/3. \tag{7.5}$$

By (7.2) and (7.5), for $s \geqq \max(s_1,s_2)$, it follows that $u(z,w) \geqq u(z^s,w^s)+2\varepsilon/3 \geqq u(z,w)+\varepsilon/3$, which is a contradiction. Therefore, it must be that $(z',w') \in g \circ f(x,x)$ and $g \circ f$ is upper semi-continuous, which was to be proved. This lemma is due to Khan and Mitra (1984).

Since Δ is compact and convex and F maps Δ into convex subsets, the Kakutani fixed point theorem [Berge (1963, p. 174)] implies there is (k,k) such that $(k,k) \in F(k,k)$. We will show that (k,k) is a stationary path supported by proportional price vectors. It may be seen that (k,k) maximizes utility over $f(k,k)$. In any case, there is (k,w) that does and, by definition of f, $w \geqq k$. Then by free disposal (S3), (k,k) also maximizes utility over $f(k,k)$.

The derivation of price supports for (k,k) parallels that of Section 6. Define the set $V = \{v | v = \rho w - z$, for some $(z,w) \in D\}$. By free disposal, (S3), and the existence of an expansible stock, (S4), $(\rho-1)k \in$ interior V. For $v \in V$, let $D_v = \{z,w) \in D | \rho w - z \geqq v\}$. D_v is bounded for any $v \in V$ by an argument parallel to that given in Section 6 for $\rho = 1$. Define $\phi(v) = \sup u(x,y)$ for $(x,y) \in D_v$, $v \in V$. The sup is attained as before. Let $W = \{(u,v) | u \leqq \phi(v), v \in V\}$. W is convex and interior $W \neq \phi$. Let $\bar{v} = (\rho-1)k$ and $\bar{u} = \phi(\bar{v})$. Then $(\bar{u},\bar{v})$ is a boundary point of W. Thus by a separation theorem for convex sets [Berge (1963, p. 163)] there is $(\pi,r) \in E_{n+1}$ and $(\pi,r) \neq 0$, such that $\pi u + rv \leqq \pi\bar{u} + (\rho-1)rk$ for all $(u,v) \in W$. Since v is unbounded below by (S3), $r \geqq 0$ must hold. Suppose $\pi = 0$. Then $rv \leqq (\rho-1)rk$ for all $v \in V$. However, $(\rho-1)k$ is interior to V, so $r = 0$. Thus $\pi \neq 0$, and we may choose (π,r) so that $\pi = 1$. Then $u + rv \leqq \bar{u} + (\rho-1)rk$ for all $(u,v) \in W$. Let $q = \rho r$. Using the definition of v we obtain [Khan and Mitra (1984)]:

Lemma 7.2

There is $(k,k) \in D$ and $q \geqq 0$ such that $u(z,w) + qw - \rho^{-1}qz \leqq \bar{u} + qk - \rho^{-1}qk$ for all $(z,w) \in D$, where $\bar{u} = u(k,k)$.

This extends similar results, arrived at independently by Flynn (1980) and McKenzie (1982).

Consider the path $\{k_t\}$, $t = 0,1,\ldots$, where $k_t = k$, all t, and the vectors q and k satisfy Lemma 7.2. Then the price path $\{p_t\}$, $t = 0,1,\ldots$, where $p_t = \rho^t q$ supports the utility function $u_t = \rho^t u$ in the sense of (4.11). It is clear that $p_t \to 0$ and k_t is bounded over t, or (W2′) holds. Thus by Theorem 5.3, the path $\{k_t\}$ is optimal. An examination of the proof of Theorem 5.2 shows that (W2′) will also replace (W2) there. Then given (W2′), it is unnecessary to use (W3) to show that $p_T(k_T - k'_T) \to 0$, and (5.6) is established directly. Thus Theorem 5.2 is valid with (W2′) replacing both (W2) and (W3), and an optimal path exists from any $x \in K$, that is, from any x for which $V_0(x) > -\infty$, where this function is defined relative to the stationary optimal path.

We may also show that (4.10) holds for p_t so that they are full Weitzman prices. Consider

$$\rho^t u(k,k) + (\rho^t - \rho^{t-1})qk = \rho^t u(k_{t-1},k_t) + \rho^t qk_t - \rho^{t-1}qk_{t-1} + \delta_t(k_{t-1},k_t), \tag{7.6}$$

where $\{k_t\}$ is a path from k_0 and $\delta_t(k_{t-1},k_t) \geqq 0$. Summing (7.6) gives

$$\sum_1^T \rho^t u + (\rho^T - 1)qk = \sum_1^T \rho^t u_t + \rho^T qk_T - qk_0 + \sum_1^T \delta_t. \tag{7.7}$$

Since $k'_t = k$ is an optimal path from k and k_t is an arbitrary path from k_0, in the limit (7.7) justifies

$$V_0(k) - qk \geqq V_0(k_0) - qk_0,$$

which establishes (4.10).

In this case it is not difficult to show that the set K of capital stocks x, with well defined values $V_0(x)$ relative to the stationary optimal path $k_t = k$, includes all sustainable stocks. If x is sustainable, that is, $(x,x) \in D$, then one feasible path from x is $k_t = x$, $t = 0,1,\ldots$. This implies

$$V_0(x) \geqq \sum_1^\infty \rho^t u(x,x) = \frac{\rho}{1-\rho} u(x,x),$$

so $V_0(x) > -\infty$ holds. Thus we have:

Theorem 7.1

If in addition to assumptions (I) and (II) we accept assumptions (S1)–(S4), there is a stationary optimal path $k_t = k$, supported by price vectors $p_t = \rho^t q$ in the sense of (4.10) and (4.11), where $q \geqq 0$. Also there is an optimal path from any sustainable stock.

According to Theorem 7.1, under the conditions assumed, there always exists a stationary optimal path $k_t = k$ supported by a price sequence $p_t = \rho^t q$, that is, by proportional prices. We may also show that any stationary optimal path (k, k) has proportional price supports if $(k, k) \in$ interior D. Since $k_t = k$, $t = 0, 1, \ldots,$ is an optimal path interior to D it satisfies the hypothesis of Lemma 4.1. Thus a sequence of support prices $\{p_t\}$, $t = 0, 1, \ldots,$ exists, and $p_t \geqq 0$ by free disposal. Consider prices (p, q) that support $u(k, k)$, that is,

$$u(x, y) - u(k, k) \leqq p(x - k) - q(y - k) \quad \text{for all} \quad (x, y) \in D. \tag{7.8}$$

Since $(k, k) \in$ interior D, it is immediate that β exists such that $|p| < \beta$ and $|q| < \beta$ must hold.

By the support property we have

$$\rho^t u(x, y) + p_t y - p_{t-1} x \leqq \rho^t u(k, k) + p_t k - p_{t-1} k,$$

Dividing through by ρ^t gives

$$u(x, y) - u(k, k) \leqq \frac{p_{t-1}}{\rho^t}(x - k) - \frac{p_t}{\rho^t}(y - k), \qquad t = 1, 2, \ldots. \tag{7.9}$$

Averaging the first $T + 1$ inequalities (7.9) gives

$$u(x, y) - u(k, k) \leqq \rho^{-1} P_T (x - k) - Q_T (y - k), \tag{7.10}$$

where

$$P_T = \frac{1}{T+1}\left(p_0 + \rho^{-1} p_1 + \cdots + \rho^{-T} p_T\right),$$

and

$$Q_T = \frac{1}{T+1}\left(\rho^{-1} p_1 + p^{-2} p_2 + \cdots + p^{-T-1} p_{T+1}\right)$$
$$= P_T + \frac{1}{T+1}\left(\rho^{-T-1} p_{T+1} - p_0\right).$$

Since $|P_T| < \beta$ for all T by (7.8), there is a subsequence $\{T_i\}$, $i = 1, 2, \ldots$, such that $P_{T_i} \to q \geqq 0$. Then Q_{T_i} also converges to q, and we obtain from (7.10)

$$u(x, y) - u(k, k) \leqq \rho^{-1} q(x - k) - q(y - k). \tag{7.11}$$

Thus the price sequence $\{p_t'\}$, where $p_t' = \rho^t q$, gives proportional support prices for $k_t = k$. This argument is due to Sutherland (1970). We have shown:

Theorem 7.2

The path $\{k_t\}$, $k_t = k$, $t = 0, 1, \ldots$, where $(k, k) \in$ interior D, is an optimal path given (S1)–(S4), if and only if there are support prices $\{p_t\}$ where $p_t = \rho^t q$, $q \geqq 0$, which satisfy (4.11).

It is implied by Theorem 7.2 that $u(k, k)$ maximizes $u(x, y)$ subject to $\rho y - x \geqq (\rho - 1)k = v$. Consider $y \geqq \rho^{-1}(x + v)$, $k = \rho^{-1}(k + v)$. Substituting in (7.11), we have

$$u(x, y) - u(k, k) \leqq \rho^{-1} q(x - k) - q(\rho^{-1} x - \rho^{-1} k),$$

or $u(x, y) - u(k, k) \leqq 0$. However, it is clear from Theorem 7.1 and the proof of Lemma 7.2 that if (k, k) maximizes $u(x, y)$ subject to $\rho y - x \geqq (\rho - 1)k$, $k_t = k$ is optimal. Then we have the:

Corollary

The path $\{k_t\}$, $k_t = k$, $t = 0, 1, \ldots$, where $(k, k) \in$ interior D, is an optimal path given (S1)–(S4) if and only if $u(k, k)$ maximizes $u(x, y)$ for $\rho y - x \geqq (\rho - 1)k$.

The necessity parts of Theorem 7.2 and the Corollary also apply to the stationary model, since the same arguments are valid. However, for sufficiency assumption (G5) would be needed, that u is strictly concave at (k, k).

8. Convergence of optimal paths

There are three general methods available for proving the convergence of optimal paths. A very simple method may be used when the utility function is uniformly concave, in a certain sense, along an optimal path. This method makes direct use of the fact that a chord of the graph of the utility function lies entirely below the graph. On the other hand, we use an alternative method when uniform concavity does not hold. This dual approach is used based upon the support prices. This approach has been referred to as the method of "value loss", since it is the accumulation of shortfalls in values of input–output combinations along one path relative to another at the other's support prices that eventually contradicts optimality. However, it is not first-order value losses that force convergence. They are fully accounted for over a segment of the optimal paths by the differences in value of initial and terminal stocks. Rather the work is done by second-order value losses due to concavity. Thus our arguments are closely related to the problem of the second variation in calculus of variations. This analogy may be illuminating to students of the calculus. However, it should be kept in mind that turnpike theory compares paths starting from different points both of which are optimal relative to their starting points. This is unlike the classical problems of calculus of variations. Finally a method is available based on the treatment of the first-order conditions for optimality as a set of difference equations that define a transformation of the paths of accumulation into a Banach space. This approach will be examined in Section 10.

Let $\{k_t\}$ and $\{k'_t\}$ be two optimal paths for $t=0,1,\ldots$, where k_0 and k'_0 may differ. Assume (I), (II), and $k_0 \in$ relative interior K_0, and suppose $k'_0 \in K_0$, that is, $V_0(k'_0) > -\infty$, when utility is normalized so that $u(k_t, k_{t+1}) = 0$, all t. The primal approach to convergence considers a path that is halfway between $\{k_t\}$ and $\{k'_t\}$, that is, $\{k''_t\}$, where $k''_t = \frac{1}{2}(k_t + k'_t)$. By convexity, $k''_t \in K_t$ for all t. Assume *uniform strict concavity* of u_t along $\{k_t\}$ in the primal sense that $|(x, y)-(k_{t-1}, k_t)| > \varepsilon > 0$ implies there is $\delta > 0$, independent of t, such that

$$u_t\left(\tfrac{1}{2}(x + k_{t-1}, y + k_t)\right) \geqq \tfrac{1}{2}\left(u_t(x, y) + u_t(k_{t-1}, k_t)\right) + \delta. \tag{8.1}$$

Applying (8.1) to $\{k_t\}$, $\{k'_t\}$, suppose the distance between the paths exceeds ε, $s(T)$ often by time T. Put $u_t(k_{t-1}, k_t) = 0$ for all t. Then

$$\sum_1^T u_t(k''_{t-1}, k''_t) \geqq \tfrac{1}{2} \sum_1^T u_t(k'_{t-1}, k'_t) + s(T)\delta. \tag{8.2}$$

If $s(T) \to \infty$ as $T \to \infty$, $\sum_1^T u''_t \to \infty$ and $V_0(k''_0) = \infty$. Since $k_0 \in$ relative interior K_0, $V_0(k_0) = \infty$ would be implied as we saw in Section 4 in contradiction to $V_0(k_0) = 0$ by the normalization. More exactly we may prove [Jeanjean (1974) and McKenzie (1982)]:

Theorem 8.1

Let $\{k_t\}$, $\{k'_t\}$, $t=0,1,\ldots$, be optimal paths and assume (I) and (II), and $k_0 \in$ relative interior K_0. Assume uniform strict concavity of u_t along $\{k_t\}$. Suppose $k'_0 \in K_0$. Then for any $\varepsilon > 0$ there is a number $N(\varepsilon)$ such that $|k'_t - k_t| > \varepsilon$ can hold for at most $N(\varepsilon)$ periods.

To find $N(\varepsilon)$ let $\bar{k} \in K_0$, where $k_0 = \alpha\bar{k} + (1-\alpha)k_0''$ for some α, $0 < \alpha < 1$. Then, by concavity of V_0,

$$0 = V_0(k_0) \geqq \alpha V_0(\bar{k}) + (1-\alpha)V_0(k_0''),$$

or

$$V_0(k_0'') \leqq \frac{\alpha}{\alpha - 1} V_0(\bar{k}).$$

At the same time,

$$V_0(k_0'') \geqq \tfrac{1}{2}\left(V_0(k_0) + V_0(k_0')\right) + N(\varepsilon)\delta = \tfrac{1}{2}V_0(k_0') + N(\varepsilon)\delta.$$

Thus

$$N(\varepsilon) \leqq \delta^{-1}\left(\frac{\alpha}{\alpha - 1} V_0(\bar{k}) - \frac{1}{2}V_0(k_0')\right),$$

which may be seen to be non-negative. This proves the theorem.

In the stationary model of Section 6, where $u_t = u$, $K_t = K$, for all t, uniform strict concavity at the (k, k) of assumption (G5) is immediate, and Theorem 8.1 implies that all expansible stocks lead to optimal paths that converge to the stock k of the optimal stationary path. This result was first proved in a model with more than one sector by Atsumi (1965), using the value loss approach.

If the hypothesis of Theorem 8.1 is strengthened by including the first part of assumption (III), so that support prices may be shown to exist, the dual approach may be used to draw the conclusion of the theorem [McKenzie (1976) and Hieber (1981)]. In this case it is convenient to use a dual notion of uniform value loss. The definition of *uniform value loss* along (k_t, p_t) is that $|(x, y)-(k_{t-1}, k_t)| > \varepsilon > 0$ implies that $\delta_{t+1}(x, y) > \delta$ for all t. Since this notion is weaker than the primal notion of uniform strict concavity, the two versions of Theorem 8.1 have no simple order of strength.

The role of uniform strict concavity in the value loss approach is to provide uniform value loss when $(x, y) \neq (k_t, k_{t+1})$. The value loss in period $t+1$ for capital stocks (x, y) relative to the path $\{k_\tau\}$, supported by prices $\{p_\tau\}$, was defined in Section 5 by

$$u_{t+1}(k_t, k_{t+1}) + p_{t+1}k_{t+1} - p_t k_t = u_{t+1}(x, y) + p_{t+1}y - p_t x + \delta_{t+1}(x, y). \tag{8.3}$$

From (4.11) the value loss $\delta_{t+1}(x, y)$ is well defined and non-negative for all

$(x, y) \in D_{t+1}$. If strict concavity holds, it is also positive for $(x, y) \neq (k_t, k_{t+1})$. Indeed, by the same proof used for Lemma 6.2, we obtain:

Lemma 8.1

If u_{t+1} satisfies (I) and is strictly concave at (k_t, k_{t+1}), for any $\varepsilon > 0$ there is $\delta > 0$ such that $|x - k_t| > \varepsilon$ implies $\delta_{t+1}(x, y) > \delta$, for any $(x, y) \in D_{t+1}$.

Let us consider two paths $\{k_t\}$ and $\{k'_t\}$, $t = 0,1,\ldots$, that are optimal where k_0 and k'_0 may differ. Assume (I) and (II), and (III) for $\{k_t\}$ and $\{k'_t\}$. Suppose $V_0(k'_0) > -\infty$ when utility is normalized on $\{k_t\}$, or $k'_0 \in K_0$. Also $V'_0(k_0) > -\infty$ when utility is normalized on $\{k'_t\}$, or $k_0 \in K'_0$. Let u_t be the utility function normalized on $\{k_t\}$ and u'_t the utility function normalized on $\{k'_t\}$. Then

$$V_0(k'_0) = \sum_{\tau=1}^{t} u_\tau(k'_{\tau-1}, k'_\tau) + V_t(k'_t),$$

and similarly for $V'_0(k_0)$. Since $u_\tau(k'_{\tau-1}, k'_\tau) + u'_\tau(k_{\tau-1}, k_\tau) = 0$, it follows that $V_t(k'_t) + V'_t(k_t) = V_0(k'_0) + V'_0(k_0)$, for all t. If $\sum_{\tau=1}^{t} u_\tau(k'_{\tau-1}, k'_\tau)$ converges, $V_0(k'_0) + V'_0(k_0) = 0$. Also support prices exist for both paths by Lemma 4.1.

The definition of the value losses in (8.3) gives symmetrical expressions for the two paths,

$$u_t(k_{t-1}, k_t) + p_t k_t - p_{t-1} k_{t-1} = u_t(k'_{t-1}, k'_t) + p_t k'_t - p_{t-1} k'_{t-1} + \delta_t, \tag{8.4}$$

$$u_t(k_{t-1}, k_t) + p'_t k_t - p'_{t-1} k_{t-1} = u_t(k'_{t-1}, k'_t) + p'_t k'_t - p'_{t-1} k'_{t-1} - \delta'_t. \tag{8.5}$$

In these formulae, $\delta_t = \delta_t(k'_{t-1}, k'_t)$, and $\delta'_t = \delta'_t(k_{t-1}, k_t)$. The prices and thus the size of value losses are independent of the normalization of u_t. Subtracting (8.5) from (8.4) gives

$$(p'_t - p_t)(k'_t - k_t) - (p'_{t-1} - p_{t-1})(k'_{t-1} - k_{t-1}) = \delta_t + \delta'_t. \tag{8.6}$$

Let $L_p(t) = (p'_t - p_t)(k'_t - k_t)$.

We may apply the support of the value function according to (4.10) to obtain

$$V_t(k_t) - p_t k_t = V_t(k'_t) - p_t k'_t + \lambda_t, \tag{8.7}$$

$$V'_t(k_t) - p'_t k_t = V'_t(k'_t) - p'_t k'_t - \lambda'_t, \tag{8.8}$$

where $\lambda_t \geqq 0$, $\lambda'_t \geqq 0$. Subtracting (8.7) from (8.8) and using $V_t(k'_t) + V'_t(k_t) = V_0(k'_0) + V'_0(k_0)$, as well as $V_t(k_t) = V'_t(k'_t) = 0$, gives

$$(p'_t - p_t)(k'_t - k_t) = -V_0(k'_0) - V'_0(k_0) - \lambda_t - \lambda'_t. \tag{8.9}$$

Since δ_t, δ'_t, λ_t, and λ'_t are non-negative, $L_p(t)$ is monotone increasing and bounded above. This line of argument leads once more to the conclusion of Theorem 8.1 with the condition that assumption (III) holds for both paths and uniform value loss holds along one of them. From (8.6) and (8.9) we may derive $\lambda_t + \lambda'_t = \lambda'_0 + \lambda_0 - \sum_{\tau=1}^{t}(\delta_\tau + \delta'_\tau)$. Therefore, to avoid contradiction, the number of periods $N(\varepsilon)$, when $|k'_t - k_t| > \varepsilon$, cannot exceed $(\lambda_0 + \lambda'_0)/\delta = -(L_p(0) + V_0(k'_0) + V'_0(k_0))/\delta$.

We have proved:

Theorem 8.2

Let $\{k_t\}$, $\{k'_t\}$, $t = 0, 1, \ldots$, be optimal paths. Assume (I), (II), and (III) for both paths and assume $k'_0 \in K_0$ and $k_0 \in K'_0$. Then support prices $\{p_t\}$ and $\{p'_t\}$ exist for $\{k_t\}$ and $\{k'_t\}$, respectively. Assume uniform value loss for either (p_t, k_t) or (p'_t, k'_t). Then for any $\varepsilon > 0$ there is a number $N(\varepsilon)$ such that $|k'_t - k_t| > \varepsilon$ can hold for at most $N(\varepsilon)$ periods.

However, Theorems 8.1 and 8.2 do not apply to objective functions that discriminate systematically against the future. The simplest of these, and one often used, is $u_t(x, y) = \rho^t u(x, y)$ where $0 < \rho < 1$ and u is a function independent of time that satisfies assumptions (I) and (II). Make assumptions (G2)–(G5). Then (G1), or $u_t = u$ and $D_t = D$, all t, implies by Theorem 6.1 that an optimal stationary path $k_t = k$ exists supported by price vectors $p^t = p$. From the proof we find that k satisfies $u(k, k) \geqq u(x, x)$ for $(x, x) \in D$. Moreover, strict concavity at (k, k), provided by (G5), implies that k satisfying this maximizing condition is unique. Also assume $(k, k) \in$ interior D.

If ρ is now introduced, that is, the utility function $u_t(x, y) = \rho^t u(x, y)$, $0 < \rho < 1$, is defined, for ρ sufficiently near 1, (G2)–(G4) will imply (S2)–(S4). Then for such a ρ, Theorem 7.1 implies that a stationary optimal path $k_t = k^\rho$ exists. From the proof, using Lemma 7.2, k^ρ satisfies $u(k^\rho, k^\rho) \geqq u(x, y)$ for all $(x, y) \in D$ such that $\rho y - x \geqq (\rho - 1)k^\rho$. Let $V(x)$ be the value function in the stationary model with $u_t = u$ and $u(k, k) = 0$. Let $K = \{x | V(x) > -\infty\}$. Assumptions (S3) and (S4) imply that K has an interior.

For each value of ρ, $\rho' < \rho < 1$, choose k^ρ satisfying the condition of Lemma 7.2. With assumptions (I), (II), and (G2)–(G5) we may prove:

Lemma 8.2

For any $\varepsilon > 0$ there is ρ' such that $|k^\rho - k| < \varepsilon$ holds for the stationary optimal path, $k_t = k^\rho$, when $1 > \rho > \rho'$.

By Theorem 7.1, assumptions (I), (II), and (S1)–(S4) imply that a stationary optimal path $k_t = k^\rho$ exists. But for ρ' near 1 these assumptions are implied. Then, as mentioned above, such a path exists where $u(k^\rho, k^\rho)$ maximizes $u(z, w)$ over all (z, w) that satisfy $\rho w - z \geqq (1 - \rho)k^\rho$, that is, over $(z, w) \in f(k^\rho, k^\rho)$. Let

$\rho^s \to 1$, where $\rho' < \rho^s < 1$ and $s = 1,2,\ldots$. Since $|k^\rho| < \zeta$ by (G2), there is a subsequence (preserve notation) such that $k^{\rho^s} \to \bar{k}$. Let $f(\rho, x, x)$ be defined in the same way as $f(x, x)$ in Section 7. Then by an argument parallel to that for Lemma 7.1, $g \circ f(\rho, x, x)$ is upper semi-continuous in (ρ, x, x). In other words, $u(k^{\rho^s}, k^{\rho^s})$ maximal over $f(\rho^s, k^{\rho^s}, k^{\rho^s})$ implies that $u(\bar{k}, \bar{k})$ is maximal over $f(1, k^1, k^1)$. Since $f(1, k^1, k^1)$ contains all $(x, x) \in D$, from Lemma 6.1 and the proof of Theorem 6.1 we find that $k_t = \bar{k}$ is a stationary optimal path when $\rho = 1$. Strict concavity of u near (k, k) from assumption (G5), implies that a stationary optimal path k that satisfies $u(k, k) \geqq u(x, x)$ for $(x, x) \in D$ is unique. Thus $\bar{k} = k$ and the original sequence $k^{\rho^s} \to k$. Since $k^{\rho^s} \to k$ for an arbitrary sequence, the convergence is uniform. We may conclude that $k^\rho \to k$ as $\rho \to 1$ and the lemma is proved.

With this preparation we can develop a turnpike theorem for the quasi-stationary model [Cass and Shell (1976)]. Substitute $\rho^t u$ for u_t in (8.4) and (8.5) and multiply through by ρ^{-t}. Define current prices by $q_t = \rho^{-t} p_t$. Then we have

$$u(k_{t-1}, k_t) + q_t k_t - \rho^{-1} q_{t-1} k_{t-1}$$

$$= u(k'_{t-1}, k'_t) + q_t k'_t - \rho^{-1} q_{t-1} k'_{t-1} + \rho^{-t}\delta_t, \tag{8.10}$$

in place of (8.4) and a similar equation in place of (8.5). For each ρ, $0 < \rho < 1$, k^ρ is chosen to satisfy the condition of Lemma 7.2. Let $k'_t = k^\rho$, all t, where k^ρ is the capital stock of the stationary optimal path and $p'_t = \rho^t q^\rho$ are the Weitzman prices provided by Theorem 7.1. Let $\{k_t(\rho)\}$ be an optimal path from $k_0 \in$ interior K and $p_t(\rho) = \rho^t q_t(\rho)$ the Weitzman support prices. The existence of the Weitzman support prices follows from Lemma 4.1, since (G3) and (G4) imply that the set S of sustainable stocks has an interior. But $S \subset K_0$ and $S \subset P_t \cap K_t$, for all t, when K_0 and K_t are defined relative to any path $\{k_t(\rho)\}$. If u is strictly concave near k^ρ for any $\varepsilon > 0$, $|k_{t-1}(\rho) - k^\rho| > \varepsilon$ implies there is $\delta > 0$ such that the value loss suffered by $k_{t-1}(\rho)$ at prices $\rho^t q^\rho$ is $\delta_t^\rho > \delta$. In formula (8.6) put $\delta_t = \delta_t(\rho)$, the value loss suffered by (k^ρ, k^ρ) at prices $p_t(\rho)$, and $\delta'_t = \delta_t^\rho$. This gives

$$(q_t(\rho) - q^\rho)(k_t(\rho) - k^\rho) - \rho^{-1}(q_{t-1}(\rho) - q^\rho)(k_{t-1}(\rho) - k^\rho)$$

$$= \rho^{-t}(\delta_t(\rho) + \delta_t^\rho), \tag{8.11}$$

for all $t \geqq 1$. Assumption (G5) implies that a neighborhood U of (k, k) exists within which u is strictly concave. Suppose ρ' is chosen near enough to 1 so that every $(k^\rho, k^\rho) \in U$ for $1 > \rho > \rho'$. This is possible by Lemma 8.2. Then for

$1>\rho>\rho'$ and any $\varepsilon>0$ there is $\delta>0$ such that $|k_{t-1}(\rho)-k^{\rho}|>\varepsilon$ implies $\rho^{-t}(\delta_t(\rho)+\delta_t^{\rho})>\delta$.

Suppose that the initial prices $q_0(\rho)$ for the path $k_t(\rho)$, and the prices q^{ρ} that support the stationary optimal path are bounded for $1>\rho>\rho'$. Then $(q_0(\rho)-q^{\rho})(k_0-k^{\rho})$ is bounded for these ρ, and ρ' may be selected near enough to 1 to imply for $1>\rho>\rho'$,

$$(\rho^{-1}-1)(q_0(\rho)-q^{\rho})(k_0-k^{\rho})>-\delta/2. \tag{8.12}$$

Let $L_c^{\rho}(t)=(q_t(\rho)-q^{\rho})(k_t(\rho)-k^{\rho})$. Adding (8.11) and (8.12) gives $L_c^{\rho}(1)-L_c^{\rho}(0)>\delta/2$. Then $(\rho^{-1}-1)L_c^{\rho}(1)>-\delta/2$ also holds. Provided $|k_{\tau}(\rho)-k^{\rho}|>\varepsilon$, for $0<\tau<t$, we may apply induction to obtain $L_c^{\rho}(t)-L_c^{\rho}(t-1)>\delta/2$, uniformly for $1>\rho>\rho'$, or

$$(q_t(\rho)-q^{\rho})(k_t(\rho)-k^{\rho})-(q_{t-1}(\rho)-q^{\rho})(k_{t-1}(\rho)-k^{\rho})>\delta/2, \tag{8.13}$$

Since feasible paths are bounded by (G2), utility is bounded above. Therefore, discounted sums converge to finite values or $-\infty$, and $V_t(k_t')+V_t'(k_t)=0$, by the reversal of the normalization. Therefore, if we multiply through by ρ^{-t}, (8.9) becomes

$$(q_t(\rho)-q^{\rho})(k_t(\rho)-k^{\rho})=-\rho^{-t}(\lambda_t+\lambda_t')\leqq 0. \tag{8.14}$$

We will see that $L_c^{\rho}(t)=(q_t(\rho)-q^{\rho})(k_t(\rho)-k^{\rho})$ may serve in place of $L_p(t)$ to prove a turnpike theorem, in the sense of convergence to a neighborhood of k^{ρ}, rather than to k^{ρ} itself.

Let R be the set of $\rho<1$ such that (S4) is satisfied. Note that (S4) is satisfied for $\rho'>\rho$ if it is satisfied for ρ. First, we must show

Lemma 8.3

The prices $q_0(\rho)$ and q^{ρ} are bounded for $\rho\in R$.

Let $k_t=k$ be the stationary optimal path provided by Theorem 6.1 and $p_t=p$ the corresponding support prices. Maintain the normalization $u(k,k)=0$. Let V_0^{ρ} be the value function at $t=0$ when ρ is the discount factor. We will show that $V_0^{\rho}(x)$ is bounded for $x\in k$ over $\rho\in R$. Let $\{k_t\}$ be an arbitrary path from x. The relation (6.1) gives

$$0\geqq u(k_{t-1},k_t)+p(k_t-k_{t-1}). \tag{8.15}$$

Multiplying through by ρ^t and summing from $t=1$ to T, we have

$$\sum_1^T \rho^t u(k_{t-1}, k_t) \leqq \rho p k_0 + \sum_1^{T-1} \rho^t(\rho-1) p k_t - \rho^T p k_T \leqq px, \tag{8.15}$$

or $V_0^\rho(x)$ is bounded above independently of ρ. On the other hand, $x \in K$ implies $V(x) > -\infty$. That is, there is a path $\{k_t'\}$ with $k_0' = x$ for which $\liminf \sum_1^T u_t(k_{t-1}', k_t')$, as $T \to \infty$, is finite. Then the argument of Theorem 8.1 implies that for any $\varepsilon > 0$ there is $N(\varepsilon)$ such that $|k_t' - k| > \varepsilon$ for no more than $N(\varepsilon)$ periods. But $(k, k) \in$ interior D implies that $(k_t', k) \in D$ for k_t' sufficiently near k. Thus we may assume $k_t' = k$ and $u_t(k_{t-1}', k_t') = 0$ for all $t > T$. Then

$$\sum_1^t \rho^\tau u_\tau(k_{\tau-1}', k_\tau') = \sum_1^T \rho^\tau u_\tau(k_{\tau-1}', k_\tau'),$$

for all $t > T$. This implies that $V_0^\rho(x)$ is bounded below over $\rho \in R$.

Consider the support formula

$$V_0^\rho(k_0) - q_0(\rho)\cdot k_0 \geqq V_0^\rho(x) - q_0(\rho)\cdot x, \tag{8.16}$$

implied by (4.10) for k_0 and $x \in k$. Choose x so that $(k_0 - x)_i = \varepsilon > 0$, all i. Set $\bar{q}_0(\rho) = q_0(\rho)/|q_0(\rho)|$. If $|q_0(\rho)|$ is unbounded over R, for some $\bar{\rho} \in R$ there is a sequence $\rho^s \to \bar{\rho}$, $s = 1, 2, \ldots$, such that $|q_0(\rho^s)| \to \infty$, and $\bar{q}_0(\rho^s) \to \bar{q}_0 \neq 0$. Then (8.16) implies

$$-\bar{q}_0\cdot k_0 \geqq -\bar{q}_0\cdot x.$$

Since $\bar{q}_0 \geqq 0$ by free disposal, this contradicts the choice of x. Therefore, $q_0(\rho)$ is bounded for $\rho \in R$.

We must now bound q^ρ. According to Lemma 7.2

$$u(k^\rho, k^\rho) + \rho q^\rho k^\rho - q^\rho k^\rho \geqq u(z, w) + \rho q^\rho w - q^\rho z, \tag{8.17}$$

for all $(z, w) \in D$. If q^ρ is unbounded for $\rho \in R$, there is a sequence $\rho^s \to \bar{p} \in R$ such that $|q^{\rho^s}|$ is unbounded and $k^{\rho^s} \to \bar{k}$. Choosing a subsequence and normalizing as before, we obtain in the limit, as a consequence of (8.17),

$$(\rho - 1)\bar{q}\bar{k} \geqq \bar{q}(\rho w - z), \quad \bar{q} \neq 0,$$

for all $(z, w) \in D$. This is a contradiction since (S3) and (S4) imply that $(\rho - 1)\bar{k}$ is interior to the set $\{\rho w - z | (z, w) \in D\}$. Thus q^ρ is bounded for $\rho \in R$.

Since V_0^ρ is concave and finite in K, it is a continuous function of y in the interior of K. Therefore, it is bounded in any compact subset of the interior of K, for example, over the set $U = \{y \| k - y| \leqq \varepsilon/2\}$. Then it is immediate from the proof that the bound on $q_0(\rho)$ for $\rho \in R$, given by the lemma, is uniform for k_0 in U. We have:

Corollary

The prices $q_0(\rho)$ are uniformly bounded for k_0 in a sufficiently small neighborhood of k and $\rho \in R$.

The corollary implies that the support prices $q_t(\rho)$ for any path lying in a small neighborhood of k are bounded as $t \to \infty$, since the $q_t(\rho)$ are possible choices of $q_0(\rho)$ for $k_0 = k_t$.

We may now prove the neighborhood turnpike theorem. A similar theorem for the case of continuous time has been proved by Nishimura (1979).

Theorem 8.3

Assume (I) and (II). Let $u_t = \rho^t u$ and $D_t = D$. Assume (G2)–(G5). Also assume that the point (k, k) of (G5) lies in interior D. Let $\{k_t\}$ be an optimal path where $k_0 \in$ interior K. Let $\{k_t'\}$, $k_t' = k^\rho$, all t, be a stationary optimal path given by Theorem 7.1. Then for any $\varepsilon > 0$, there is $\rho(\varepsilon)$ and $N(\varepsilon)$ such that $1 > \rho > \rho(\varepsilon)$ implies $|k_t - k^\rho| < \varepsilon$ holds for all $t > N(\varepsilon)$.

Since the prices $q_0(\rho)$ and q^ρ are bounded for ρ near 1 by Lemma 8.3, the argument leading to (8.13) may be applied for an arbitrary $\varepsilon > 0$. Let $N > -2L_c^\rho(0)/\delta$. Then (8.13) and (8.14) are inconsistent unless $|k_t(\rho) - k^\rho| < \varepsilon$ for some $t < N$. The choice of N is independent of ρ so long as $1 > \rho > \rho'$. This shows the optimal path must approach k^ρ at least once [see the "visit lemma" of Scheinkman (1976)]. However, we will show that there is a neighborhood of k^ρ in which the path remains thereafter.

By Lemma 8.2 and the assumption that $(k, k) \in$ interior D, it is possible to choose ρ' so that $(k^\rho, k^\rho) \in C \subset$ interior D for $1 > \rho > \rho'$, where C is compact. Then any y sufficiently near k^ρ will have $(y, y) \in$ interior D, which implies $y \in$ interior K, for any ρ with $1 > \rho > \rho'$. Let $U_\varepsilon = \{y \| y - k^\rho| \leqq \varepsilon\}$. Choose ε so small that $U_\varepsilon \subset$ interior K for $(k^\rho, k^\rho) \in C$. Then, by the Corollary to Lemma 8.3, the prices $q_t(\rho)$ are bounded for $k_t(\rho) \in U_\varepsilon$ uniformly for ρ with $1 > \rho > \rho'$.

We may suppose that ρ' has been chosen so that $k_t(\rho)$ lies in the neighborhood U_ε for some $t < N$ for any path ρ with $1 > \rho > \rho'$. Uniformly bounded prices for $1 > \rho > \rho'$ and $k_t(\rho) \in U_\varepsilon$ imply that L exists such that $0 \geqq L_c^\rho(t) > \rho L$ for $k_t(\rho) \in U_\varepsilon$. Then if $k_t(\rho) \in U_\varepsilon$, it follows from (8.11) that $L_c^\rho(t+1) \geqq L$. However, $-L_c^\rho(t)$ is seen from (8.7) and (8.14) to be the sum of the remainder terms in the supports of the value function, thus, using strict concavity, $|k_t(\rho) - k^\rho| \to 0$

as $L \to 0$. Then for any $\varepsilon' > 0$, ε may be chosen so small, that is, L so near 0, that $L_c^\rho(t+1) \geqq L$ implies $|k_{t+1}(\rho) - k^\rho| < \varepsilon'$. This follows from the strict concavity of u near (k^ρ, k^ρ) and thus of V_{t+1}^ρ, for ρ near 1. If $k_{t+1}(\rho)$ is outside U_ε, it is implied by (8.11) that $L_c(t+2) - \rho^{-1}L_c(t+1) \geqq \delta$ for some $\delta > 0$. Also ρ' may be chosen near enough to 1 so that $(\rho^{-1} - 1)L > -\delta/2$ for $1 > \rho > \rho'$. Then from $L_c^\rho(t+1) \geqq L$ we have

$$L_c^\rho(t+2) - L_c^\rho(t+1) \geqq \delta + (\rho^{-1} - 1)L > \delta/2. \tag{8.18}$$

This implies $L_c(t+2) \geqq \rho^{-1}L$ also holds and $|k_{t+2}(\rho) - k^\rho| < \varepsilon'$.

Let $U_{\varepsilon'} = \{x \| x - k^\rho| < \varepsilon'\}$. Then (8.18) may be used again to imply that $k_{t+\tau}(\rho) \in U_{\varepsilon'}$ for $\tau \geqq 2$, so long as $k_{t+\tau-1}(\rho)$ does not lie in U_ε. But $k_{t+\tau}(\rho)$ must eventually re-enter U_ε, or $L_c^\rho(t+\tau)$ will become positive, which is impossible. A repetition of the argument shows that $k_{t+\tau}(\rho)$ remains in $U_{\varepsilon'}$ again. Thus $k_{t+\tau}(\rho)$ can never leave $U_{\varepsilon'}$ and the theorem is proved.

Theorem 8.3 is weaker than Theorem 8.1 where $\rho = 1$, since it is not asserted that ρ can be chosen so that $k_t(\rho)$ converges asymptotically to k^ρ. Indeed, there may be other optimal stationary paths interior to $U_{\varepsilon'}$ and cyclical paths as well [see Benhabib and Nishimura (1978)]. However, the assumption (G5) may be strengthened to give asymptotic convergence for ρ sufficiently near 1. Suppose that u has continuous second partial derivatives at (k, k) and the Hessian of u is negative definite there. Then ρ may be chosen near enough to 1 so that

$$Q(\rho) = \begin{bmatrix} \rho u_{11} & \rho u_{12} \\ u_{21} & u_{22} \end{bmatrix},$$

evaluated at (k^ρ, k^ρ), is negative quasi-definite for (k^ρ, k^ρ) in a neighborhood W of (k, k). Since the neighborhood W expands as $\rho \to 1$, while $k^\rho \to k$ as $\rho \to 1$, ρ may be chosen near enough to 1 to bring (k^ρ, k^ρ) inside W. But $Q(\rho)$ negative definite implies that (8.13) will hold for some $\delta > 0$ for any $\varepsilon > 0$.

Indeed, write the left-hand side of (8.13) as

$$L_c^\rho(t) - L_c^\rho(t-1) = -(u_2^t - u_2^\rho)(k_t - k^\rho) - (\rho u_1^t - \rho u_1^\rho)(k_{t-1} - k^\rho), \tag{8.19}$$

where $u_2^t = (\partial/\partial y)u(k_{t-1}, y)]_{y=k_t}$, $u_2^\rho = (\partial/\partial y)u(k^\rho, y)]_{y=k^\rho}$, and similarly for u_1^t, and u_1^ρ. We may express (8.19) in a small neighborhood of (k^ρ, k^ρ) as

$$\begin{aligned} &(u_{21}^\rho(k_{t-1} - k^\rho) + u_{22}^\rho(k_t - k^\rho))(k_t - k^\rho) + (\rho u_{11}^\rho(k_{t-1} - k^\rho) \\ &\quad + \rho u_{12}^\rho(k_t - k^\rho))(k_{t-1} - k^\rho) + o(\varepsilon^2) \\ &= L_c^\rho(t-1) - L_c^\rho(t). \end{aligned} \tag{8.20}$$

where $o(\varepsilon^2)$ is of order higher than the second in ε and $\varepsilon = |(k_{t-1}, k_t) - (k^\rho, k^\rho)|$. If $Q(\rho)$ is negative quasi-definite, (8.20) implies

$$L_c^\rho(t) - L_c^\rho(t-1) \geqq -\lambda\varepsilon^2 - o(\varepsilon^2), \tag{8.21}$$

where λ is the characteristic root of $\frac{1}{2}(Q^T(\rho) + Q(\rho))$ of maximal absolute value. Thus $\varepsilon' > 0$ may be chosen so that $-\lambda\varepsilon^2 - o(\varepsilon^2) > -\frac{1}{2}\lambda\varepsilon^2$ for all $0 < \varepsilon < \varepsilon'$. But from Theorem 8.3, $\bar{\rho}$ may be chosen near enough to 1 so that $|k_t(\rho) - k^\rho| < \varepsilon'$ for all $t > N(\varepsilon')$ and all ρ with $1 \geqq \rho \geqq \bar{\rho}$. We may also choose $\bar{\rho}$ so that $Q(\rho)$ is uniformly negative quasi-definite for $1 \geqq \rho \geqq \bar{\rho}$, that is, (8.21) holds for given λ for all ρ with $1 \geqq \rho \geqq \bar{\rho}$. To avoid contradicting (8.15), $|k_t(\rho) - k^\rho| \to 0$ must hold. Indeed, for any $\varepsilon > 0$ there is $N_1(\varepsilon)$ such that $|k_t(\rho) - k^\rho| < \varepsilon$ for $t > N_1(\varepsilon)$ when $1 \geqq \rho \geqq \bar{\rho}$.

We have proved:

Theorem 8.4

If in addition to the hypotheses of Theorem 8.3, u has continuous second partial derivatives at the optimal stationary path (k, k) of Assumption (G5) and the Hessian of u is negative definite at (k, k), there is $\bar{\rho}$ such that $1 > \rho > \bar{\rho}$ implies for any $\varepsilon > 0$ there is $N(\varepsilon)$ such that $|k_t - k^\rho| < \varepsilon$ for all $t > N(\varepsilon)$, where $\{k_t\}$ is any optimal path satisfying $k_0 \in$ interior K.

The fact that k_0 is assumed interior to E_+^n is not a restriction, since the capital stock space can be chosen differently for each t, so long as all stocks are included which can appear in that period given the initial stocks [McKenzie (1976)]. Of course, the requirements that k_0 be interior to K and (k, k) be interior to D *are* substantive restrictions.

Theorem 8.4 extends the classical theorem for $\rho = 1$ to the case $\rho < 1$ and sufficiently near 1. A result of this type was first obtained by Scheinkman (1976). A similar result was obtained by Brock and Scheinkman (1978). Theorem 8.3 may be extended to utility functions u_t that depend more generally on time where a uniform concavity condition can be obtained in a way analogous to the move from $\rho^t u$ to u. Suppose there exist numbers $\rho_\tau > 0$ such that $\tilde{u}_t = \prod_1^t \rho_\tau^{-1} u_t$ is uniformly strictly concave along $\{k_t\}$. Then the argument leading to Theorem 8.3 can be retraced in this broader context [McKenzie (1976)]. A particular case would be that of variable discount factors, or $u_t = \prod_1^t \rho_\tau u$ so that $\tilde{u}_t = u$ for all t.

In the special case of the stationary model another type of turnpike theorem was established in the course of proving Theorem 6.2. It was shown there that even without strict concavity of u near a point (k, k) where sustainable utility is maximized, if this point is unique [Assumption (G5′)], the average input–output vector of a maximal path $(1/T)\sum_1^T(k_{t-1}, k_t)$ converges to (k, k). Brock (1970) refers to this behavior of maximal paths as an average turnpike property. The circumstances that underlie the average turnpike property become clearer when a general analysis of asymptotic behavior of maximal paths is made using the notion of the von Neumann facet in the following section.

The asymptotic properties of optimal paths in the continuous time model have been investigated along lines similar to those of this section, in particular, by Cass and Shell (1976) and Brock and Scheinkman (1976).

9. The von Neumann facet

Although the support prices were found for maximal paths in Section 4 with utility functions that were only assumed to be concave, the turnpike theorems that have been proved so far have used stronger assumptions involving strict concavity, at least at an optimal path. Strict concavity is used to provide value losses $\delta_t(x, y) > 0$ whenever $(x, y) \neq (k_{t-1}, k_t)$ for an optimal path $\{k_t\}$. However, if the basis for a value loss argument exists in terms of uniformity of concavity over time, it will still be true that paths must behave asymptotically to eliminate the value loss. This means that asymptotically optimal paths must be supported by the same prices. If we define a facet as the set of $(x, y) \in D_t$ that are supported by a particular price vector (p, q), the elimination of value losses will require that the input–output vectors of optimal paths eventually approach the same facets. Thus a weaker form of convergence will continue to hold. This convergence may, in fact, lead to a turnpike in the original sense when the facets have an appropriate structure. This is a generalization of the turnpike theorems to the case where utility may not be strictly concave, and value losses do not necessarily appear off the turnpike.

The case of non-strictly concave utility is not really a borderline case in terms of the economic problem. Suppose that the extensive model has neo-classical production functions with homogeneous labor input and no net joint products. That is, if (x, y) is an input–output vector for the jth industry, $x_i \geqq y_i$ for $i \neq j$. Output is divided between consumption and terminal stocks. Let utility be a strictly concave function of consumption. Yet the reduced model cannot have a strictly concave utility function in terms of initial and terminal stocks. A flat piece of the graph of $u_t(x, y)$, and thus a non-trivial facet, will be generated by the variations in activity levels which are consistent with the labor supply and with the consumption vector c that underlies $u_t(x, y)$. The possible variations will be significant whenever the variations of the input–output vector can be absorbed by the initial and terminal stocks without varying either c or the total labor supply. If stocks are depleted from use so that an activity from each industry must be used to obtain $y \geqq x$ for $(x, y) \in D_t$, the dimension of the facet will be at least $n-1$ if stocks are maintained somewhere on it. To this extent input–output changes can be made to fall on the accumulation program without losing efficiency by varying activity levels for activities in use.

Define $F_t(p,q)$ as all $(x,y) \in D_t$ such that

$$u_t(x,y) + qy - px = \sup(u_t(z,w) + qw - pz), \tag{9.1}$$

over $(z,w) \in D_t$. Concavity and closedness of u_t implies that $F_t(p,q)$ is a closed convex subset of D_t. Also F_t is an upper semi-continuous correspondence from $E_{t-1} \times E_t$ to the non-negative orthant of $E_{t-1} \times E_t$. Let

$$\mathrm{d}((z,w), F_t) = \min|(z,w) - (x,y)|,$$

for $(x,y) \in F_t$. We reformulate the value loss result as [McKenzie (1968)]:

Lemma 9.1

Let u_t satisfy assumptions (I) and (II). Let $F_t(p,q) \neq \phi$ be a facet of D_t. For any $\eta > 0$, $\varepsilon > 0$ there is $\delta > 0$ such that $|z| < \eta$ and $(z,w) \in D_t$ implies $\delta(z,w) > \delta$ for $\mathrm{d}((z,w), F_t) > \varepsilon$.

Consider a sequence (z^s, w^s) that violates the conclusion, that is, $|z^s| < \eta$, $\mathrm{d}((z^s,w^s), F_t) > \varepsilon$, but $\delta_t(z^s,w^s) < \delta^s$ where $\delta^s \to 0$. By assumption (II) w^s is also bounded, so there is a convergent subsequence whose limit $(\bar{z},\bar{w})$ satisfies $\delta_t(\bar{z},\bar{w}) = 0$, $|\bar{z}| \leqq \eta$, and $\mathrm{d}((\bar{z},\bar{w}), F_t) \geqq \varepsilon$. However, $\delta_t(\bar{z},\bar{w}) = 0$ implies $(\bar{z},\bar{w}) \in F_t$, which is a contradiction.

Lemma 9.1 may be used to prove a theorem which is the analog of Theorem 8.1. Suppose that $\{k_t\}$, $t = 0,1,\ldots,$ is a maximal path and F_t is a sequence of facets where $(k_{t-1}, k_t) \in F_t$ for all t. Such a sequence is defined by the sequence of support prices $\{p_t\}$ guaranteed by Lemma 4.1. It is not unreasonable, in view of bounded land and labor services, to assume F_t to be bounded, even uniformly over time. Let us assume further that the value loss off F_t is uniform over t in the sense that η, ε, and δ may be chosen independently of t in Lemma 9.1. Let K_0 be the set of initial stocks with well defined values when u_t is normalized so that $u_t(k_{t-1}, k_t) = 0$, all t. Then the analog of the dual argument for Theorem 8.2 will prove convergence of maximal paths $\{k_t'\}$, from initial stocks $k_0' \in K_0$, to the facet sequence $\{F_t\}$. The argument for Theorem 8.1 is also valid for a primal version of convergence to the sequence of facets. If strict concavity holds, $F_t = \{k_{t-1}, k_t\}$ and the original theorems are true. We may state:

Theorem 9.1

Let $\{k_t\}$, $\{k_t'\}$, $t = 0,1,\ldots,$ be maximal paths and assume (I) and (II), and (III) for both paths. Let $\{p_t\}$ support $\{k_t\}$ and let $\{F_t\}$ be the corresponding facet sequence. Assume $k_0' \in K_0$ and $k_0 \in K_0'$, and there is uniform value loss along

$\{F_t\}$. Then for any $\varepsilon > 0$ there is $N(\varepsilon)$ such that $d((k'_{t-1}, k'_t), F_t) > \varepsilon$ can hold for at most $N(\varepsilon)$ periods.

One case to which Theorem 9.1 applies is the stationary model of Section 6 with the strict concavity assumption (G5) omitted. Lemma 6.1 is valid since it does not use (G5). Thus there exists $p \geqq 0$ such that $u(x, y) + py - px \leqq \bar{u}$ for all $(x, y) \in D$ where $\bar{u} = \max u(x, x)$ for $(x, x) \in D$. The price vector p defines a facet $F(p, p)$. We may prove [Peleg (1973)]:

Lemma 9.2

Under assumptions (G1)–(G4) there is $(k, k) \in D$ such that $u(k, k) \geqq u(x, x)$, for all $(x, x) \in D$ and $k_t = k$, $t = 0, 1, \ldots,$ is a maximal path.

Let $C = \{(x, x) | u(x, x) = \bar{u}\}$. By assumptions (I) and (G2), C is compact. Then there is $(k, k) \in C$ such that

$$pk \leqq px \quad \text{for all} \quad (x, x) \in C. \tag{9.2}$$

Suppose $k_t = k$ is not maximal. Then there is a path $\{k'_t\}$ and $T > 0$ such that $k'_0 = k$ and

$$\sum_1^t \left(u(k'_{t-1}, k'_t) - \bar{u} \right) > \varepsilon > 0, \tag{9.3}$$

for all $t > T$. Consider

$$(x_{t-1}, y_t) = \frac{1}{t} \sum_1^t (k'_{\tau-1}, k'_\tau).$$

Convexity of D implies $(x_{t-1}, y_t) \in D$. Since k'_t is bounded by (G2), there is a point of accumulation $(\bar{k}, \bar{k})$ of the sequence (x_{t-1}, y_t). Moreover,

$$u(x_{t-1}, y_t) \geqq \frac{1}{t} \sum_1^t u(k'_{\tau-1}, k'_\tau),$$

by concavity of u. Then (9.3) and assumption (I) imply $u(\bar{k}, \bar{k}) \geqq \bar{u}$, and

$(\bar{k}, \bar{k}) \in C$. By Lemma 6.1, and (9.3), for all $t > T$,

$$\varepsilon < \sum_{1}^{t} \left(u\left(k'_{\tau-1}, k'_{\tau}\right) - \bar{u} \right) \leqq \sum_{1}^{t} p\left(k'_{\tau-1} - k'_{\tau}\right) = pk - pk'_{t}. \tag{9.4}$$

Since $y_t = (1/t)\Sigma_1^t k'_\tau$ and $y_t \to \bar{k}$, (9.4) implies $pk > p\bar{k} + \varepsilon$. This contradicts (9.2), since $(\bar{k}, \bar{k}) \in C$. Therefore, $k_t = k$ is a maximal path.

The set C and, in particular, the maximal path $k_t = k$, lies on the facet $F(p, p)$. We may set $F_t = F(p, p)$ in Theorem 9.1 and derive the convergence of k'_t to $F(p, p)$. In a similar way, Theorem 8.3 may be given a facet generalization where k^ρ is replaced by the facet $F^\rho = F(q^\rho, \rho q^\rho)$ on which (k^ρ, k^ρ) lies. $F(p, p)$ or F^ρ will be referred to as a von Neumann facet. The argument for Lemma 8.2 now proves that (k^ρ, k^ρ) converges to the compact set C, as $\rho \to 1$. Note that in the proof of Lemma 8.3 the support prices given by Lemma 6.1 are used, but the stationary optimal path plays no role. Therefore, the boundedness of $q_0(\rho)$ and q^ρ follows for the present case just as before. However, two new assumptions are needed.

(F1) The unique support prices for all points of the von Neumann facet containing (k^ρ, k^ρ) are $(q^\rho, \rho q^\rho)$.

(F2) For any $\xi > 0$, $\varepsilon > 0$, there is $\delta > 0$, such that $|x| < \xi$ and $\mathrm{d}((x, y), F(q^\rho, \rho q^\rho)) > \varepsilon$ implies $\delta(x, y) > \delta$, uniformly for ρ near 1.

Assumption (F1) implies directly that $L_c^\rho(t) = (q_t(\rho) - q^\rho)(k_t(\rho) - k^\rho)$ is 0 for $(k_t(\rho), k_{t+1}(\rho)) \in F(q^\rho, \rho q^\rho)$. By assumption (G2), $k_t(\rho)$ is bounded independently of ρ by the maximum of $|k_0(\rho)|$ and ζ. Then assumption (F1) implies that $L_t(\rho)$ is also near 0 for $(k_t(\rho), k_{t+1}(\rho))$ near $F(q^\rho, \rho q^\rho)$, since in that case $q_t(\rho)$ will be near q^ρ. Assumption (F1) is needed because k_t need not be near k^ρ although $(k_t(\rho), k_{t+1}(\rho))$ is near $F(q^\rho, \rho q^\rho)$. Assumption (F2) provides the value losses that lead to $L_{t+1}(\rho) - \rho^{-1} L_t(\rho) \geqq \delta > 0$ for some δ when $(k_t(\rho), k_{t+1}(\rho))$ is outside an ε-neighborhood of $F(q^\rho, \rho q^\rho)$.

These facts allow the proof of a neighborhood turnpike theorem for the von Neumann facet which is the analogue of Theorem 8.3 [details may be found in McKenzie (1983)]. We have:

Theorem 9.2

Assume (I) and (II). Let $u_t = \rho^t u$, $D_t = D$, and assume (G2), (G3), (G4), and F(1), F(2). Let $\{k_t\}$ be an optimal path where $k_0 \in$ interior K. Let $\{k'_t\}$, $k'_t = k^\rho$, all t, be a stationary optimal path, and $p'_t = \rho^t q^\rho$ the support prices, given by Theorem 7.1. Then for any $\varepsilon > 0$, there is $\rho(\varepsilon)$ and $N(\varepsilon)$ such that $1 > \rho > \rho(\varepsilon)$ implies $\mathrm{d}((k_{t-1}, k_t, F(q^\rho, \rho q^\rho)) > \varepsilon$ holds for no more than $N(\varepsilon)$ periods.

The principal change that must be made in the proof of Theorem 8.3 to obtain the proof of Theorem 9.2 is to replace the condition $|k_t(\rho) - k^\rho| > \varepsilon$ by the condition $d((k_{t-1}(\rho), k_t(\rho), F(q^\rho, \rho q^\rho)) > \varepsilon$ whenever lower bounds are being deduced for value losses $\delta_t(\rho)+\delta_t^\rho$. Then the elimination of value losses to avoid contradiction with (8.14) forces convergence to the facet $F(q^\rho, \rho q^\rho)$ rather than to the path $k_t' = k^\rho$. The part of the earlier proof that required (k^ρ, k^ρ) to enter a strictly concave neighborhood of (k, k) is no longer needed, since it is no longer necessary to establish $F(q^\rho, \rho q^\rho) = (k^\rho, k^\rho)$.

Given some assumptions on the structure of the facets F_t of Theorem 9.1 to which the (k_{t-1}, k_t) belong, and which are defined by the (p_{t-1}, p_t), it may be that paths that remain close to the F_t for a long time must approach each other. This can be seen most easily for stationary models where one of the price supported paths is a maximal stationary path supported by constant current prices so that $F_t = F$ for all t.

Choose points in the facet F which affinely span the smallest flat containing F, say (x^i, y^i), $i=1,\ldots,r$, where the dimension of F is $r-1 \leqq 2n$. Then any point $(z, w) \in F$ can be expressed as $\sum_1^r \alpha_i(x^i, y^i)$ where $\sum \alpha_i = 1$. If $\{k_t\}$ is a path on F, we have $\{k_{t-1}, k_t\} = \sum_1^r \alpha_i^t(x^i, y^i)$, and $(k_t, k_{t+1}) = \sum_1^r \alpha_i^{t+1}(x^i, y^i)$, or $\sum_1^r \alpha_i^t y^i = \sum_1^r \alpha_i^{t+1} x^i$. Suppose that $r = n+1$ and A and B are square matrices with columns

$$\begin{pmatrix} x^i \\ 1 \end{pmatrix} \quad \text{and} \quad \begin{pmatrix} y^i \\ 1 \end{pmatrix},$$

respectively. Then $t \geqq 0$, the equation $B\alpha^t = A\alpha^{t+1}$ must be satisfied for some vectors α^t and α^{t+1} if (k_{t-1}, k_t) and (k_t, k_{t+1}) lie on F. If A is non-singular, this may be written

$$\alpha^{t+1} = A^{-1}B\alpha^t. \tag{9.2}$$

Suppose $A^{-1}B$ has only one characteristic root λ with absolute value one and this root is simple. Then $\lambda = 1$, since α must solve (9.2) where

$$\sum_1^r \alpha_i(x^i, y^i) = (k, k),$$

and k is the capital stock vector of the stationary maximal path. The last rows of A and B imply that $\sum_{i=1}^r \alpha_i(j) = 0$ if $\lambda_j \neq 1$ and $\sum_{i=1}^r \alpha_i(1) = 1$ for $\lambda_1 = 1$. The path $k_t = k$ is optimal for the quasi-stationary case by Theorem 5.3. We will show later that it is optimal for the stationary case as well. If we make assumption (G2) of the stationary model of Section 6 that sustainable stocks are bounded, $|k_t|$ is bounded by a number ζ. Then for any path $\{k_t\}$ on F, $k_t \to k$ must hold [McKenzie (1968)]. This is easily seen if the characteristic roots are all simple, so the characteristic vectors span the complexification of the r-dimensional Euclidean space [Hirsch and Smale (1974, pp. 64–65)]. Then $\alpha^t = \sum_1^r \beta_j \lambda_j^t \alpha(j)$ where $\alpha(j)$ is the characteristic vector associated with λ_j and β_j is a given number, possibly complex. Also $\{k_t\}$ on F implies $\sum \alpha_i^t = 1$, so $\beta_1 = 1$. If $|\lambda_i| > 1$, $\beta_j = 0$ must hold, or else α^t and thus k_t is unbounded as $t \to \infty$. If $|\lambda_i| < 1$, $\lambda_i^t \to 0$ as $t \to \infty$. Thus $\alpha^t \to \alpha(1)$ and $k_t \to k$. By an extension of this argument the same convergence property will be shown to hold for any path that converges to F. Then the convergence of maximal paths will once again be established.

In the case where $(k, k) \in$ interior F relative to the smallest flat that contains F we may prove $r \leqq n+1$. The proof of Lemma 6.1 implies that $u + pv = \bar{u}$ for a vector $p \geqq 0$ for every $v = (y - x)$ and $(x, y) \in F$. Thus all (u, v) corresponding to points in F lie in a flat of dimension less than or equal to n in E^{n+1} and $(\bar{u}, 0)$ is expressible as an affine combination with non-zero coefficients of r affinely independent vectors of W, $(u^i, v^i) = (u(x^i, y^i),\ y^i - x^i)$, where $r \leqq n+1$. This means $(u(k, k), k, k)$ is the same affine combination of r affinely independent vectors $(u(x^i, y^i), x^i, y^i)$ of the graph of u. Consequently the dimension of F is at least $r-1$. We will show that the dimension of F is exactly $r-1$. Let $(k, k) = \sum_{i=1}^r \alpha_i (x^i, y^i)$. Then $0 = \sum_{i=1}^r \alpha_i v^i$. Suppose there were $(x, y) \in F$ which was affinely independent of the (x^i, y^i). Let $v = (y - x)$. Then $v = \sum_{i=1}^r \beta_i v^i$ for some β_i where $\sum_1^r \beta_i = 1$, or $v^j = -\beta_j^{-1} \beta_i v^i + \beta_j^{-1} v$, and $0 = \sum_{i \neq j} (\alpha_i - \alpha_j \beta_j^{-1} \beta_i) v^i + \alpha_j \beta_j^{-1} v$. This implies there is (k', k') which is the same linear combination of the (x^i, y^i), $i \neq j$, and (x, y). Moreover, the assumption that (x, y) is affinely independent of the (x^i, y^i) implies that $k' \neq k$. But $(k', k') \in S$ where S is the smallest flat containing F, since S and F are convex, and (k, k) is interior to F, there is a point (k'', k'') on the line segment joining (k', k') and (k, k), which lies in F. This contradicts the uniqueness of the stationary optimal path. Thus no such $(x, y) \in F$ can exist, or the dimension of F is $r-1$. If $r = n+1$, matrices A and B will have the same number of rows as columns and except for coincidence their columns will be linearly independent. If the model is neoclassical, small perturbations of the processes will eliminate characteristic roots of absolute value one except for the root one which is present by construction. Finally, when u is piecewise linear, the graph of u is polyhedral and $(k, k) \in$ relative interior F holds by definition. See Morishima (1969, chs. 10 and 13), for a careful discussion of the polyhedral case.

We may say that the structure of the von Neumann facet F is *stable* if, for any $\varepsilon > 0$, there is T such that every solution α^t of the difference equation (9.2), for which $(A\alpha^t, B\alpha^t) = (k_t', k_{t+1}') \in F$ for $t \geqq 0$, satisfies $|k_t' - k| < \varepsilon$ for some k for all $t > T$ [Inada (1964)]. The case outlined in the last paragraph is an example of a stable facet. Suppose that a bounded path $\{k_t\}$ converges to F, but that $\{k_t\}$ does not converge to k. Choose a sequence of neighborhoods U^s of F defined by $U^s = \{(x, y) | \mathrm{d}((x, y), F) < \varepsilon^s > 0\}$ where $\varepsilon^s \to 0$. Let t_s be a sequence of times such that $(k_t, k_{t+1}) \in U^s$ for $t \geqq t_s$. This is possible since (k_t, k_{t+1}) converges to F by assumption. Consider the sequence of paths $\{k_\tau^s\}$, $\tau = 0, 1, \ldots,$ where $k_\tau^s = k_{t_s + \tau}$. Since the $\{k_\tau^s\}$ are bounded, we may use the Cantor process to choose a subsequence converging to a path $\{k_\tau'\}$. Then assumption (I) and F closed imply that $(k'_\tau, k'_{\tau+1}) \in F$ for all $\tau \geqq 0$. If $\{k_\tau\}$ does not converge to k, given $T \geqq 0$ the times t_s may be chosen so that $|k^s_T - k| > \varepsilon > 0$ for all s. This implies that there exist paths beginning at time $\tau = 0$ on F that lie outside an ε-neighborhood of k at time $\tau = T$ where T may be set arbitrarily large, in contradiction to the stability of F. We may prove:

Theorem 9.3

If a path $\{k_t\}$ in a stationary model, satisfying assumptions (I), (II), and (G2), converges to the von Neumann facet F and the structure of F is stable, then $k_t \to k$ where k is the capital stock vector of the stationary optimal path. If $\{k_t\}$ is a maximal path, it is optimal.

The convergence has been shown. However, the argument for optimality leading to Theorem 5.1 only uses (W1) and (W2), which are met here, together with the turnpike property which was proved using (W3). Since in the present case the turnpike property is established, optimality follows for $\{k_t\}$. The facet F where A is non-singular and $A^{-1}B$ has a unique characteristic root with absolute value one, which is simple and equal to one, gives a particular case for Theorem 9.3. A condition which is equivalent to stability is that the stationary path on F be unique and there be no cyclic paths on F of constant amplitude [McKenzie (1968)].

In the quasi-stationary case the analogue of Theorem 9.3 is not useful since Theorem 9.2 only gives a neighborhood theorem, or Liapounov stability, not asymptotic stability, for the von Neumann facet. However, the neighborhood counterpart of Theorem 9.3 can be proved for quasi-stationary models. First, when there is a unique optimal stationary stock, that is, $C = \{k\}$, for $\rho = 1$, Lemma 8.2 remains valid and $k^\rho \to k$ as $\rho \to 1$. Furthermore, it is easily seen that $\rho \to 1$ implies that $F^\rho \to F$ where F is the von Neumann facet for $\rho = 1$. This allows us to prove a neighborhood version of Theorem 9.3 [see McKenzie (1983)].

Theorem 9.4

In addition to the hypothesis of Theorem 9.2, assume that the von Neumann facet F for $\rho = 1$ is stable. Then for any $\varepsilon > 0$ there is $\rho(\varepsilon)$ and $T(\varepsilon)$ such that $\rho(\varepsilon) < \rho < 1$ implies that $(k_t - k^\rho) < \varepsilon$ holds for $\tau > T(\varepsilon)$.

The proof of this theorem is entirely parallel to the proof of Theorem 9.3, except that use must be made of a sequence of paths $\{k_t(\rho^s)\}$ from $k_0(\rho^s) = k_0$, since no single path need converge either to F or to F^ρ. Indeed, suppose the theorem were false. Then there are sequences $\rho^s \to 1$, $\varepsilon^s \to 0$, and τ^s, $s = 1, 2, \ldots$, such that $(k_t(\rho^s), k_{t+1}(\rho^s))$ lies in the ε^s-neighborhood of F for $t > \tau^s$, but for which there is $t^s = \tau^s + T$ and $|k_{t^s}(\rho^s), k^{\rho^s}| > \varepsilon$, where T may be arbitrarily large. Let $h^s_t = k_{t^s+t}$. Since the paths $\{h^s_t\}$ lie in a bounded set, we may choose, by a Cantor process, a subsequence converging to a sequence $\{h_t\}$, $t = 0, 1, \ldots$. As before, the limit path lies in F but $|h_T - k| > \varepsilon$. Since this construction is possible for T arbitrarily large the stability of F is violated. Thus no such sequences can exist, or $k_t(\rho)$ eventually remains in an ε-neighborhood of k, where ε may be chosen arbitrarily small if ρ is then chosen near enough to 1.

It is also possible to prove an asymptotic theorem in the case of a non-trivial von Neumann facet and ρ near 1 if differentiability is assumed in the manner of Theorem 8.4. Of course, the presence of the facet implies that the second differential of u cannot be negative definite at (k, k). However, it can be negative semi-definite and negative definite in the subspace S of E^{2n} defined by $S = \{(z, w)|(z, w)\cdot((x, y)-(k, k)) = 0$, for all $(x, y) \in F\}$. That is, $[u_{ij}]$ evaluated at (k, k) is negative definite on the orthogonal complement of $(F-(k, k))$. The asymptotic convergence to k is proved by appeal to the local stability theorem of Scheinkman (1976). Since Theorem 9.4 brings the path $k_t(\rho)$ into a small neighborhood of k^ρ for ρ near 1, local asymptotic stability of k^ρ completes the argument. The local argument uses a linear approximation (12.8) to the Euler equations (12.1) and a regularity assumption for the stable manifold of the linear approximation. The theorem and its proof may be found in McKenzie (1983).

III. Comparative Statics and Dynamics

10. Differentiable utility

If we assume, in addition to concavity and closedness of u_t, differentiability of u_t with respect to capital stocks, a new method of proving the turnpike theorem becomes available, due to Araujo and Scheinkman (1977), that does not depend on the condition that ρ be near 1. Differentiability also facilitates comparative studies analogous to the comparative statics of general equilibrium theory. The special assumptions which are used to obtain the results are the analogues in the dynamic setting of the familiar assumptions of comparative statics and stability theory for general equilibrium, that is, a dominant diagonal or negative definiteness for the appropriate Jacobian matrix [see Arrow and Hahn (1971, ch. 12)]. Negative definiteness is almost equivalent in the differentiable context to the value loss assumptions of the last section. However, the dominant diagonal assumption for the Jacobian matrix is independent of negative definiteness. The concavity of utility that is crucial for calculus of variations, and maximum theory in general, is still needed. This should not be surprising since the conditions of Weierstrass and Legendre in calculus of variations, which imply local concavity of utility with respect to rates of change, are *necessary* conditions along an optimal path.

Because of the differentiability of u_t we do not need to appeal to Section 4 for support prices since the derivatives of u_t take their place. Recall that a path of accumulation is optimal if it catches up to every alternative path from the same initial stocks. If $\{k_t\}$ is a path with (k_t, k_{t+1}) interior to D, consider alternative paths $\{k'_t\}$ where $k'_t = k_t$ for $t \neq \tau$ and $k'_\tau = x > 0$. Then $\{k_t\}$ catches up to $\{k'_t\}$ if and only if $u_\tau(k_{\tau-1}, x) + u_{\tau+1}(x, k_{\tau+1}) \leqq u_\tau(k_{\tau-1}, k_\tau) + u_{\tau+1}(k_\tau, k_{\tau+1})$. The differentiability assumption implies that this condition will be violated for an appropriate choice of x and τ unless

$$u_2^t(k_{t-1}, k_t) + u_1^{t+1}(k_t, k_{t+1}) = 0, \tag{10.1}$$

for all t, where u_1^t denotes the vector of derivatives of u_t with respect to initial stocks and u_2^t the vector of derivatives with respect to terminal stocks. Thus (10.1) is a necessary condition for optimal paths and corresponds to the Euler condition of the calculus of variations.

We assume:

(I′) The utility functions $u_t(x, y) = \rho^t u(x, y)$, where $0 < \rho < 1$ and $u(x, y)$ is concave and closed on the convex set D, contained in the non-negative orthant of E^{2n}. Interior $D \neq \phi$ and u has continuous second partial derivatives in the interior of D.

For the sake of simplicity we make our argument in terms of the quasi-stationary case $u_t = \rho^t u$, $0 < \rho < 1$, although the argument can be given in a general form applying to utility functions that depend on time in more complicated ways, reflecting changes in taste and technology [McKenzie (1977)]. Let $\{k_t\}$, $t = 0, 1, \ldots$, be a path satisfying (10.1) for $u_t = \rho^t u$, where the distance of the path from the boundary of $D_t = D$ is at least $\varepsilon > 0$ in all periods. Represent an arbitrary path $\{k'_t\}$ by $\{z_t\}$ where $z_t = k'_t - k_t$, and rewrite (10.1), after dividing through by ρ^t, as

$$v_2(z_{t-1}, z_t) + \rho v_1(z_t, z_{t+1}) = 0, \tag{10.2}$$

for all t, setting $v(z_{t-1}, z_t) = u(k'_{t-1}, k'_t)$ for all t. We will refer to $\{z_t\}$ also as a path. For a given $0 < \beta < 1$, let $x_t = \beta^{-t} z_t$. Let G_z be the set of paths $\{z_t\}$ with $\beta^{-t}|z_t| < \varepsilon/2$ for all t. Let G_x be the corresponding set of sequences $\{x_t\}$. Then G_x is contained in the Banach space l_∞^n of bounded sequences of vectors in E_n. The norm $|x|_\infty$ of $x \in l_\infty^n = \sup|x_t|$ over $t \geqq 0$, where $|x_t|$ is the Euclidean norm. The set G_x is not empty since it contains 0. By the assumption that (k'_{t-1}, k'_t) is bounded interior to D, G_x has a non-empty interior in l_∞^n.

We define a function ξ by

$$\xi_t(x_0, x) = \beta^{-t} v_2(\beta^{t-1} x_{t-1}, \beta^t x_t) + \beta^{-t} \rho v_1(\beta^t x_t, \beta^{t+1} x_{t+1}),$$

$t = 1, 2, \ldots$, where $x = \{x_t\}$, $t = 1, 2, \ldots$. Then $\xi(0,0)_t = 0$ for all t, by the first-order condition (10.2) for an optimum. If v has second partial derivatives at (0,0) that are bounded and uniformly continuous over t, $\xi_t(x_0, x)$ is bounded over t for small ε. Thus ξ maps G_x into l^n_∞.

We will say that a path $\{k_t\}$ is *smooth* if it satisfies the Euler equation (10.1) and is bounded away from the boundary of D. Then u has second partial derivatives that are bounded from ∞ over t and uniformly continuous along $\{k_t\}$. It is possible to show for smooth paths that the derivative $D_x\xi$ at (0,0) is given by

$$(D_x\xi h)_1 = (v^1_{22} + \rho v^2_{11})h_1 + \beta\rho v^2_{12}h_2,$$

$$(D_x\xi h)_t = \beta^{-1}v^t_{21}h_{t-1} + (v^t_{22} + \rho v^{t+1}_{11})h_t + \beta\rho v^{t+1}_{12}h_{t+1}, \tag{10.3}$$

for $t = 2,3,\ldots,$ where $h \in l^n_\infty$, $v^t_{ij} = v_{ij}(z_{t-1}, z_t)$, and the partial derivatives are evaluated at (0,0). Also $D_x\xi$ is continuous at (0,0) [see Araujo and Scheinkman (1977)]. We may represent $D_x\xi$ as an infinite matrix, or

$$D_x\xi = \begin{bmatrix} v^1_{22} + \rho v^2_{11} & \beta\rho v^2_{12} & 0 & & \\ \beta^{-1}v^2_{21} & v^2_{22} + \rho v^3_{11} & \beta\rho v^3_{12} & & \\ & \beta^{-1}v^3_{21} & v^3_{22} + \rho v^4_{11} & \beta\rho v^4_{12} & \\ 0 & & & \ddots & \end{bmatrix}. \tag{10.4}$$

If $|\cdot|_1$ is a norm on R^n, for a matrix argument $|\cdot|_1$ indicates the corresponding operator norm, that is $|M_{ij}|_1$ is $\sup|M_{ij}y|_1$ for $y \in E^n$, $|y|_1 = 1$. Given any norm on R^n, an infinite matrix M formed of $n \times n$ blocks M_{ij}, with M_{ii} invertible, is said to have *dominant diagonal blocks* if $\sup|M_{ii}^{-1}|_1 < \infty$ over i and $\sup\sum_{j \neq i} |M_{ii}^{-1}M_{ij}|_1 = \delta < 1$ over i. M defines a transformation of l^n_∞ into l^n_∞ when $\sum_j|M_{ij}|_1$ is bounded over i. The boundedness of the second partial derivatives of u imply this condition for $D_x\xi$ if $\{k_t\}$ is smooth. The matrix M is said to be *invertible* if it defines a linear homeomorphism of l^n_∞ onto l^n_∞ [Dieudonné (1960, p. 45)]. We may show:

Lemma 10.1

If an infinite matrix M that maps l^n_∞ into l^n_∞ has dominant diagonal blocks, it is invertible.

Since M is bounded on the unit ball in l^n_∞, it is a continuous linear map. Let M_1 be the matrix of diagonal blocks M_{ii} with 0's elsewhere. Since $|M_{ii}^{-1}|$ is bounded over i by the assumption of dominant diagonal blocks, M_1^{-1} exists and is continuous. Let $M_2 = M_1^{-1}M$. Then the dominant diagonal assumption implies, for some norm $|\cdot|_1$,

$$|M_2 - I|_1 = \sup_i \sum_{j \neq i} |M_{ii}^{-1}M_{ij}|_1 = \delta < 1.$$

Since $M_2 = I - (I - M_2)$, formally $M_2^{-1} = I + (I - M_2) + (I - M_2)^2 + \cdots$. But the Neumann series on the right-hand side converges, so M_2 has a continuous inverse over l_∞^n. Thus $M = M_1 M_2$ has a continuous inverse over l_∞^n.

Assume that $D_x\xi$ has dominant diagonal blocks at (0,0), which corresponds to $k_t' = k_t$, for all t. This condition will hold for β sufficiently near 1 if it holds for $\beta = 1$. Then $D_x\xi$ is a linear homeomorphism of l_∞^n onto l_∞^n. Also $\xi(x_0, x)$ maps a neighborhood of (0,0) in $E^n \times l_\infty^n$ into l_∞^n with $\xi(0,0) = 0$. We may apply the implicit function theorem [Dieudonné (1960, p. 265)] to obtain a continuous function $\psi(x_0)$ valid in a neighborhood of $x_0 = 0$ such that $\xi(x_0, \psi(x_0)) = 0$ where $\psi(x_0)$ has continuous derivatives and $\psi(0) = 0$.

The continuity of ψ implies that $|x_0|_1$ may be chosen small enough to put x near 0, that is, $\sup|x_t|_1 < \varepsilon$ over t for small positive ε. Then $z_t = \beta^t x_t$ for $\beta < 1$ implies that z_t converges exponentially to 0, that is k_t' converges exponentially to k_t. We note that for ε sufficiently small $\{k_t'\}$ is also a smooth path. This proves:

Lemma 10.2

If (k_0, k) is a path of accumulation that is smooth and the Jacobian of the map ξ, derived from the Euler equation (10.1), has dominant diagonal blocks, there is a neighborhood W of k_0 such that $k_0' \in W$ implies there is a smooth path $\{k_t'\}$, $t = 0, 1, \ldots$, and $k_t' \to k_t$ exponentially as $t \to \infty$.

In order to derive a local turnpike theorem from Lemma 10.2, it is only necessary to show that the paths $\{k_t'\}$ derived there are optimal paths from k_0' near k_0. An additional assumption is needed, which in the quasi-stationary case can take the form of (S2) or (G2), introduced in Section 6. The effect of (S2) is to bound any path. We can prove:

Lemma 10.3

If assumptions (I′), (II), and (S2) hold, any path that satisfies the Euler equation and is bounded interior to D is smooth, and any infinite smooth path is optimal.

All that is needed to give smoothness for an Euler path that is bounded away from the boundary of D is that its second partial derivatives be bounded. However, smoothness is immediate by continuity of the derivatives if the path is confined to a compact subset of D. But this follows from (S2) and the fact that the path is bounded interior to D.

To show optimality for smooth paths observe that concavity of u implies for $\{k_t'\}$, $t=0,1,\ldots$,

$$\rho^{t+1}u(k_t', k_{t+1}') - \rho^{t+1}u_2^{t+1}k_{t+1}' - \rho^{t+1}u_1^{t+1}k_t'$$
$$\geqq \rho^{t+1}u(x, y) - \rho^{t+1}u_2^{t+1}y - \rho^{t+1}u_1^{t+1}x, \tag{10.5}$$

for $(x, y) \in D$, where $u_2^{t+1} = u_2(k_t', k_{t+1}')$, for example. By the Euler equation (10.1), $u_2(k_{t-1}', k_t') = -\rho u_1(k_t', k_{t+1}')$ for a smooth path $\{k_t'\}$. Thus $\rho^t u_2^t = -\rho^{t+1}u_1^{t+1}$ in (10.5). Let $p_t = -\rho^t u_2^t$, $t = 0,1,\ldots$. Then (10.5) implies (W1) and smoothness implies the second part of (W2′). Since (S2) implies the first part of (W2′), $\{k_t'\}$ is optimal by Theorem 5.3.

Together Lemmas 10.2 and 10.3 imply, except for uniqueness:

Theorem 10.1

Suppose $\{k_t\}$, $t=0,1,\ldots$, is a path that is smooth and satisfies the dominant diagonal condition, and assumptions (I′), (II), (G2) are met by the utility function. Then every capital stock k_0' near k_0 initiates a unique optimal path and this path converges exponentially to $\{k_t\}$.

To see that the optimal path is unique, suppose there were a second optimal path $\{k_t''\}$ with $k_0'' = k_0'$. Consider a path $\{\bar{k}_t\}$ with $\bar{k}_t = \alpha k_t' + (1-\alpha)k_t''$, $0 < \alpha < 1$. Then

$$\alpha \sum_1^T u_t(k_{t-1}', k_t') + (1-\alpha)\sum_1^T u_t(k_{t-1}'', k_t'') = \sum_1^T u_t(\bar{k}_{t-1}, \bar{k}_t) - \sum_1^T \varepsilon_t,$$

where $\varepsilon_t \geqq 0$. Thus the optimality of $\{k_t'\}$ and $\{k_t''\}$ implies that $\varepsilon_t = 0$, all t. For α sufficiently small $\{\bar{k}_t\}$ lies in a small neighborhood of $\{k_t'\}$ and thus of $\{k_t\}$. Since $\{\bar{k}_t\}$ is also optimal from k_0' it must satisfy the Euler equation. However, by the implicit function theorem the solution of the Euler equation in a small neighborhood of $\{k_t\}$ is unique. Thus $k_t'' = k_t'$ for all t.

Theorem 10.1 is a local turnpike result. However, it may be used to prove a global theorem. Let C be the set of capital stocks that initiate smooth paths at $t=0$ along which the dominant diagonal condition is met. Theorem 10.1 implies that these paths are optimal. By assumption, C is not empty. If $\{k_t\}$ is a smooth optimal path satisfying the dominant diagonal condition, Theorem 10.1 implies that k_0' in a small neighborhood of k_0 initiates a smooth path $\{k_t'\}$ that converges to $\{k_t\}$. Moreover, the uniform continuity of the second partial derivatives near the path $\{k_t\}$ implies that the dominant diagonal condition is also met by $\{k_t'\}$ when the neighborhood is chosen small enough. Thus we may consider the maximal connected component C_0 of C that contains k_0.

Let S be the subset of C_0 such that the optimal path from $w \in S$ converges exponentially to $\{k_t\}$. If $w \in S$, Theorem 10.1 implies there is a neighborhood of w which is also in S. Let $\{k_t'\}$ be the optimal path from w and let y lie in this neighborhood. Then there is a path $\{k_t''\}$ from y, and $\beta < 1$, for which $|k_t - k_t''| \leqq |k_t - k_t'| + |k_t' - k_t''| \leqq \beta^t |k_0 - w| + \beta^t |w - y|$, so k_t'' also converges exponentially to k_t as $t \to \infty$. Thus S is open.

Now suppose that $x \in$ boundary S and $x \in C_0$. Since $x \in C_0$, Theorem 10.1 applies and there is $y \in S$ near x such that the path $\{k_t''\}$ optimal from y converges exponentially to the optimal path $\{k_t'\}$ that departs from x. But $y \in S$ implies that k_t'' converges exponentially to k_t. Therefore, k_t' must converge exponentially to k_t, or S is closed in C_0. But C_0 is a connected set so $S = C_0$. We have proved a global result.

Theorem 10.2

Suppose $\{k_t\}$, $t = 0, 1, \ldots,$ is a path that is smooth and satisfies the dominant diagonal condition, and assumptions (I′), (II), and (S2) are met by the utility function. Let C be the set of capital stocks, at $t = 0$, that initiate smooth paths satisfying the dominant diagonal condition. Let C_0 be the maximal connected component of C that contains k_0. Then $x \in C_0$ implies there is a unique optimal path $\{k_t'\}$ with $k_0' = x$ and $k_t' \to k_t$ at an exponential rate, as $t \to \infty$.

The crucial feature of the argument leading to the turnpike result is the invertibility of the derivative of the Euler functions. This derivative was used to define a transformation of l_∞^n into l_∞^n. Sometimes, however, other Banach spaces may be more effective. For example, if assumptions are made like those in Section 8 to support a value loss argument, the appropriate space is Hilbert space l_2^n. The invertibility lemma follows if the derivative is negative definite. Consideration of the matrix representation (10.4) shows that the derivative is negative definite if the matrix

$$\begin{bmatrix} \rho u_{11}^t & \rho u_{12}^t \\ u_{21}^t & u_{22}^t \end{bmatrix}$$

is negative quasi-definite uniformly over the path. This is implied by uniformity over interior $D \cap W$, for $W = \{(x, y) \| x| < \max(|k_0|, \zeta)\}$ where ζ is from (S2). It may be shown [Brock and Scheinkman (1978)] that this condition is almost equivalent to the value loss conditions (8.14) and (8.21) needed for the turnpike results in Section 8 when u is twice continuously differentiable. Thus the method of this section is very powerful for interior paths when u is twice continuously differentiable.

The arguments used here like those in Section 9 are not limited to the quasi-stationary case. With minor complications they can be adapted to utility functions $u_t(x, y)$ which depend on time in the way described in Section 2 [McKenzie (1977)].

11. Comparative dynamics for optimal paths

By use of the infinite Jacobian matrix of the first-order conditions (the discrete Euler conditions) for an optimal path it is possible to derive comparative dynamic results for the differentiable model [Araujo–Scheinkman (1979)]. These are analogous to the comparative static results proved in general equilibrium theory and use the same assumptions adapted to the infinite case. The Jacobian matrix is shown to be negative definite, or it is assumed to have dominant diagonal blocks with certain sign patterns for diagonal and off-diagonal blocks. The parameters that shift demand between the numéraire and other goods in general equilibrium are replaced by the discount factor or the initial stocks in the dynamic case of optimal growth.

Let $\{k_t\}$, $t = 0,1,\ldots$, be an optimal path. Let $z_t = k'_t - k_t$, and $z = \{z_t\}$, $t = 1,2,\ldots$. Define $\zeta(z_0, z, \rho)$ for $0 < \rho < 1$ by

$$\zeta_t(z_0, z, \rho) = v_2(z_{t-1}, z_t) + \rho v_1(z_t, z_{t+1}), \tag{11.1}$$

where $v(z_{t-1}, z_t) = u(k'_{t-1}, k'_t)$ for all t. If $\{k_t\}$ is a smooth path and B_z is the set of paths $\{z_t\}$ with $|z_t| < \varepsilon$, for small ε, ζ maps B_z into l^n_∞. Similarly if H_z is the set of paths $\{z_t\}$ with $\sum_1^\infty |z_t|^2 < \infty$ and $|z_t| < \varepsilon$, for small ε, ζ maps H_z into l^n_2. In the first case ζ maps a neighborhood of 0 in l^n_∞ into l^n_∞, and in the second case ζ maps a neighborhood of 0 in l^n_2 into l^n_2.

As in Section 10, under assumption (I′) for a smooth path $\{k_t\}$, $D_z\zeta(0,0,\rho)$ can be represented in either space by an infinite matrix,

$$D_z\zeta = \begin{bmatrix} v^1_{22} + \rho v^2_{11} & \rho v^2_{12} & & & 0 \\ v^2_{21} & v^2_{22} + \rho v^3_{11} & \rho v^3_{12} & & \\ & v^3_{21} & v^3_{22} + \rho v^4_{11} & \rho v^4_{12} & \\ & & & \ddots & \\ 0 & & & & \end{bmatrix}. \tag{11.2}$$

In this expression $v^t_{ij} = v_{ij}(k_{t-1}, k_t)$. Suppose the quadratic bilinear form $h^T(D_z\zeta)h$ is negative definite, that is, $h^T(D_z\zeta)h < -\varepsilon\sum_1^\infty |h_t|^2$, for all $h \in l^n_2$ and some $\varepsilon > 0$. Then $D_z\zeta$ is invertible on l^n_2 [Araujo and Scheinkman (1977, p. 619)].

It is clear from the representation (11.2) that $D_z\zeta$ will be negative definite if

$$\begin{bmatrix} \rho v_{11}^t & \rho v_{12}^t \\ v_{21}^t & v_{22}^t \end{bmatrix}$$

is negative quasi-definite, uniformly with respect to t along the path $\{k_t\}$. At the stationary optimal path, $k_t = k$, all t, $D_z\zeta$ is negative definite if and only if

$$\begin{bmatrix} \rho v_{11} & \rho v_{12} \\ v_{21} & v_{22} \end{bmatrix}$$

is negative quasi-definite, where $v_{ij} = u_{ij}(k, k)$. Also, from Lemma 10.1, $D_z\zeta$ is invertible over l_∞^n when the dominant diagonal condition is met. These are the two conditions which have been shown to imply a turnpike theorem. As in the general equilibrium tâtonnement, there is a close relationship between conditions which imply stability and conditions which allow comparison of equilibrium paths.

As for $D_z\xi$ in Section 10, the invertibility of $D_z\zeta$ allows the implicit function theorem to be applied to obtain a function $\phi(z_0, \rho')$, defined in some small neighborhood of $(0, \rho)$, such that $\zeta(z_0, \phi(z_0, \rho'), \rho') = 0$. Also $\phi(z_0, \rho')$ is differentiable and the derivatives are given by

$$\begin{aligned} D_{z_0}\phi(z_0, \rho') &= -[D_z\zeta(z_0, \phi(z_0, \rho'), \rho')]^{-1} \cdot D_{z_0}\zeta(z_0, \phi(z_0, \rho'), \rho'), \\ D_\rho\phi &= -[D_z\zeta]^{-1} \cdot D_\rho\zeta, \end{aligned} \tag{11.3}$$

[Dieudonné (1960, p. 265)]. We first show:

Lemma 11.1

If $D_z\zeta$ is invertible on l_∞^n, assumption (I′) implies

$$\sum_1^\infty \rho^{t+1}[\mathrm{d}z/\mathrm{d}\rho]_t [D_\rho\zeta]_t \geqq 0, \tag{11.4}$$

where $D_z\zeta$ and $D_\rho\zeta$ are evaluated at $(z_0, \phi(z_0, \rho'), \rho')$, $\mathrm{d}z/\mathrm{d}\rho = D_\rho\phi(z_0, \rho')$, and (z_0, ρ') is sufficiently near $(0, \rho)$.

Since $\mathrm{d}z/\mathrm{d}\rho = -[D_z\zeta]^{-1}\cdot D_\rho\zeta$ by (11.3), we obtain $D_z\zeta\cdot \mathrm{d}z/\mathrm{d}\rho = -D_\rho\zeta$. Therefore,

$$\sum_1^\infty \rho^{t+1}[\mathrm{d}z/\mathrm{d}\rho]_t[D_z\zeta\cdot \mathrm{d}z/\mathrm{d}\rho]_t = -\sum_1^\infty \rho^{t+1}[\mathrm{d}z/\mathrm{d}\rho]_t[D_\rho\zeta]_t. \tag{11.5}$$

The left-hand side of (11.5) is equal to $[\mathrm{d}z/\mathrm{d}\rho]^T A(\mathrm{d}z/\mathrm{d}\rho)$, where A is equal to the matrix obtained from $D_z\zeta$ when the tth row is multiplied by ρ^{t+1}. However, it is easily seen from (11.2) that A is negative semi-definite if

$$\begin{bmatrix} v_{11}^t & v_{12}^t \\ v_{21}^t & v_{22}^t \end{bmatrix}$$

is negative semi-definite, which is implied by the concavity of v. The concavity of v is immediate from the concavity of u given by assumption (I′). Also the convergence of the sums in (11.5) follows from the fact that the derivatives belong to l_∞^n and $0<\rho<1$. This completes the proof of the lemma. Of course, $D_z\zeta$ is invertible on l_∞^n when the path $\{k_t\}$ is smooth and the dominant diagonal condition holds.

From (11.1) we observe that $[D_\rho\zeta(0,0,\rho)]_t = v_1(0,0) = u_1(k_t, k_{t+1}) = -\rho^{-1}u_2(k_{t-1}, k_t)$. Thus $([D_\rho\zeta]_t, \rho[D_\rho\zeta]_{t+1})$ supports $u(k_t, k_{t+1})$ in the sense of (4.11) by virtue of the concavity of u. Put $p_t = \rho^{t+1}[D_\rho\zeta]_t$, $t=1,2,\ldots$, and $p_0 = u_1(k_0, k_1)$. Then $\{p_t\}$, $t=0,1,\ldots$, satisfies (4.11). Moreover, by the differentiability of u, these supports are unique, so they must satisfy (4.10) as well by Lemma 4.1. Since $\mathrm{d}z/\mathrm{d}\rho = \mathrm{d}k/\mathrm{d}\rho$, the conclusion of Lemma 11.1 may be written $\sum_1^\infty p_t(\mathrm{d}k_t/\mathrm{d}\rho) \geqq 0$, or an increase in the discount factor for utility cannot reduce the present value of the stream of capital stocks at the support prices.

The conclusion of Lemma 11.1 holds equally well when $D_z\zeta$ is invertible on l_2^n by the same argument. As mentioned earlier $D_z\zeta$ will be invertible for a smooth path under assumption (I′) if the quadratic bilinear form $h^T(D_z\zeta)h$ is negative definite, that is, if

$$\begin{bmatrix} \rho v_{11}^t & \rho v_{12}^t \\ v_{21}^t & v_{22}^t \end{bmatrix}$$

is uniformly negative quasi-definite with respect to t. Indeed, in this case it is unnecessary to multiply by ρ^{t+1}. Current prices may be used, that is, putting

$q_t = \rho^{-t} p_t$,

$$\rho \sum_1^\infty [\mathrm{d}z/\mathrm{d}\rho]_t [D_\rho \zeta]_t = \sum_1^\infty q_t [\mathrm{d}z/\mathrm{d}\rho]_t > 0 \tag{11.6}$$

will hold. On the other hand, the economic meaning of a sum of current values is not clear.

The results so far are not intrinsic to the stationary model. However, for the stationary model when a stationary optimal path exists that is interior, Araujo and Scheinkman (1979) have shown a more intimate connection between stability and Lemma 11.1. If the linear approximation to the Euler equations, as a system of difference equations, is asymptotically stable at the stationary optimal path, and also the optimal path $\{k_t\}$ converges to the stationary optimal path, then the Jacobian matrix $D_z\zeta$ along this path is invertible on l_∞^n, and the consequence (11.4) may be drawn. A path that converges in this fashion is said to satisfy a strong global turnpike condition.

The foregoing discussion may be collected in:

Theorem 11.1

Assume (I′), (II), and (S2), and let $\{k_t\}$, $t = 0, 1, \ldots$, be a smooth optimal path. Let $\{p^t\}$ be the unique support prices for $\{k_t\}$. Then $\sum_1^\infty p_t(\mathrm{d}k_t/\mathrm{d}\rho) \geqq 0$ if any of the following conditions hold:

(1) The Jacobian $D_z\zeta(0, 0, \rho)$ has dominant diagonal blocks along $\{k_t\}$, where $z_t = k_t' - k_t$.

(2) The matrix

$$\begin{bmatrix} \rho u_{11}^t & \rho u_{12}^t \\ u_{21}^t & u_{22}^t \end{bmatrix}$$

is negative quasi-definite along $\{k_t\}$, uniformly with respect to t.

(3) The path $\{k_t\}$ satisfies a strong global turnpike condition. If condition (2) holds, the inequality is strict and p_t may be replaced with $q_t = \rho^{-t} p_t$.

It should be noted that Theorem 11.1 does not make a comparison of stationary optimal paths. Even when $k_t = k$, $p_t = p$, for all t, we cannot expect $\mathrm{d}k_t/\mathrm{d}\rho$ to be constant with respect to t. A shift in ρ to ρ' will lead to a new *stationary* optimal path $k_t' = k'$, and the new *optimal* path from k will converge to k'.

Comparative dynamic results may also be obtained when the initial stocks vary if appropriate assumptions are made on the signs of elements of the Jacobian. These assumptions will be sufficient to sign the inverse of the Jacobian matrix just as in the static case of general equilibrium theory. The crucial mathematical tool is a generalization to infinite dimensions of the theorem on non-negative inverses for Leontief type matrices. Araujo and Scheinkman (1979) proved:

Lemma 11.2

Let M be an infinite matrix written as a collection of $n \times n$ blocks M_{ij}, $i, j = 1, 2, \ldots$, with $\sup \sum_{j=1}^{\infty} |M_{ij}| < \infty$, over i. If M has dominant diagonal blocks and $M_{ii}^{-1} \leqq 0$, $M_{ij} \geqq 0$, for $i \neq j$, then $M^{-1} \leqq 0$.

As in proving Lemma 10.1 let M_1 be the matrix of diagonal blocks M_{ii} with 0's elsewhere. Let $M_2 = M_1^{-1} M$. As before $M_2^{-1} = I + (I - M_2) + (I - M_2)^2 + \cdots$, since $M_1^{-1} \leqq 0$ and $M_{ij} \geqq 0$ for $i \neq j$, $I - M_2 \geqq 0$. Thus $M_2^{-1} \geqq 0$ and $M^{-1} = M_2^{-1} M_1^{-1} \leqq 0$.

The condition $M_{ii}^{-1} \leqq 0$ will be satisfied by the theorem for Leontief matrices [McKenzie (1960)] if M_{ii} has quasi-dominant diagonal elements, either by rows or columns, that are negative, and the off-diagonal elements are non-negative. A square matrix A has *quasi-dominant diagonal elements* by rows if there exist numbers $d_i > 0$ such that $d_i |a_{ii}| > \sum_j d_j |a_{ij}|$ for all i, and *mutatis mutandis* for columns.

Assume that the Jacobian matrix (11.2) of the Euler conditions $D_z \zeta$ satisfies the conditions of Lemma 11.2 on an optimal path. That is, $(v_{22}^t + \rho v_{11}^{t+1})^{-1} \leqq 0$ and $v_{12}^{t+1} \geqq 0$, $v_{21}^{t+1} \geqq 0$, for $t = 1, 2, \ldots$. Then if $[D_z \zeta]^{-1}$ exists, it will satisfy $[D_z \zeta]^{-1} \leqq 0$. However, by (11.3), $D_{z_0} \phi(z_0, \rho') = -[D_z \zeta]^{-1} \cdot D_{z_0} \zeta$. From (11.1), $[D_{z_0} \zeta]_1 = v_{21}(z_0, z_1)$, and $[D_{z_0} \zeta]_t = 0$, for $t > 1$. Thus, by assumption, $D_{z_0} \zeta \geqq 0$, and finally $[D_{z_{i0}} \phi(z_0, \rho')]_t = \mathrm{d}k_t / \mathrm{d}k_{i0} \geqq 0$ for all i and t. The effect of increasing any initial stock is to cause all subsequent stocks along the optimal path to increase or remain constant. This justifies:

Theorem 11.2

Assume (I′), (II), and (G2), and let $\{k_t\}$, $t = 0, 1, \ldots$, be a smooth optimal path. Suppose the matrix $D_z \zeta(0, 0, \rho)$ has dominant diagonal blocks, and the sign conditions $(u_{22}^t + \rho u_{11}^{t+1})^{-1} \leqq 0$, $u_{12}^{t+1} \geqq 0$, $u_{21}^{t+1} \geqq 0$, $t = 1, 2, \ldots$, are met where $u_{ij}^t = u_{ij}^t(k_{t-1}, k_t)$. Then $\mathrm{d}k_t / \mathrm{d}k_{i0} \geqq 0$ for all i and all t.

12. Comparative statics of stationary states

Comparative statics is confined to the stationary or quasi-stationary model and compares stationary optimal paths which correspond to different values of the discount factor or other parameters of the model. Our interest will lie in the quasi-stationary model where the following assumption holds:

(I″) The utility function $u_t = \rho^t u$ for $0 < \rho \leqq 1$ and u is concave and closed over D which is a convex set contained in the non-negative orthant of E^{2n}. Interior $D \neq \phi$. Also there is a stationary optimal path $\{k_t\}$ interior to D with $k_t = k$, where u has continuous second partial derivatives at (k, k) and the Hessian matrix of u at (k, k) is negative definite.

We will be concerned with the effect of small changes in parameters for stationary optimal paths whose input–output vectors (k, k) are interior to D.

Let $k_t = k$ be a stationary optimal path where (k, k) is interior to D. Then the first-order conditions for optimality (10.1) imply

$$u_2(k, k) + \rho u_1(k, k) = 0. \tag{12.1}$$

As noted in Section 11, if the matrix

$$Q(\rho) = \begin{bmatrix} \rho u_{11} & \rho u_{12} \\ u_{21} & u_{22} \end{bmatrix},$$

evaluated at (k, k), is negative quasi-definite the local turnpike theorem holds, that is, for any capital stocks in a small neighborhood of k, the unique optimal path converges to $k_t = k$, as $t \to \infty$. Then we may say that the stationary optimal path $k_t = k$ is locally stable.

The Jacobian matrix of (12.1) with respect to k is $J(\rho) = u_{21} + u_{22} + \rho u_{11} + \rho u_{12}$. If this matrix is non-singular, the implicit function theorem may be applied to (12.1). That is, if (k', ρ') satisfy (12.1) for $0 < \rho' \leqq 1$ and $(k', k') \in$ interior D, there is a unique differentiable function $k(\rho)$ such that $(k(\rho), \rho)$ satisfy (12.1) for ρ near ρ', and $k(\rho') = k'$. Let $q(\rho) = \rho u_1(k(\rho), k(\rho))$ we may consider the inequality

$$q(\rho') \cdot \mathrm{d}k(\rho)/\mathrm{d}\rho|_{\rho=\rho'} > 0. \tag{12.2}$$

If (12.2) is satisfied at $(k(\rho'), \rho')$) Burmeister and Turnovsky (1972) say that the model is *regular* at $(k(\rho'), \rho')$). Regularity means that an increase in the discount factor leads to an increase in the value of capital for a stationary optimal path when prices are held constant.

If the necessary condition (12.1) for optimality with $k = k(\rho)$ is totally differentiated with respect to ρ, we obtain

$$(u_{21} + u_{22} + \rho u_{11} + \rho u_{12})(\mathrm{d}k/\mathrm{d}\rho) + u_1 = 0,$$

or

$$J(\rho)(\mathrm{d}k/\mathrm{d}\rho) = -u_1 = -\rho^{-1}q, \tag{12.3}$$

where the functions are evaluated at $(k(\rho), k(\rho))$, and $k = k(\rho)$. Multiplying (12.3) by $\mathrm{d}k/\mathrm{d}\rho$ on the left gives

$$(\mathrm{d}k/\mathrm{d}\rho)^T J(\rho)(\mathrm{d}k/\mathrm{d}\rho) = -\rho^{-1}q(\mathrm{d}k/\mathrm{d}\rho). \tag{12.4}$$

But if $Q(\rho)$ is negative quasi-definite, so is $J(\rho)$ and (12.4) implies (12.2). Thus we have:

Theorem 12.1

Under assumption (I″), the sufficient condition for local stability of a stationary optimal path, $Q(\rho)$ negative quasi-definite, implies that the stationary optimal path is regular.

Another condition that implies $J(\rho)$ negative quasi-definite and thus establishes regularity is $u_{21} + u_{22}$ negative quasi-definite. Put $J(\rho, \alpha) = u_{21} + u_{22} + \alpha u_{11} + \alpha u_{12}$, evaluated at $k(\rho)$. Then $J(\rho, \alpha)$ is negative quasi-definite when $\alpha = 0$, and $J(\rho, \alpha)$ is negative definite for $\alpha = 1$. Since $J(\rho) = \rho(J(\rho, 1)) + (1-\rho)(J(\rho, 0))$, $J(\rho)$ is negative quasi-definite for $0 < \rho < 1$ [Dasgupta and McKenzie (1983)]. The analogous condition in the continuous time model is shown by Magill (1977) to imply stability for that model. If we write $x_{t+1} = k_{t+1} - k_t$ and $U(k_t, x_{t+1}) = u(k_t, k_{t+1})$, then $u_{21} + u_{22} = U_{21}$. Thus $u_{21} + u_{22}$ is the effect on the marginal utility cost of investment of an increase in the initial stocks.

According to Theorem 10.1 local asymptotic stability holds around an optimal path if the assumption of dominant diagonal blocks is met along this path. For a stationary optimal path the dominant diagonal assumption for the infinite Jacobian matrix $D_x\xi(x_0, x)$ with $\beta = 1$ is reduced to

$$|(u_{22} + \rho u_{11})^{-1}u_{21}| + |(u_{22} + \rho u_{11})^{-1}\rho u_{12}| < 1, \tag{12.5}$$

since the non-zero blocks of each row are the same, with $u_{ij}^t = u_{ij}(k, k)$, all t, and $i, j = 1, 2$. Recall that $|\cdot|$ for a matrix argument denotes the operator norm, that is, $|A| = \sup|Ax|$ for $|x| = 1$. As in the case of market tâtonnement, the dominant diagonal assumption is not effective by itself but requires a supplement, for example, symmetry or sign restrictions on the elements. Assume, as in the dynamic case of Theorem 11.2, that $(u_{22} + \rho u_{11})^{-1} \leqq 0$ and $u_{12} = u_{21}^T \geqq 0$. From (12.3) we have

$$\mathrm{d}k/\mathrm{d}\rho = -\left(I + (u_{22} + \rho u_{11})^{-1}(u_{21} + \rho u_{12})\right)^{-1}(u_{22} + \rho u_{11})^{-1}\rho^{-1}q. \tag{12.6}$$

By the proof of Lemma 11.2 and the sign assumptions $(I+(u_{22}+\rho u_{11})^{-1}(u_{21}+\rho u_{12}))^{-1}$ exists and is non-negative. Therefore $\mathrm{d}k/\mathrm{d}\rho = Mq$, where $M \geqq 0$. Since free disposal implies that q is non-negative, we have $\mathrm{d}k_i/\mathrm{d}\rho \geqq 0$, for all i, or an increase in the discount factor cannot lead to a decrease in any capital stock. This justifies:

Theorem 12.2

Under assumption (I″), if the dominant diagonal block condition holds and $(u_{22}+\rho u_{11})^{-1} \leqq 0$, $u_{12} \geqq 0$, along the stationary optimal path $k(\rho)$, then $\mathrm{d}k_i(\rho)/\mathrm{d}\rho \geqq 0$, all i.

We may say that *weak regularity* holds if $\geqq$ replaces $>$ in (12.2). Then Theorem 12.2 implies weak regularity of the stationary optimal path but it is much stronger than weak regularity. Also the condition $(u_{22}+\rho u_{11})^{-1} \leqq 0$ is implied, as noted earlier, if $(u_{22}+\rho u_{11})$ has quasi-dominant diagonal elements, by row or by column, and the off-diagonal elements are non-negative.

The relation between stability and regularity, illustrated by Theorems 12.1 and 12.2, seems to be typical, that is, sufficient conditions for local stability often imply regularity of the stationary optimal path, whether the discrete time or the continuous time model is used. For the continuous time model additional examples may be found in Brock (1976). Results of this type illustrate the *Correspondence Principle* of Samuelson, that "the problem of stability of equilibrium is intimately tied up with the problem of deriving fruitful theorems in comparative statics" [Samuelson (1947)]. Also see Burmeister and Long (1977).

For further examples of the Correspondence Principle we may consider the autonomous difference equation of second order

$$u_2(k_{t-1}, k_t) + \rho u_1(k_t, k_{t+1}) = 0. \tag{12.7}$$

This is the form taken by the necessary condition of optimality (10.1) for a stationary model. It is approximated in a small neighborhood of the stationary optimal path $k_t = k$ by the linear equation

$$u_{21}z_{t-1} + (u_{22}+\rho u_{11})z_t + \rho u_{12}z_{t+1} = 0, \tag{12.8}$$

where $z_t = k_t - k$. The characteristic equation of (12.8) is

$$\det(u_{21} + (u_{22}+\rho u_{11})\lambda + \rho u_{12}\lambda^2) = 0, \tag{12.9}$$

where $\det A$ is the determinant of A. Suppose (12.9) has n roots of absolute value less than 1 and $\det u_{21} \neq 0$. We may appeal to the argument of Scheinkman (1976, pp. 25–26) which is given for the case $\rho = 1$, but also applies when $\rho < 1$, to conclude that these assumptions imply the local turnpike theorem for (k, k), or local stability for optimal paths near (k, k). We will show that, if u_{21} is also symmetric, (k, k) is regular [Dasgupta and McKenzie (1983)].

Rewrite (12.9) as

$$\det(A + B\lambda + \rho A\lambda^2) = 0. \tag{12.10}$$

The proof that (k, k) is regular depends on:

Lemma 12.1

If A is non-singular and symmetric, the characteristic roots of $B^{-1}A$ are less than $1/(1+\rho)$ in absolute value, if and only if there are n roots of (12.10) with absolute value less than 1.

Since $-B$ is positive definite and A is symmetric there is a non-singular matrix Q such that $Q^TBQ = -I$ and $Q^TAQ = -R$ where $R = Q^{-1}(B^{-1}A)Q$ is a diagonal matrix with the characteristic roots of $B^{-1}A$ on the diagonal. Also R is real. See Gantmacher (1960, p. 310). Since A is non-singular, the diagonal elements of r_i of R are non-zero.

(12.10) is equivalent to $\det(Q^T(A + B\lambda + \rho A\lambda^2)Q) = 0$ or $\det(\lambda I + (1 + \rho\lambda^2)R) = 0$. Thus the roots of (12.10) are the roots of the equations

$$\lambda + (1 + \rho\lambda^2)r_i = 0, \qquad i = 1, \ldots, n, \tag{12.11}$$

where repeated roots are counted. The discriminant of (12.11) is $(1 - 4\rho r_i^2)$. Thus the roots of the ith equation are real if $|r_i| \leqq 1/(1+\rho)$.

Suppose all $|r_i| < 1/(1+\rho)$. Then all roots are real. Also $\det A \neq 0$ implies $\lambda = 0$ is not a root of (12.10). Then (12.11) implies

$$\rho|\lambda| + 1/|\lambda| = 1/|r_i| > 1 + \rho. \tag{12.12}$$

This gives $(|\lambda| - 1) > (|\lambda| - 1)(1/\rho|\lambda|)$. Thus $|\lambda| > 1$ implies $1/\rho|\lambda| < 1$. But substitution in (12.11) shows that λ a root implies $1/\rho\lambda$ is the other root. Therefore, one of the roots has absolute value less than 1. Since this is true for all i, there must be n roots λ_i of (12.10) with $|\lambda_i| < 1$.

On the other hand, suppose there are n roots λ_i of (12.10) with $|\lambda_i| < 1$. If the roots λ_i are real, the equation in (12.12) implies that $|r_i| = 1/(1+\rho)$ for $|\lambda_i| = 1$. Also the derivative of the left-hand side of this equation with respect to $|\lambda|$ is negative for $|\lambda| \leqq 1$. Then $|r_i| < 1/(1+\rho)$ for $|\lambda_i| < 1$. If a root λ_i is complex, it follows that $1/\rho\lambda_i = \bar{\lambda}_i$, or $|\lambda_i| = 1/\rho > 1$ in contradiction to the hypothesis. Thus $|r_i| < 1/(1+\rho)$ holds for all i.

To show regularity, assume there are n roots λ_i of (12.10) with $|\lambda_i|<1$. By Lemma 12.1 if r is any root of $B^{-1}A$, then $-1/(1+\rho)<r<1/(1+\rho)$. But $-B$ positive definite and A symmetric implies $\min_i r_i \leqq x^TAx/x^TBx \leqq \max_i r_i$, when $x \neq 0$, and r_i, $i=1,\dots,n$, are the roots of $B^{-1}A$ [Gantmacher (1960, p. 319)]. Thus, $|x^TAx/x^TBx|<1/(1+\rho)$, or, since B is negative definite, $x^T(A+B+\rho A)x = x^T(u_{21}+u_{22}+\rho u_{11}+\rho u_{12})x<0$ for $x \neq 0$. By (12.4) this implies regularity for (k,k).

It is easily seen [Araujo and Scheinkman (1977, p. 611)] that if (12.8) has $n+1$ roots of absolute value larger than 1, the stationary optimal path cannot be stable. Thus our result may be stated.

Theorem 12.3

Assume (I″) and let (12.8) represent the Euler equations linearized about the stationary optimal path $k_t = k$. If (12.8) has no roots of absolute value equal to 1, and u_{12} is non-singular and symmetric, the stationary optimal path is regular if it is locally stable.

For a symmetric matrix the operator norm defined by the Euclidean norm is equal to the maximum of the absolute values of the characteristic roots [see Araujo and Scheinkman (1977, p. 607)]. Let $|\cdot|_1$ be the matrix defined by the norm on R^n given by $|x|_1=(-xBx)^{1/2}$. Then $\max|B^{-1}Ax|_1$ over $|x|_1=1$ equals $\max|E^{-1}(B^{-1}A)Ey|$ over $|y|=1$, where $-B=EE^T$. However, $E^{-1}(B^{-1}A)E$ is symmetric, and it has the same characteristic roots as $B^{-1}A$. This means that the dominant diagonal block condition for (12.8) is met if $(1+\rho)B^{-1}A$ has the absolute value of all its characteristic roots less than 1, which by Lemma 12.1 is implied by local stability. On the other hand, the dominant diagonal condition implies local stability by Lemma 10.2. Thus Lemma 12.1 has:

Corollary

If u_{12} is non-singular and symmetric, the stationary optimal path is locally stable, if and only if (12.8) has a dominant diagonal block.

The symmetry of u_{12} is equivalent to the symmetry of $U_{21}=u_{21}+u_{22}$. The implications of the symmetry condition that corresponds to $U_{12}=U_{21}$ in the continuous time model have been extensively explored by Magill and Scheinkman (1979). They prove that the sufficient condition for regularity which corresponds to $J(\rho)$ negative definite implies local stability in the continuous case. However, Dasgupta (1982) has shown by a counterexample that $J(\rho)$ negative definite does not imply stability in the discrete case.

If symmetry is strengthened to separability of $U(k,0)$, that is, $U_{21}=u_{21}+u_{22}=0$, the stability assumption of Theorem 11.1 becomes unnecessary. With this assumption $J(\rho)=\rho(u_{11}+u_{12})$. Since $J(1)$ is negative definite by assumption

(I″), and in this case $J(\rho)=\rho J(1)$, $J(\rho)$ is negative definite also, which implies regularity by (12.4). However, we may also prove stability under the assumption of separability.

Consider $(u_{22}+\rho u_{11})^{-1}J(\rho)=I-(1+\rho)(u_{22}+\rho u_{11})^{-1}u_{22}=I+(1+\rho)B^{-1}A$. By a theorem of Arrow (1974, p. 200) if X is positive quasi-definite and M is symmetric, the real parts of th roots of XM have the same sign distribution as the real parts of the roots of M. In this case $M=-J(\rho)$ and $X=-(u_{22}+\rho u_{11})^{-1}$. Since M is positive definite, the roots of $I+(1+\rho)B^{-1}A=XM$ have positive real parts. Indeed, the roots are positive since the fact that B is definite and A is symmetric implies that the roots of $B^{-1}A$ are real [Gantmacher (1960, p. 310)]. Let r be a root of $B^{-1}A$, then $1+(1+\rho)r>0$. But $B^{-1}A$ has negative roots by the same result of Arrow, since it is the product of a positive definite matrix and a negative definite matrix and XM and MX have the same roots. Therefore, $0<-r_i<1/(1+\rho)$ for all i and (12.9) has n roots with absolute values less than 1 by Lemma 12.1. By the result of Scheinkman this implies that (k,k) is locally stable, since $\det u_{21}=-\det u_{22}\neq 0$ by assumption (I″). We have proved [Dasgupta and McKenzie (1983)]:

Theorem 12.4

Assume (I″). If $u_{21}+u_{22}=0$, or $U(k,0)$ is separable on a stationary optimal path $k_t=k$, the stationary optimal path is locally stable.

In the continuous time model global stability of an interior stationary optimal path has been proved under the assumption of separability by Scheinkman (1978).

It is an implication of regularity that the utility achieved on the stationary optimal path increases with the discount factor ρ. Indeed, if $k_t=k(\rho')$ is a stationary optimal path and $k(\rho)$ satisfies (12.1) near ρ', putting $u^*(\rho)=u(k(\rho),k(\rho))$, we have

$$\mathrm{d}u^*/\mathrm{d}\rho=(u_1+u_2)(\mathrm{d}k(\rho)/\mathrm{d}\rho)|_{\rho=\rho'}. \tag{12.13}$$

Since $u_2=-\rho u_1$ by (12.1), (12.13) implies $\mathrm{d}u^*/\mathrm{d}\rho>0$ at $\rho=\rho'$ if and only if

$$(1-\rho')u_1(\mathrm{d}k(\rho)/\mathrm{d}\rho)=(\rho'^{-1}-1)q(\mathrm{d}k(\rho)/\mathrm{d}\rho)>0, \tag{12.14}$$

where the derivatives are evaluated at $\rho=\rho'$. If $0<\rho'<1$, the inequality (12.14) is implied by regularity. Following Burmeister and Turnovsky (1972), we may refer to a stationary optimal path that satisfies $\mathrm{d}u^*/\mathrm{d}\rho>0$ as *non-paradoxical*. Thus we have the result:

Theorem 12.5

Under assumption (I″), if a stationary optimal path is regular, it is non-paradoxical.

The Jacobian matrix of the necessary condition (12.1) may also be used to study the question of global uniqueness [Brock (1973)]. Let us say that a stationary optimal path $k_t = k$ is interior if $(k, k) \in \text{int}\, D$. Theorem 7.2, together with the remark that follows its corollary, implies that (12.1) is necessary and sufficient for a stationary path that is interior to be optimal when $0 < \rho \leqq 1$. Thus the number of solutions to (12.1) for given ρ and the number of stationary optimal paths for ρ that are interior to D are the same. Also from the remark following the corollary to Theorem 7.2, for $\rho = 1$ the input–output vector of an interior stationary optimal path maximizes $u(x, y)$ over $(x, y) \in D$ such that $y - x \geqq 0$. Because the Hessian is negative definite by assumption (I″), the maximum is achieved at a unique point. Thus the interior stationary optimal path is unique for $\rho = 1$. The capital stock of this path is also the unique solution of (12.1) for $\rho = 1$.

Write $G(\rho, x) = u_2(x, x) + \rho u_1(x, x)$ for $(x, x) \in D$ and $0 < \rho \leqq 1$. Let C be a convex subset of the diagonal of $E^n \times E^n$ which contains the input–output vector (k, k) of a stationary optimal path for $\rho = 1$. Assume that C is open relative to the diagonal, and the closure $\bar{C}$ lies in the interior of D. It follows from Lemma 8.2 that for some ρ' such that $0 < \rho' < 1$, all ρ such that $\rho' \leqq \rho \leqq 1$ have the property that the solutions $k(\rho)$ of (12.1) satisfy $(k(\rho), k(\rho)) \in C$. Let C_1 be the projection of C on the first component of the product $E^n \times E^n$. Since C_1 is a convex open subset of Euclidean space, it is an oriented differentiable manifold and its closure $\bar{C}_1$ is a manifold with boundary. For ρ given, $G(\rho, x)$ defines a vector field on $\bar{C}_1$, which for any value of ρ with $\rho' \leqq \rho \leqq 1$ has no zeros on the boundary of $\bar{C}_1$.

Since G is a continuously differentiable function of ρ and ρ varies between ρ' and 1, the vector fields on $\bar{C}_1$ are smoothly homotopic to one another. Therefore, the degree of the vector field on the boundary of C_1 is invariant [Milnor (1965, p. 28)]. Let $G_x(x, \rho)$ be the derivative of G with respect to x. The degree of the vector field on boundary C_1 equals the sum of the signs of $\det G_x(\rho, x)$ over all $x = k$ such that $G(\rho, k) = 0$ [Milnor (1965, pp. 36–37)]. But for $\rho = 1$, there is only one such k, and $\det G_x(1, x) = (-1)^n$ at this k since the Hessian matrix of u is negative definite there by assumption (I″). Thus the degree of the vector field on boundary $\bar{C}_1 = (-1)^n$. Assume that the sign of $\det G_x(\rho, x)$ does not change over the zeros of the field for $\rho' < \rho < 1$. Since the sum of signs must have absolute value 1, there can be only one zero for G at such ρ, or equivalently only one stationary optimal path interior to D.

At a zero of G, the derivative $G_x(\rho, x)$ is the Jacobian matrix $J(\rho, x)$ of (12.1). Thus we have proved:

Theorem 12.6

Let C be defined as above. There is $\rho' < 1$ such that all stationary optimal paths $k_t(\rho) = k(\rho)$ with $(k(\rho), k(\rho)) \in \text{interior}\ D$, for any ρ with $\rho' \leqq \rho \leqq 1$, satisfy

$(k(\rho), k(\rho)) \in C$. Under assumption (I″) if the sign of det $J(\rho, k(\rho))$ is constant for each ρ over the stocks $k(\rho)$ of stationary optimal paths with $(k(\rho), k(\rho)) \in C$, there is only one such path for each ρ.

This theorem was first proved, in a neo-classical model, by Benhabib and Nishimura (1979).

Sufficient conditions for stability are also useful for comparative statics when parameters of the utility function, other than ρ, are varied. Let $u_t(x, y) = \rho^t u(x, y, \alpha)$, where α is a vector of m parameters of the current utility function. Differentiating (12.1) totally with respect to α gives

$$(u_{21} + u_{22} + \rho u_{11} + \rho u_{12})(\mathrm{d}k/\mathrm{d}\alpha) = -u_{2\alpha} - \rho u_{1\alpha}. \tag{12.15}$$

Multiplying (12.15) on the left by $(\mathrm{d}k/\mathrm{d}\alpha)^T$, an $m \times n$ matrix, we have

$$(\mathrm{d}k/\mathrm{d}\alpha)^T(u_{21} + u_{22} + \rho u_{11} + \rho u_{12})(\mathrm{d}k/\mathrm{d}\alpha) = -(\mathrm{d}k/\mathrm{d}\alpha)^T(u_{2\alpha} + \rho u_{1\alpha}). \tag{12.16}$$

The sufficient condition for local stability of the stationary optimal path $k_t(\rho) = k$ that $Q(\rho)$ be negative quasi-definite at $(k(\rho), k(\rho))$ implies that $(u_{21} + u_{22} + \rho u_{11} + \rho u_{12})$ is negative quasi-definite. Then $(\mathrm{d}k/\mathrm{d}\alpha)^T(u_{2\alpha} + \rho u_{1\alpha})$ is positive quasi-definite. In applications to particular problems, for example, investment of the firm with adjustment cost, this result may be fruitful [Brock (1976)].

Similarly, the sufficient condition (12.5) for local stability of $k_t(\rho) = k$, that is, a dominant diagonal for the infinite Jacobian matrix, may be applied to

$$\begin{aligned}\mathrm{d}k/\mathrm{d}\alpha = &-\left(I + (u_{22} + \rho u_{11})^{-1}(u_{21} + \rho u_{12})\right)^{-1} \\ &\times (u_{22} + \rho u_{11})^{-1} \cdot (u_{2\alpha} + \rho u_{1\alpha}).\end{aligned} \tag{12.17}$$

As before, the matrix $-(I + (u_{22} + \rho u_{11})^{-1}(u_{21} + \rho u_{12}))^{-1}(u_{22} + \rho u_{11})^{-1}$ is non-negative when the sign assumptions $(u_{22} + \rho u_{11})^{-1} \leqq 0$, $u_{21} = u_{12}^T \geqq 0$, are made. In applications this result may also be useful.

As an example, consider a simple model of investment by the firm with adjustment costs, related to the continuous model of Treadway (1971). Let $\pi(k_t, k_{t+1}, \alpha) = f(k_t, k_{t+1}) - \alpha(k_{t+1} - k_t)$, where f represents gross revenue, after maximizing on current spending for variable inputs, and α is a vector of prices for new capital goods. The presence of k_{t+1} as an argument of f is a consequence of the adjustment costs of capital expansion incurred within the firm. The firm's objective is to maximize $\sum_{t=1}^{\infty} \rho^t \pi(k_t, k_{t+1}, \alpha)$ given some initial stocks k_0. Prices are formed on competitive markets and are expected to remain constant, while $\rho = 1/(1+r)$, where r is the interest rate, also expected to remain constant.

Differentiating $\pi(k,k,\alpha)$, where $k_t(\rho)=k$ is a stationary optimal path for the utility function $u_t=\rho^t\pi$, we have $\pi_2=f_2-\alpha$ and $\pi_1=f_1+\alpha$. Then $\pi_{2\alpha}=-I$ and $\pi_{1\alpha}=I$. Substituting in the right-hand side of (12.15) and solving, $\mathrm{d}k/\mathrm{d}\alpha=(1-\rho)(J(\rho))^{-1}$. If we make the assumption, sufficient for local stability and regularity of $k_t(\rho)$, that $Q(\rho)$ evaluated at (k,k) and thus $J(\rho,k)$, is negative quasi-definite, we see from (12.16) that $\mathrm{d}k/\mathrm{d}\alpha$ is negative quasi-definite. Then the equilibrium demand for each capital stock is decreasing with respect to its own price, so long as the interest rate is positive.

We may also apply the assumption of dominant diagonal blocks, with $(\pi_{22}+\rho\pi_{11})^{-1}\leqq 0$ and $\pi_{12}=\pi_{21}^T\geqq 0$, which is also sufficient for local stability of $k_t(\rho)=k$. Using again that $\pi_{2\alpha}+\rho\pi_{1\alpha}=-(1-\rho)I$, we infer from (12.17) that $\mathrm{d}k/\mathrm{d}\alpha\leqq 0$. This is what one would expect by analogy to timeless production where factors are normally gross complements. See Rader (1968). We may finally note that $\mathrm{d}k/\mathrm{d}\alpha$ negative quasi-definite in the adjustment cost model also follows from the other assumptions that we used to establish regularity, since the arguments proceeded by way of negative quasi-definite $J(\rho,k)$. The assumption of symmetry and local stability in Theorem 12.3, the assumption of separability, and the assumption of $U_{21}(=u_{21}+u_{22})$ negative quasi-definite are cases in point. Another model of investment, used by Lucas (1967), satisfies the separability assumption. It has been studied in this context by Scheinkman (1978).

13. Comparative dynamics near stationary states

In the neighborhood of stationary states some further results of comparative dynamics are available. In the quasi-stationary model it is convenient to write ζ as a function of k_0, $k=(k_1,k_2,\ldots)$, and ρ and give a definition equivalent to (11.1),

$$\zeta_t(k_0,k,\rho)=u_2(k_{t-1},k_t)+\rho u_1(k_t,k_{t+1}). \tag{13.1}$$

Then $D_k\zeta$ is given by (11.2) if u_{ij} replaces v_{ij} everywhere. Let $k_t=k(\rho)$, $t=0,1,\ldots$, be a stationary optimal path and let $D_\rho k=(Dk(\rho),Dk(\rho),\ldots)$ where $Dk(\rho)=\mathrm{d}k(\rho)/\mathrm{d}\rho$. From (11.2) and (12.3) we have

$$\begin{aligned} D_k\zeta(k_0,k,\rho)D_\rho k&=[J(\rho)Dk(\rho)-u_{21}Dk(\rho),J(\rho)Dk(\rho),\ldots]\\ &=-(u_1-u_{21}Dk(\rho),u_1,\ldots). \end{aligned} \tag{13.2}$$

However, from (13.1) we obtain

$$D_\rho\zeta(k_0,k,\rho)=(u_1,u_1,\ldots). \tag{13.3}$$

Therefore, substituting in (13.2),

$$D_k\zeta(k_0,k,\rho)(D_\rho k) = -D_\rho\zeta(k_0,k,\rho)-(u_{21}Dk(\rho),0,0,\ldots). \tag{13.4}$$

If $D_k\zeta$ is invertible, (13.4) implies

$$D_\rho k = -(D_k\zeta)^{-1}D_\rho\zeta-(D_k\zeta)^{-1}(u_{21}Dk(\rho),0,0,\ldots). \tag{13.5}$$

But the implicit function theorem implies that in a small neighborhood of (k_0,k,ρ) there is a continuously differentiable function $\psi(k_0',\rho')$ which maps a neighborhood of (k_0,ρ) into l_∞^n such that $[\zeta(k_0',\psi(k_0',\rho'),\rho']_t=0$, all t, and $\psi(k_0,\rho)=k$. Furthermore, as in (11.3),

$$D_\rho\psi(k_0',\rho') = -\left(D_k\zeta(k_0',\psi(k_0',\rho'),\rho')\right)^{-1}\cdot D_\rho\zeta(k_0',\psi(k_0',\rho'),\rho'). \tag{13.6}$$

Thus, combining (13.5) and (13.6), we obtain an expression for the variation of the path $k_t=k(\rho)$ with ρ,

$$D_\rho\psi(k_0,\rho) = D_\rho k + D_k\zeta(k_0,k,\rho)^{-1}\cdot(u_{21}Dk(\rho),0,0,\ldots). \tag{13.7}$$

Moreover, since the derivative $D_\rho\psi(k_0',\rho')$ is continuous in its arguments, $D_\rho\psi(k_0',\rho')$ converges to $D_\rho\psi(k_0,\rho)$ as $k_0'\to k_0=k(\rho)$ and $\rho'\to\rho$.

Similarly the variation of the path with respect to the initial stock is given by

$$D_{k_0}\psi(k_0',\rho') = -\left(D_k\zeta(k_0',\psi(k_0',\rho'),\rho')\right)^{-1}\cdot D_{k_0}\zeta(k_0',\psi(k_0',\rho'),\rho'),$$

so that

$$D_{k_0}\psi(k_0,\rho) = -(D_k\zeta(k_0,k,\rho))^{-1}(u_{21},0,0,\ldots), \tag{13.8}$$

where $k_0=[\psi(k_0,\rho)]_t=k(\rho)$, all t, and $k=(k(\rho),k(\rho),\ldots)$. From (13.7) and (13.8) we derive

$$D_\rho\psi(k_0,\rho) = D_\rho k - D_{k_0}\psi(k_0,\rho)\cdot Dk(\rho). \tag{13.9}$$

The preceding argument is apparently special to ρ as a parameter because of the part played by u_1. However, the fact that $Dk(\rho)=u_1$ need not be introduced. If the utility function depends on a parameter α, which may be a vector, it may be written as before $u(x, y, \alpha)$. Then ζ may be defined by

$$\zeta_t(k_0, k, \rho, \alpha) = u_2(k_{t-1}, k_t, \alpha) + \rho u_1(k_t, k_{t+1}, \alpha). \tag{13.10}$$

The earlier arguments can now be made with $k(\rho, \alpha)$ replacing $k(\rho)$, $D_\alpha k(\rho, \alpha)$ replacing $Dk(\rho)$, $D_\alpha \zeta$ replacing $D_\rho \zeta$, and $D_\alpha(u_2 + \rho u_1)$ replacing u_1. Then in place of (13.7) we derive

$$D_\alpha \psi(k_0, \rho, \alpha) = D_\alpha k + D_k \zeta(k_0, k, \rho, \alpha)^{-1} \cdot (u_{21} Dk(\rho, \alpha), 0, 0, \ldots), \tag{13.11}$$

and in place of (13.9) we derive

$$D_\alpha \psi(k_0, \rho, \alpha) = D_\alpha k - D_{k_0} \psi(k_0, \rho, \alpha) \cdot D_\alpha k(\rho, \alpha), \tag{13.12}$$

where $D_\alpha k = (D_\alpha k(\rho, \alpha), D_\alpha k(\rho, \alpha), \ldots)$ and $D_\alpha k(\rho, \alpha)$ is the variation of an optimal stationary state associated with ρ and α from a change in α. If α is an m vector, $D_\alpha k(\rho, \alpha)$ may be represented by an $n \times m$ matrix while $D_{k_0}\psi$ may be represented conformably by a matrix of n columns arranged in $n \times n$ blocks indexed by t. On the other hand, $D_\alpha k$ and $D_\alpha \psi$ are representable by a matrix of m columns arranged in $n \times m$ blocks indexed by t.

On the basis of the preceding arguments we may state [Dasgupta and McKenzie (1983)]:

Theorem 13.1

Let $k_t = k(\rho, \alpha)$, $t = 0, 1, 2, \ldots$, be a stationary optimal path. Assume that $D_k \zeta(k_0, k, \rho, \alpha)$ is invertible. Then for (k'_0, ρ', α') near (k_0, ρ, α), there is a unique optimal path $\psi(k'_0, \rho', \alpha')$ with $[\psi(k'_0, \rho', \alpha')]_0 = k'_0$, and $D_\rho \psi(k'_0, \rho', \alpha')$ and $D_\alpha \psi(k'_0, \rho', \alpha')$ converge to the expressions given in (13.7) and (13.11) as $(k'_0, \rho', \alpha') \to (k_0, \rho, \alpha)$. Also $D_{k_0} \psi(k'_0, \rho', \alpha')$ converges to the analog of the expression in (13.8) when α is introduced as a parameter.

However, to make use of these expressions we need to evaluate $(D_k \zeta(k_0, k, \rho, \alpha))^{-1} D_\alpha k(\rho, \alpha)$ and $(D_k \zeta(k_0, k, \rho, \alpha))^{-1} D_\rho k(\rho, \alpha)$. Replacing v by u in (11.2) we observe that

$$[D_k \zeta(k_0, k, \rho, \alpha) \cdot z]_t = u_{21} z_{t-1} + (u_{22} + \rho u_{11}) z_t + \rho u_{12} z_{t+1}, \tag{13.13}$$

where $z_0 = 0$ and $u_{ij} = u_{ij}(k(\rho, \alpha), k(\rho, \alpha))$. Equate $D_k \zeta(k_0, k, \rho, \alpha) \cdot z$ to $a \in l^n_\infty$. If $|u_{12}| \neq 0$, (13.13) may be written

$$b_t = z_{t+1} - N z_t - M z_{t-1}, \tag{13.14}$$

where $M = -\rho^{-1} u_{12}^{-1} u_{21}$, $N = -u_{12}^{-1}(\rho^{-1} u_{22} + u_{11})$, and $\rho^{-1} u_{12}^{-1} a_t = b_t$. A solution z_t, $t = 0,1,2,\dots$, of (13.14) with $z_0 = 0$ satisfies $z_t = [(D_k\zeta)^{-1}\cdot a]_t$. Thus solutions of (13.14) with appropriate values for b_t will provide explicit formulae for the expressions in (13.7) and (13.11). Araujo and Scheinkman have shown that solutions exist for any $b \in l_\infty^n$ if the Euler equations (12.8) have no roots λ with $|\lambda| = 1$ and $k(\rho)$ is locally stable.

It is helpful to transform (13.14) into a first-order system. Define $h_t(z_{t-1}, z_t)$ and $d_t = (0, b_t)$, so that $d_t \in R^{2n}$. Then (13.14) may be written

$$d_t = h_{t+1} - Hh_t, \tag{13.15}$$

where

$$H = \begin{bmatrix} 0 & I \\ M & N \end{bmatrix}.$$

Assuming that (12.8) has no roots of unit norm and $k(\rho)$ is locally stable, there are n characteristic roots λ of H with $|\lambda| < 1$. Thus there is a non-singular real matrix P such that

$$P^{-1}HP = L = \begin{bmatrix} L_1 & 0 \\ 0 & L_2 \end{bmatrix}, \tag{13.16}$$

where L_1 and L_2 are $n \times n$, and the characteristic roots of L_1 have absolute value less than 1 and the characteristic roots of L_2 have absolute value greater than 1. After transformation by P the system of equations (13.15) become

$$e_t = y_{t+1} - Ly_t, \tag{13.17}$$

where

$$e_t = P^{-1}d_t \quad \text{and} \quad y_t = P^{-1}h_t.$$

From (13.7) and (13.11) we find that in the case where ρ varies $a = (u_{21}D_\rho k(\rho, \alpha), 0, 0, \dots)$ and $b = -(MD_\rho k(\rho, \alpha), 0, 0, \dots)$. Therefore, the first set of equations (13.15) appears as

$$\begin{pmatrix} z_1 \\ z_2 \end{pmatrix} - \begin{bmatrix} 0 & I \\ M & N \end{bmatrix}\begin{pmatrix} 0 \\ z_1 \end{pmatrix} = \begin{pmatrix} 0 \\ MD_\rho k(\rho, \alpha) \end{pmatrix}. \tag{13.18}$$

These equations are equivalent to

$$\begin{pmatrix} z_1 \\ z_2 \end{pmatrix} - \begin{bmatrix} 0 & I \\ M & N \end{bmatrix} \begin{pmatrix} -D_\rho k(\rho,\alpha) \\ z_1 \end{pmatrix} = 0. \tag{13.19}$$

When (13.19) is transformed by P, it becomes

$$\begin{pmatrix} y_1^1 \\ y_2^1 \end{pmatrix} - \begin{bmatrix} L_1 & 0 \\ 0 & L_2 \end{bmatrix} \begin{pmatrix} y_1^0 \\ y_2^0 \end{pmatrix} = 0. \tag{13.20}$$

It is clear from (13.17) that $y \in l_\infty^n$ implies $y_2^0 = 0$. Thus it must be that

$$P\begin{pmatrix} y_1^0 \\ 0 \end{pmatrix} = \begin{pmatrix} -D_\rho k(\rho,\alpha) \\ z_1 \end{pmatrix} \quad \text{and} \quad P\begin{pmatrix} y_1^1 \\ 0 \end{pmatrix} = \begin{pmatrix} z_1 \\ z_2 \end{pmatrix}. \tag{13.21}$$

Putting

$$P = \begin{bmatrix} P_{11} & P_{12} \\ P_{22} & P_{21} \end{bmatrix},$$

(13.21) implies

$$P_{11} y_1^0 = -D_\rho k(\rho,\alpha) \quad \text{and} \quad P_{21} y_1^0 = z_1.$$

It has been proved by Scheinkman (1976, pp. 25–26) that P_{11} is non-singular if $\rho = 1$. However, the same argument is effective for $0 < \rho < 1$. Then

$$z_1 = -P_{21} P_{11}^{-1} D_\rho k(\rho,\alpha). \tag{13.22}$$

Also from (13.21), $z_1 = P_{11} y_1^1$, and from (13.20), $y_1^1 = L_1 y_1^0 = -L_1 P_{11}^{-1} D_\rho k(\rho,\alpha)$. Thus $z_1 = -P_{11} L_1 P_{11}^{-1} D_\rho k(\rho,\alpha)$. Indeed, it follows that

$$z_t = -P_{11} L_1^t P_{11}^{-1} D_\rho k(\rho,\alpha), \quad \text{all} \quad t \geqq 1. \tag{13.23}$$

Note that $L_1 = P_{11}^{-1} P_{21}$ from (13.22) and (13.23).

We may now use (13.7) and (13.23) to derive, for $k_0 = k(\rho)$,

$$\left[D_\rho \psi(k_0,\rho,\alpha)\right]_t = D_\rho k(\rho,\alpha) - P_{11} L_1^t P_{11}^{-1} D_\rho k(\rho,\alpha), \tag{13.24}$$

which is valid for $t \geqq 1$. Thus $[D_\rho \psi(k_0,\rho)]_t = 0$ for $t = 0$ and converges to $D_\rho k(\rho,\alpha)$ as $t \to \infty$. The argument for the derivative with respect to α proceeds in the same way to give the analogous formula

$$[D_\alpha \psi(k_0, \rho, \alpha)]_t = D_\alpha k(\rho, \alpha) - P_{11} L_1^t P_{11}^{-1} Dk(\rho, \alpha). \tag{13.25}$$

Moreover, we have from (12.3) that $D_\rho k(\rho, \alpha) = (J(\rho, \alpha)^{-1} u_1(k(\rho, \alpha), k(\rho, \alpha))$ if $J(\rho, \alpha)$ is non-singular. However, $J(\rho)$ singular implies $\lambda = 1$ is a root of (12.8). Thus assuming that (12.8) has no roots of unit norm we may derive from (13.24) an explicit expression for the variation of path with ρ when $k_0 = k(\rho, \alpha)$, that is

$$[D_\rho \psi(k_0, \rho, \alpha)]_t = -P_{11}(I - L_1^t) P_{11}^{-1} (J(\rho, \alpha))^{-1} u_1. \tag{13.26}$$

Also comparison of (13.26) with (13.9) shows that

$$[D_{k_0} \psi(k_0, \rho, \alpha)]_t = P_{11} L_1^t P_{11}^{-1}. \tag{13.27}$$

On the basis of the foregoing argument we may assert [Dasgupta and McKenzie (1983)]:

Theorem 13.2

If the path $k_t = k(\rho, \alpha)$, $t = 0, 1, 2, \ldots$, is locally stable, u_{12} is non-singular, the Euler equations (12.8) have no roots of unit norm, and assumption (I″) holds, the variation of the optimal path for $k_0 = k(\rho, \alpha)$ with respect to ρ, α, and k_0 is given by expressions (13.24) and (13.26), (13.25), and (13.27), respectively. Moreover, the variations for optimal paths where the parameter values are ρ' and α', and initial stocks are k_0', converge to these expressions as $(k_0', \rho', \alpha') \to (k_0, \rho, \alpha)$.

If it is assumed that u_{12} is symmetric, as before, sharper results may be reached. Let Q be the matrix that appears in the proof of Lemma 12.1. We will need:

Lemma 13.1

On the hypothesis of Theorem 12.4, if also u_{12} is symmetric, $P_{11} = Q$.

The symmetry of u_{12} implies that M in (12.16) equals $-\rho^{-1} I$. Also $N = -\rho^{-1} u_{12}^{-1}(u_{22} + \rho u_{11})$. Let R be the diagonal matrix with the roots of (12.11) on the diagonal. Then

$$\begin{bmatrix} Q^{-1} & 0 \\ 0 & Q^{-1} \end{bmatrix} \begin{bmatrix} 0 & I \\ -\rho^{-1} I & N \end{bmatrix} \begin{bmatrix} Q & 0 \\ 0 & Q \end{bmatrix} = \begin{bmatrix} 0 & I \\ -\rho^{-1} I & -\rho^{-1} R^{-1} \end{bmatrix}. \tag{13.28}$$

Let $\delta^i \in R^n$ satisfy $\delta^i_j = 0,\ j \neq i$, and $\delta^i_i = 1$. Consider

$$\begin{bmatrix} 0 & I \\ -\rho^{-1}I & -\rho^{-1}R^{-1} \end{bmatrix}\begin{pmatrix} \delta^i \\ \lambda\delta^i \end{pmatrix} = \begin{pmatrix} \lambda\delta^i \\ -\rho^{-1}(1+r_i^{-1}\lambda)\delta^i \end{pmatrix}. \tag{13.29}$$

If λ is chosen to be a root of (12.11), $1 + r_i^{-1}\lambda = -\rho\lambda^2$, so the right side of (13.29) is

$$\lambda\begin{pmatrix} \delta^i \\ \lambda\delta^i \end{pmatrix}.$$

In other words,

$$\begin{pmatrix} \delta^i \\ \lambda\delta^i \end{pmatrix}$$

is a characteristic vector of

$$\begin{bmatrix} 0 & I \\ -\rho^{-1}I & -\rho^{-1}R^{-1} \end{bmatrix},$$

for the characteristic root λ, and the matrix of characteristic vectors may be written

$$T = \begin{bmatrix} I & I \\ L_1 & L_2 \end{bmatrix},$$

where L_1 and L_2 are diagonal matrices with the characteristic roots of (12.10) on the diagonal. Then

$$T^{-1}\begin{bmatrix} 0 & I \\ -\rho^{-1}I & -\rho^{-1}R^{-1} \end{bmatrix}T = \begin{bmatrix} L_1 & 0 \\ 0 & L_2 \end{bmatrix}.$$

Finally, from (13.16) and (13.28),

$$P = \begin{bmatrix} Q & 0 \\ 0 & Q \end{bmatrix}\begin{bmatrix} I & I \\ L_1 & L_2 \end{bmatrix} = \begin{bmatrix} Q & Q \\ QL_1 & QL_2 \end{bmatrix}, \quad \text{so} \quad P_{11} = Q.$$

In accord with the notion of regularity for stationary optimal paths defined in Section 12, optimal paths which satisfy $\sum_1^\infty P_t(\mathrm{d}k_t/\mathrm{d}\rho) \geqq 0$, as in Theorem 11.1, may be said to be *dynamically regular*. Then we will say that an optimal path k_t,

$t=0,1,\ldots,$ has *strong dynamic regularity* if $P_t(\mathrm{d}k_t/\mathrm{d}\rho)>0$ for all $t>0$. With the help of Lemma 8.2 we may prove [Dasgupta and McKenzie (1983)]:

Theorem 13.3

Under the hypothesis of Theorem 13.2, if u_{12} is symmetric, optimal paths from k_0 in a sufficiently small neighborhood of $k(\rho,\alpha)$ have strong dynamic regularity.

We apply Lemma 12.1, and the properties of Q, and Lemma 13.1 to (13.26). Write $J(\rho,\alpha)=B+(1+\rho)A$ where $A=u_{12}$ and $B=u_{22}+\rho u_{11}$. Then

$$\begin{aligned}(J(\rho,\alpha))^{-1}&=(A^{-1}B+(1+\rho)I)^{-1}A^{-1}\\&=-(QR^{-1}Q^{-1}+(1+\rho)I)^{-1}QR^{-1}Q^{T}\\&=-Q(R^{-1}+(1+\rho)I)^{-1}R^{-1}Q^{T}.\end{aligned}$$

Thus (13.26) may be written, using Lemma 13.1,

$$\left[D_\rho\psi(k_0,\rho,\alpha)\right]=Q(I-L_1^t)(I+(1+\rho)R)^{-1}Q^Tu_1 \tag{13.28}$$

for $k_0=k(\rho,\alpha)$. Since L_1 is diagonal with λ_i on the diagonal and $|\lambda_i|<1$, and R is diagonal with r_i on the diagonal and $|r_i|<1/(1+\rho)$ by Lemma 12.1, the matrix on the right-hand side of (13.28) is positive definite. Therefore, $u_1[D_\rho\psi(k_0,\rho,\alpha)]_t>0$ and the theorem follows for k_0 near $k(\rho,\alpha)$ by Theorem 12.3. We may note that this argument also provides an alternative proof of Theorem 12.3, since $(J(\rho,\alpha))^{-1}$ is shown to be negative definite. Otani (1982) derives results in the continuous time model parallel to Theorem 13.3.

Formula (13.25) may be applied to the adjustment cost model described in Section 12. Then α represents the prices of new capital goods. Solving equations (12.15) for $\mathrm{d}k/\mathrm{d}\alpha\equiv D_\alpha k(\rho,\alpha)$ gives $D_\alpha K(\rho,\alpha)=(1-\rho)(J(\rho))^{-1}$ as before. Substituting this expression in (13.25), we obtain

$$\left[D_\alpha\psi(k_0,\rho,\alpha)\right]_t=P_{11}(I-L_1^t)P_{11}^{-1}(1-\rho)(J(\rho))^{-1}. \tag{13.29}$$

Thus by the argument leading to Theorem 13.3 $[D_\alpha\psi(k_0,\rho,\alpha)]_t$ is negative definite when u_{12} is symmetric. Consequently an increase in the price of any capital good reduces its stock along an optimal path of accumulation from k_0 near $k(\rho,\alpha)$. Also investment, which equals $[D_\alpha\psi]_{t+1}-[D_\alpha\psi]_t$, is reduced along the path in the component corresponding to a capital good whose price increases. This application to the adjustment cost problem corresponds to results proved by Mortensen (1973) in a continuous time model of investment with adjustment costs.

References

Araujo, A. and J. A. Scheinkman (1977), "Smoothness, comparative dynamics, and the turnpike property", Econometrica, 45:601–620.

Araujo, A. and J. A. Scheinkman (1979), "Notes on comparative dynamics", in: J. R. Green and J. A. Scheinkman, eds., General Equilibrium, Growth and Trade. New York: Academic Press.

Arrow, Kenneth J. (1974), "Stability independent of adjustment speed", in: G. Horwich and P. A. Samuelson, eds., Trade, Stability, and Macroeconomics. New York: Academic Press.

Arrow, Kenneth J. and Frank H. Hahn (1971), General Competitive Analysis. San Francisco: Holden-Day.

Atsumi, Hiroshi (1965), "Neoclassical growth and the efficient program of capital accumulation", Review of Economic Studies, 32:127–136.

Becker, Robert A. (1980), "On the long-run steady state in a simple dynamic model of equilibrium with heterogeneous households", Quarterly Journal of Economics, 94:375–382.

Bellman, Richard (1957), Dynamic Programming. Princeton: Princeton University Press.

Benhabib, Jess and Kazuo Nishimura (1979), "The Hopf bifurcation and the existence and stability of closed orbits in multi-sector models of optimal economic growth", Journal of Economic Theory, 21:421–444.

Benhabib, Jess and Kazuo Nishimura (1979), "On the uniqueness of steady states in an economy with heterogeneous capital goods", International Economic Review, 20:59–82.

Benveniste, Lawrence M. and José Scheinkman (1982), "Duality theory for dynamic optimization models: The continuous time case", Journal of Economic Theory, 27:1–19.

Berge, Claude (1963), Topological Spaces. London: Oliver and Boyd.

Bewley, Truman (1982), "An integration of equilibrium theory and turnpike theory", Journal of Mathematical Economics, 10:233–268.

Bliss, Gilbert A. (1965), Calculus of Variations. LaSalle: Open Court Publishing Company.

Brock, William A. (1970), "On existence of weakly maximal programmes in a multi-sector economy", Review of Economic Studies, 37:275–280.

Brock, William A. (1973), "Some results on the uniqueness of steady states in multisector models of optimum growth when future utilities are discounted", International Economic Review, 14:535–559.

Brock, William A. (1974), "Money and growth: The case of long-run perfect foresight", International Economic Review, 15:750–777.

Brock, William A. (1976), "Applications of recent results on the asymptotic stability of optimal control to the problem of comparing long run equilibria", Unpublished paper. Chicago.

Brock, William A. (1979), "An integration of stochastic growth theory and the theory of finance, Part I: The growth model", in: J. R. Green and J. A. Scheinkman, eds., General Equilibrium, Growth and Trade. New York: Academic Press.

Brock, William A. (1980), "An integration of stochastic growth theory and the theory of finance", in: J. J. McCall, ed., The Economics of Uncertainty. Chicago: University of Chicago Press.

Brock, William A. and Mukul Majumdar (1978), "Global asymptotic stability results for multi-sector models of optimal growth under uncertainty when future utilities are discounted", Journal of Economic Theory, 18:225–243.

Brock, William A. and Leonard J. Mirman (1976), "Optimal economic growth and uncertainty: The discounted case", Journal of Economic Theory, 12:479–513.

Brock, William A. and José Scheinkman (1976), "Global asymptotic stability of optimal control systems with applications to the theory of economic growth", Journal of Economic Theory, 12:164–190.

Brock, William A. and José Scheinkman (1978), "On the long-run behavior of a competitive firm", in: G. Schwödiauer, ed., Equilibrium and Disequilibrium in Economic Theory. Dordrecht: D. Reidel.

Burmeister, Edwin and Ngo van Long, (1977), "On some unresolved questions of capital theory", Quarterly Journal of Economics, 91:289–314.

Burmeister, Edwin and Stephen J. Turnovsky (1972), "Capital deepening response in an economy with heterogeneous capital goods", American Economic Review, 62:842–853.

Cass, David (1966), "Optimum growth in an aggregative model of capital accumulation: A turnpike theorem", Econometrica, 34:833–850.

Cass, David and Karl Shell (1976), "The structure and stability of competitive dynamical systems", Journal of Economic Theory, 12:31–70.

Dasgupta, Swapan (1982), "A local analysis of stability and regularity of stationary states in discrete symmetric optimal accumulation models", Journal of Economic Theory, forthcoming.

Dasgupta, Swapan and Lionel W. McKenzie (1983), "The comparative statics and dynamics of stationary states", in a festschrift for Leonid Hurwicz, forthcoming.

Dechert, W. Davis and Kasuo Nishimura (1983), "A complete characterization of optimal growth paths in an aggregated model with a non-concave production function", Journal of Economic Theory, 31: 332–354.

Dieudonné, Jean Alexandre (1960), Foundations of Modern Analysis. New York: Academic Press.

Dorfman, Robert, Paul Samuelson and Robert Solow (1958), Linear Programming and Economic Analysis. New York: McGraw-Hill.

Evstigneev, I. V. (1974), "Optimal stochastic programs and their stimulating prices", in: J. Łoś and M. W. Łoś, eds., Mathematical Models in Economics. Amsterdam: North-Holland.

Flynn, James (1980), "The existence of optimal invariant stocks in a multi-sector economy", Review of Economic Studies, 47:809–812.

Gale, David (1967), "On optimal development in a multi-sector economy", Review of Economic Studies, 34:1–18.

Gantmacher, F. R. (1960), The Theory of Matrices, Vol. I. New York: Chelsea.

Hammond, Peter (1975), "Agreeable plans with many capital goods", Review of Economic Studies, 42:1–14.

Hammond, Peter and James A. Mirrlees (1973), "Agreeable plans", in: J. A. Mirrlees and N. H. Stern, eds., Models of Economic Growth. New York: Wiley.

Hestenes, Magnus R. (1966), Calculus of Variations and Optimal Control Theory. New York: Wiley.

Hieber, Gerhard (1981), On Optimal Growth Paths and Variable Technology. Cambridge: Oelgeschlager, Gunn, and Hain.

Hirsch, Morris W. and Stephen Smale (1974), Differential equations, dynamic systems, and linear algebra. New York: Academic Press.

Inada, Ken-ichi (1964), "Some structural characteristics of turnpike theorems", Review of Economic Studies, 31:43–58.

Intriligator, Michael D. (1971), Mathematical Optimization and Economic Theory. Englewood Cliffs: Prentice-Hall.

Iwai, Katsuhito (1972), "Optimal economic growth and stationary ordinal utility: A Fisherian approach", Journal of Economic Theory, 5:121–151.

Jeanjean, Patrick (1974), "Optimal development programs under uncertainty: The undiscounted case", Journal of Economic Theory, 7:66–92.

Kahn, M. Ali and Tapan Mitra (1984), "On the existence of a stationary optimal stock for a multi-sector economy: A primal approach", unpublished paper. Ithaca: Cornell University.

Koopmans, Tjalling C. (1965), "The concept of optimal economic growth", in: The Econometric Approach to Development Planning, Pontificae Academiae Scientiarum Scripta Varia, No. 28. Amsterdam: North-Holland.

Kuhn, Harold W. and A. W. Tucker (1951), "Non-linear programming", in: J. Neyman, ed., Proceedings of the Second Berkeley Symposium on Mathematics, Statistics, and Probability. Berkeley: University of California.

Lucas, Robert E., Jr. (1967), "Optimal investment policy and the flexible accelerator", International Economic Review, 8:78–85.

Magill, Michael J. P. (1977), "Some new results on the local stability of the process of capital accumulation", Journal of Economic Theory, 15:174–210.

Magill, Michael J. P. and José A. Scheinkman (1979), "Stability of regular equilibria and the correspondence principle for symmetric variational problems", International Economic Review, 20:297–315.

Majumdar, Mukul and Tapan Mitra (1982), "Intertemporal allocation with a non-convex technology: The aggregative framework", Journal of Economic Theory, 27:101–136.

Malinvaud, Edmond (1953), "Capital accumulation and efficient allocation of resources", Econometrica, 21:233–268.

McKenzie, Lionel W. (1960), "Matrices with dominant diagonals and economic theory", in: K. J.

Arrow, S. Karlin and P. Suppes, eds., Mathematical Methods in the Social Sciences, 1959. Stanford: Stanford University Press.

McKenzie, Lionel W. (1963), "Turnpike theorems for a generalized Leontief model", Econometrica, 31:165–180.

McKenzie, Lionel W. (1968), "Accumulation programs of maximum utility and the von Neumann facet", in: J. N. Wolfe, ed., Value, Capital and Growth. Edinburgh: Edinburgh University Press.

McKenzie, Lionel W. (1971), "Capital accumulation optimal in the final state", in: G. Bruckmann and W. Weber, eds., Contributions to the von Neumann Growth Model. Vienna: Springer-Verlag.

McKenzie, Lionel W. (1974), "Turnpike theorems with technology and welfare function variable", in: J. Łoś and M. W. Łoś, eds., Mathematical Models in Economics. New York: American Elsevier.

McKenzie, Lionel W. (1976), "Turnpike theory", Econometrica, 44:841–865.

McKenzie, Lionel W. (1977), "A new route to the turnpike", in: R. Henn and O. Moeschlin, eds., Mathematical Economics and Game Theory. New York: Springer-Verlag.

McKenzie, Lionel W. and Makoto Yano (1980), "Turnpike theory: Some corrections", Econometrica, 48:1839–1840.

McKenzie, Lionel W. (1982), "A primal route to the turnpike and Liapounov stability", Journal of Economic Theory, 27:194–209.

McKenzie, Lionel W. (1983), "Turnpike theory, discounted utility, and the von Neumann facet", Journal of Economic Theory, 30:330–352.

Milnor, John (1965), Topology from the Differentiable Viewpoint. Charlottesville: University Press of Virginia.

Morishima, Michio (1961), "Proof of a turnpike theorem: 'The no joint production case'", Review of Economic Studies, 28:89–97.

Morishima, Michio (1969), Theory of Economic Growth. Oxford.

Mortensen, D. T. (1973), Generalized costs of adjustment and dynamic factor demand theory", Econometrica, 41:657–665.

Nishimura, K. G. (1983), "A new concept of stability and dynamical economic systems", Journal of Economic Dynamics and Control, 6:25–40.

Otani, Kyoshi (1982), "Explicit formulae of comparative dynamics", International Economic Review, 23:411–419.

Peleg, Bezalel (1973), "A weakly maximal golden-rule program for a multi-sector economy", International Economic Review, 14:574–579.

Peleg, Bezalel and Harl E. Ryder, Jr. (1974), "The modified golden rule of a multi-sector economy", Journal of Mathematical Economics, 1:193–198.

Peleg, Bezalel and Itzhak Zilcha (1977), "On competitive prices for optimal consumption plans, II", SIAM Journal of Applied Mathematics, 32:627–630.

Pontryagin, L. S. and Associates (1962), The Mathematical Theory of Optimal Processes. New York: Interscience.

Rader, Trout (1968), "Normally factor inputs are never gross substitutes", Journal of Political Economy, 76:38–43.

Radner, Roy (1961), "Paths of economic growth that are optimal with regard only to final states", Review of Economic Studies, 28:98–104.

Ramsey, Frank (1928), "A mathematical theory of savings", Economic Journal, 38:543–559.

Samuelson, Paul A. (1947), Foundations of Economic Analysis. Cambridge: Harvard University Press.

Samuelson, Paul A. (1965), "A catenary turnpike theorem involving consumption and the golden rule", American Economic Review, 55:486–496.

Samuelson, Paul A. (1971), "Turnpike theorems even though tastes are intertemporally dependent", Western Economic Journal, 9:21–26.

Scheinkman, José (1976), "On optimal steady states of *n*-sector growth models when utility is discounted", Journal of Economic Theory, 12:11–20.

Scheinkman, José (1978), "Stability of separable Hamiltonians and investment theory", Review of Economic Studies, 45:559–570.

Skiba, A. K. (1978), "Optimal growth with a convex–concave production function", Econometrica, 46:527–540.

Starrett, David A. (1972), "Fundamental nonconvexities in the theory of externalities", Journal of Economic Theory, 2:180–199.

Sutherland, W. R. S. (1970), "On optimal development in a multi-sectoral economy: The discounted case", Review of Economic Studies, 37:585–589.

Takekuma, Shin-ichi (1982), "A support price theorem for the continuous time model of capital accumulation", Econometrica, 50:427–442.

Treadway, Arthur B. (1971), "The rational multivariate flexible accelerator", Econometrica, 39:845–855.

Weitzman, M. L. (1973), "Duality theory for infinite horizon convex models", Management Science, 19:783–789.

von Weizsäcker, C. C. (1965), "Existence of optimal programs of accumulation for an infinite time horizon", Review of Economic Studies, 32:85–104.

Yano, Makoto (1981), "Trade, capital accumulation, and competitive market", Ph.D. thesis. University of Rochester.

PART 5

MATHEMATICAL APPROACHES TO ECONOMIC ORGANIZATION AND PLANNING

Chapter 27

ORGANIZATION DESIGN*

THOMAS A. MARSCHAK

University of California, Berkeley

1. Introduction

The study of organizations has taken two directions in recent economic research. First, an organization is a productive unit. It transforms resources – members' time and informational equipment of various sorts – into certain outputs, namely actions that yield some sort of desired commodity, called *payoff*. A major theoretical and descriptive task is to characterize the technology of such productive units and to discover what distinguishes well designed organizations from poorly designed ones. The task is analogous in spirit to the modeling of a complex physical or chemical technology so as to find those resource combinations which produce the technology's outputs efficiently. Accomplishing the task leads to the ranking of possible *designs* for an organization. We shall use that term loosely for the moment. A design specifies, informally speaking, who does what when.

Second, one particular class of organizations has been studied intensively, namely economies, whose members are producers and consumers. The members spend some of their time in observing their changing local environments, in transmitting messages to other members, in computating, in storing and retrieving information. These efforts produce the resource-allocation actions followed in the economy. One class of designs for such organizations are called price mechanisms and have traditionally received central attention. But more recently, possible designs for economies have been studied abstractly and the general term "resource-allocation mechanisms" has come into use. Competitive mechanisms using prices are examples, but so are planning mechanisms in which various sorts of directives are transmitted by a center.

The work on resource-allocation mechanisms has so far been obliged to take a very incomplete view of the issues facing a *designer* of mechanisms who has to

*This chapter was to have been prepared by Jacob Marschak. Before his death (July 1977), he had prepared an outline. A few initial pages were found as well. The present essay is in no sense an attempt to reconstruct what he intended to write. It is, however, unified by the concept of a design composed of "tasks" (or "processors"). This concept is central to his final publication [J. Marschak (1979)] and, judging by the outline, was to have unified his writing of the present chapter.

Handbook of Mathematical Economics, vol. III, edited by K.J. Arrow and M.D. Intriligator
© 1986, Elsevier Science Publishers B.V. (North-Holland)

choose between two candidates. Such a designer would want to know, for each of two proposed resource-allocation mechanisms, its *net* performance over time: the final commodities available for consumption as the economy passes through any given sequence of environments and responds to them with the mechanism's resource-allocating actions, *after* allowing for the resources used in the operation of the mechanism itself. This would require a complete characterization of the technology of a mechanism's operation. So far, only fragmentary and extremely simple models of technology have been explored.

Another issue arises both in the general study of efficient organization design and in the analysis of resource-allocation mechanisms for economies. That is the question of incentives. If the organization members are humans and not programmable robots, they may fail to follow the instructions specified by a given design because they do not want to. When is it the case that the benefits which members receive as the mechanism generates actions make each member prefer following the designer's instructions to violating them? For organizations which earn a transferable payoff – a commodity desired by all members and divisible among them – it may be possible to award a portion of payoff to each member in such a way that each member finds it in his own interest to follow a given design.[1] The designer may view such rewards as part of a design's cost. It is then payoff less rewards less the other costs of operating a design which determine the designer's ranking of alternative designs. The incentive problem is discussed extensively in Chapter 28, largely in connection with resource-allocation mechanisms. We shall omit it, in order to keep the present survey within manageable bounds.

Fortunately, one *can* usefully study the efficiency of designs, separately from the incentive issue. One assumes, in effect, that members *are* programmable robots. A design which has been found to be promising from the efficiency point of view can then be studied further to see if its incentive properties are acceptable – to see whether some system of rewards can induce humans to behave as the design's robots behave.

The present survey takes the point of view of a designer of organizations who is to choose among alternative designs for the organization, given certain data which determine the designs available as well as the payoffs and the costs associated with each available design. Typically, the performance of the design will depend on uncertain external events. The generally accepted contemporary view as to ideal behavior for a designer – or any other chooser under uncertainty – is that he be an expected-utility maximizer who attaches suitable personal probabilities to the uncertain events. The difficulty with requiring such uncompromising rationality is that – *unless the designer's utility function is sharply restricted* – the study of designs with regard to efficiency becomes irrelevant. Without restricting the utility function – or, possibly, making very specific assumptions about the

[1]An example is studied in J. Marschak (1977).

probabilities – it becomes irrelevant to compare designs with respect to the amount of "output", as measured by *expected payoff*, which can be achieved for given bundles of costly inputs (communication effort, suitably measured; observing effort; computing effort; and so forth).

Suppose utility, defined on payoff and on these input quantities, is increasing in the former and decreasing in the latter. Suppose design A dominates design B: A achieves at least as high an expected payoff as B and requires no more of any input with the strict inequality holding with regard to one or more of these magnitudes. Then it is *not* true that for all utility functions and all probability distributions, the expected-utility maximizing designer must prefer A to B.[2,3] If, moreover, both payoff and all input costs can be measured in dollars, then only for a linear utility function – i.e. only for a risk-neutral designer – is it true that "A dominates B with respect to expected payoff and input costs" implies "A has higher expected utility than B".

Yet much of the work which has been done, including work on resource-allocating designs for economies, can be viewed (as we shall see) as work of the efficiency sort. There seems to be agreement that the work is interesting and ought to continue. Indeed, this seems often to be the only sort of work presently feasible. But one *cannot* take the view that such work is relevant for *any* expected-utility maximizing designer.

How then, without restricting utility, can one motivate the efficiency approach? The most promising answer is to appeal to *bounded rationality* as the appropriate standard of behavior for a designer. Designing an organization is perhaps the most complex decision making that one can study. If ever there were a task for which the unlimited rationality of the expected-utility maximizer is too ambitious a standard, it is the design of organizations. But "bounded rationality" in the present state of the discussion [Simon (1972)] is still only an informal guide, permitting a wide choice of models.[4] The discussion provides one with useful suggestions – e.g., that a boundedly rational decision maker revises his "aspiration level" in accordance with the observed difficulty of achieving the level's previous value. But beyond this, the modeler has no accepted and specific ground rules.

What would be crucial as a foundation for further work in the efficiency of organization design is a set of simple, appealing, and modest axioms for a boundedly rational decision maker which respect his non-neutrality towards risk

[2] For a proof, see J. Marschak (1971).

[3] For this to be true with non-linear utility it would suffice for utility to be the sum of a possibly nonlinear function of payoff plus several other possibly non-linear functions, each defined on a different input. Such separability seems unreasonable in many settings. In ideal models both costly inputs and payoff would be measurable in dollars, and such separability of the utility-function would be ruled out. But even in models which fall short of the ideal the "preferential independence" [Keeney (1972)] required for separability is highly implausible.

[4] See, for example, Radner (1975) and Radner and Rothschild (1976).

and yet imply that when one design dominates another with regard to expected payoff and costly inputs, then the dominating design is preferred.

No such axioms are presently available. In their absence, not only expected payoff, but also other "output" measures – or *gross performance* measures, as we shall sometimes call them – have a legitimate claim on the attention of the student of organization design who is interested in efficiency. One of these is *minimum* payoff or maximum distance from a payoff maximizing action, and some initial research using such a measure is summarized below (Section 4.3).

Our central concept in the present survey will be a simple sort of design – called a *one-step* design, to be developed in Section 2. We shall be able to interpret major recent work as a contribution to the efficiency study of one-step designs. Section 3 so interprets work in the theory of teams. Section 4 similarly interprets work in the theory of adjustment processes (including certain work on resource-allocation mechanisms) and suggests some possible new directions. Section 5 interprets briefly some general issues in organization design, notably "centralization versus decentralization".

2. One-step designs

2.1. General concepts

We consider a designer of an organization. He is given

(1) a set E of possible *environments* e,

(2) a set A of possible organization *actions* a,

(3) a set R of possible *results* or *payoffs* r,

(4) a *result function* or *payoff function* ρ from $A \times E$ to R.

He is to choose what we shall call a *design*. He has *beliefs* about the elements of E, expressed as probabilities, as well as preferences among the pairs (r, c), where c denotes the *cost* of a contemplated design. For each environment in E, the design yields a pair (r, c). If the designer obeys suitable axioms [J. Marschak and Radner (1971, ch. 2)], then there will be a *utility function* on the pairs (r, c) such that he prefers the first of two designs to the second if and only if expected utility over the pairs (r, c) which the design yields is higher for the first design than for the second. If a payoff r is a quantity of a desired commodity and the cost c is a quantity of the same commodity, then the utility function is defined on the variable $r - c$.

"Nature" chooses, at the start of each *time period*, an element e of E. In each time period, the designed organization must choose an element a of A, thereby generating an element $r=\rho(a,e)$ of R. Each element a of A is a vector $a=(a_1,\dots,a_l)$. We shall call the l components of the organization's action "the values assigned to its l *attributes*". The attributes comprise the l-element set L.

We list now the elements of a *one-step design*, the object our designer is to choose. The adjective "one-step" will be explained shortly. For ease of exposition, we confine the definition to the *finite* case – i.e., unless otherwise indicated the sets E, A, R and the sets comprising a design are all finite.

First, there has to be chosen a set N of organization members. Let N contain n members.

Second, for each member $i\in N$, there has to be chosen a *task* $T^i=(\bar{X}^i, X^i, \Gamma^i)$, where[5] $\bar{X}^i$ is a finite non-empty set of *inputs* $\bar{x}^i$; X^i is a finite non-empty set of *outputs* x^i; and Γ^i is a matrix of conditional probabilities $\Pr(x^i|\bar{x}^i)$. If Γ^i contains only zeros and ones, then Γ^i is called a *noiseless* task. To define the sets $\bar{X}^i$, X^i, the designer specifies the *observing*, *message-receiving*, *message-sending*, and *action-taking* in which i engages. To be precise, he must choose for every member i

(a) a partitioning $\mathscr{P}^i$ of the environment set E. Member i, observing the environment, determines in which of the sets in $\mathscr{P}^i$ the environment lies (there is a finite number of these sets). If $\mathscr{P}^i$ contains only E itself, then i does no observing at all.

(b) a subset L^i of the action-attribute set L, and a finite set A_{L^i} whose typical element a_{L^i} specifies a value for every attribute in L^i. Here $\bigcup_i L^i = L$; $\bigcap_i^{L^i} L^i$ is empty (only one member is responsible for a given attribute); and A_{L^i} is a subset of the projection of A with regard to the attributes in L^i. If L^i is empty, that means that member i has no direct responsibility for any attribute of the organization's action.

(c) For each $j\neq i$, a finite set M^{ij} of possible *messages to* j. For all i, j in N with $i\neq j$, the set M^{ij} is a subset of the same set, called a *language*. If M^{ij} is empty, then i never sends messages to j.

A member's inputs are observations or messages or both; his outputs are messages or values of action attributes or both. For every i, then, we define

$$\bar{X}^i=\mathscr{P}^i\times\prod_{k\neq i}M^{ki},\qquad X^i=\prod_{j\neq i}M^{ij}\times A_{L^i}.$$

(Thus, one component of an input is a set – the subset of E in which i knows the environment to lie.) Once we have specified the matrix $\Gamma^i=\Pr(x^i|\bar{x}^i)$, we have

[5] The alternative term "processor" has been used for this triple in engineering and computer-science literature.

also specified various associated probabilities. In particular, we have specified the conditional probability distribution of certain *components* of i's output given the value of certain *components* of i's input – for example, the probability distribution of the message j sends to k, given a value of the message j receives from i. We can then also speak, in the usual way, of a certain component of i's output as stochastically *independent* of (or dependent on) a certain component of i's input – for example, the message i sends to j^* is independent of the message i receives from k^*.

Now, in a *one-step* design a member i never sends to $j \neq i$ a message which is itself (stochastically) based, either directly or indirectly (via intermediate members) on a message sent *by* j. Formally, we shall say that *j sends message to k based on messages received from i* if for some $\tilde{m}^{jk}$ in M^{jk}, and for some inputs $\bar{x}^{*j}, \bar{x}^{**j}$ both in $\bar{X}^j$ and differing in only *one component*, namely the message received by j from i, we have

$$\Pr(\tilde{m}^{jk}|\bar{x}^{*j}) \neq \Pr(\tilde{m}^{jk}|\bar{x}^{**j}).$$

We shall say that *r receives messages influenced by s* if there is some sequence of members $q_1, \ldots, q_H$ distinct from each other and from r and s, such that

q_1 sends messages to q_2 based on messages received from s,
q_2 sends messages to q_3 based on messages received from q_1,
⋮
q_{H-1} sends messages to q_H based on messages received from q_{H-2},
q_H sends messages to r based on messages received from q_{H-1}.

The design defined by the triples $\{(\bar{X}^i, X^i, \Gamma^i)\}_{i \in N}$ is then a *one-step* design *if no member i sends to $k \neq i$ messages based on received messages, where those received messages are influenced by i himself.*

We can now interpret the operation of a one-step design – a design which also meets our condition on the sets L^i and our condition that $\bar{X}^i, X^i$ are non-empty – so that, given a period's new environment the design generates new actions in response to it, and so that the probability distribution of these actions can (in principle) be computed. To do so we assume that *there is no noise in the acquisition of inputs and the adjustment of action attributes to their chosen values.* Member i knows exactly the set in $\mathscr{P}^i$ to which the period's environment belongs, knows exactly the messages sent to him, and the value of a_{L^i} which he chooses becomes in fact the value taken by the organization. In Section 2.2.3 below, we consider the case of intervening noise.

Each member adopts the following procedure: "Send a message to a given member and choose a value for a given action attribute as soon as you receive all those input components which determine the probability distribution of those output components." Members initiate the period's action-generating process by observing the environment. Each observing member then adjusts those action attributes whose values depend only on his observation, and sends those messages which are based only on his observation. He next sends those messages and adjusts those attributes which depend only on his observation *and* on the messages received in the first interchange of messages. He thereupon sends those messages and adjusts those attributes which depend only on this observation and on the messages received in the first two interchanges. And so on until he has selected all components of his output.

Fulfillment of the one-step condition implies that every action attribute will be adjusted. It cannot happen that in order to adjust an attribute, i must wait for a message from j which does not arrive because it depends (indirectly) on a message i is to send, and that is based in turn on the message i awaits.

A design, finally, will be said to *cover E with regard to A* or to be a *design on E with regard to A* if for every environment in E the design generates values of the action attributes which define an action[6] in A.

A one-step design covering E with regard to A is then a triple

$$\left(N, \{M^{ij}\}_{i \neq j, i \in N, j \in N}, \{\mathscr{P}^i, A_{L^i}, \Gamma^i\}_{i \in N}\right),$$

fulfilling all the conditions we have stated. Given an environment in E, the design implies a conditional probability distribution over the actions in A. If the design is noiseless (i.e. each of its tasks is noiseless), then it assigns a unique action to every environment.

The concept of design just given can be generalized in a number of ways while still preserving the one-step property. In particular, (1) finiteness could be dropped and (2) *memory* could be added. Some components of an output could be sent to a member i's memory, to be retrieved as an input component, in a subsequent period. We shall refer to such generalizations as needed, when our simple concept of design is not adequate to interpret a topic in the survey. In particular, memory will be added in Section 4.1.

2.2. *Models of technology and costs*

The designer, with his beliefs about E, can proceed to compare two proposed one-step designs if he knows something about their *costs*. A design's cost reflects the effort required to carry out each task (transforming inputs into outputs); to

[6] If the set A is the cartesian product of its l projections (with respect to the l attributes), then every design covers E with regard to A.

acquire each task's inputs (by observing the environment, by transmitting messages); and to execute the actions whose current value is an output of some tasks. We consider several possible approaches to the modeling of a design's technology and its costs.

2.2.1. *Probability-free fixed cost models of technology*

In one approach, one *ignores* the fact that some environments occur more frequently than others, some messages are sent more frequently than others, and some actions taken more frequently than others. One portrays a technology in which a separate "detector" is required for each of the possible values of every component of a task's input. Such a detector is a device which at any moment is in a "no" or a "yes" state and at the start of a period is in the "no" state. As the period's value for an input component is determined, the device corresponding to that value takes the "yes" state, while all the other devices, corresponding to other values, continue to take the "no" state. Similarly, for each possible value of every component of a task's output there is a "selector", a device which is always either in a "yes" state or a "no" state and is in a "no" state at the start of a period. As the period's value for an output component is obtained by member i, in response to the states of the input detectors, the corresponding selector is put into the "yes" state and those corresponding to other values continue to take the "no" state.

A fixed *investment* has to be made in the required number of detectors and selectors and the design's cost depends on this investment alone. The detectors and selectors may experience different degrees of "wear and tear", since some take the "yes" state more often than others. But that, in the model, has no effect on cost.

In the simplest model of this class, cost is given by some function whose arguments are the number of members in N and the number of elements in the sets $\mathscr{P}^i$, M^{ij}, A_{L^i}; the function is increasing in each of these arguments. Possibly, in addition, cost goes up rapidly, for each *fixed* i, as the sizes of the sets $\mathscr{P}^i$, $(M^{ij}\}_{j\neq i}$, A_{L^i} go up, so that the designer can strike a balance between the number of members and the size of their tasks.

A more elaborate probability-free model would require more than the counting of elements of sets. It might, for example, consider some partitionings $\mathscr{P}^i$ more costly than others. It might describe i's observing of the environment as the asking of a sequence of binary questions which ends when the correct set in $\mathscr{P}^i$ has been found. The (probability-free) "observing cost" might then be an increasing function of the largest number of questions which could ever be asked.

Under a probability-free model of technology and cost, a designer will never be interested in a design with noisy tasks. He is charged nothing extra for noiselessness and can therefore only gain – with regard *both* to expected utility and expected payoff – by confining his attention to noiseless designs in which a well-chosen output always follows a given input.

More elaborate probability-free models would, however, allow for differences in *computational complexity*. Some noiseless tasks require relatively great effort to perform in a given time. The required output is a complicated function of the components of the input. Even though one confines oneself to the finite case, so that the function is a finite "table", some tables are easier to search through to determine the proper output than others. There is substantial theoretical literature on computational complexity, but so far it has not taken a form that appears useful for the cost comparison of specific designs. We shall comment briefly later (Section 4.4) on the theory of sequential finite-state machines, which suggests one approach to ranking tasks according to computational difficulty.

2.2.2. *Probabilistic but noiseless technologies which exploit frequency differences*

Consider next models of a technology which are to be used for a noiseless design (all tasks are noiseless) and for which there is no noise in the acquisition of inputs and the execution of actions. The technology pays special attention to the more frequently observed environments, more frequently sent messages, and more frequently chosen actions.

Each task receives its inputs and issues its outputs through various noiseless *devices*[7] – devices which observe the environment, transmit messages, and execute actions, adjusting each attribute to its chosen level.[8] Randomness of the messages received and actions taken springs then, from the randomness of the environment. Suppose Nature chooses the successive environments out of E, in successive time periods, in a serially independent manner. Given, then, an unchanging probability distribution on E, one can, in principle, compute for a given design the probability that member i has to send to j a particular message m^{ij} in the typical time period.

[7]Formally, we could calll each such device a task. Thus, transmission of a message which is i's output to j, for whom it is an input, could be a "task", whose inputs are messages sent by i and whose outputs are messages actually received by j. If transmission is noisy, that means the newly defined task is a noisy one. For modeling purposes it seems useful, however, to reserve the "task" concept for the n (human) members, and to use the word "device" to describe transmitting "hardware" which links tasks to each other, observing hardware and action-executing hardware. Of course, the "hardware" might have some human components (e.g. messengers).

[8]Given the set E and a partitioning $\mathscr{P}^i$ on E, a noiseless observing device always correctly informs i as to the set in $\mathscr{P}^i$ to which the current environment belongs. For a noisy device there is at least one set S in $\mathscr{P}^i$, containing a subset of positive probability measure, such that for the environments in this subset, the device indicates (to i) with positive probability, a set other than S. Of course, one can formally redefine E so that the device's noise itself becomes part of the typical element of E. Relative to the redefined set E, the device is then noiseless. This exercise seems unlikely to be useful as modeling. One would like those objects which are *given* to the designer – the sets E, A, and the result function ρ – to be independent of a particular device, which is just one of many devices choosable by the designer.

We want now to model the technology of message transmission from i to j. One possible transmission device sends certain *symbols* from i to j and is just "large" enough to send a fixed number, say f^{ij}, of symbols per *time unit*, which we shall take to be our "period". The cost of the device is an increasing function of f^{ij}. The original Noiseless Coding Theorem of Information Theory[9] [Shannon (1949)] then tells us that given the probabilities of the messages m^{ij}, there is a greatest lower bound to those values of f^{ij} which are capable – using a suitable *coding* of messages into symbols – of "keeping up" with the stream of messages i is required to send to j in successive time periods as nature picks successive environments. A good code assigns longer symbol sequences to infrequent messages. Using such codes, there is a greatest lower bound to those values of f^{ij} such that with probability one the backlog of messages waiting to be transmitted never exceeds a constant bound. If there are t symbols, the greatest lower bound is given by the entropy

$$-\sum_{m^{ij}\in M^{ij}} \Pr(m^{ij})\log_t \Pr(m^{ij}).$$

This is the greatest lower bound to the average number of symbols needed per message. Hence it is the greatest lower bound to the number of symbols per time unit which the device must be capable of transmitting if member i is to keep up with the stream of messages which he has to send to j, since i has to send a new message to j in each time unit.

Suppose now that member i is permitted to accumulate the messages to be sent to j in *blocks*. Each block is coded into a sequence of symbols which are decoded by the receiver. The code has the property that the start of a new message can always be recognized. By making the block size long enough, one can bring as close to the lower bound just given as one wants the symbols-per-time unit capability required in order to process, with a bounded backlog, the stream of messages which i has to send to j as Nature chooses its stream of environments in successive time periods.

The difficulty is that if a block is to contain more than one message (from i to j), then several time periods, with their successive environments, must pass. Hence the action which each of these environments generates is not taken until

[9] The term Information Theory usually refers to the work of Shannon and his successors. It deals with the properties of certain models of transmission. The more recent term "Information Economics" was originally applied [e.g., J. Marschak (1971) and papers cited there] to studies of the value of information to a single decision maker or possibly (like some of the work surveyed here) groups of decision makers. In these studies cost either plays no role or else simple assumptions about cost are made, *not* necessarily those made in transmission models of the Shannon type. Even more recently, the term "Economics of Information" has come to be used (somewhat confusingly) for a still different area: the working of markets whose members respond to various signals which, at some cost, they can observe.

several time periods after the environment has ceased to prevail. We have assumed so far that the payoff $\rho(a, e)$ collected in a period depends *only* on the environment e prevailing and the action a taken *in that period*. If we continue to assume that, and if *successive environments are indeed serially independent*, then the delayed actions generated through message-block accumulation can be no better – with regard both to expected utility and expected payoff – than a well-chosen constant action repeated in every time period. In that case, the greatest lower bound of the Noiseless Coding Theorem can be of no interest to the designer. What becomes relevant instead is the lowest symbols-per-time-unit capability which – using the best possible code – permits i to send every message to j that he may be obliged to send and to do so within a certain fixed time interval. The relevant fixed time interval is, in general, less than the time period between environments. One has to allow for the time required for coding and decoding as well as the time required for members' observing, action-taking, and other transmissions (between other pairs of members) which are required by the design to precede or follow the transmission from i to j.

Given, then, the fixed interval T during which the transmission from i to j must be completed ($T<1$, if we continue to take the period between environments as the time unit), the required symbol-per-time-unit capability of a t-symbol transmission device is Q/T symbols per time unit, where Q is simply the smallest integer satisfying

$$t^Q \leqq \text{number of elements in } M^{ij}.$$

(Each message in M^{ij} is coded into a distinct sequence of symbols and there are as few unused sequences as possible.)

The Noiseless Coding Theorem becomes relevant if we continue to assume environments to be serially independent but modify our original set of objects given to the designer. Let $\hat{a}(e)$ denote the action in A which a (feasible) design generates in response to a period's new environment. Replace the payoff function ρ by a function w defined on quadruples (a, e, e', T) with a in A, e and e' in E, and T a positive integer. Suppose a proposed design, yielding $\hat{a}(e)$, takes T time periods to generate an action in response to an environment, where "time period" means, as before, the interval between successive environments. Then if the environment sequence is $e_1, e_2, \dots, e_t, \dots,$ the payoff collected by the organization in the typical time period t is $w(\hat{a}(e_{t-T}), e_{t-T}, e_t, T)$.

High values of w are desired by the designer and w is decreasing in T, that is, smaller delays yield more payoff. The arguments of w include the current environment e_t – the environment prevailing when the design finally generates an action in response to e_{t-T}. This reflects the fact that delay is undesirable because of the elapsed time itself ("impatience") and because the action $\hat{a}(e_{T-t})$ is to some extent obsolete when the environment has become e_t. Consider, as an example, a design whose purpose is to fill current "orders", which are one component of the current environment. The generated action is the fulfillment of an order, namely, the shipment of commodities to certain consumers. A long delay is undesirable for consumer and designer (impatience). But, in addition, the delivered commodities may have become, to some extent, inappropriate in view of new tastes (a further component of a current environment) when delivery finally occurs.

If the function w replaces the original payoff function ρ, then in choosing among designs, the designer has to balance the transmission-cost saving due to accumulating blocks of messages (and so allowing a low symbols-per-time-unit capability) against the lower values of w resulting from longer delay. Needless to say, the trade-off is a complicated one, and does not appear to have been worked out, even for simple examples.

The relevance of coding theorems to transmission costs in the case of serially *dependent* environments is dealt with in Section 2.2.3 below.

Aside from message transmission, the other efforts required by a design may be performed in a manner that exploits frequency differences. Unlike the case of message transmission, there are no standard models, exploiting frequency differences, of observing, action-taking, and computing. But models can be constructed. In the case of observing, for example, a possible probability-free model was discussed briefly earlier: cost depends on the largest number of binary questions ever needed to locate the environment in a set of $\mathscr{P}^i$. In a frequency-exploiting model of observing, one would consider cost to change from period to period and to be an increasing function of the number of binary questions which need to be asked in that period in order to locate the environment.

Given the sets in $\mathscr{P}^i$, many question-asking schemes (algorithms) are possible. A partial catalogue of schemes has been explored by H. Oniki (1974a, 1974b). One could simply proceed down a given list of sets and ask for each whether or not it contains the environment, stopping as soon as the answer is "yes". One could arbitrarily divide the sets into two groups, equal in number of elements, or differing by one, then similarly divide each of those into two sub-groups, and so forth, letting the resulting binary tree (two branches at each node) guide the questioning. Or one can arrange the sets in $\mathscr{P}^i$ along many other binary trees (e.g. trees in which groups are divided into two sub-groups far from equal in size). If a probability is attached to each set in $\mathscr{P}^i$, then one natural criterion for choosing among these algorithms is the expected number of questions that need to be asked until the environment is located. The argument establishing the Noiseless Coding Theorem tells us a greatest lower bound for this expected value as one passes over the possible algorithms, namely, the entropy

$$H = -\sum_{S \in \mathscr{P}^i} \Pr(S) \log \Pr(S).$$

A further theorem of Shannon tells us that the least upper bound for the expected value is $H+1$. There exists, moreover, an algorithm [Huffman (1962)] for constructing all optimal binary trees – all binary trees which minimize the expected value of the number of questions asked.

This approach rests on a model of the observing technology wherein all questions are equally difficult to answer. In some settings this may, of course, be quite unrealistic; some parts of the set E may be harder to "scan" than others, or, to use another terminology, some aspects of the current environment may be harder to measure than others. Clearly, one has to begin modeling specific real organizational observing tasks before one can begin to judge the usefulness of the binary-tree model or any other model.

There have not yet been attempts to construct frequency-dependent models of action-taking as such. One could, presumably, try to capture the idea that frequent *changes* of action are costly as well as the idea that the instructions issued to an action taker (which are the outputs of certain tasks) ought to be made "simpler" for more frequently taken actions than for less frequently taken ones.

2.2.3. *Noisy models*

In these models it is costly to *diminish* the noise in a task – to replace $(\overline{X}^i, X^i, \Gamma^i)$ by $(\overline{X}^i, X^i, \tilde{\Gamma}^i)$, where $\tilde{\Gamma}^i$ is, in some appropriate sense, closer than Γ^i to a noiseless zero–one matrix. Similarly, in these models, the acquisition of inputs through noisy transmission and noisy observation is cheaper than through noiseless transmission and observation; the carrying out of intended actions with error is cheaper than without error.

Consider first the devices used for message transmission from i to j. Suppose environments are serially independent, so that the successive messages i is required to send to j are also serially independent. Suppose that i can be provided with a device for transmission to j. The device transmits symbols ("0" and "1"). But it is not completely reliable; it is characterized by a probability matrix Δ:

		Symbol received	
		1	0
	1		$p 1-p$
Symbol sent	0	q	$1-q$

Suppose one can choose both the matrix Δ and the device's speed (in symbols-per-time unit). The device is cheaper the less its speed and the "further" is Δ from the identity matrix.

The device is used as follows. Member i is given, in each time unit (the time period between environments), a message in M^{ij} to be sent to j. Suppose there are R elements in M^{ij} and their probabilities in every time period are $s_1,\ldots,s_R$. Member i takes the messages to be sent, accumulates them into suitable blocks, codes each block into sequences of zeros and ones. The sequence is sent over the noisy device and received (in somewhat distorted form) by j who decodes it; j, that is to say, assigns to the received zero/one sequence a sequence of messages in M^{ij}, namely, the sequence with highest posterior probability.

For any n-tuple of probabilities $(p_1,\ldots,p_n)$ with $0 \leqq p_i \leqq 1$, $i=1,\ldots,n$, and $\sum_i p_i = 1$, let $H(p_1,\ldots,p_n)$ denote the entropy $-\sum_i p_i \log p_i$. Now let

$$c(\Delta) = \max_{0 \leqq \alpha \leqq 1} \Bigg[H(\alpha, 1-\alpha) - (p\alpha + q(1-\alpha)) H\left(\frac{p\alpha}{p\alpha + q(1-\alpha)}, \frac{q(1-\alpha)}{p\alpha + q(1-\alpha)}\right) - [(1-p)\alpha + (1-q)(1-\alpha)] H\left(\frac{(1-p)\alpha}{(1-p)\alpha + (1-q)(1-\alpha)}, \frac{(1-q)(1-\alpha)}{(1-p)\alpha + (1-q)(1-\alpha)}\right)\Bigg].$$

We can interpret $C(\Delta)$ (the "channel capacity" associated with Δ) as the largest average uncertainty reduction which a noisy device characterized by the matrix Δ is capable of achieving, where entropy measures uncertainty and the average is taken over the possible symbols received. The first H-term measures a receiver's uncertainty about the sender's choice of zero or one before he has received a symbol and knows only the probabilities that the sender sent zero or one, which are, respectively, α and $1-\alpha$. The last two H-terms measure the receiver's reduced uncertainty about the sender's choice given that he has received a particular symbol. Now *if*

$$c(\Delta) > H(p_1,\ldots,p_R),$$

then the *error probability*, i.e. the probability that a message decoded by i differs from the message which i would have obtained from j if Δ were the identity matrix, can be made arbitrarily small by choosing sufficiently long blocks of messages. Moreover, this can be done so that the average number of symbols (zeros and ones) required to code a message exceeds $c(\Delta)$ by as little as desired. Hence, provided one accepts sufficiently long block accumulations, any device transmitting $c(\Delta)$ symbols (zeros or ones) per time unit (time period) will suffice to keep up with the stream of messages that i has to send to j.

Error probability is, of course, a peculiar criterion, since it weighs all errors equally, even though some lead to lower payoffs than others. A 1960 generalization of the result [Shannon (1960)] allows for a more appealing criterion, namely, "fidelity", that is, the expected value of some "benefit" function – a function of the message sent by i and the message recognized (after transmission and decoding) by j. In the generalized result, expected benefit can be made as close as desired to the highest expected benefit achievable for a noiseless device (for which Δ is the identity matrix) and this can still be done with a symbols-per-time-unit speed as close as desired to $C(\Delta)$. The result extends from binary devices to devices with any number of symbols.

Even the generalized result is remote from the assessment of a design's cost when a "good" collection of transmission devices is chosen for the design. For the result to be relevant in the case of serially independent environments, we again need to introduce delay penalties. One has to balance the cost saving due to noisiness *and* the saving due to slowness of the devices used in transmission between *all* pairs (i, j) not only against each other but also against the penalty due to delay. Then in extremely simple cases the generalized result might guide one in achieving the balance.

If one does not accept delay but requires every message to be sent in the time period in which the prevailing environment is the one that gave rise to the message, then, as in the noiseless-transmission case, the coding theorems of Information Theory have no relevance. Once may use codings and noisy-transmission devices, but the speed of a device (in symbols-per-time-unit) must be high enough to permit the longest possible coded message. It must be the speed Q defined in Section 2.2.2 above.

Once a complete set of transmission, observing, and action-taking devices are in place – both noisy ones and noiseless ones – so that a design's tasks can be carried out, a conditional probability distribution on the actions in A is determined for each environment in E. To find it, one has to interpret properly the inputs and outputs of each task. Thus, suppose each input in $\bar{X}^j$ includes among its components a message in M^{ij}; if i transmits to j through a noisy device, then a typical value of this input component is j's best guess, after decoding (his maximum-likelihood guess) as to the message in M^{ij} which i intended to send. The probability distribution on A for a given e in E depends, possibly in a complex way, on the probabilities characterizing every device used for input acquisition and action-taking. But it *is* determined once those probabilities are specified. Given a probability distribution on E as well, the designer who is to choose between two designs and does not accept delay can, in principle, rank them according to highest attainable expected utility, where utility is defined on payoff and cost. Cost consists of the cost of an array of (possibly noisy) devices capable of acquiring the design's inputs and executing its actions plus the cost of performing the design's tasks (assigning outputs to inputs). A design's highest attainable expected utility is its expected utility for the best possible choice of devices.

Alternatively, one can take the less ambitious bounded-rationality or linear-utility viewpoints described in the Introduction, and could try to study, for a given design, the trade-off between expected payoff and the various elements of cost – e.g., noisiness of devices and of tasks, speed of transmission, sizes of the finite sets defining the design.

What of the costs associated with the members' tasks themselves? If a task $T^i = (\bar{X}^i, X^i, \Gamma^i)$ is noisy, one approach to modeling its cost is simply to select some measure of dispersion and to suppose that *given the sets* $\bar{X}^i$ *and* X^i, the lower is the average value of this measure for the probabilities $\Pr(x^i|\bar{x}^i)$ – where the average is taken over all inputs $\bar{x}^i$ – the costlier is the task. The costliest possible task for a given pair $(\bar{X}^i, X^i)$ is that for which Γ^i is noiseless. This is performed by a totally reliable member. A less reliable, more "confused" member, who is, perhaps, less well trained or less gifted, may be acquired (purchased) instead. He is able to carry out – given the sets $\bar{X}^i$, X^i – only a noisy task. From this point of view, then, a particular acquired member is part of a design; to acquire him, and to assign him the sets $\bar{X}^i$, X^i, is to acquire the matrix Γ^i. It might then be reasonable to let the cost associated with the triple $(\bar{X}^i, X^i, \Gamma^i)$ depend on the sizes of $\bar{X}^i$ and X^i and on the average dispersion of Γ^i.

Note that one possible dispersion measure is, once again, average entropy, i.e.

$$\sum_{\bar{x}^i \in \bar{X}^i} \Pr(\bar{x}^i)\left[-\sum_{x^i \in X^i} \Pr(x^i|\bar{x}^i)\log \Pr(x^i|\bar{x}^i)\right].$$

But the grounds for using entropy in this context are *not* those of the transmission model. Member i selects outputs in X^i in response to inputs in $\bar{X}^i$ so nothing need be transmitted. Average entropy is merely one possible measure of i's "confusedness" or "reliability".

2.2.4. *Models in which cost varies from period to period and depends on the period's input – output pairs*

In models of this sort the cost of a task T^i depends on more than the sizes of $\bar{X}^i$, X^i and the "reliability" expressed in Γ^i. Some inputs are harder to detect than others in the time available and some outputs harder to select than others. There is, then, a cost $c(\bar{x}^i, x^i)$ attached to every pair $(\bar{x}^i, x^i)$ with $\bar{x}^i \in \bar{X}^i$ and $x^i \in X^i$; and the cost is incurred in any period in which i assigns the output $\bar{x}^i$ to the input x^i. A design's cost varies from period to period. An example of a "cheap" output of member i might be a "null message" sent to j ("silence"). This, of course, does tell j something about i's inputs, but the act of forming it ("doing nothing") may, in some technologies, require little effort.

Such an approach to cost may, of course, be combined with some of the other approaches considered, so that the total cost of a design is composed of a variable part and a fixed part.

2.2.5. *Costs: The case of serially dependent environments*

To conclude our general discussion of costs, two final remarks need to be made about the case of serially dependent environments. First, if the environments are serially dependent – if they are, for example, the successive states of a Markov chain – then the coding theorems of Information Theory are again relevant to the modeling of transmission costs provided delay is accepted. But the application of the theorems is more complicated, since the probability distribution on M^{ij} changes from period to period. One possibility is to require the transmission device to have a sufficient speed (in symbols-per-time-unit) to keep up with the stream of messages if the probability distribution *were* stationary and were the "worst" of the distributions that could, in fact, prevail – if such a "worst" distribution exists.

Second, to allow for the undesirability of delay one now need *not* redefine payoff as in Section 2.2. Payoff can remain, as in our original formulation, a quantity collected in each period depending only on that period's action and that period's environment. Since successive environments are now correlated, an action which the design generates in response to a period's environment may not, on the average, be inappropriate to the environment of some periods later; in that later period, the delayed action may generate more payoff, on the average, than would the best constant action (repeated in all periods, regardless of present or past environments). Hence, for many plausible stochastic processes generating successive environments, expected payoff decreases as delay increases, even where current payoff depends only on current environment and current action. The designer is interested, then, in balancing the low cost of slow devices against the lower payoff due to the delay these devices causes.

3. Contributions to organization design, interpreted in the framework of one-step designs: Information structures and decision rules in teams

3.1. *Information structures and designs*

In the theory of teams [J. Marschak and Radner (1971)], one is given a collection N of n members, an environment set E (not necessarily a finite set) with elements e, a set A of possible action n-tuples $a = (a_1, \dots, a_n)$, and a payoff function ρ on $E \times A$. Let A_i, $i \in N$, denote the projection of A with respect to the ith coordinate of a. For each $i \in N$, a_i (an attribute of a in our previous terminology) is called *member i's action*. Only *two* objects are to be chosen: (1) an *information structure* (Y, η), where Y is an n-tuple of *signal sets* $(Y_1,\dots,Y_n)$ and $\eta=(\eta_1,\dots,\eta_n)$ is an n-tuple of functions η_i from E to Y_i; (2) a feasible *team decision rule* $\delta=(\delta_1,\dots,\delta_n)$, where δ_i (member i's decision rule) is a function from Y_i to A_i such that for all $(y_1,\dots,y_n)$ in $Y_1\times\cdots\times Y_n$, $(\delta_1[\eta_1(y_1)],\dots,\delta_n[\eta_n(y_n)])$ lies in A. The latter feasibility requirement is trivially satisfied for all δ if $A=A_1\times A_2\times\cdots\times A_n$.

Suppose an expected-utility maximizing designer is given two pairs: $[(\bar{Y},\bar{\eta}),\bar{\delta}]$ which has a cost $\bar{c}$, and $[(\bar{\bar{Y}},\bar{\bar{\eta}}),\bar{\bar{\delta}}]$ which has a cost $\bar{\bar{c}}$. Suppose he has a probability distribution on E and a utility function u defined on payoff–cost pairs. Suppose there is a sequence of periods in each of which an environment e is drawn from E according to the given probability distribution. Then the designer prefers the first pair if and only if the expected value of

$$u\left(\rho\left(\left[\bar{\delta}_1(\bar{\eta}_1(e)),\dots,\bar{\delta}_n(\bar{\eta}_n(e))\right],e\right),\bar{c}\right)$$

exceeds the expected value of

$$u\left(\rho\left(\left[\bar{\bar{\delta}}_1(\bar{\bar{\eta}}_1(e)),\dots,\bar{\bar{\delta}}_n(\bar{\bar{\eta}}_n(e))\right],e\right),\bar{\bar{c}}\right).$$

But most of the work which has been done in the theory of teams deals with a simpler issue. That is to find, for a given interesting information structure, the "best" decision rule, where "best" means expected-payoff maximizing and not expected-utility maximizing, with utility defined on payoff and cost. Given a structure (Y,η), one seeks the rule $\hat{\delta}$ such that for all other rules δ the expected value of $\rho(\delta_1(\eta_1(e)),\dots,\delta_n(\eta_n(e)),e)$ is not greater than the expected value of $\rho([\hat{\delta}_1(\eta_1(e))],\dots,[\hat{\delta}_n(\eta_n(e))],e)$. Such work has to appeal to the linear utility or the bounded-rationality viewpoints sketched in the Introduction. Studies which compare information structures with regard to highest attainable expected payoff seem likely, in any case to continue. They appear far more tractable than full-scale studies of highest attainable expected utility for non-linear utility functions and specific cost assumptions.

If we have found, for a given team of n members, an information structure and an expected-payoff maximizing team decision rule, have we then also defined a one-step design? No, for if we only write down an information structure and a decision rule, we say nothing about the observing effort and the message-sending which occurs and about who performs these tasks. Only action-taking is discussed: member i adjusts the value of a_i. Suppose, however, one *adds* that e is an n-tuple $(e_1,\dots,e_n)$ and that member i always observes e_i, his "local" characteristic. (In our previous terminology, i has a partitioning $\mathscr{P}^i$ whose typical set has the form $\{e: e_i=\bar{e}_i\}$.) A designer is given, in other words, some sort of "natural" association between action and local observation. He may choose from a set of available information structures (Y, η) which *enrich* the information (about the entire environment e) of at least one member k. To be more precise: for member k, for such a structure (Y, η), and for all elements $\bar{e}$ of some subset $\bar{E}$ of E which has positive probability measure, it is *not* true that

$$\{e : \eta_k(e) = \eta_k(\bar{e})\} \supseteq \{e : e_k = \bar{e}_k\}. \tag{3.1}$$

Now suppose further that every such member k receives the additional information through a single $(n-1)$-tuple of messages received, all at once, from the other members. Clearly, there will be some $(n-1)$-tuple which suffices, if nothing else the $(n-1)$-tuple $(e_1, \ldots, e_{k-1}, e_{k+1}, \ldots, e_n)$. Then we have all the elements of a noiseless one-step design which achieves a given information structure (Y, η) and a given decision rule δ. Member k's inputs are his own observation and the messages received from others. His outputs are the messages he sends (which depend only on his observations) and a value of a_k, which depends on his own observation and the messages received and equals $\delta_k[\eta_k(e)]$. But some one-step designs achieving the structure (Y, η) are wasteful. For each member i, one would like to search among all $(n-1)$-tuples of functions $\{\varphi_{ik}\}_{i,k \in N, i \neq k}$, where φ_{ik} is defined on the possible values of e_k and there is some function r_k such that for every e in E,

$$\eta_i(e) = r_i[e_i, \varphi_{i,1}(e_1), \ldots, \varphi_{i,i-1}(e_{i-1}), \varphi_{i,i+1}(e_{i+1}), \ldots, \varphi_{i,n}(e_n)]. \tag{3.2}$$

Among all such pairs $(r_i, \{\varphi_{ik}\}_{i \neq k})$ one is interested in those which are economical with regard to transmission (from k to i) of the messages $\varphi_{ik}(e_i)$ and with regard to the difficulty of computing the functions φ_{ik} and r_i. If E is finite, then the approaches to measuring noiseless transmission costs which were discussed above are relevant again. If E is a continuum, and the range of φ_{ik} is a continuum as well, then one may require that φ_{ik} satisfy appropriate smoothness conditions and treat the dimension of the range of φ_{ik} as a measure of transmission cost. Work of this sort, with a different motivation than the achieving of a given information structure for a team, is summarized below in Section 4.2. As for computational difficulty (complexity), that, as remarked earlier, is a subject still largely unexplored by economists concerned with organization design.

An information structure for a team, as defined so far, is noiseless relative to a given set E of environments. Member i generally does not have complete information about the environment but he always receives the same information – namely, $\eta_i(e)$ – about a given environment e.

One can study instead information structures which are noisy relative to a given E. In the simplest sort of noisy structure, member i obtains for each e in E, a signal $\eta_i(e)+\lambda_i$, where $\eta_i(e)$ is a real number and λ_i is a real random variable[10] with a probability distribution F_i. As before, a decision rule δ is to be chosen. Assume again that e is an n-tuple $(e_1,\dots,e_n)$. But now the "local" information which member i always has in any case – regardless of the information structure – is $e_i+\mu_i$, where μ_i is a random variable with probability distribution G_i. The analogue of (3.1) is then the condition that for at least one member k the conditional distribution of e given $\eta_i(e)+\lambda_i$ is not the same as the conditional distribution given $e_i+\mu_i$. If this condition is met, then we seek a (noisy) one-step design which achieves the given noisy information structure. To do so we seek for each i [analogously to (3.2)] a function r_i and an $(n-1)$-tuple $\{(\varphi_{ik}, D_{ik})\}_{i,k\in N,\ i\neq k}$, where φ_{ik} is a function and D_{ik} is the probability distribution of a random variable γ_{ik}, such that for each i in N and each $\bar{e}$ in E the distribution of $r_i[\bar{e}_i+\mu_i, \varphi_{i,1}(\bar{e}_1+\mu_1)+\gamma_{i,1},\dots,\varphi_{i,i-1}(\bar{e}_{i-1}+\mu_{i-1})+\gamma_{i,i-1}, \varphi_{i,i+1}(\bar{e}_{i+1}+\mu_{i+1})+\gamma_{i,i+1},\dots,\varphi_{i,n}(\bar{e}_n+\mu_n)+\gamma_{i,n}]$ is the same as the distribution of $\eta_i(\bar{e})+\lambda_i$. Such $(n-1)$-tuples, together with the distributions G_i describing the members' noisy observing, define a noisy one-step design achieving the required noisy information structure. As before, the design's action outputs may be assigned to inputs so as to express a chosen team decision rule.

Some investigations which have used the framework of the theory of teams have, in effect, proceeded in the reverse direction from that just described. They have considered a particular one-step design which is of interest because of its historic role in certain discussions and have then studied the information structure which the design achieves. The studies by Groves, Radner, and others of a "Lange–Lerner" price mechanism and rival mechanisms in a certain class of teams are of this sort. Below, in Section 3.4, we briefly consider these studies.

3.2. *Finding best expected payoff for a given structure*

The work done so far in the theory of teams has not concerned itself with the design needed to achieve a structure and with the design's costs. That is natural, since it is difficult enough to find, as the existing work does, decision rules which are best, in the expected-payoff sense, for some interesting information structure and payoff function. A main tool for this purpose is the "person-by-person satisfactoriness" theorem. Suppose that $A=A_1\times\cdots\times A_n$. If a decision rule $\delta=(\delta_1,\dots,\delta_n)$ is best for a given structure (Y,η) with regard to the expected

[10] Of course, as remarked earlier in connection with designs, one can redefine the environment set so that the noise λ_i becomes part of the environment; relative to the new set the information structure is again noiseless. But doing so obscures the special form taken by the analogues of (3.1) and (3.2) for a noisy structure. From a modeling point of view, it may be useful to distinguish between the aspects of the environment which no transmission or observing devices can affect, and the aspects which are simply properties of the devices the designer chooses.

value of payoff $\rho(a, e)$, then clearly it must be true that each rule δ_i is best given the other rules $\{\delta_k\}_{k \neq i}$. That is to say, for every y_i in Y_i such that the event $\eta_i(e) = y_i$ has positive probability, $\delta_i(y_i)$ must equal that element $\bar{a}_i$ of A_i which maximizes with respect to a_i, the conditional expected value of $\rho([\delta_1(\eta_1(e)), \dots, \delta_{i-1}(\eta_{i-1}(e)), a_i, \delta_{i+1}(\eta_{i+1}(e)), \dots, \delta_n(\eta_n(e))], e)$ given that $\eta_i(e) = y_i$. The theorem says that if ρ is strictly concave and differentiable in its arguments, then such "person-by-person satisfactoriness" is also *sufficient* for δ to be a best rule".[11]

3.3. *The quadratic case*

If the function ρ is quadratic and concave, then the condition that δ_i be best against the other rules gives, for each y_i, an equation which is linear in the best rules $\hat{\delta}_1, \dots, \hat{\delta}_{i-1}, \hat{\delta}_{i+1}, \dots, \hat{\delta}_n$. To be precise, suppose the payoff function $\rho(a, e)$ has the form

$$e_0 + 2a'\mu(e) - a'Q(e)a,$$

where e_0 is a constant n-vector, μ is a vector-valued function of e, and Q is a matrix-valued function with $Q(e)$ always a positive definite $n \times n$ matrix. Member i's signal $\eta_i(e)$ tells him something about $\mu(e)$ and $Q(e)$ and hence about the team action which ought to be taken to maximize payoff. For a given information structure (Y, η) and for any $y_i \in Y_i$, the person-by-person satisfactoriness condition is (after differentiating the conditional expected payoff given $\eta_i(e) = y_i$ and setting equal to zero)

$$\delta_i(y_i)\mathscr{E}(Q_{ii}|\eta_i(e) = y_1) + \sum_{j \neq i} \mathscr{E}\big(\delta_j(\eta_j(e))|\eta_i(e) = y_i\big) = \mathscr{E}(\mu_i|\eta_i(e) = y_i), \tag{3.3}$$

where $\mathscr{E}$ denotes expectation.

For the quadratic case there are several procedures for finding explicitly a best δ given (Y, η), or at least finding the expected payoff under a best δ:

(1) One can, for some information structures, guess, using intuition, at a decision rule δ which seems likely to satisfy (3.3). If one can then show that it does, one has shown it to be a best rule.

[11] The theorem generalizes to the case in which the action a has to lie, for every e, in a convex set $A(e)$ which need not be the Cartesian product of its n projections. In the generalized theorem, "$(\bar{\delta}_1, \dots, \bar{\delta}_n)$ is person-by-person satisfactory" means, for example, that for every e, $\bar{a}_1 = \bar{\delta}_1(\eta_1(e))$ maximizes the conditional expected value of $\rho([a_1, \bar{\delta}_2(\eta_2(e)), \dots, \bar{\delta}_n(\eta_n(e))], e)$ on the set of those member-1 actions a_1 for which $[a_1, \bar{\delta}_2(\eta_2(e)), \dots, \bar{\delta}_n(\eta_n(e))] \in A(e)$.

(2) One can assume that E is the n-dimensional real space and that the probability distribution on E is normal. If $y_i = \eta_i(e)$ is a real vector, also normally distributed, and if Q is constant, then for the $(\delta_1, \ldots, \delta_n)$ which satisfies (3.3) and is therefore best, δ_i is a linear function of the components of y_i. Each coefficient of δ_i is a function of the parameters of the distribution of $\mu(e)$ and the constants Q and e_0.

(3) If Q is constant, then for certain information structures one can show, using (3.3), that for each y_i, $\hat{\delta}_i(y_i)$ (with $\hat{\delta}$ best) satisfies a linear equation in $\hat{\delta}_i(y_i)$ *only*. This occurs, in particular, for a structure wherein each member i knows some function $\zeta_i(e)$, with $\zeta_1(e), \ldots, \zeta_n(e)$ independently distributed, and also knows a vector $\tau_1(e) = (\tau_1(e), \ldots, \tau_n(e))$, where $\tau_k(e)$ is member k's "report" about $\zeta_k(e)$; the "report" partitions E more coarsely than does $\zeta_k(e)$.

(4) If Q is constant, then without further assumptions it is straightforward to show that if $\hat{\delta}$ is best for a structure (Y, η), then (using some compact and obvious vector notation)

$$\mathscr{E}\hat{\delta}[\eta(e)] = Q^{-1}\mathscr{E}\mu(e);$$

at the same time, "value of information" – the amount by which best expected payoff for (Y, η) exceeds best expected payoff for the "no-information" structure (where, for every i, $\eta_i(e)$ is a constant signal) – is given by

$$\mathscr{E}[\hat{\delta}(\eta(e))]\mu(e) - [\mathscr{E}\hat{\delta}(\eta(e))]'\mathscr{E}\mu(e).$$

For a number of structures, this permits calculation of the "value of information" [and hence of the best expected payoff for (Y, η)] *without* explicitly computing a best decision rule.

Using one or another of these four approaches, a variety of information structures have been explored for the quadratic case[12] [Chapter 7 of J. Marschak and Radner (1971)]. Some of these results can be interpreted (in accordance with the central concern of the present survey) as tracing the effect on best expected payoff as one varies one or another element of an information structure's cost – or the cost, rather, of a design achieving that structure.

(a) There is a family of structures, called [in J. Marschak and Radner (1971)] "dissemination of independent information" and just summarized above in connection with the third approach to calculating best decision rules. Consider moving from a structure in which a particular member j does not send to any

[12] Chu (1976a, 1976b) considers, for the quadratic case, information structures in which member i knows the linear combination $\sigma_{k=1}^{n} h_{ik}\mu_k(e)$. One might regard the *rank* of the matrix (h_{ik}) as an indicator of its cost. An algorithm is given which converges to that matrix among all those of given rank for which best expected payoff [with $\mu_1(e), \ldots, \mu_n(e)$ independent and normally distributed] is highest.

others a report of $\zeta_j(e)$ (his "local" observation), to a structure which is the same except that j now sends such a report to everyone. It turns out that the resulting improvement in best expected payoff is independent of the size and composition of that group of other members $r \neq j$ who send reports in both structures. So if allowing a new "reporter" has a constant cost [if communicating $\zeta_j(e)$ to the others is equally costly for all j], then if the designer wants expected payoff minus cost to be large, all of the members for whom improvement in expected payoff exceeds the reporting cost should be instructed to report.

(b) Interesting classes of structures are suggested by the term "management by exception". In one such class, the variable $\mu_j(e)$ is assumed "locally" observable by j and to take values on the real line. To define the information structure, a part of the real line – a set R_j – is chosen for each j and is called j's "exception set". If and only if j finds $\mu_j(e)$ to lie in the exception set, he communicates it to some "central agent" who relays it to all the other members whose observations are also exceptional (or, equivalently, who computes the action all those members ought to take and tells each of them what that action is). Then $\eta_j(e)$ equals $\mu_j(e)$ if $\mu_j(e) \notin R_j$ and otherwise equals $\{\mu_k(e): k \in J(e)\}$, where $J(e)$ is the set $\{k \in N: \mu_k(e) \in R_k\}$. If the variables $\mu_j(e)$ are assumed independently distributed and $Q(e)$ a constant, then a best decision rule is found [by making a good guess and verifying that the rule satisfies (3.3)]. When, for an environment $\bar{e}$, j's observation $\mu_j(\bar{e})$ is not exceptional, the best rule tells him to take the action which would be best under that information structure wherein each member k knows only $\mu_k(e)$ for every e. When j's observation $\mu_j(\bar{e})$ is exceptional, the best rule tells him to take the action which would be best for that structure wherein, for every e, all members in $J(\bar{e})$ known each other's observations [i.e. for any e each knows $\{\mu_k(e): k \in J(\bar{e})\}$].

For given probabilities $\{p_j\}_{j \in N}$, where $p_j = \Pr[\mu_j(e) \in R_j]$, the sets R_j can be chosen so that best expected payoff is a maximum. Doing so, letting $p_j = p$ for all j, and letting Q take a special form – namely, $q_{ii} = 1$, $q_{ij} = q$, $i \neq j$ – one can trace (for fixed n) the effect of increasing p on best expected payoff. As p goes toward 1, best expected payoff increases but does so more and more slowly. Suppose the relevant technology is of the general sort in which a different cost is attached to each of a task's input–output pairs and one pays for each pair when it is used (Section 2.2.4 above). In the present setting, suppose a certain cost is incurred whenever member j reports an exception but not otherwise. If one assumes the cost per report to be identical for all members and all reports, and independent of the number of members reporting, and if the designer wants the expected value of payoff minus cost to be high, then there is a best value of p. Beyond this value, the improvement in expected payoff due to more frequent reporting of exception is less than the increase in expected cost.

(c) The effect of increasing or decreasing error in observing and in the transmission of certain messages can be examined. One can study, for example, a structure in which complete information about e is accumulated (by some central agent). Assume Q constant; the central agent knows $(\mu_1(e),\dots,\mu_n(e))$. He computes the payoff-maximizing action $\hat{a}(e)=(\hat{a}_1(e),\dots,\hat{a}_n(e))$ and sends $\hat{a}_i(e)$ to i. That transmission, however, is subject to error, so that member i ends up knowing $y_i=\eta_i(e)=\hat{a}_i(e)+\varepsilon_i(e)$. Assuming all the $\hat{a}_i(e)$, $\varepsilon_i(e)$ to be independently normally distributed with zero means, a best decision rule can be computed (using the second of the above four approaches). Consider the central agent to be an $(n+1)$st member of the team. In one model of technology we may attribute error to "confusedness" or "unreliability" of member $n+1$ himself. Suppose that the more dispersed is the variable $\varepsilon_i(e)$, the cheaper the design, where we measure dispersion by the variance of the $\varepsilon_i(e)$. If all the variables $\varepsilon_i(e)$ have a common variance, a designer can balance the cost of low dispersion against the resulting improvement in expected payoff.

A similar analysis can be performed with regard to observing error. This time, the central agent receives erroneous messages $\mu_i(e)+\varepsilon_i$ as to the individual observations $\mu_i(e)$. He computes the action $\tilde{a}(e)=(\tilde{a}_1(e),\dots,\tilde{a}_n(e))$ which would be payoff maximizing if $\mu(e)$ were to equal $(\mu_1(e)+\varepsilon_1,\dots,\mu_n(e)+\varepsilon_n)$. He sends $\tilde{a}_i(e)$, without error, to member i. Now the cost of reducing observing error can be balanced against best expected payoff.

(d) The *size* of the team can be considered an element of cost and the effect of varying it on best expected payoff traced for a specific family of information structures. To do so, one has to specify how the payoff function and the probability distribution on E vary as n varies. One may, for example, assume that Q is constant and has ones on the diagonal and q everywhere off the diagonal [with $-1/(n-1)<q<1$, which assures positive definiteness]. One can assume the variables $\mu_i(e)$ to be independent of each other for all team sizes, and to be drawn, for all team sizes, from the same probability distribution with zero mean and, for every n, an $n\times n$ variance–covariance matrix with diagonal elements 1 and off-diagonal elements σ. Under such assumptions, then, the "returns to scale" issue can be investigated. For many structures, best expected payoff increases as n increases but for some of these structures one has "decreasing returns" (each additional member adds less to best expected payoff than the previously added member), for others increasing returns; and for still others constant returns. For some structures, the limiting behavior of best expected payoff, as n increases without limit, can be conveniently and suggestively studied.

(e) One can check certain general conjectures as to the desirability of certain properties of structures or designs as one changes certain aspects of the payoff function. Such properties may be associated, in particular, with the term "decentralization", discussed in more detail in Section 5 below. For the present, suppose we consider a decentralized information structure to be one in which member i knows only certain "local" or "private" information. Suppose, in particular, that it is the structure in which i knows $\mu_i(e)$. One can then study the improvement in payoff as one moves away from the structure toward "more centralized" structures [wherein at least some member i knows more than $\mu_i(e)$]. One may be interested in the way the improvement changes as (1) one varies the payoff function so as to increase the "interaction" between members – the sensitivity, with respect to changes in j's action, of i's marginal contribution to payoff – as measured, say, by $\partial^2\rho/\partial a_i \partial a_j$; (2) one varies the environmental probability distribution so that i's local observation $\mu_i(e)$ becomes more strongly correlated with j's local observation $\mu_j(e)$. More specifically, let Q again be constant with ones on the diagonal and q off it. Then $|q|$ measures the strength of interaction between any two members. It is found – as intuition suggests – that as $|q|$ rises above zero, there is an increase in the "penalty" due to decentralization, i.e. in the improvement in payoff when "local information only" is replaced by, for example, complete pooling of local information, or management by exception, or pooling of local information (for every e) among groups of members.[13] Similarly, if one lets $\mathscr{E}\mu_i = 0$ and lets $\mathscr{E}\mu_i\mu_j = 1$ when $i = j$ and $\sigma > 0$ otherwise (all i, j in N), then for those "non-decentralized" structures which have been studied in this connection, the improvement in payoff as one substitutes that structure for "local information only" goes down as σ goes up. Again, this is as intuition suggests.

3.4. *Studies of a resource-allocation team*

A number of studies have dealt with a team of $n+1$ members composed of one "center" and n local "managers". Each manager i is to use a resource assigned to him by the center and is to choose a value of a local decision variable L_i. An environment is $e = (k, \mu_1, \ldots, \mu_n)$, where k is the total availability of the resource and μ_i is a parameter defining a local production function f. A team action is the vector $a = (k_1, \ldots, k_n)$ where k_i is i's assigned share of the centrally allocated resource. The arguments of the production function f are L_i and k_i. The team's payoff is $\rho(a, e) = \sum_{i=1}^{n} f(L_i, k, \mu_i)$, provided $\sum_{i=1}^{n} k_i \leq k$. [In a generalized version, Groves and Radner (1972), k and k_i are vectors.]

The first studies [Radner (1972) and Groves and Radner (1972)] were inspired by the classic claims made for a "price" mechanism, wherein the center adjusts $(k_1, \ldots, k_n)$. Each adjustment is made after receiving from each manager i a "profit-maximizing demand" message – i.e. a quantity $\hat{k}_i(p, \mu_i)$ which would maximize $\max_{L_i} f(k_i, L_i, \mu_i) - pk_i$, where p is a price announced at each step by the center and adjusted in response to the previous step's excess demand. While the process may converge, under suitable assumptions on f, to a payoff-maximiz-

[13] But these results fail to have analogues, when one studies quadratic payoff functions and designs which carry out successive steps of certain adjustment processes. See Section 4.4 below.

ing value of $((k_1, L_1),\ldots,(k_n, L_n))$ it would, in practice, have to be stopped after a finite number of steps, say, at the Tth step. At that point, the information structure to be studied is defined: for each manager the center knows his original local observation (k for the center, μ_i for each manager i) together with the *accumulation* of messages received in the preceding steps of the process (the profit-maximizing demand messages for the center, the prices for each manager). In the classic discussion (of tatonnement processes), the action finally taken at the Tth step is in fact based on the messages most recently received and *not* on the entire accumulation of messages: the center, for example, allocates k in accordance with the final profit-maximizing demand messages (each manager's share equals his final profit-maximizing demand if these demands do not exceed k and otherwise falls short of his final demand, in some suitable way). Nevertheless, one can ignore the costs of memory and can seek the best decision rule assigning, at the final step, a value of $(k_1,\ldots, k_n)$ to the center's accumulated information and a value of L_i to each manager's accumulated information.

This was done, to begin with for the case of a *single* step ($T=1$), with production functions f quadratic and the parameter μ_i a pair (μ_{ik}, μ_{iL}):

$$f(k_i, L_i, \mu_i) = 2\mu_{ik}k_i + 2\mu_{iL}L_i - k_i^2 - L_i^2 - 2qk_iL_i.$$

The information structure, labeled "one-stage Lange–Lerner" (OSLL), is intended to capture the interchange of a "price mechanism" which terminates after one step. The center announces a price and the managers respond with profit-maximizing demands. The center's choice of price can only depend on k, since that is all the center knows at the start. Once the center knows what the profit-maximizing demands are, he can uniquely deduce, for each i, the quantity $\nu_i = \mu_{ik} - q\mu_{iL}$. It is then argued that nothing is lost if one says that the center in fact sends k *itself* to the managers and each manager sends ν_i; in the OSLL structure (Y, η), then, manager i, $i=1,\ldots,n$, knows $\eta_i(e) = (\mu_i, k)$ and the center (member $n+1$) knows $\eta_{n+1}(e) = (k, \nu_1,\ldots, \nu_n)$. Informally speaking, this viewpoint might rest on a model of technology in which only transmission is costly, all real numbers (and vectors of real numbers) can be transmitted exactly, and the dimension of the real message sent from i to j determines the cost of that transmission.

Note that given the OSLL structure, each manager has to choose a current value of L_i in ignorance of other managers' production functions, of the values other managers give their local decision variables, and of the quantity of the centrally allocated resource which he and others will receive. A team decision rule tells i what value of L_i to choose.

Assume all the random variables μ_{ik}, μ_{iL}, k to be independently distributed and assume that the distribution of μ_{ik} is the same for every i, as is the distribution of μ_{iL}. It is then verified that for the OSLL structure a certain team decision rule in fact satisfies that version of the person-by-person-satisfactoriness condition which is appropriate[14] when the decisions are to fulfill a linear constraint like $\sum k_i \leq k$. Hence, in view of the concavity and differentiability of ρ, that decision rule is best. The best expected payoff so obtained is then compared with the best expected payoff for several other structures.

In particular, it is found that the best expected payoff is not increased if each manager i sends not ν_i – which is a "contraction" of his local information – but rather sends the vector $\mu_i = (\mu_{ik}, \mu_{iL})$ itself, so that $\eta_{n+1}(e) = (k, \mu_1, \dots, \mu_n)$, while $\eta_i(e) = (\mu_i, k)$, $i = 1, \dots, n$, as before. One virtue of the "price mechanism" is then demonstrated: given the center's "price" message (here taken to be k itself), each manager's "profit-maximizing demand" response (here taken to be ν_i) cannot be improved upon. If there were allowed to be a *second* price announcement by the center, then the best possible (or "full-communication") payoff – i.e. $\max_a \rho(a, e)$ – could be attained for every e, since $(k, \nu_1, \dots, \nu_n)$ gives the center enough information to compute exactly the optimal shadow price (that price for which the profit-maximizing demands would comprise the optimal allocation). That is a property of the quadratic payoff function and is not in general true for other concave payoff functions.

A surprisingly rich assortment of further structures have been studied for the quadratic case – in a slightly generalized version – by Welch (1980). In the generalized version, there is no separate "center" member and each manager i has an endowment w_i of the "team" resource. Each manager has to choose a value of his local variable L_i and a final allocation k_i of the team resource, which may be more or less than the random variable w_i. For a feasible decision rule $\sum_{i=1}^{n} k_i = \sum_{i=1}^{n} w_i$. A linear function of i's information plays a role in several of the structures studied, namely, $\gamma_i = \nu_i - (1 - q^2)w_i$, where ν_i is defined as before.

Under the structure called "bilateral sampling", member i knows γ_{i-1}, γ_i, and γ_{i+1}. Under "pass the mean" he knows $(\bar{\gamma}_{i-1}, \gamma_i, \gamma_{i+1})$, where $\bar{\gamma}_i = (1/i)\sum_{j=1}^{i} \gamma_j$. Under the "Feldman Round",[15] he knows γ_i, γ_{i+1}, and p_{i-1}, where $p_i = \gamma_i / 2^{i-1} + \sum_{j=2}^{i} \gamma_j / 2^{i-j+1}$. For each of these structures, the optimal decision rule for i can be interpreted as a profit-maximizing rule for a certain shadow price π_i, whose definition varies from structure to structure.

In this study, as well as in the Groves–Radner studies, it is instructive to examine the performance – that is, the limit of the best expected payoff – of the various structures as the number of managers increases without limit. Given the assumed form for ρ, and the assumptions concerning identical distributions of the random variables, the move from an r-manager to an $(r+1)$-manager team is well defined. It may be possible to compute not only the limit of best

[14] See footnote 11.
[15] Inspired by Feldman (1973).

expected payoff but also the path of best expected payoff for an initial range of team sizes. In the Welch study it is found, for example, that "bilateral sampling" performs worse than "pass the mean" for small team sizes, but after $n = 8$, it quickly becomes far superior and stays so in the limit.

The Radner–Groves studies find that in the limit the penalty of the OSLL structure – the amount by which its best expected payoff falls short of best expected payoff under full communication – goes to zero as n increases without limit. One might view this fact as another virtue of "price-like" mechanisms. The results has been generalized [Arrow and Radner (1978)] to the case of a general concave function f.

Arrow and Radner (1978) find a similar asymptotic result for the case of general concave production functions f and for an information structure which can *not* be interpreted as a price mechanism. They study the structure wherein each manager conveys to the center a complete description of the manager's current function f. The center is to allocate each resource in a list of fixed resources (whose availability he knows) among the managers, and bases the allocation on full information. Each manager,, however, knows only his own function f and has to chose a value of his local decision variable in ignorance of other managers' functions and in ignorance of the total resource availabilities. As the number of managers increase without limit, the best expected payoff attainable under this structure converges to the best expected payoff attainable under completely share full information (i.e. the best expected payoff attainable when all managers as well as the center know all production functions as well as the resource availabilities). To put it simply, local ignorance becomes less and less damaging as the number of managers grows.

A similar asymptotic result has been shown by Groves and Hart (1982) to hold for a structure which it is again difficult to identify with a price mechanism but which appears considerably more appealing, with regard to informational costs, than the Arrow–Radner structure, with its full communication of production functions to the center. Groves and Hart study an "uninformed demand" information structure. Each manager sends to the center a demand for the centrally allocated resource (or for each of several centrally allocated resources). The demand is *not* based on any "price" message from the center but only on the manager's local information (his current production function). If the total demands do not exceed the current central resource availability then the demands are met exactly. If they do exceed it, then one of several rationing schemes are used. Formally, the information structure tells each manager only his own production function and tells the center the manager's demands (manager i's demand is some function of i's current production function), as well as the central resource availability (or availabilities). A team decision rule for this structure tells each manager what value to choose for his local decision variable and tells the center how much of the central resource (or resources) to allocate to each manager – it tells the center, that is to say, to fulfill the managers' demands when that is feasible and to apply some rationing scheme when it is not. To achieve the paper's asymptotic results it is not necessary to compute a team decision rule which is best (expected-payoff-maximizing) for each such uninformed-demand information structure, but only to study particular interesting rationing schemes and the team decision rules associated with them. It turns out that for some uninformed-demand structures and some choices of rationing schemes, as the number of managers increases without limit, the output of each manager (the value taken by his function f) converges almost surely to the highest attainable output (i.e. the output attainable when all decisions are based on fully shared complete information). In particular, an extremely simple rationing scheme suffices – a scheme in which the managers can be viewed as arriving, in an arbitrary sequence, at the pool of fixed resources. Each arriving manager takes what he wants from the pool until the pool is exhausted.

From the viewpoint of a designer who wants to compare one-step designs whose costs are explicit, these results – asymptotic or otherwise – about the merits of "price" mechanisms (in their single-interchange-of-messages form) are interesting but quite incomplete. Such a designer would really like to know, in weighing the classic claims for price mechanisms as a guide for choice among designs, whether "price" designs extract good performance (expected payoff) from the effort required to run them; or whether other designs would extract more from the same effort. One very special form of this question could in fact be studied using the tools of the Theory of Teams (i.e. the techniques which sometimes permit computation of best decision rules and best expected payoff).

In the Radner–Groves problem, one could confine attention to finite designs and could permit manager i to impose a "grid" on the set E_i of possible values of (μ_{ik}, μ_{iL}). He imposes on E_i, that is to say, a partitioning $\mathscr{P}^i$ composed of B_i sets. Similarly, the center partitions E_{n+1}, the set of possible k's, according to $\mathscr{P}^{n+1}$, composed of B_{n+1} sets. Consider a technology in which a design's cost is increasing in B_i, $i=1,\dots,n+1$, and depends on nothing else. Let E_i, $i=1,\dots,n$, be the non-negative quadrant of real two-space and let E_{n+1} be the non-negative real line. One can, in particular, consider a "$(\overline{B}_1,\dots,\overline{B}_n,\overline{B}_{n+1})$-grid" finite version of the OSLL structure, wherein the center knows, for each manager i, that the true value of ν_i lies in one of $\overline{B}_i$ sets, and each manager knows that the true value of k lies in one of $\overline{B}_{n+1}$ sets. The best decision rules for this structure can be found. They are obtained in a simple way from the best rules for the original "continuum" OSLL structure by substituting, respectively, $\mathscr{E}(\nu_i|\mu_i \in S_i)$ for $\mathscr{E}\nu_i$ and $\mathscr{E}(k|k \in T)$ for $\mathscr{E}k$, where S_i is a set in $\mathscr{P}^i$ and T is a set in $\mathscr{P}^{n+1}$. A similar substitution yields best expected payoff. One can then choose the $\overline{B}_i$ sets for each i, $i=1,\dots,n+1$, so that best expected payoff is not less than for any other $(n+1)$-tuple of partitionings, where the ith partitioning is a $\overline{B}_i$-fold partitioning, $i=1,\dots,n+1$. Finally, one can ask whether this best $(\overline{B}_1,\dots,\overline{B}_{n+1})$-grid finite version of the OSLL structure makes the best use of its effort. Is there any other structure, under which member i, $i=1,\ldots,n$, conveys to member $n+1$ one of $\bar{B}_i$ possible signals about his local environment and member $n+1$ sends out one of $\bar{B}_{n+1}$ signals about his, for which best expected payoff is higher? The answer is not yet known.[16]

3.5. *The polyhedral case*

One of the distressingly few attempts to model real organizations in the framework of the theory of teams is a study of a sales organization by McGuire (1963). In the simplest model there considered, each of n salesmen decides, in a given period, on an order, a_i, to be centrally produced and delivered to that salesman's location. Salesman i knows a current price, e_i (a random variable), at which an unlimited amount can be sold in his location. The unit production cost for $\sum a_i$, the sum of salesmen's orders, is 1 if $\sum a_i \leqq c$ (c, say, is a "normal-shift capacity"), but it is $1+k$ for any excess of $\sum a_i$ over c (k is, say, the extra unit cost of an "overtime" shift). Team payoff is then

$$\begin{aligned}\rho(a,e) &= \sum_{i=1}^{n} a_i(e_i-1) - k\max\left(0, \sum_{i=1}^{n} a_i - c\right)\\ &= \min\left(\sum_{i=1}^{n} a_i(e_i-1), \sum_{i=1}^{n} a_i(e_1-1-k)+ck\right),\end{aligned}$$

and A is non-negative real n-space. One may be interested, for example, in the structure (η, Y) for which $\eta_i(e)=e_i$ and in comparing it with the "centralized" structure for which $\eta_i(e)=e$. Consider a generalized form of the problem: $\rho(a,e)=\min(\rho_1(a,e),\ldots,\rho_g(a,e))$, where $\rho_k(a,e)$ is linear in the vector a for $k=1,\ldots,g$.

Suppose, further, that for any e, the (real-valued) actions are required to satisfy not only $a_i \leqq 0, i\in N$, but also linear constraints depending on e, namely,

$$\sum_{i=1}^{n} \lambda_i(e)a_i \leqq \mu_s(e), \qquad s=1,\ldots,S.$$

If E is finite then, not surprisingly, the best decision rules for any information structure (η, Y) can be found by solving an associated linear programming problem [J. Marschak and Radner (1971, ch. 5)].

[16] More ambitiously, one can add a further "effort" dimension, namely, "fineness of action implementation", to be discussed below in Section 4.3. One studies a finite version of the OSLL structure in which not only the numbers $\bar{B}_i$ are fixed, but also the number of possible team actions, ans asks whether any other finite structure characterized by the same $\bar{B}_i$'s can do better when the number of permitted team actions is kept the same.

Handbook of Mathematical Economics, vol. III, edited by K.J. Arrow and M.D. Intriligator
© 1985, Elsevier Science Publishers B.V. (North-Holland)

4. Contributions to organization design: Adjustment processes

4.1. General concepts and background

An adjustment process is a system of difference or differential equations in n variables, each variable associated with one of the n members – comprising the set N – of an organization which has to take actions in response to a changing environment. The variable associated with member i is a vector of messages sent to other members or possibly to i himself (a stored piece of information). In general, the equation associated with a member has the organization's environment as a parameter. In important cases the parameter is not the complete environment but rather the aspect of which that member has "private" knowledge – knowledge he gains through his own observation of the environment and *not* through messages received from others. The organization's action is an n-tuple, whose ith coordinates is called "member i's action". There may be a natural association between a member's action and his private knowledge: whoever is in charge of a certain coordinate of the organization's action "automatically", or very cheaply, has private knowledge of a certain aspect of the environment. Such "cospecialization of action and observation", as it has been called [J. Marschak and R. Radner (1971, ch. 4)], is a technological fact and partly determines the costs of carrying out an adjustment process.

We shall confine attention to difference-equation processes. Suppose again that new environments occur regularly, one time period apart. The first step of the adjustment process follows observation of the new environment. In practice, some finite number of steps – say, T – would have to be carried out. Following the Tth step, the organization takes an action which is a function of the values taken by the members' variables at the Tth step. This action is, then, a response to that environment which preceded the first step.[17]

Given an n-member organization with environment set E and action set A, a *temporally homogeneous* adjustment process[18] is the quadruple

$$\pi = \left(\mathcal{M}, m_0, f = (f^1,\dots,f^n), h\right).$$

Here $\mathcal{M}$ is a set called a *language*; m_0, an *initial message*, lies in $\mathcal{M}^{(n)}$, the n-fold Cartesian product of $\mathcal{M}$; f^i is, for every i in N, a function from $E \times \mathcal{M}^{(n)}$ to $\mathcal{M}$; and h – called the *outcome function* – is from $E \times \mathcal{M}^{(n)}$ to A, an action set. The quadruple defines, for any e in E, the difference equation system

$$m_t^i = f^i(e, m_{t-1}), \qquad t > 1, \quad \text{all } i \text{ in } N,$$

where $m_{t-1} \in \mathcal{M}^{(n)}$ is the n-tuple $(m_{t-1}^1,\dots,m_{t-1}^n)$.

[17] One may wish to add the requirement that T be small enough so that the action is taken before a new environment occurs.

[18] We consider here a variant of the formulation first given by Hurwicz (1960).

In a temporally non-homogeneous process, which we shall not consider, each function f^i would have a further argument, namely, the interger t. The variable m_t^i can be interpreted as a message formed by i at step t and sent to any member j who needs it in order to form his next message m_j^{t+1}. If T steps of the process are carried out following the environment e, then $h(m_T, E)$ becomes the organization's action in response to e.

Now suppose we are given n partitionings $\mathscr{P}^i$ on E, $i \in N$. Let the sets in $\mathscr{P}^i$ be indexed by the variable e_i, called the ith *environmental characteristic*. Then we call the process $(\mathscr{M}, m_0, f, h)$ *privacy-preserving relative to* $\{\mathscr{P}^i\}_{i \in N}$ if for every i, there is a function $\tilde{f}^i$ such that if e lies in the set of $\mathscr{P}^i$ indexed by $\bar{e}_i$, then

$$f^i(e, m) = \tilde{f}^i(\bar{e}_i, m), \qquad \text{all } m \text{ in } \mathscr{M}^{(n)}.$$

An action in A is the n-tuple $a = (a_1, \ldots, a_n)$. If there is an inevitable cospecialization of action and observation, so that *only* the member who takes, say, the action a_k can observe a certain environmental characteristic, then that means it is technologically impossible to operate a process, say π^*, which is privacy-preserving relative to partitionings wherein not k but some other member observes that characteristic. If such cospecialization is not inevitable but is extremely cheap, then the process π^* is not impossible but is forbiddingly expensive and perhaps not worth studying.

As for the selection of an action once a terminal message m_T has been reached, the function h determines a vector of functions $(h^1, \ldots, h^n)$ where $h^i(m, e)$, with $m \in \mathscr{M}^{(n)}$, is a value of a_i, $i \in N$. One may wish to require a similar privacy-preserving property with regard to action-choosing, i.e. for every i, there is a function $\tilde{h}^i$ such that when e is in the set of $\mathscr{P}^i$ indexed by $\bar{e}_i$, then $h^i(m, e) = \tilde{h}^i(m, \bar{e}_i)$ for all m in $\mathscr{M}^{(n)}$. This is achieved automatically if h is a function only[19] of m.

Suppose *both* f and h in the process $\pi = (\mathscr{M}, m_0, f, h)$ are privacy-preserving relative to $\{\mathscr{P}^i\}_{i \in N}$. Suppose the partitioning $\mathscr{P}^i$ describes the environmental observing done by member i. The possible actions $\hat{a}_i(e, T) = h^i(m_T, e)$ generally partition the set E more finely than the observational partitioning $\mathscr{P}^i$. The further refinement is due to the T interchanges of messages which occur. The messages which member i may receive from j are implied by the function f^i: i receives messages from $j \neq i$ if and only if there is some $\overline{m} = (\overline{m}^1, \ldots, \overline{m}^j, \ldots, \overline{m}^n) \in \mathscr{M}^{(n)}$, some $\overline{\overline{m}} = (\overline{m}^1, \ldots, \overline{\overline{m}}^j, \ldots, \overline{m}^n) \in \mathscr{M}^{(n)}$, with $\overline{\overline{m}}^j \neq \overline{m}^j$, and some $e \in E$ such that $f^i(e, \overline{m}) \neq f^i(e, \overline{\overline{m}})$. But $\hat{a}_i(e, T)$ does not depend on the entire T-step

[19]Then h is a "non-parametric" outcome function in the terminology of Hurwicz (1972). Formally, of course, one could let $\tilde{h}_i(m_{t-1}, e_i)$ be an element of the vector m_t^i. Member i, that is to say, keeps a running record (in the form of a message sent to himself) of the action he would take were the current step to be the final one, and this is, say, the last coordinate of the vector m_t^i. If the current step *is* the final one – if $t = T$ – then the action taken by i is given by the non-parametric function $h^i(m_T)$ which equals the last coordinate of m_T^i.

accumulation of messages received by i; that accumulation would generally partition E still more finely than does $\hat{a}_i(e, T)$. At each step, in the typical process, member i, knowing only e_i and m_{t-1}, is unable to reconstruct (does not remember) the sequence of messages which he has received since the first step.[20]

The abstraction just presented was inspired by the classic debates about the virtues of price or "competitive" mechanisms for resource allocation in an economy. Once classic claim was that an economy, with or without private ownership, could be operated at all – or at least could not achieve Pareto-optimality unless – the information repeatedly exchanged among its many members consists, for each commodity, of a price and an excess demand. The unthinkable alternative was some scheme which would require descriptions of members' preferences, endowments, and technologies to be gathered in a central place, where individual consumptions and productions are computed and then issued as instructions to the economy's members. Discussions since the 1930's have made clear that the choices are not quite so polarized. One can formulate schemes ("planning" mechanisms of various sorts), in which there is still a "center", which fall short of the unthinkable total centralization, but in which the center may yet receive more information about members' technologies and tastes than would be given by a classic sequence of utility- and profit-maximizing excess demands. The messages sent by the center, moreover, constrain members' actions more than prices alone constrain them in the classic scheme. To clarify the issues that arise in choosing among the rich variety of resource-allocating schemes which are in principle possible, one first needs an abstract concept fitting *any* scheme. The abstraction just given has served the purpose reasonably well in a number of studies of resource-allocating mechanisms. We shall consider several lines of study and shall relate each of them to the comparison of one-step designs.

4.2. *The equilibrium study of adjustment processes*[21]

Hurwicz's original paper (1960) presents a process, the "quasi-competitive" process, which captures, in one form, the message interchanges that precede a competitive equilibrium. Each member's message is a *set* of "resource-flow matrices"; such a matrix describes all trades and productions in the economy. To form a new message, a member i finds the set, say S_i, of those matrices which would leave him at least as well off as any matrix in the intersection of all members' previous messages. His new message is not S_i itself, but rather the smallest cone containing S_i. That is of interest because the smallest cone generally

[20] The formulation above does, however, *permit* complete (or partial) accumulation of messages. If the language $\mathscr{M}$ is sufficiently rich, i's message m_t^i could contain a complete summary of the sequence of messages i has so far received.

[21] Much of the literature dealing with this topic is surveyed in Hurwicz (1973); we shall explicitly mention here only a small part of the literature.

partitions the set of possible previous message n-tuples more coarsely than does the set S_i itself and therefore is, in a certain sense, an informationally less costly message. If the environment set (the set of possible tastes, technologies, and endowments) has classic properties, then for every initiating environment the quasi-competitive process yields *at equilibrium* an n-tuple of messages – i.e. sets of resource-flow matrices – whose intersection contains only the competitive equilibria of the economy defined by that environment.

In this approach to the study of processes, then, one looks not at the messages reached after some finite number T of steps, but rather at the *equilibrium messages*. To be precise, one now no longer defines a process as a quadruple but merely as the triple,

$$\pi = (\mathcal{M}, f, h);$$

the initial message m_0 is omitted. Then, for e in E, the equilibrium messages comprise a set $D(e) \subseteq \mathcal{M}^{(n)}$, where $m = (m^1, \dots, m^i, \dots, m^n) \in D(e)$ implies

$$f^i(e, m) = m^i, \quad \text{all} \quad i \in N.$$

The action generated by the process in response to e lies in a set, namely, the set $\mathcal{D}(e)$ of *equilibrium outcomes or actions*,

$$\mathcal{D}(e) = \{a \in A : h(m, e) = a \text{ for some } m \text{ in } D(e)\}.$$

We shall say that the process *covers E with regard to equilibria* if, for every e in E, $D(e)$ is not empty. We shall then also call it a *process on E*.

One simply lays aside the question of which equilibria, if any, will in fact be reached for specific initial messages, and the question of how long (how many steps) this might take.[22] In the spirit of classic debates about competitive resource allocation, one investigates only the achievements of a process at equilibrium. For the case of resource-allocating processes for a certain set E of economic environments, one may be interested, for example, only in processes which (1) cover E

[22] There is one process for which equilibrium is achieved in one step. This is the privacy-preserving process wherein each member i announces the characteristic e_i to the others. For this process, the language is $\mathcal{M} = \cup_i E_i$, where E_i is the set of possible values of e_i, and $f^i(e, m) = \tilde{f}^i(e_i, m) = e_i$. Following any initial message m_0, one has for all i, $m_1^i = f^i(e, m_0) = \tilde{f}^i(e_i, m_0) = e_i = \tilde{f}^i(e_i, m_1) = m_2^i$. One-step processes are considered in a general manner below.

with regard to equilibria and (2) for every e in E achieve Pareto optimality, relative to the economy defined by e, at *every* equilibrium outcome – i.e. every element of $\mathscr{D}(e)$. One can then ask whether a certain process within the class of processes satisfying (1) and (2) for some E is informationally inferior to some other process in this class – whether, for example, the quasi-competitive process, which satisfies (1) and (2) for the set E of classic economies, is inferior to any other process satisfying (1) and (2) for the same[23] set E.

For some purposes it is useful to provide the following interpretation for the equilibrium study of a process $\pi = (\mathscr{M}, f, h)$ on E which is used by an n-person organization and is privacy-preserving relative to partitionings $\{\mathscr{P}^i\}_{i \in N}$ on E with sets indexed by the variables $\{e_i\}_{i \in N}$. Let there be a *center* – an $(n+1)$st member – who announces to all the members in N (members $1, \dots, n$) an arbitrary non-repetitive sequence of trial messages $m = (m^1, \dots, m^i, \dots, m^n)$ belonging to $\mathscr{M}^{(n)}$. After the current message m is announced, every member i in N examines the current value of his private environmental characteristic e_i to see whether $\tilde{f}^i(e_i, m) = m^i$ [where, as before, $\tilde{f}^i(e_i, m) = f^i(e, m)$]. If so, member i sends a "Yes" signal to the center. If the center receives n "Yes" signals, then an equilibrium message, say $\overline{m} \in \mathscr{M}^{(n)}$, has been found; the center then computes the value of $h(\overline{m}, e)$ for the current environment and this becomes the organizational action taken in response to e. [If h determines a privacy-preserving n-tuple $h^1, \dots, h^n$, and every a in A is an n-tuple $(a_1, \dots, a_n)$ of members' actions, then each member i computes the new action $h^i(\overline{m}, e_i)$ and takes that action.] If and only if the center receives less than n "Yes" signals, a new trial message m is announced.

One ignores, then, the fact that sometimes a large number of trials may be needed to reach an equilibrium message and sometimes a small number. Doing so, one can simply treat as costly the "size" of the set $\mathscr{M}^{(n)}$ – i.e. some suitable measure of the size of the collection of messages which the center must be prepared, in the worst case, to try out. If $\mathscr{M}^{(n)}$ is finite, the size is the number of elements in $\mathscr{M}^{(n)}$. If $\mathscr{M}^{(n)}$ is a finite-dimensional vector space, then its dimension is a measure of size.

The dimension approach is followed in a number of studies of resource-allocating mechanisms for economies [Mount and Reiter (1974) and Reiter (1974a,

[23] Hurwicz's quasi-competitive process is informationally superior to the "greed" process with regard to coarseness of the partitioning on the possible message n-tuples $m_t = (m_t^1, \dots, m_t^n)$ induced, for any t and any fixed e, by the n-tuple m_{t+1}. In both processes, messages are sets of resource-flow matrices and the outcome function assigns to an n-tuple of sets another set, namely, their intersection. In the greed process, the entire set S_i, not the smallest cone containing it, is sent by i. On the other hand, the greed process achieves Pareto-optimality at equilibrium for a wider class of economies than the classic set E. It is not yet settled whether there is another process, in the class of processes satisfying (1) and (2) for the classic E, which is informationally *superior* – in the same precise sense – to the quasi-competitive processes.

1974b)]. For these mechanisms $\mathcal{M}^{(n)}$ is a continuum. These studies are well surveyed by Reiter (1977) and we shall not re-survey them here. In all of these studies, certain *smoothness* conditions are placed on the functions f and h. Without such conditions, a process with a many-dimensional $\mathcal{M}^{(n)}$ could be replaced by a process with a lower-dimensional $\mathcal{M}^{(n)}$ having the same equilibrium outcomes for any e – by coding, in a "non-smooth" way, every many-dimensional message as a one-dimensional one (e.g., n real numbers can be hidden in a single decimal number with n digits). Such smuggling of many dimensions into one dimension is felt to introduce certain additional costs, so that the apparent cheapness of the single dimension is illusory. These costs are not explicitly modeled but perhaps have to do with the fact that in practice a continuum of messages would have to be approximated by a finite collection of messages and a process with "smuggling" would be hard to approximate, since a small error in the one-dimensional message would lead to a very large error in the "smuggled" n-dimensional message obtained after decoding.

A specific technology in which such costs would arise for non-smooth processes has not so far been presented in the discussions. Instead the general question asked has been: Given a certain class E of economic environments (economies), and given that the action set A comprises the economy's possible final consumptions, what is the lowest dimension of $\mathcal{M}$, where $\mathcal{M}$ is a Euclidean space, for which there is a "smooth" process $(\mathcal{M}, f, h)$, privacy-preserving with respect to the natural partitions[24] $\mathcal{P}^i$ on E, covering E with regard to equilibria, and achieving Pareto-optimality at every equilibrium outcome?[25] In particular, is the lowest dimension that required by the competitive process, suitably defined? For the class E of classic economies, and various specific versions of smoothness, the answer to the second question has been shown to be "Yes" [Mount and Reiter (1974) and Hurwicz (1977)].

The simplest and earliest of these results is due to Hurwicz.[26] Consider an n-person exchange economy with L commodities. In the price mechanism a message consists of n L-dimensional trade vectors and $L-1$ prices. Now let $n=2$ and $L=3$ and let each of the two persons have a utility function on the commodity triples with a quadratic term and a linear term. (We shall be considering the very same economy for $n=L=2$ in Section 4.3.3 below.) It can readily be shown that if a mechanism had as its messages the two trade vectors plus *less* than 2 $(=L-1)$ auxiliary variables and if it covered the set of all possible economies (endowments and linear-quadratic utility functions), that would imply a one-to-one mapping from the message space onto an "economy-

[24] In the natural partitions, each agent knows his own endowment, preferences, and technology.

[25] If the sets $\mathcal{M}$ are not restricted to Euclidean spaces, then dimension is replaced by a more general concept.

[26] The central technique of proof first appeared in Hurwicz (1972), but the discussion there did not deal with economies. The most general version of the result sketched here appears in Hurwicz (1977).

space" of higher dimension; but such a mapping cannot be Lipschitzian. So if "smooth" is taken to mean that the mapping from messages to environments is Lipschitzian (or, more precisely, contains a Lipschitzian selection), then there exists no smooth mechanism which covers all linear-quadratic two-person exchange economies, achieves (at equilibrium) what the price mechanism achieves, and has a message space consisting of trade vectors and auxiliary variables and having a dimension lower than that of the price mechanism's message space. *A fortiori* there exists no such mechanism covering a larger class of two-person economies than the linear-quadratic class. The result extends to n persons and to message spaces more general than the trade-vectors-plus-auxiliary-variables spaces.

Results of this sort are clearly important if one wants to assess in a preliminary way the classic informational claims made for the competitive mechanism – claims long unsubstantiated since it appeared too difficult to study them with rigor. This motivation amply justifies the approach even though it is incomplete from the point of view of a *designer* of organizations. The designer would be concerned with a specific transmission, observing, and action-taking technology. He *would* be concerned by the changing number of trials required, from one environment to the next, to reach an equilibrium message. He would be concerned with costs not captured in the size of $\mathscr{M}$, e.g. the action-taking effort, perhaps measured by the size of the set of possible equilibrium outcomes (actions). The "trial message" procedure, moreover, is generally *not* a well defined one-step design in our earlier sense.

Some processes, however, *are* one-step designs. To define them – and for other purposes as well – it is first convenient to replace our definition of process by a slightly more compact one. The new definition is suggested by the "trial message" interpretation just given. We define a process on E, privacy-preserving with respect to partitionings $\{\mathscr{P}^i\}_{i\in n}$ on E, with each set in $\mathscr{P}^i$ indexed by a value of $e_i \in E_i$, as a triple

$$P = (M, g, h),$$

where M is a language (whose elements we may think of as a center's trial announcements) g is an n-tuple of functions $(g^1,\dots,g^i,\dots,g^n)$, where g^i is a function on $E_i \times M$ taking[27] two integer values, namely, zero and 1; and h is from $M \times E$ to A. For a privacy-preserving process $\pi = (\mathscr{M}, f, h)$ as defined so far, one obtains the new form $P = (M, g, \bar{h})$ by letting $M = \mathscr{M}^{(n)}$, letting

$$g^i\left[e_i,(m^1,\dots,m^i,\dots,m^n)\right] = 0 \quad \text{if} \quad \tilde{f}^i\left[e_i,(m^1,\dots,m^i,\dots,m^n)\right] = m^i,$$
$$= 1 \quad \text{otherwise},$$

[27]Again, E_i denotes the set of possible values of e_i.

and letting $\bar{h}=h$. The new form, then, simply suppresses the fact that an announced trial message is an n-tuple of individual messages. The equilibrium messages and actions for the new form are the same as those for the associated original form: for $P=(M, g, h)$, $m \in M$ is an equilibrium message for e, with characteristics $(e_1,\dots,e_n)$, if and only if $g^i(e_i, m)=0$, all i in N. We shall, through the rest of the present section and the next one (Section 4.3) consider only processes $P=(M, g, h)$ as just defined.

Now suppose that a process $P=(M, g, h)$ on E is privacy-preserving with respect to the partitionings $\{\mathscr{P}_i\}_{i \in N}$, where for each i the sets in $\mathscr{P}_i$ are indexed by the variable e_i and $E_i \equiv \{e_i\}$. We shall say that the process P is a *one-step* process if (i) there exists for each i a set T_i of non-empty sets whose union equals E_i and (ii) there exists a one-to-one mapping γ from M to $\overline{M} \equiv T_1 \times \cdots \times T_n$ such that for each i and for every m in M, $g^i(e_i, m)=0$ if and only if $e_i \in t_i$, where $(t_1,\dots,t_n)=\gamma(m)$. One can interpret such a process as follows: Person i observes his local environment e_i and determines a set t_i in T_i to which it belongs. This set is communicated to a center. The center finds that m for which $\gamma(m)=(t_1,\dots,t_n)$; then the action $h(m, e)$ is taken. *A sequence of trial announcements is not needed*. Suppose we now add an explicit statement as to who computes and takes the action $h(m, e)$. If h is privacy-preserving, then the action has n parts, each the responsibility of one member, who computes and takes that part of the action once the center has announced m; if not, then the entire action is computed and taken by the center. To carry out the process in this way is to operate a one-step design as we have defined it. In the terminology of Section 2.1, a one-step process defines a one-step design covering E with respect to A, where A is any set containing the set of actions $\{a: a=h(m, e)$ for some m in M and some e in $E\}$.

To any arbitrary process $P=(M, g, h)$, there corresponds a one-step process which we shall call the *standard form of P*; it is denoted $P^*=(M^*, g^*, h^*)$ and is privacy-preserving with respect to the same partitionings as P. A message in M^* is an *n-tuple of sets*. We have

$$M^* = \left\{(R_1,\dots,R_n): \text{for } i=1,\dots,n,\ R_i=\mu_g^i(e_i) \text{ for some } e_i \in E_i\right\}, \tag{4.1}$$

where

$$\mu_g^i(e_i) \equiv \left\{m \in M: g^i(e_i, m)=0\right\}. \tag{4.2}$$

Further, for any $\bar{m}=(R_1,\dots,R_n) \in M^*$,

$$\begin{aligned} g^{*i}(e_i, \bar{m}) &= 0 \quad \text{if} \quad R_i=\mu_g^i(e_i), \\ &= 1 \quad \text{otherwise}, \end{aligned} \tag{4.3}$$

and, for any e in E,

$$h^*(\overline{m}, e) = h(m^e, e), \tag{4.4}$$

where m^e denotes an element uniquely selected from the set $\bigcap_{i \in N} R_i$.

The standard form $P^* = (M^*, g^*, h^*)$ has the following interpretation: Member i observes the environment to determine the current value of e_i, say $\bar{e}_i$. He then sends to the center a message, which, is a *set*, namely, the set $\mu^i_g(\bar{e}_i)$ of all those messages – all those elements of the original language M – for which he would, in the *original* trial-message procedure, say "Yes" when e_i has the value $\bar{e}_i$. The center examines the n such sets received and selects a message, in the original set M, which lies in all of these n sets.[28] For the current environment this is an equilibrium message of the original process $P = (M, g, h)$ and is also an equilibrium message (with respect to g^*) of the standard-form process P^*. (If the original process covers E with respect to equilibria, then the intersection of the n sets cannot be empty.) An action is then assigned to the equilibrium message so found; it must belong to the set of equilibrium actions for e in the original process P. If we interpret P^*, then, in the sending-of-messages-to-the-center manner, and if we add an explicit statement as to who determines the equilibrium action, then we have well defined a *one-step design, for n members plus a center, which realizes the process* $P = (M, g, h)$. Given any e, the design generates as an output an action which is an equilibrium outcome for P.

As an informally sketched example, consider an adjustment process of a "Lange–Lerner" price type with n managers and a price-announcing center who allocates an organizational resource. In the original process,[29] the center announces a new price at each step t as a function of profit-maximizing excess demands received at $t-1$. Only at equilibrium are the right prices found; it is its achievement at equilibrium that makes the process worth studying and worth comparing with others which achieve the same allocation at equilibrium. In the *standard* form of the process, each manager, after observing the current environ-

[28] To see that the standard form is indeed a one-step process as defined above, we have to exhibit the mapping γ. For any message $m^* = (S_1, \ldots, S_n) \in M^*$ let $\gamma(m^*) = [\gamma_1(m^*), \ldots, \gamma_n(m^*)]$, where $\gamma_i(m^*) \equiv \{e_i : g^i(e_i, m) = 0 \text{ for all } m \in S_i\}$. The mapping is one-to-one. Suppose not. Then M^* contains an $\overline{m}^* = (\bar{S}_1, \ldots, \bar{S}_n)$ and an $\overline{\overline{m}}^* = (\bar{\bar{S}}_1, \ldots, \bar{\bar{S}}_n)$, such that for some i, $\bar{S}_i \neq \bar{\bar{S}}_i$ and $\gamma_i(\overline{m}^*) = \gamma_i(\overline{m}^{**})$. Suppose (i) $\bar{S}_i = \{m \in M : g^i(\bar{e}_i, m) = 0\}$ and (ii) $\bar{\bar{S}}_i = \{m \in M : g^i(\bar{\bar{e}}_i, m) = 0\}$. Since $\bar{S}_i \neq \bar{\bar{S}}_i$, $\exists \tilde{m} \in M$ such that (iii) $\tilde{m} \in \bar{S}_i$ but (iv) $\tilde{m} \notin \bar{\bar{S}}_i$. Now (iv) means that $g^i(\bar{\bar{e}}_i, \tilde{m}) \neq 0$; that means in turn (since $\tilde{m} \in \bar{S}_i$) that $\bar{\bar{e}}_i \notin \gamma_i(\overline{m}^*) = \{e_i : g^i(e_i, m) = 0 \text{ for all } m \text{ in } \bar{S}_i\}$. On the other hand (ii) implies that $\bar{\bar{e}}_i \in \gamma_i(\overline{\overline{m}}^*) = \{e_i : g^i(e_i, m) = 0 \text{ for all } m \text{ in } \bar{\bar{S}}_i\}$. That contradicts the statement that $\gamma_i(\overline{m}^*) = \gamma_i(\overline{m}^{**})$.

[29] The original process can be defined formally so that it has the "(M, g, h)" form. An element of M is a trial announcement of managers' local actions *and* prices. One can also define the process in the "$(\mathscr{M}, f, h)$" form. An element of $\mathscr{M}$ (a message by member i) is a *set* of prices and managers' actions; i proposes a value of his own action by announcing the set defined by that value and by all possible values of the other variables.

ment – determining his current technology – computes a *demand schedule*, giving the profit-maximizing demand he would announce, for that technology, at *every* possible price in the original process. The center receives the n demand schedules and uses them to find a price at which total excess demand would be zero (or possibly negative); he then gives each manager i an allocation equal to i's (profit-maximizing) demand at this equilibrium price. Informed of his allocation, each manager chooses a (profit-maximizing) value of whatever local action variables are in his charge. A one-step design, then, has achieved the equilibrium of the price process.

The standard form requires, in general, a richer language than the original form,[30] but since it defines a one-step design, with no mystery as to a terminal step, the assessment of its cost, and its payoffs over successive time periods, can proceed.

4.3. *Discrete processes*[31]

4.3.1. *Introduction*

The assumption that a process $P = (M, g, h)$ has a *countable* language M permits mathematically distinct approaches to the assessment of its costs and payoffs. The language may be not only countable but *discrete*. A discrete language lies in a metric space and for each element of the language there is a neighborhood containing that element but no other element. Among discrete processes those with a *finite* language are of particular interest.

In the present section we mainly consider processes with discrete languages. But some remarks apply as well to the larger class of countable processes and some only to the smaller class of finite processes. We shall refer to any process whose language is *not* countable as a *continuum* process.

From a technology-modeling point of view, one many argue that discreteness is realistic: it is not possible to send any one of a continuum of messages over a transmission device found in the real world. If the continuum is the real line, for example, then any number to be sent has to be rounded off to a pre-selected number of digits. A further reason to study countable or discrete processes is to see what results from the study of continuum processes have natural counterparts in the countable or discrete case. In particular, are there counterparts to the finding that there are no "smooth" continuum processes which achieve what the price mechanism achieves but have a message space of lower dimension?

[30] An example is given in footnote 36, below.

[31] The problems and results summarized in this section are based on joint work of L. Hurwicz and the author [Hurwicz and Marschak (1984)]. An early remark on a finite counterpart to the smoothness conditions of the "continuum" literature considered above is found in Hurwicz (1972, p. 314).

We start by considering any process $P = (M, g, h)$ which

$$\left.\begin{array}{l}\text{(a) has a countable } M\\ \text{(b) is privacy-preserving relative to partitionings } \{\mathscr{P}^i\}_{i\in N}\\ \qquad \text{on } E\text{, where each set in } \mathscr{P}^i \text{ is indexed by a value of the}\\ \qquad \text{variable } e_i\text{, whose possible values comprise the set } E_i\\ \text{(c) covers } E \text{ [for every } e \text{ in } E\text{, there is an } m \text{ in } M\\ \qquad \text{for which } g^i(e_i, m) = 0\text{, all } i]\\ \text{(d) has an outcome function } h \text{ whose domain is } M \text{ and not } M\times E\end{array}\right\}. \tag{4.5}$$

We assume further that

$$\left.\begin{array}{l}\text{the index } n\text{-tuple } (e_1,\dots,e_n) \text{ uniquely determines } e\text{, i.e.,}\\ \text{for every } n\text{-tuple } (T_1,\dots,T_n) \text{ with } T_i\in\mathscr{P}^i\text{, all } i \text{ in } N,\\ \bigcap_{i\in N}T_i \text{ is a singleton}\end{array}\right\}. \tag{4.6}$$

Now (4.6) is satisfied if the variable e is identical with the n-tuple $(e_1,\dots,e_n)$ and if

$$E = E_1 \times E_2 \times \cdots \times E_n \tag{4.7}$$

where, as before, E_i denotes the set of possible values of the variable e_i.

We consider in the rest of the present section (Section 4.3) only triples $(E, P, \{\mathscr{P}^i\}_{i\in N})$ satisfying (4.5) and (4.7).

4.3.2. *Realizing a discrete process in one step*

It will be useful to define, for any process $P = (M, g, h)$ on E,

$$\left.\begin{array}{l}\sigma_g^i(m) \equiv \{e_i\in E_i : g^i(e_i, m) = 0\}\\ \sigma_g(m) \equiv \prod_{i\in N}\sigma_g^i(m)\\ \Sigma_g^i \equiv \{S\subseteq E_i : S \text{ is non-empty; } S = \sigma_g^i(m) \text{ for some } m \text{ in } M\}\end{array}\right\}. \tag{4.8}$$

If E_i is a set in one-dimensional Euclidean space, then every set $\sigma_g(m)$ is the intersection of E with a countable union of rectangles, where each rectangle has dimension n or lower.

We turn now to a two-person organization whose environment set $E = E_1 \times E_2$ is a closed rectangle in non-negative real two-space with one corner at the origin;

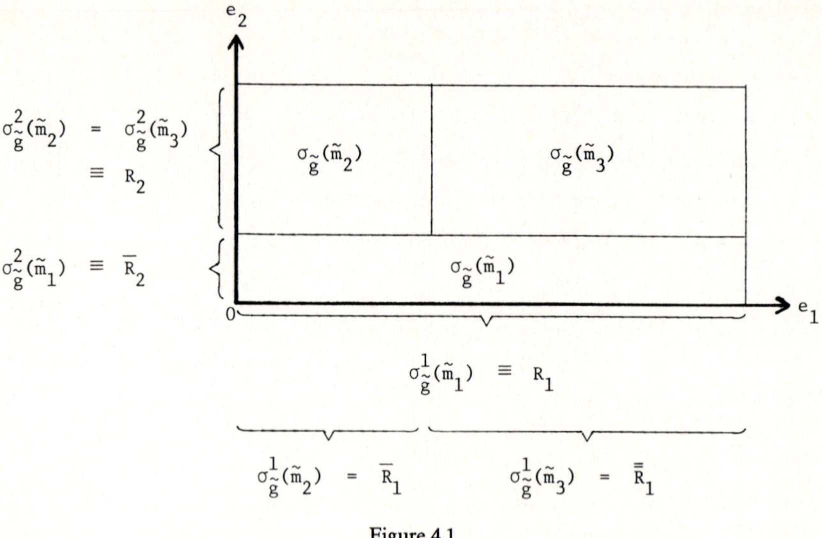

Figure 4.1

E_i $(i=1,2)$ is a closed interval of the non-negative real line, with zero its smallest element. The two figures which follow portray two privacy-preserving processes on E, i.e. two triples, $(\tilde M, \tilde g, \tilde h)$ and $(\bar M, \bar g, \bar h)$. Process $\tilde P = (\tilde M, \tilde g, \tilde h)$ has a language $\tilde M$ containing just three messages: $\tilde m_1, \tilde m_2, \tilde m_3$. Process $\bar P = (\bar M, \bar g, \bar h)$ has a language $\bar M$ with just four messages: $\bar m_1, \bar m_2, \bar m_3, \bar m_4$. Assume that process $\tilde P$ obeys the condition that "$m' \neq m''$ and $\sigma_{\tilde g}(m') \cup \sigma_{\tilde g}(m'')$ is a rectangle" implies "$\tilde h(m') \neq \tilde h(m'')$".[32] In Figure 4.1 the three interior clcsed rectangles portray the three sets $\sigma_{\tilde g}(\tilde m_1), \sigma_{\tilde g}(\tilde m_2), \sigma_{\tilde g}(\tilde m_3)$. The indicated closed intervals portray the sets $\sigma^i_{\tilde g}(\tilde m_k)$ $(i=1,2;\ k=1,2,3)$, which, for brevity, are called $R_1, \bar R_1, \bar{\bar R}_1$ and $R_2, \bar R_2$. The interior closed rectangles and closed intervals of Figure 4.2 portray the analogous sets for process $\bar P$; there the interior intervals are denoted $T_1, \bar T_1$ and $T_2, \bar T_2$. In the notation of (4.8), $\Sigma^1_{\tilde g} = \{R_1, \bar R_1, \bar{\bar R}_1\}$, $\Sigma^2_{\tilde g} = \{R_2, \bar R_2\}$; $\Sigma^1_{\bar g} = \{T_1, \bar T_1\}$, $\Sigma^2_{\bar g} = \{T_2, \bar T_2\}$.

Now we can associate with process $\bar P$ a one-step process which we shall call the *observational-report form* of the original process. In this form the language consists of *pairs of sets*, namely the four pairs $\{(T_1, T_2), (\bar T_1, T_2), (T_1, \bar T_2), (\bar T_1, \bar T_2)\}$. Person i observes in which of the closed intervals, T_i or $\bar T_i$, the current

[32] Clearly if a process $P = (M, g, h)$ has two messages, m' and m'', for which $h(m') = h(m'')$, while at the same time $\sigma_g(m') \cup \sigma_g(m'')$ is a rectangle, then the two messages can be consolidated, i.e. there exists another privacy-preserving process on E, with the same equilibrium actions as P and with a language smaller than M if M is finite.

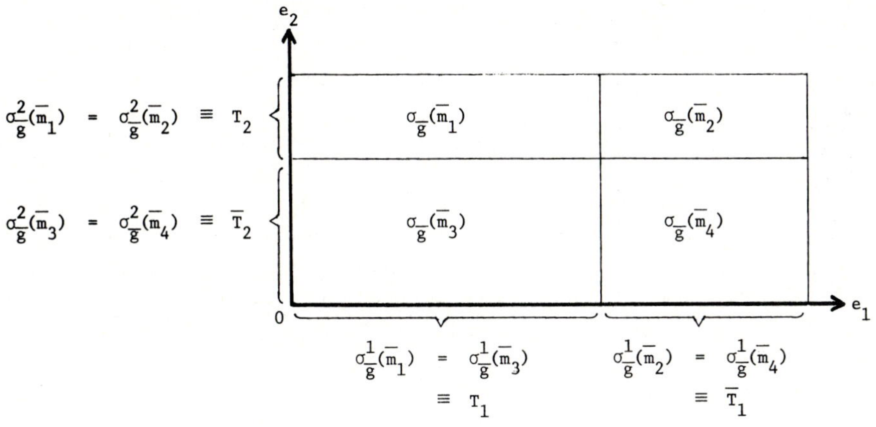

Figure 4.2

local environment e_i lies and announces that interval to the center. If e_i lies on the boundary of two intervals, he announces either one. The center next finds that m in M such that $\sigma_{\bar{g}}(m)$ is the closed rectangle which is the cartesian product of the two intervals. The action $\bar{h}(m)$ is then taken, where $\bar{h}$ is the outcome function in the original triple $(\bar{M}, \bar{g}, \bar{h})$.

For any e in E the equilibrium actions for the one-step observational-report form are exactly those of the original process, i.e. the one-step process realizes the original process $\bar{P}$. Note that the observational-report form has a four-element language, a language of the same size as that of the original process.

We may define the observational-report form for any process $P = (M, g, h)$. It is the triple $\hat{P} = (\hat{m}, \hat{g}, \hat{h})$, where

$$\left.\begin{array}{l}\text{(a) } \hat{M} = \left\{(S_1,\ldots,S_n) : S_i \in \Sigma^i_g,\ i=1,\ldots,n\right\} \\ \text{(b) } \hat{g}^i\left[(e_i,(S_1,\ldots,S_n)\right] = 0 \quad \text{if} \quad e_i \in S_i \\ \qquad\qquad\qquad\qquad\quad\ = 1 \quad \text{otherwise} \\ \text{(c) } \hat{h}\left[(S_1,\ldots,S_n)\right] = h(m), \text{ where } m \text{ is the only element of } M \\ \qquad \text{satisfying } \sigma_g(m) = \prod_{i=1}^n S_i \end{array}\right\}. \tag{4.9}$$

But the observational-report form of a given process P may not exist, since (c) may not be satisfiable for some n-tuple $(S_1,\ldots,S_n)$. The observational-report form exists [i.e. (c) can always be satisfied] if and only if for *every* n-tuple $(S_1, \ldots, S_n)$ with $S_i \in \Sigma^i_g$, all i, there is one and only one $m \in M$ such that $\sigma_g(m) = \prod_{i=1}^n S_i$. If the observational-report form of P does exist then it realizes P.

For process $\tilde{P}$ of Figure 4.1, the observational-report form, as just defined, does not exist, since e.g. there is no element m in $\tilde{M} = \{\tilde{m}_1, \tilde{m}_2, \tilde{m}_3\}$ such that $\sigma_{\tilde{g}}(\tilde{m}) = R_1 \times R_2$. It is true that we can depart from the observational-report form as just defined, and construct another one-step process which realizes process $\tilde{P}$, by *discarding* the set R_1 from the pairs in the language. For that process – call it $P^* = (M^*, g^*, h^*)$ – we have $M^* = \{(\overline{R}_1, R_2), (\overline{R}_1, \overline{R}_2), (\overline{\overline{R}}_1, R_2), (\overline{\overline{R}}_1, \overline{R}_1)\}$, while g^{*i} and h^* are defined analogously to $\hat{g}^i$, $\hat{h}$ in (a) and (c) of (4.9). But that process, while indeed a one-step process realizing $\tilde{P}$, has *four* elements in its language, whereas $\tilde{P}$ itself has only three. There is, in fact, *no* privacy-preserving one-step process with a three-element language which realizes process $\tilde{P}$ [assigns to each (e_1, e_2) in E the equilibrium actions which process $\tilde{P}$ assigns to that (e_1, e_2)]. Process $\tilde{P}$ covers E with the three closed rectangles of Figure 4.1 and assigns a distinct outcome to each rectangle. To construct a one-step process realizing process $\tilde{P}$, we must choose subsets of E_1 and E_2 such that the Cartesian product of each subset pair is contained in one of the rectangles and each point in E is contained in one of the Cartesian products. But every collection of subset-pairs with that property contains more than three pairs.[33]

Summarizing the basic differences between process $\tilde{P}$ and process $\overline{P}$:

(i) Process $\overline{P}$ has an observational-report form, as defined in (4.9), in which *every* set in Σ^i_g is a possible message sent by i. The observational-report form of $\overline{P}$ realizes $\overline{P}$ in one step and has a language exactly as large as that of $\overline{P}$ itself.

(ii) There is no one-step process which realizes process $\tilde{P}$ and has a language no larger than that of $\tilde{P}$ itself.

To put it another way, both of the original processes $\tilde{P}$ and $\overline{P}$ cover E with regard to equilibria, assigning certain action sets to points of E.[34] The assignments made in process $\tilde{P}$ might be superior to those made in process $\overline{P}$, with regard to some appropriate measure of performance,[35] even though $\tilde{P}$ has a

[33] If the collection contains three or fewer subset pairs, then for either E_1 or E_2 there is exactly one subset. Suppose that is true for E_1. Then there is no subset S of E_2 such that $E_1 \times S$ is contained in the top left rectangle. Suppose it is true for E_2. Then there is no subset S of E_1 such that $E_2 \times S$ is contained in the bottom rectangle.

[34] To points on the boundary of two rectangles a set containing more than one action is assigned. For process $\overline{P}$, a point on the boundary shared by the two lower rectangles is assigned the action set $\{\overline{h}(\overline{m}_3), \overline{h}(\overline{m}_4)\}$.

[35] See Section 4.3.3 below.

smaller language. But unless the language size is increased, the equilibrium actions of process $\tilde{P}$ cannot be realized in one step; it requires a sequence of trial announcements – sometimes as many as three – before an equilibrium message is found. The equilibrium actions of process $\overline{P}$, on the other hand, *can* be realized in one step, by transmission of observational reports to the center, without changing the language size.[36]

Now what *general* property of a process, going beyond the finite Euclidean space of the preceding two illustrative processes, makes it behave like process $\overline{P}$ and not like process $\tilde{P}$? The answer is provided by the following definition; the definition abstracts from the visual intuition that a process like $\overline{P}$ imposes a *grid* on E while a process like $\tilde{P}$ does not:

Definition

A process $P = (M, g, h)$ on E is said to have a *grid structure* if

$$\left.\begin{array}{l}\text{for every } i \text{ in } N \text{ no set } S \text{ in } \Sigma^i_g \text{ is contained in the union}\\ \text{of some sets in } \Sigma^i_g, \text{ all of them distinct from } S.\end{array}\right\}. \tag{4.10}$$

(Equivalently, P has a grid structure if for every i in N and every S in Σ^i_g, there exists an element $x \in S$ such that $x \notin \overline{S}$ for all $\overline{S} \in \Sigma^i_g$ with $\overline{S} \neq S$.)

Clearly process P has a grid structure and process $\tilde{P}$ does not. If E is a subset of a metric space, then a process $P = (M, g, h)$ on E has a *non-overlapping* grid structure if for all i any two sets S, S' in Σ^i_g are either disjoint or have only boundary points in common, where every neighborhood of a point in the boundary of the two sets contains points in both sets. Process $\overline{P}$ has a non-overlapping grid structure. The discussion which now follows would be greatly simplified if we confined attention to non-overlapping grid-structure processes on subsets of metric spaces. But the question of when a one-step realization is possible and what language size this requires is important enough to merit a more general treatment.

[36] For both processes $\tilde{P}$ and $\overline{P}$ there is another one-step form, namely the standard form, defined in (4.1) to (4.4). But that requires a larger language than the observational report form. Thus in the standard form of process $\tilde{P}$ there are three possible messages for person 1: (i) the message $(\tilde{m}_1, \tilde{m}_2)$ to be sent if e_1 lies in the interior of $\overline{R}_1$ [i.e. for such an e_1, the set $\mu^1_{\tilde{g}}(e_1)$, defined in (4.2), is $\{\tilde{m}_1, \tilde{m}_2\}$]; (ii) the message $(\tilde{m}_1, \tilde{m}_3)$ if e_1 lies in the interior of $\overline{\overline{R}}_1$; (iii) the message $(\tilde{m}_1, \tilde{m}_2, \tilde{m}_3)$ if e_1 lies at the boundary of $\overline{R}_1$ and $\overline{\overline{R}}_1$. Similarly there are three possible messages for person 2. So the language of the standard form of $\tilde{P}$ has nine elements.

It can be shown[37] that if a process $P=(M, g, h)$ on E has the grid structure, then

$$\left.\begin{array}{ll} \text{(a)} & \text{for every } n\text{-tuple } (S_1,\ldots,S_n) \text{ with } S_i \text{ in } \Sigma_g^i\text{, there is an} \\ & m \text{ in } M \text{ for which } \sigma_g(m)=\prod_{i\in N} S_i \\ \text{(b)} & \text{for every } \overline{m} \text{ in } M \text{ with } \sigma_g(\overline{m}) \text{ non-empty there is an } n\text{-tuple} \\ & (S_1,\ldots,S_n) \text{ with } S_i \text{ in } \Sigma_g^i \text{ and } \sigma_g(\overline{m})=\prod_{i\in N} S_i \\ \text{(c)} & \text{if } M \text{ is finite and contains no } m \text{ for which } \sigma_g(m) \text{ is empty,} \\ & \text{then the number of } n\text{-tuples } (S_1,\ldots,S_n) \text{ with } S_i \text{ in } \Sigma_g^i\text{,} \\ & i=1,\ldots,n\text{, equals the number of elements in } M \end{array}\right\}. \tag{4.11}$$

A process which obeys (4.11) need not have the grid structure. But if it obeys (4.11) together with the condition of *non-redundancy* then it must have the grid structure. A process $P=(M, g, h)$ satisfies this condition if M contains no redundant messages. A message $\overline{m}\in M$ is redundant if $\sigma_g(\overline{m})$ is non-empty and is contained in $\bigcup_{m\in\overline{M}}\sigma_g(m)$, where $\overline{M}\subset M$ and $\overline{m}\notin\overline{M}$. (If a process has redundant message there exists a second process with the same equilibrium actions for every e as the first process and no redundant messages.)

[37] Proposition (b) of (4.11) follows immediately from the definition of σ_g. The proof of proposition (a) is as follows, for the case $n=2$. Suppose $S_1=\sigma_g^1(\overline{m})$, $S_2=\sigma_g^2(\overline{\overline{m}})$, and both sets are non-empty. If both S_1 and S_2 contain only one element or if Σ_1 or Σ_2 contains only one set, then the assertion of (a) follows trivially. Suppose (without losing generality) that S_1 has more than one element and that Σ_1 contains more than one set. Then, in view of (4.10), there exists a point $e^*=(e_1^*, e_2^*)$ in $S_1\times S_2$ such that

$$\text{“}e_1^*\in\sigma_g^1(\tilde{m})\text{”} \quad \text{implies} \quad \text{“}\sigma_g^1(\tilde{m})=\sigma_g^1(\overline{m})\text{”}. \tag{*}$$

[Suppose not. Then for every (e_1, e_2) in $S_1\times S_2$ there exists an m such that $e_1\in\sigma_g^1(m)$ and $\sigma_g^1(m)\neq\sigma_g^1(\overline{m})$. That is to say, for every (e_1, e_2) in $S_1\times S_2$ there exists $\tilde{S}_1\in\Sigma_g^1$ with $\tilde{S}_1\neq S_1$, such that $e_1\in\tilde{S}_1$ and $e_1\in S_1$. That means S_1 is contained in the union of some sets in Σ_g^1, all distinct from S_1, which contradicts (4.10).]

Since the process covers E with regard to equilibria there exists an m^* with $e^*\in\sigma_g(m^*)$. We shall show in two steps, that $\sigma_g(m^*)=S_1\times S_2$.

Step I. Let $\tilde{e}\neq e^*$ belong to $\sigma_g(m^*)$; in particular (without losing generality) suppose (α) $\tilde{e}_1\notin S_1=\sigma_g^1(\overline{m})$. We have ($\beta$) $\tilde{e}_1\in\sigma_g^1(m^*)$; ($\gamma$) $e_1^*\in\sigma_g^1(\overline{m})$; ($\delta$) $e_i^*\in\sigma_g^1(m^*)$. Now one of three possibilities must hold: (a) $\sigma_g^1(m^*)=\sigma_g^1(\overline{m})$, which contradicts ($\alpha$), ($\beta$); (b) $\sigma_g^1(m^*)\cap\sigma_g^1(\overline{m})$ is empty, which contradicts (γ), (δ); or (c) $\sigma_g^1(m^*)\neq\sigma_g^1(\overline{m})$, but $\sigma_g^1(m^*)\cap\sigma_g^1(\overline{m})$ is not empty. But (c), together with (γ) and (δ), contradict (*). Hence $\tilde{e}_1\in S_1$. An identical argument establishes $\tilde{e}_2\in S_2$.

Step II. Suppose $\hat{e}\neq e^*$ belongs to $S_1\times S_2$. Suppose $\hat{e}\notin\sigma_g(m^*)$; in particular, suppose (without losing generality) that (λ) $\hat{e}_1\notin\sigma_g^1(m^*)$. We also have ($\mu$) $\hat{e}_1\in\sigma_g^1(\overline{m})$, in addition to ($\gamma$) and ($\delta$) of step I. Of the three possibilities listed in step I, (a) contradicts (λ), (μ); and (b) and (c) are ruled out as in step I. Hence, $\hat{e}_1\in\sigma_g^1(m^*)$. An identical argument establishes $\hat{e}_2\in\sigma_g^2(m^*)$.

Proposition (c) of (4.11) follows directly from (a) and (b).

If a process $P = (M, g, h)$ is non-redundant and has the grid structure then it obeys, moreover, a strengthened form of (4.11), in which (a) becomes "for every n-tuple $(S_1, \ldots, S_n)$ with S_i in Σ^i_g there is one and only one m in M for which $\sigma_g(m) = \prod_{i \in N} S_i$". The strengthened form is needed if the observational-report form of P is to exist, i.e. if the outcome function $\hat{h}$ in (4.9) is to be well-defined. A non-redundant finite-language grid-structure process $P = (m, g, h)$ has then [in view of (4.11) in its strengthened form] an observational-report form which realizes P and has a language the same size as M.

Finally, if a process $P = (M, g, h)$ with a finite language M is non-redundant, if "$m \neq m'$ and $\sigma_g(m) \cup \sigma_g(m')$ is a Cartesian product" implies "$h(m) \neq h(m')$", and if the process lacks the grid structure, then there does not exist a one-step process which realizes P and has a language not larger than M. The argument used above to establish this fact for the illustrative process $\tilde{P}$ can be generalized.

4.3.3. *Informational efficiency of discrete processes*

We turn now to the informational efficiency of discrete processes. Suppose the set A of possible actions lies in a Euclidean space. An organization has to choose a point in A in response to an environment e in a set E. It contemplates doing so by using a proposed discrete process on E. We shall use a new measure – not expected payoff – of the process's "gross" performance. The measure is well-defined for many infinite-language discrete processes but will be particularly useful for the study of finite-language processes. Let $\phi(e)$ denote, for any e in E, a uniquely selected payoff-maximizing action in A. (We confine attention to cases where a payoff-maximizing action exists for any e in E.) Our measure of gross performance of a process $P = (M, g, h)$ satisfying (4.5) will be the "maximum possible error" of the process P or, as we shall call it for brevity, the "error of the process P". This is

$$\varepsilon_\phi(P) \equiv \sup_{m \in M} \sup_{e \in \sigma_g(m)} \|h(m) - \phi(e)\|,$$

where, as before,

$$\sigma_g(m) = \{e = (e_1, \ldots, e_n) : g^i(e_i, m) = 0, \text{ all } i \in N\},$$

and the symbols $\|\;\|$ denote Euclidean distance. Study of this measure is confined to processes $\tilde{P}$ for which ε_ϕ exists (a sufficient condition is that h be bounded on M). The error $\varepsilon_\phi(P)$ is, loosely speaking, the largest possible distance between the action which the organization "should" take [i.e. $\phi(e)$] and the action actually taken when the process P attains an equilibrium message. We shall call ϕ an *optimality function*.

Consider now the following question about a given process P. Is there any other process P^* on E with $\varepsilon_\phi(P^*) < \varepsilon_\phi(P)$, with the informational costs of P^* not higher than those of P and with at least one of these inequalities strict? If not, the process P is *efficient*. If there is no such P^* among the processes in a given *class* of processes, then P is efficient within that class.

Studying such an efficiency question for a minimax performance measure like ε_ϕ is, of course, not accepted lightly by someone who regards expected-utility maximization with personal probabilities as the only proper procedure for a designer or any other decision maker. As the remarks in Section 1 emphasized, however, such a viewpoint must *also* reject the efficiency study of designs with regard to expected payoff. Both lines of study are equally "illegitimate", yet it is such lines of study, not uncompromising expected-utility comparisons (with utility defined on payoff and cost), that one can feasibly pursue.

The performance measure ε_ϕ, in the proposed efficiency study of processes, is itself subject to criticism, since it ignores the payoffs earned by the actions which might be in force before the process generates an equilibrium action. If the process has a one-step realization, the criticism is not serious.

On the cost side, progress requires some bold assumptions. In the spirit of existing equilibrium studies of processes we ignore all costs incurred in the steps which precede equilibrium. Again the omission is not serious if the process has a one-step realization; it then becomes a one-step design in our earlier sense.

Two cost elements will be considered: (i) the size, or for infinite-language processes, the "fineness", of the language M, and (ii) the size or fineness of the process's *action set* $h(M) \equiv \{a : a = h(m)$ for some m in $M\}$. If a process $P = (M, g, h)$ has the grid-structure, then it can be realized by the one-step observational-report form, and the size or fineness of M is a measure of "observing effort" or "the precision of observational reporting", since M and g determine the sets $\sigma_g^i(m)$ in which person i seeks to locate the current environment e_i. The size or fineness of the action set $h(M)$ is a measure of implementing or action-taking effort. In some cases it may also measure some aspects of computing effort: an outcome function h which partitions M coarsely may be easier to compute than one which partitions M finely. Beyond that, however, no explicit measure of computing effort will be suggested. That may create difficulties, as the "price-mechanism" result summarized in Section 4.3.4 below illustrates. Useful measures of computing effort remain a major unmet challenge in the study of adjustment process and the study of designs in general.

We shall not present a general definition of the two "fineness" measures which are to be used when language and outcome set are infinite. Some natural measures will emerge in the illustrations of Section 4.3.4.

We now narrow the discussion to the case of an optimality function ϕ which takes its values in one-dimensional real space and to processes whose action sets lie in one-dimensional real space.

Three basic propositions are important tools in the construction of discrete privacy-preserving processes on a set $E = E_1 \times \cdots \times E_n$ whose actions are points on the real line and which are informationally efficient with regard to a real-valued ϕ.

Proposition 4.1 (optimality of the "closest-to-midcontour" outcome function) Let $P = (M, g, h)$ be a process on E for which $\varepsilon_\phi(P)$ exists. Let A denote the action set $h(M)$. Let $P^* = (M, g, h^*)$ be another process on E, where h^* is defined as follows:

$$\left.\begin{array}{l}\text{for any } m \in M^*,\ h^*(m) \text{ is the smallest element of a set} \\ \text{which has either one or two elements, namely the set} \\ \left\{\bar{a} \in A : |\bar{a} - d_\phi(m)| \leqq |a - d_\phi(m)|, \text{ all } a \text{ in } A\right\}, \text{ where} \\ d_\phi(m) \equiv \tfrac{1}{2}\left| \inf_{e \in \sigma_g(m)} \phi(e) + \sup_{e \in \sigma_g(m)} \phi(e) \right| \end{array}\right\}. \tag{4.12}$$

Then $\varepsilon_\phi(P^*)$ exists and $\varepsilon_\phi(P^*) \leqq \varepsilon_\phi(P)$.

The proposition says that a process for which the error ε_ϕ exists can be improved with regard to error, or at least not damaged, if the process's outcome rule is changed so that the action set remains the same but the rule for assigning actions becomes the following: to each of the sets $\sigma_g(m)$ the rule assigns[38] that action (in the unchanged action set) which is closest to the "midcontour" action for $\sigma_g(m)$, with ties being broken in favor of the smaller action. The "midcontour" action for $\sigma_g(m)$ is that action which is midway between the infinum of the function ϕ on the set $\sigma_g(m)$ and the supremum of ϕ on that set. Given any triple (A, M, ϕ) we shall call the function h^* defined in (4.12) the *closest-to-midcontour function for* (A, M, ϕ).

Proof of the proposition is straightforward.

For the next proposition call a process $\pi = (M, g, h)$ on E *ϕ-connected* whenever we have (i) e', e'', e''' in E; (ii) $\bar{m} \in \mu_g(e')$, $\bar{m} \in \mu_g(e''')$; (iii) $\phi(e') \leqq \phi(e'') \leqq \phi(e''')$. Then we also have $\bar{m} \in \mu_g(e'')$. [Here $\mu_g(e)$, following the definition in (4.2), denotes $\{m \in M: g^i(m, e_i) = 0, \text{ all } i\}$.] The proposition says that if a process which lacks the property is modified so that it displays the property, then the error ε_ϕ is not increased. Specifically:

Proposition 4.2 (Optimality of "ϕ-connectedness")

Suppose $e', e'', e''' \in E$ and $\phi(e') \leqq \phi(e'') \leqq \phi(e''')$. Suppose $P = (M, g, h)$ is a (not ϕ-connected) process on E such that for some $\bar{m} \in M$, $\bar{m} \in \mu_g(e')$, $\bar{m} \in$

[38] Since an outcome rule assigns an action to each m in M, it also assigns an action to each set $\sigma_g(m)$.

$\mu_g(e''')$. Let $P^* = (M^*, g^*, h^*)$ be another process on E such that $M^* = M$, $h^* = h$,

$$u_{g^*}(e) = \mu_g(e), \quad \text{for all } e \text{ in } E \setminus \{e''\},$$

and

$$\mu_{g^*}(e'') = \{\overline{m}\}.$$

Then

$$\varepsilon_\phi(P^*) \leqq \varepsilon_\phi(P).$$

The proposition implies, in particular, that if for every $i = 1, \ldots, n$, E_i is a subset of the real line, and if ϕ is non-decreasing or non-increasing in each of its arguments then we can confine attention, in searching for *grid-structure* processes with low error, to grid-structure processes (M, g, h) for which every set $\sigma_g(m)$ is the intersection of E with a single rectangle of dimension n or lower.

To state the third proposition some definitions are needed. For a given process $P = (M, g, h)$ on E, let α, β be, respectively, the infinum and the supremum of ϕ on E. Assume that $]\alpha, \beta[\subseteq \phi(E)$. For m in M, define $u_m \equiv \inf\{\phi(e): e \in \sigma_g(m)\}$, $v_m \equiv \sup\{\phi(e): e \in \sigma_g(m)\}$. For $\alpha \leqq X < Y \leqq \beta$, define the set B^m_{XY} to be the empty set if $u_m \geqq Y$ or $v_m \leqq X$, and the set $\{e: e \in \sigma_g(m);\ X \leqq \phi(\theta) \leqq Y\}$ otherwise. Thus B^m_{XY} comprises those environments which are in $\sigma_g(m)$ and have ϕ-values in $[X, Y]$, except that if such an environment, say e^*, lies also in $\sigma_g(\overline{m})$ for some $\overline{m} \neq m$, then e^* is assigned to B^m_{XY} or $B^{\overline{m}}_{XY}$ according to whether $\sigma_g(\overline{m})$ or $\sigma_g(m)$ contains environments whose ϕ-values are in $]\phi(X), \phi(Y)[$.

For $\alpha \leqq X < Y \leqq \beta$ and $A \subset \boldsymbol{R}$, let $E_A(X, Y) \equiv \sup\{|h_A(m) - \phi(e)|: m \in M;$ B^m_{XY} non-empty; $e \in B^m_{XY}\}$, where h_A denotes the closest-to-midcontour function for (A, M, ϕ). (Thus, for the action set A, $E_A(X, Y)$ is the supremum of the errors on the entire "belt" of environments whose ϕ-values are in $[X, Y]$, where environments lying in several distinct sets $\sigma_g(m)$ are deemed to belong to the belt or not in accordance with the preceding definition.) For m in M and $\alpha \leqq X < Y \leqq \beta$, $A \subset \boldsymbol{R}$ and B^m_{XY} non-empty, define $F_{mA}(X, Y) \equiv \sup\{|h_A(m) - \phi(e)|: e \in B^m_{XY}\}$. Call the set $\sigma_g(m)$ "A-critical on (X, Y)" if B^m_{XY} is not empty and $F_{mA}(X, Y) = E_A(X, Y)$. [Thus $\sigma_g(m)$ is A-critical on (X, Y) if it contains points (environments) in the "(X, Y)-belt" and for one of these points, say e^*, the error $|h_A(m) - \phi(e^*)|$ equals the supremum of the errors over all points in the (X, Y)-belt.] Finally, we say that the finite set A has the "no-alien property for (A, M, ϕ)" if for two successive elements r, s of A (i.e., $r < s$ and $]r, s[\cap A$ is empty), the action $h_A(m)$ equals r or s if $\sigma_g(m)$ is A-critical on (r, s). [Thus if $\sigma_g(m)$ is A-critical on the "belt" defined by two successive elements of A, then m is assigned one of those elements (by the closest-to-midcontour function h_A) rather than some "alien" element.]

Then Proposition 4.3 is as follows:

Proposition 4.3 (optimality of "equal-error" action k-tuples among all action k-tuples when ϕ is bounded)

Let ϕ be bounded from above and below, i.e. there exist $\alpha, \beta \in \boldsymbol{R}$ such that $\inf\{\phi(e): e \in E\} = \alpha$, $\sup\{\phi(e): e \in E\} = \beta$. Let ϕ, E satisfy $]\alpha, \beta[\subseteq \phi(E)$. Let $P = (M, g, h)$ be any process on E with $h(M) = A$, where A has k elements. Let $P^* = (M, g, h^*)$ be another process on E, where:

(i) $h^*(M) = A^*$ and h^* is the closest-to-midcontour function for (M, A^*, ϕ);
(ii) A^* has k distinct elements, namely $a_1, \ldots, a_k$, which are ordered, without loss of generality, so that $a_1 < a_2 < \cdots < a_k$;
(iii) A^* has the no-alien property for (a, M, ϕ); and
(iv) A^* satisfies the "equal-error" condition

$$\begin{aligned} E_{A^*}(\alpha, a_1) &= E_{A^*}(a_1, a_2) \\ &= E_{A^*}(a_2, a_3) \\ &\vdots \\ &= E_{A^*}(a_{k-1}, a_k) \\ &= E_{A^*}(a_k, \beta). \end{aligned} \tag{4.13}$$

Then $\varepsilon_\phi(P^*) \leqq \varepsilon_\phi(P)$.

The proposition says that if ϕ is bounded then a process with a k-element action set can always be improved, or at least not damaged, if the action set becomes a k-tuple with the "equal-error" and no-alien properties, and the closest-to-midcontour outcome function is used. The proposition rests on a somewhat intricate argument.

Propositions 4.1 and 4.3 tell us, then, that if ϕ is bounded, then a search among processes with a k-element action set in order to find a process with a low error ε_ϕ can be confined to those processes for which the outcome rule is the closest-to-midcontour rule, and the action k-tuple has the no-alien and equal-error properties.[39]

[39] Each of the three propositions has a counterpart for the case of actions which lie in the real space of dimension greater than one.

One can, in particular, study the case of a linear ϕ on a compact E in a finite Euclidean space. Consider the class of k-action grid-structure processes on E, where the grid is constrained to consist of cubes of the same size [each cube being a set $\sigma_g(m)$]. An exact error-minimizing action k-tuple for such processes (a k-tuple with the equal-error and no-alien properties) has been found as a function of ϕ, E, and cube size.

4.3.4. *Illustrative applications to the study of resource-allocating mechanisms*

Suppose persons 1 and 2 comprise a two-person two-commodity exchange economy. Let x and z be symbols associated with the two commodities. For person i, endowments of the two commodities are w_{x_i}, w_{z_i}. Additions to endowments (net trades) are denoted x_i, z_i. Utility as a function of net holdings is

$$u_i = \alpha_i(w_{x_i} + x_i) - \tfrac{1}{2}\beta_i(w_{x_i} + x_i)^2 + z_i + w_{z_i}$$

(where $\alpha_i > 0$, $\beta_i > 0$), provided

$$\partial u_i/\partial x_i = \alpha_i - \beta_i(w_{x_i} + x_i) \geqq 0. \tag{4.14}$$

Now define $e_{11} = \alpha_1 - \beta_1 w_{x_i}$, $e_{21} = \alpha_2 - \beta_2 w_{x_2}$, $e_{12} = \beta_1 > 0$, $e_{22} = \beta_2 > 0$, $e_1 = (e_{11}, e_{12})$, $e_2 = (e_{21}, e_{22})$, $e = (e_1, e_2)$. Then (4.14) becomes

$$x_i \leqq e_{i1}/e_{i2}. \tag{4.14$'$}$$

The pair of consumptions $[(w_{x_1} + x_1, w_{z_1} + z_1), (w_{x_2} + x_2, w_{z_2} + z_2)]$ is an interior Pareto optimum if at that pair (4.14′) holds for $i = 1,2$ and

$$\begin{aligned} x_1 &= (\alpha_1 - \alpha_2 - \beta_1 w_{x_1} + \beta_2 w_{x_2})/(\beta_1 + \beta_2) \\ &= (e_{11} - e_{21})/(e_{12} + e_{22}), \end{aligned} \tag{4.15}$$

$$x_2 = -x_1, \tag{4.16}$$

$$z_1 + z_2 = 0, \tag{4.17}$$

$$z_1 + w_{z_1} \geqq 0, \qquad i = 1,2. \tag{4.18}$$

Now let a set E of pairs $e = (e_1, e_2)$ be given. Consider the following continuum "price" mechanism $\bar{P} = (\bar{M}, \bar{g}, \bar{h})$ on E:

$$\begin{aligned} \bar{M} = \{(p, x)\colon p &= (e_{11}e_{22} + e_{21}e_{12})/(e_{12} + e_{22}), \\ x &= (e_{11} - e_{21})/(e_{12} + e_{22}) \text{ for some } e \in E\}, \end{aligned}$$

$$\bar{g}^1 = e_{11} - e_{12}x - p, \qquad \bar{g}^2 = e_{21} + e_{22}x - p,$$

$$\bar{h}(p, x) = x.$$

Here p is a price and x is a proposed value of the net trade x_1. For every $e \in E$, the unique equilibrium outcome (the unique element of the set $h[\mu_g(e)]$) is $x_e \equiv (e_{11} - e_{21})/(e_{12} + e_{22})$. The inequality (4.14′) is satisfied for $x_1 = x_e$ and $x_2 = -x_e$ provided

$$e_{22}e_{11} + e_{21}e_{12} \geqq 0. \tag{4.19}$$

We consider only E such that (4.19) is satisfied for all $e \in E$.

We are interested in the informational efficiency of *discrete versions of the mechanism (process)* $\bar{P}$, as well as the efficiency of other processes on E. An appropriate optimality function for this purpose would be one which assigns to each $e \in E$, the set of interior Pareto optima for the economies defined by e. Fortunately, however, we do not have to deal with optimality functions which are set-valued and not point-valued. For it happens that in the economy considered, with its utility functions linear in z_i, *every* interior Pareto-optimal trade vector (x_1, x_2, z_1, z_2) for the economies defined by a given e has the same value of x_1 (and hence also the same value of $x_2 = -x_1$), namely $x_1 = (e_{11} - e_{21})/(e_{12} + e_{22})$. Any (z_1, z_2) satisfying (4.17) and (4.18), combined with $x_1 = (e_{11} - e_{21})/(e_{12} + e_{22})$ is also, for any $e \in E$, the unique equilibrium outcome of the continuum price mechanism $\bar{P}$. Therefore, if one chooses $\bar{P}$ out of those privacy-preserving continuum processes on E which achieve a Pareto-optimum at equilibrium, then one imposes no "bias" in favor of certain Pareto optima: the x-components of an optimum are unique and the process $\bar{P}$ leaves open the specification of the z-components.

Hence, in judging a certain discrete process on E, it is natural to use the point-valued optimality function $\phi(e) = (e_{11} - e_{21})/(e_{12} + e_{22})$, and to compare the maximum distance between that process's equilibrium actions and $\phi(e)$, with the corresponding maximum distance for other processes on E.

Judging the information efficiency of discrete versions of the price process $\bar{P}$ is a task which is very different when the environment set is compact than when it is unbounded. We consider both compact and unbounded examples.

A compact two-parameter set of economies. Consider first the economies defined by the following compact set E:

$$E = \{e = [(e_{11}, e_{12}), (e_{21}, e_{22})] : e_{12} = e_{22} = 1,\ 0 \leqq e_{11} \leqq 1,\ 0 \leqq e_{21} \leqq 1\}.$$

For every $e \in E$, (4.19) holds. It will save notation in studying this E, to let the symbols e_1, e_2 temporarily take new meanings: let e_1 now denote e_{11}, let e_2 denote e_{21}, let $e = (e_1, e_2)$ and let E now denote the set $\{e = (e_1, e_2) : 0 \leqq e_1 \leqq$

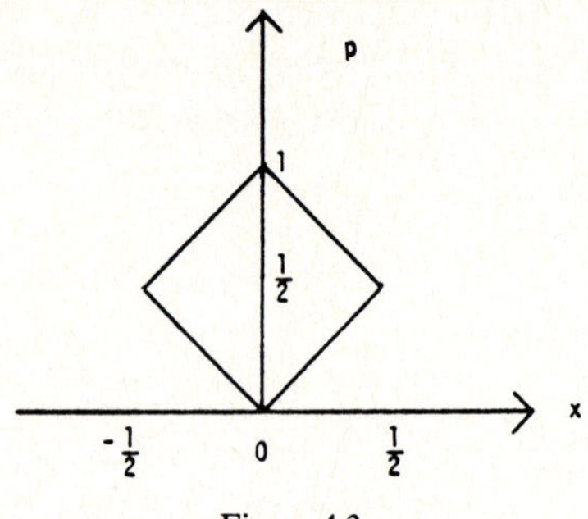

Figure 4.3

$1, 0 \leqq e_2 \leqq 1\}$. Our optimality function becomes $\phi(e) = \frac{1}{2}(e_1 - e_2)$. The continuum price mechanism $\overline{P} = (\overline{M}, \bar{g}, \bar{h})$, redefined so that it is now a mechanism on the redefined E, has

$$\overline{M} = \{(p, x): p \geqq 0;\ p + x - e_1 = 0,\ p - x - e_2 = 0$$
$$\text{for some } (e_1, e_2) \in E\}$$
$$= \{(p, x): p \geqq 0, |x| \leqq p \leqq 1 - |x|, -\tfrac{1}{2} \leqq x \leqq \tfrac{1}{2}\},$$
$$\bar{g}^1 = p + x - e_1, \qquad \bar{g}^2 = p - x - e_2,$$
$$\bar{h}(p, x) = x.$$

The language $\overline{M}$, that is to say, consists of all points in the rotated square shown below (Figure 4.3).

We consider now a discrete approximation, called $\overline{\overline{P}}$, to the continuum price process $\overline{P}$. The approximation is a process whose language is a *subset* of M^*, where M^* denotes the points in M which are obtained when one imposes on M a uniform lattice whose points are spaced a distance of $\frac{1}{6}$ apart, with the origin being a lattice point. The set M^* has 25 elements, namely the points shown in Figure 4.4.

In the approximating process $\overline{\overline{P}}$, there is for every $e_i \in E_i$, a *surrogate value* e_i' at a distance of not more than $\frac{1}{6}$ from e_i. Person i observes e_i and finds the surrogate value e_i'. (For $e_i = \frac{1}{2}$ or $\frac{2}{3}$, there are two surrogates.) Person i says Yes for a given message m if he would have said Yes for that m and for e_i' in the continuum process $\overline{P}$. Any "tighter" choice of uniformly placed surrogates than those which are within $\frac{1}{6}$ of any given e_i *would not permit coverage* of E: for some surrogates there would be no lattice point among those in Figure 4.4 such that in the continuum process i would say Yes to that surrogate at that lattice point.

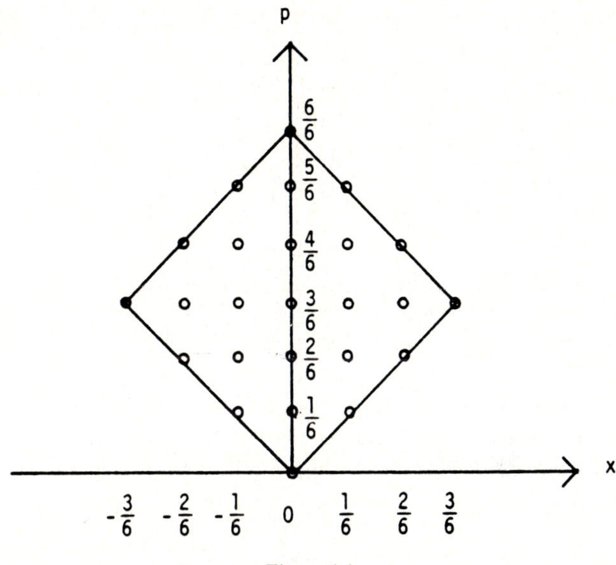

Figure 4.4

Formally, for the approximating discrete price process $\bar{\bar{P}} = (\bar{\bar{M}}, \bar{\bar{g}}, \bar{\bar{h}})$, we have

(i) $\bar{\bar{M}}^i = \{(\tfrac{3}{6}, -\tfrac{2}{6}), (\tfrac{4}{6}, -\tfrac{1}{6}), (\tfrac{5}{6}, 0), (\tfrac{2}{6}, -\tfrac{1}{6}), (\tfrac{3}{6}, 0),$
$(\tfrac{4}{6}, \tfrac{1}{6}), (\tfrac{1}{6}, 0), (\tfrac{2}{6}, \tfrac{1}{6}), (\tfrac{3}{6}, \tfrac{2}{6})\},$

(ii) $\bar{\bar{g}}(m, e_i) = 0$ if $\bar{g}^i(m, r) = 0$ for some r in $\rho(e_i)$,
$= 1$ otherwise, $i = 1, 2$,

where

$$\begin{aligned}
\rho(e_i) &= \{\tfrac{1}{6}\}, & 0 \leqq e_i < \tfrac{1}{3}, \\
&= \{\tfrac{1}{6}, \tfrac{1}{2}\}, & e_i = \tfrac{1}{3}, \\
&= \{\tfrac{1}{2}\}, & \tfrac{1}{3} < e_i < \tfrac{2}{3}, \\
&= \{\tfrac{1}{2}, \tfrac{5}{6}\}, & e_i = \tfrac{2}{3}, \\
&= \{\tfrac{5}{6}\}, & \tfrac{2}{3} < e_i \leqq 1,
\end{aligned}$$

(iii) $\bar{\bar{h}}(m) = \bar{h}(m)$.

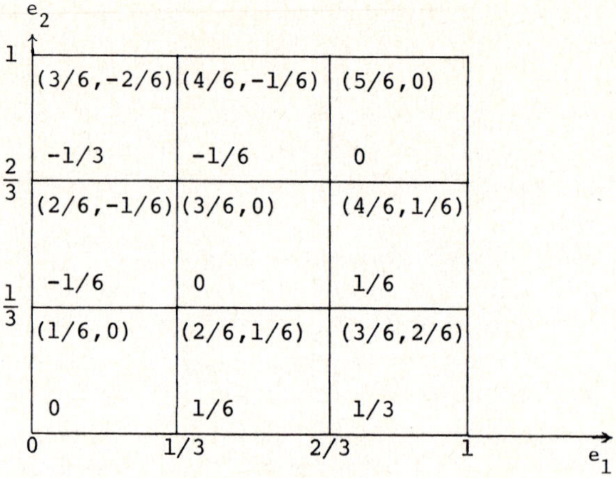

e_2	0 to $1/3$	$1/3$ to $2/3$	$2/3$ to 1
$\frac{2}{3}$ to 1	(3/6,-2/6) -1/3	(4/6,-1/6) -1/6	(5/6,0) 0
$\frac{1}{3}$ to $\frac{2}{3}$	(2/6,-1/6) -1/6	(3/6,0) 0	(4/6,1/6) 1/6
0 to $\frac{1}{3}$	(1/6,0) 0	(2/6,1/6) 1/6	(3/6,2/6) 1/3

e_1

Figure 4.5

Thus in the approximation $\bar{\bar{P}}$, the sets in $\{\sigma_{\bar{\bar{g}}}(m): m \in \bar{\bar{M}}\}$ are the nine squares in Figure 4.5. Inside each square $\sigma_{\bar{\bar{\pi}}}(m)$ is written $m = (p, x)$ (at the top) and the outcome $\bar{\bar{h}}(m) = x$ (at the bottom). The example illustrates what can be stated in a precise manner, namely, a general procedure for obtaining a discrete process which approximates a given continuum process whose language is a subset of a Euclidean space. As in the example, the approximation imposes a uniform lattice on the continuum language.

The process $\bar{\bar{P}}$ is a grid-structure process with a nine-message language $\bar{\bar{M}}$, and a *five*-element action set $\bar{\bar{h}}(\bar{\bar{M}}) = \{-\frac{1}{3}, -\frac{1}{6}, 0, \frac{1}{6}, \frac{1}{3}\}$. The five actions correspond to the values of ϕ along five contours (lines of constant ϕ); the center of each square is on one of these contours. Each square, in other words, is assigned its *midcontour*, which, as we know from Proposition 4.1, is the best possible outcome that can be assigned to the square. It is easily checked that for our optimality function $\phi = \frac{1}{2}(e_1 - e_2)$, $\varepsilon_\phi(\bar{\bar{P}}) = \frac{1}{6}$. The maximum error of $\frac{1}{6}$ occurs at the *corner points* of the squares. At $e = (\frac{1}{3}, \frac{2}{3})$ for example, an equilibrium message [one of four in $\mu_{\bar{\bar{g}}}(e)$] is $m = (p, x) = (\frac{3}{6}, 0)$. For this message, the action is $x = \bar{\bar{h}}(m) = x = 0$. But the optimal action is $\phi(e) = \frac{1}{2}(\frac{1}{3} - \frac{2}{3}) = -\frac{1}{6}$.

The result concerning linear ϕ and compact E, alluded to at the end of Section 4.3.3 above, shows that if a grid-structure process on our unit-square E has 9 messages and 5 actions, then it cannot have a maximum error less than $\frac{1}{6}$. So $\bar{\bar{P}}$ is informationally efficient in the class of grid-structure processes on E. That is to say there is no grid-structure process on E with not more than 9 messages, not more than 5 actions, an error ε_ϕ not more than $\frac{1}{6}$, and one of these inequalities a strict inequality.[40]

From another point of view, however, the discrete price process $\bar{\bar{P}}$ is not efficient, since only 9 of the 25 points in M^* are used in the approximating $\bar{\bar{M}}$. Suppose that in "purchasing" the lattice on $\overline{M}$ whose points are $\frac{1}{6}$ apart, one has paid for a 25-element language. *Given this capability*, is there a mechanism which improves upon the maximum error of $\frac{1}{6}$ achieved by $\bar{\bar{P}}$ and does so without requiring more than 5 actions, the number required by $\bar{\bar{P}}$? The answer is that there exists a grid-structure process which achieves this improvement. An error of $\frac{3}{20}$ (which is less than $\frac{1}{6}$) is achieved by a grid-structure process $\tilde{P} = (M^*, \tilde{g}, \tilde{h})$ on E, portrayed in Figure 4.6, whose language has 25 elements. The sets $\sigma_{\tilde{g}}(m)$ are squares and the 5 actions are $\{-\frac{7}{20}, -\frac{4}{20}, -\frac{1}{20}, \frac{3}{20}, \frac{7}{20}\}$.

These results generalize to any discrete process on E which approximates the continuum process $\overline{P}$ in the manner illustrated and whose language is obtained by imposing a lattice of arbitrary fineness on the continuum $\overline{M}$. If one counts only *the lattice points actually used*, then the discrete price process is informationally efficient, at least within the class of grid-structure processes. If one supposes that both the used and the unused lattice points are available as messages, then a process on E with no more messages, no more actions and a smaller error than the price process can be constructed.

A compact three-parameter set of economies. Consider now a new E. This time let $e_1 = (e_{11}, e_{12})$ have its initial meaning: let e_{22}, as initially defined, equal one; let e_2 denote e_{21} as initially defined. We consider the class E of three-parameter economies, where E is the non-negative unit cube,

$$E = \{[(e_{11}, e_{12}), e_2] : 0 \leqq e_{11} \leqq 1, 0 \leqq e_{12} \leqq 1, 0 \leqq e_2 \leqq 1\}.$$

The continuum price mechanism $\overline{P} = (\overline{M}, \bar{g}, \bar{h})$ on E has, as its language $\overline{M}$, the

[40] Moreover, it is conjectured, but so far unproved, that if one wants to divide the unit square into t non-overlapping rectangles so that the *maximum perimeter* of the rectangles is minimized, then one can confine one's search to t rectangles of *equal* perimeter. By Proposition 4.1 the best possible outcome to assign to a rectangle is its midcontour. But for $\phi = \frac{1}{2}(e_1 - e_2)$, the maximum error on a rectangle of sides a, b, when the midcontour outcome is assigned to the rectangle is $(a + b)/4$ (this error occurs at a corner). Finding an error-minimizing mechanism which uses t messages is therefore the same as dividing the unit square into t non-overlapping rectangles so as to minimize the maximum perimeter. *If the conjecture is true*, then it is easy to show that when $t = r^2$, r an integer, then the maximum perimeter is minimized if the unit square E is divided into t *equal squares*. Since $9 = 3^2$, then *if the conjecture were true*, it would follow that the equal-square mechanism of Figure 4.5, in which each square is assigned an outcome equal to its midcontour, is a best mechanism (with regard to error) among all 9-message 5-outcome processes, i.e. that the discrete "price" mechanism $\bar{\bar{P}}$ is efficient in the class of *all* discrete processes on E, not only in the class of grid-structure processes.

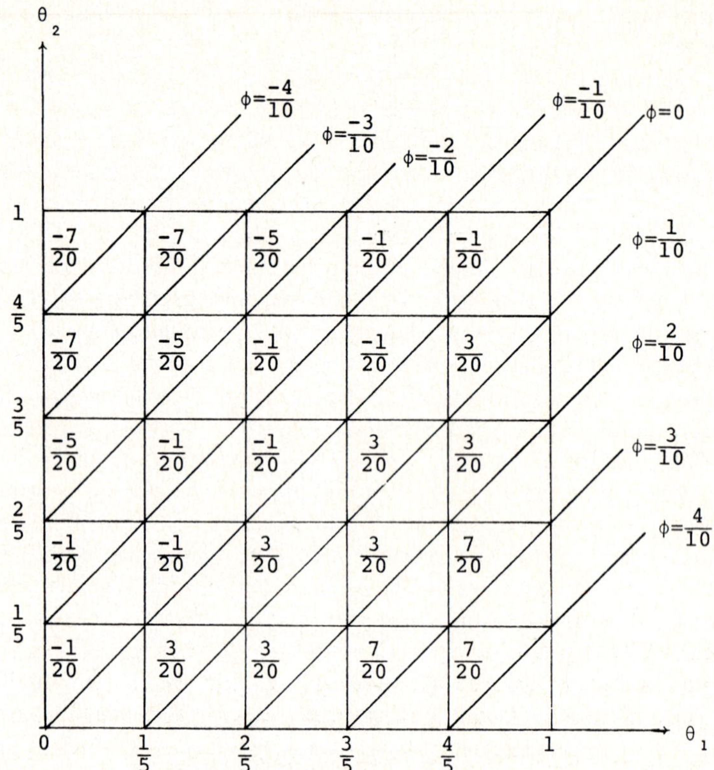

Figure 4.6. In every square $\sigma_{\bar{\bar{g}}}(m)$, the outcome $\bar{\bar{h}}(m)$ is written. The maximum error in every square occurs either at the "northwest" or at the "southeast" corner of the square and is exactly 3/20.

set of messages which are equilibrium messages for some e in E. It is straightforward to verify that this set $\overline{M}$ is the parallelogram in Figure 4.7. (It contains the rotated square of Figure 4.3 since the E of the preceding two-parameter example is contained in the new E.)

Suppose as before that a uniform lattice is placed on the parallelogram E, generating a set M^* of lattice points. Suppose that, $\overline{\overline{P}}$, our discrete approximation to $\overline{P}$, is to use as its language a subset $\overline{\overline{M}}$ of M^*. Let the discrete approximation $\overline{\overline{P}} = (\overline{\overline{M}}, \bar{\bar{g}}, \bar{\bar{h}})$ be defined quite analogously to that of the preceding example, with "surrogates" playing the same role. It is again a grid-structure process. The surrogates are now given by the centers of uniform *cubes* of side V. The cubes cover E. To be an acceptable approximation to $\overline{P}$, however, the process $\overline{\overline{P}}$ must cover all points in E with regard to equilibria. The fineness of the lattice

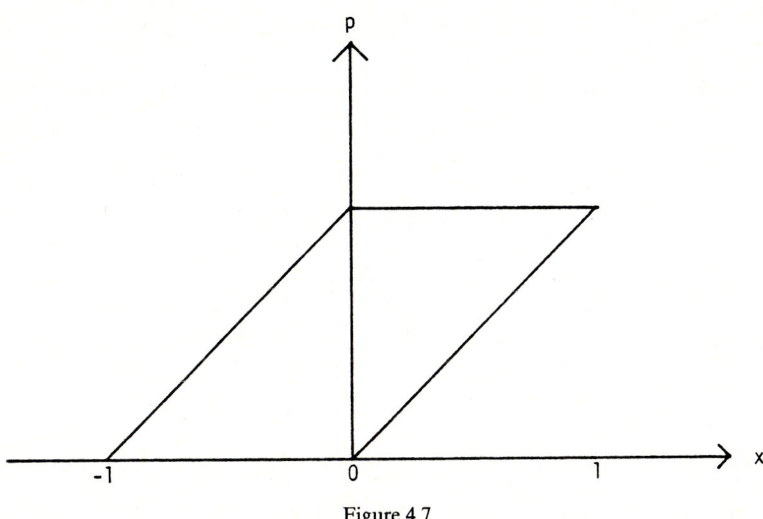

Figure 4.7

placed on $\overline{M}$ must be consistent with the "tightness" of the surrogates, i.e. with the number V. That is to say, for each of person i's surrogates there must be a lattice point in M^* such that in the continuum process i would say Yes if his local environment were that surrogate and the message were that lattice point.

It can be shown that if $V = 1/t$, $t > 0$ an integer, then:

(i) For any lattice on M which is consistent with $1/t$, the process $\overline{\overline{P}} = (\overline{\overline{M}}, \overline{\overline{g}}, \overline{\overline{h}})$ will not use all the lattice points in M^* [i.e. for some m in M^*, $\sigma_{\overline{\overline{g}}}(m)$ is empty].

(ii) One can construct another discrete process (a grid-structure process) whose language has the same number of elements as the lattice points *actually used*, whose action set is no larger than that of $\overline{\overline{P}}$, and which yet has a lower error ε_ϕ. So the price process $\overline{\overline{P}}$ is inefficient: one can do better without increasing either language or action set.

It is perhaps surprising that the three-parameter case should prove less "favorable" to the price process than the two-parameter case. One might have expected that the price process displays its advantages the more strongly the larger the number of parameters determining an environment. In fact a result similar to that just summarized, a result which we shall not sketch here, can be obtained for the *four*-parameter case (with $\delta \leqq e_{22} \leqq 1$, where $\delta > 0$ is arbitrarily small and with e_{11}, e_{12}, e_{21} each taking values $\geqq 0$ and $\leqq 1$). On the other hand, perhaps two, three, and four are all "small" numbers and perhaps it takes an example with a much larger number of parameters to illustrate the informational virtues of a discrete price process.

Note finally that the price process $\bar{\bar{P}}$ follows one of several possible "styles" in which one might approximate the continuum price process. Another style ("rounding off" the functions g^i) is considered in the unbounded example which now follows. Whether a "round-off" approximation to the price process is inefficient for bounded sets of economies remains open.

A two-parameter unbounded set of economies. Now let $E=\{e=[(e_{11},e_{12}),(e_{21},e_{22})]: e_{12}=e_{22}=1,\ e_{11}+e_{22}\geqq 0\}$. Again let e_1 denote e_{11} and let e_2 denote e_{21}. We shall consider a discrete process $\bar{\bar{P}}$ on E which approximates the continuum price process $\bar{P}$ in a different way than the "parameter surrogate" processes just considered. For the process $\bar{P}=(\bar{M},\bar{g},\bar{h})$ on E we have

$$\bar{M}=\{(p,x): p+x=e_1,\ p-x=e_2 \text{ for some } e_1, e_2 \text{ with } e_1+e_2\geqq 0\}.$$

Now impose on this $\bar{M}$ a lattice in which the distance between the x-coordinates of the lattice points is not necessarily the same as the distance between p-coordinates. Denote the lattice $L_{\rho\tau}$, where $\rho>0, \tau>0$. The lattice is the set of points $\{(\bar{p},\bar{x}): \bar{p}=l\rho,\ \bar{x}=k\tau$ for some integers $l,k\}$.

The language of the discrete approximating process $\bar{\bar{P}}=(\bar{\bar{M}},\bar{\bar{g}},\bar{\bar{h}})$ is $\bar{\bar{M}}=L_{\rho\tau}\cap\bar{M}$. Further for $i=1,2$ and $m=(\bar{p},\bar{x})$ in $\bar{\bar{M}}$,

$$\begin{aligned}\bar{\bar{g}}^i(m,e_i) &=0 \quad \text{if } |\bar{g}^i(m,e_i)|\leqq\delta_i,\\ &=1 \quad \text{otherwise,}\end{aligned}$$

and

$$\bar{\bar{h}}[(\bar{p},\bar{x})]=\bar{x}.$$

The numbers δ_1,δ_2 are positive. In this style of approximation, person i says Yes (for a given "local environment" e_i) to a lattice-point message $(\bar{p},\bar{x})$ if he would be "within δ_i" of saying Yes in the continuum process, i.e., if g^i takes a value within δ_i of zero. This corresponds, for example, to rounding off g^i to a given number of digits (decimal, binary, or in general, t-nary) and saying Yes if the rounded off value is zero. The *round-off numbers* δ_1,δ_2 are to be chosen so as to *permit coverage of* E. That means they must lie in the set $S_{\rho\tau}\equiv\{(\delta_1,\delta_2)$: for every $(e_1,e_2)\in E$, there exists $(\bar{p},\bar{x})$ in $\bar{\bar{M}}$ such that $|\bar{p}+\bar{x}-e_1|\leqq\delta_1$, $|\bar{p}-\bar{x}-e_2|\leqq\delta_2\}$ (recall that $g^1[(p,x),e_1]=p+x-e_1$, $g^2[(p,x),e_2]=p-x-e_2$).

Suppose a (δ_1, δ_2) in $S_{\rho\tau}$ has been chosen. Then for a given (e_1, e_2), the lattice point $(\bar{p}, \bar{x})$ is an equilibrium message if

$$\bar{p} + \bar{x} - e_1 = \varepsilon_1, \quad \bar{p} - \bar{x} - e_2 = \varepsilon_2, \qquad |\varepsilon_1| \leqq \delta_1, \quad |\varepsilon_2| \leqq \delta_2.$$

That means that $\bar{x} = (\frac{1}{2})(e_1 - e_2) + \frac{1}{2}(\varepsilon_1 - \varepsilon_2)$ will be the action taken. But the optimal action for (e_1, e_2) is $\phi[(e_1, e_2)] = \frac{1}{2}(e_1 - e_2)$. The supremum of the distances between the equilibrium action and the optimal action, i.e. of the possible values of $|\frac{1}{2}(\varepsilon_1 - \varepsilon_2)| = \frac{1}{2}|\varepsilon_1 + \varepsilon_2|$, is then $\frac{1}{2}(\delta_1 + \delta_2)$. That is to say

$$\varepsilon_\phi(\bar{\bar{P}}) = (\tfrac{1}{2})(\delta_1 + \delta_2). \tag{4.20}$$

To judge the proposed discrete price process fairly, we must select a pair of round-off numbers (δ_1, δ_2) in $S_{\rho\tau}$ for which $\delta_1 + \delta_2$ (and hence ε_ϕ) is a minimum.

Note that for a given pair (ρ, τ) of lattice finenesses, the action set of the discrete price process is the set $T_\tau \equiv \{x : x = k\tau$ for some integer $k\}$. The informational efficiency question is therefore the following: For given ρ, τ does there exist a process $\tilde{P} = (\tilde{M}, \tilde{g}, \tilde{h})$ on E, with $\tilde{M} \subseteq L_{\rho\tau}$ and with action set $\tilde{h}(\tilde{M}) = T_\tau$ such that $\varepsilon_\phi(\tilde{P}) < \varepsilon_\phi(\bar{\bar{P}})$, where $\bar{\bar{P}}$ is defined for a (δ_1, δ_2) which minimizes $\delta_1 + \delta_2$ on $\delta_{\rho\tau}$? If the answer is no, then $\tilde{P}$ is informationally efficient. The answer for $\rho < 1$, is, however, "Yes". From one point of view, in fact, it is a rather strong "Yes". Specifically, for every rival process $\tilde{P}$, the action set is the set T_τ. Hence we immediately have a lower bound to $\varepsilon_\phi(\tilde{P})$, namely $\tau/2$. [To see this, pick an arbitrary integer k. Let $\bar{e} = (\bar{e}_1, \bar{e}_2) \in E$ satisfy $\frac{1}{2}(\bar{e}_1 - \bar{e}_2) = k\tau + \tau/2$. Then $|\phi(\bar{e}) - x| \geqq \tau/2$ for all $x \in T_\tau$. So we have $\varepsilon_\phi(\tilde{P}) \geqq \inf_{x \in T_\tau} |\phi(\bar{e}) - x| \geqq \tau/2$.] The error of the discrete price process [for a best choice of δ_1, δ_2) in $S_{\rho\tau}$] exceeds the lower bound $\tau/2$. But for any $\lambda > 0$ and any lattice $L_{\rho\tau}$ with $0 < \rho < 1$, however fine, we can construct a process $\tilde{P}$ for which $\varepsilon_\phi(\tilde{P}) = \tau/2 + \lambda$. That is to say, for a given lattice $L_{\rho\tau}$ *there is a family of discrete processes which are not the price process* $\bar{\bar{P}}$ *but have the same language and action set as the price process* $\bar{\bar{P}}$; *by choosing an appropriate member of this family we obtain an error as close as desired to the lower bound* $\tau/2$.

To establish the result one proceeds as follows. The set $S_{\rho\tau}$ turns out to be very difficult to compute for arbitrary ρ, τ. It is a bit easier to compute the not larger set $\tilde{S}_{\rho\tau} \equiv \{(\delta_1, \delta_2)$: for every $(e_1, e_2) \in E$, there exist integers l, k such that $|l\rho + k\tau - e_1| \leqq \delta_1$, $|l\rho - k\tau - e_2| \leqq \delta_2\}$. [Here a weaker requirement is imposed on (δ_1, δ_2) than is the case for $S_{\rho\tau}$, since negative "prices" $l\rho$ are now permitted.] Let the expression $a\tilde{S}_{bc}$ denote the set $\{(\bar{\delta}_1, \bar{\delta}_2) : \bar{\delta}_1 = a\delta_1,\ \bar{\delta}_2 = a\delta_2$ for some (δ_1, δ_2) in $\tilde{S}_{bc}\}$. One shows easily that

$$\tilde{S}_{\rho\tau} = \tau S_{\rho/\tau, 1}, \tag{4.21}$$

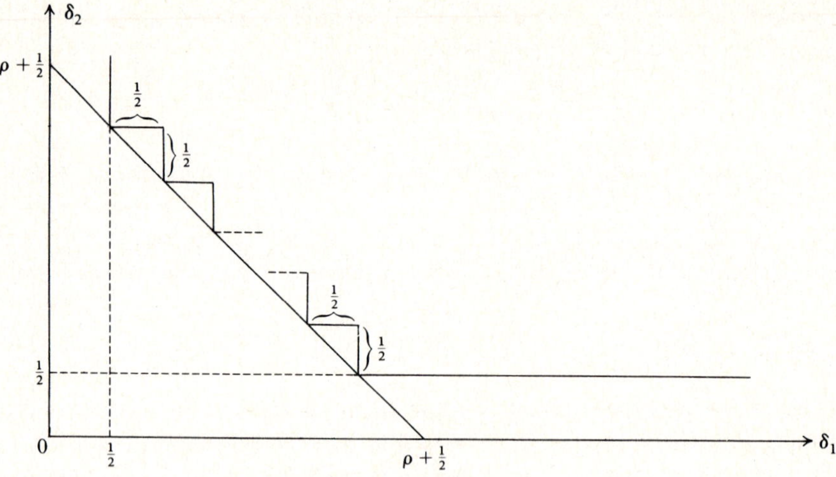

Figure 4.8

$$\tilde{S}_{1/\rho,1} = (1/\rho)\tilde{S}_{\rho 1} \quad \text{or equivalently} \quad \tilde{S}_{\rho 1} = \rho \tilde{S}_{1/\rho,1}. \tag{4.22}$$

Now for $\rho \geqq 1$, $\tilde{S}_{\rho 1}$ consists of all points on or above a certain "staircase" graph in the (δ_1, δ_2)-space.[41] The set $S_{\rho 1}$ contains points for which $\delta_1 + \delta_2 = \rho + \frac{1}{2}$ but no points for which $\delta_1 + \delta_2 < \rho + \frac{1}{2}$. Consequently, in view of (4.22), if $\rho < 1$ $(1/\rho \geqq 1)$ then some of the points in $\tilde{S}_{\rho 1} = \rho S_{1/\rho,1}$ lie on the line $\delta_1 + \delta_2 = \rho(1/\rho + \frac{1}{2}) = 1 + \rho/2$ and none have a value of $\delta_1 + \delta_2$ less than that. Hence, in view of (4.21),

$$\left.\begin{array}{l} \text{if} \quad \rho \leqq \tau \quad (\rho/\tau \leqq 1), \\ \text{then the minimum of } \delta_1 + \delta_2 \text{ on} \quad \tilde{S}_{\rho\tau} = \tau \tilde{S}_{\rho/\tau,1} \\ \text{is} \quad \tau(1 + (\rho/\tau))/2 = \tau + \rho/2 \end{array}\right\}. \tag{4.23}$$

Now let $\rho \leqq \tau$. Since $S_{\rho\tau} \subseteq \tilde{S}_{\rho\tau}$ and since $\tau + \rho/2$ is the minimum of $\delta_1 + \delta_2$ on $\tilde{S}_{\rho\tau}$, every (δ_1, δ_2) in $S_{\rho\tau}$ satisfies $\delta_1 + \delta_2 \geqq \tau + \rho/2$. But then, in view of (4.20), the error of the discrete process $\bar{P}$, using any (δ_1, δ_2) in $S_{\rho\tau}$, is $\varepsilon_\phi(\bar{P}) = \frac{1}{2}(\delta_1 + \delta_2) \geqq \tau/2 + \rho/4$ and so exceeds the lower bound $\tau/2$.

[41] For $\rho \geqq 1$ an integer, $\tilde{S}_{\rho 1}$ is the set of points on or above the "staircase" shown in Figure 4.8. For $\rho = a/b > 1$, with a, b relatively prime integers, $b \neq 1$, the role of the lines $\delta_i = \frac{1}{2}$ in the figure is played by the lines $\delta_i = 1/2b$; the staircase has some steps resting on the line $\delta_1 + \delta_2 = \rho + \frac{1}{2}$, but other steps (to the left and to the right of these) rest on certain lines $\delta_1 + \delta_2 = J$, where $J > \rho + \frac{1}{2}$. For $\rho > 1$ irrational, the set $\tilde{S}_{\rho 1}$ changes but still contains points for which $\delta_1 + \delta_2 = \rho + \frac{1}{2}$ and no points for which $\delta_1 + \delta_2 < \rho + \frac{1}{2}$.

Consider now the following rival process $P_\eta = (\tilde{M}, g_\eta, \tilde{h})$ on E, defined as follows for any number $\eta > 1$:

(i) $\tilde{M} = L_{\rho\tau}$,

(ii) for $m = (\pi, x)$ in $\tilde{M}$,

$$g_\eta^1(e_1, m) = 0 \quad \text{if} \quad |\pi/\eta + x - e_1| \leqq \tilde{\delta}_1, \qquad = 1 \quad \text{otherwise,} \tag{4.24}$$

$$g_\eta^2(e_2, m) = 0 \quad \text{if} \quad |\pi/\eta - x - e_2| \leqq \tilde{\delta}_2, \qquad = 1 \quad \text{otherwise,} \tag{4.25}$$

(iii) $\tilde{h}[(\pi, x)] = x$,

where $(\tilde{\delta}_1, \tilde{\delta})$ are chosen so that coverage is achieved for all e in E [i.e. so that for every $e = (e_1, e_2)$, there exists (π, x) in $\tilde{M}$ satisfying (4.24) and (4.25)].

One can interpret the process P_η as follows: Person i, observing e_i, and an announced message (π, x), first multiplies π by the positive constant $1/\eta < 1$. He then interprets $(\pi/\eta, x)$ as though it were a message $(\bar{p}, \bar{x})$ in the discrete price process $\overline{\overline{P}}$ (with $\delta_1 = \tilde{\delta}_1$, $\delta_2 = \tilde{\delta}_2$) and determines whether to say Yes or No exactly as in the price process. A "price" may now, however, be negative, since the language is now the full lattice $L_{\rho\tau}$, whereas in $\overline{\overline{P}}$ it is $\overline{\overline{M}} = L_{\rho\tau} \cap \overline{M}$, which excludes negative prices. In effect, person i *rescales* the announced price, multiplying it by $1/\eta$ to do so.

Now for any $e = (e_1, e_2)$, the equilibrium action is $\frac{1}{2}(e_1 - e_2) + \frac{1}{2}(\varepsilon_1 - \varepsilon_2)$, where $|\varepsilon_1| \leqq \tilde{\delta}_1$, $|\varepsilon_2| \leqq \tilde{\delta}_2$. Since $\phi(e) = \frac{1}{2}(e_1 - e_2)$, the error $\varepsilon_\phi(P_\eta)$ is (quite analogously to the price process $\overline{\overline{P}}$) the supremum of the possible values of $\frac{1}{2}|\varepsilon_1 + \varepsilon_2|$. That equals $\frac{1}{2}(\tilde{\delta}_1 + \tilde{\delta}_2)$.

Clearly the set of prices $(\tilde{\delta}_1, \tilde{\delta}_2)$ which achieve coverage of E for the process P_η is the set $\tilde{S}_{\rho/\eta, \tau}$, where $\rho/\eta < \tau$ (since $\rho < \tau$ and $\eta > 1$). In view of (4.23), the best choice of $(\tilde{\delta}_1, \tilde{\delta}_2)$ in this set (a pair for which $\tilde{\delta}_1 + \tilde{\delta}_2$ is a minimum) is a pair for which $\tilde{\delta}_1 + \tilde{\delta}_2 = \tau + (\rho/\eta)/2 = \tau + \rho/2\eta$. Then the error of the rival process P_η becomes

$$\varepsilon_\phi(P_\eta) = \tau/2 + \rho/4\eta.$$

By choosing η suitably large, this error can be made as close to the lower bound $\tau/2$ as desired. In particular we obtain an error smaller than that of the discrete price process $\overline{\overline{P}}$, even though our language is the same lattice (though with negative prices now included) as the language of the price process and even though the action set remains the same as well.[42]

[42] It seems very likely that the same result holds if negative prices are excluded in the rival process; but that has not yet been established.

A similar result can be obtained (though it is more difficult) when E is redefined so that $e_{12}=\beta_1$, $e_{22}=\beta_2$ and β_1, β_2 are arbitrary positive constants.

There are several possible ways to interpret the result. One is that the "correct" way to approximate the continuum price mechanism, using a "round-off" approach, is to admit a rescaling operation, so that P_η becomes the "correct" approximation. From this point of view, one has a "favorable" result about the price process, an analogue of the result in the continuum literature. For there (as summarized briefly above in Section 4.1), if one studies the price mechanism on the two-person linear-quadratic exchange economy just considered, one finds that no smooth (Lipschitzian) process can achieve that mechanism's equilibria with a message space of dimension less than that of the price mechanism: that dimension is a lower bound. The price mechanism achieves the lower bound. In our result, a suitably "rescaled" price mechanism comes as close as desired to the lower bound on the error permitted by a given lattice, i.e. by a given language and action fineness. So, in a sense, admitting the rescaling operation is the analogue of requiring smoothness in the continuum approach.

In another interpretation, one would say that "rescaling" is a somewhat alien operation, nowhere suggested in any discussion of the price mechanism, and that P_η is not, therefore, a "price" process. The result would then be rather "unfavorable" to the price mechanism, *unless* one feels that the rescaling operation has its own cost, namely a *computing* cost, not well captured in language fineness or action fineness. But in fact any process which is a rival to the price process will have computations (in its g its h or both) which are distinct from the computations of the price process. So if one takes this point of view one has first to supply a general measure of computing cost, preferably as workable a measure as language fineness or size and action-set fineness or size. *Such a measure remains elusive.*

In a third interpretation, one would say that the rescaling result shows that one should not in fact study infinite-language mechanisms on unbounded sets of economies but only finite-language processes on bounded (though arbitrarily large) sets. The rescaling phenomenon just illustrated depends on the non-finiteness of the language: having, so to speak, "paid for" an infinite language, one can use it, without extra charge, for a rescaled process.

A three- or four-parameter unbounded set of economies. If one now redefines E so that e_{12} or e_{22} or both becomes a free unbounded parameter instead of a constant, then a discrete process approximating the price process in the preceding "round-off" style *cannot* be constructed. It turns out that *no* pair of round-off numbers (δ_1, δ_2) achieves coverage of such a three-or-more-parameter unbounded E. If one wants to cover E with a price process, one has to adopt another style of approximation, for example the "parameter surrogate" style of the preceding compact-environment illustrations. Whether a "rescaling" result similar to the one just given holds with regard to such a price process is not yet settled.

The price-mechanism investigations just summarized have dealt with very small classes of economies, with each member identified by a small number of parameters. (The same small class, however, also plays, as we have seen, a crucial role in the continuum literature.) It remains to be seen whether analogous results can be obtained for both bounded and unbounded sets of many-parameter economies.

4.3.5. *A conjecture about the informational efficiency of discrete price mechanisms*

A possibly distant goal is the settling of a certain general conjecture as to whether a discrete price mechanism makes good use of the informational resources it requires. Suppose that a class $E = E_1 \times E_2 \times \cdots \times E_n$ of n-person k-commodity exchange economies e is given. Every e is an n-tuple $(e_1, \ldots, e_n)$, where $e_i \in E_i$ is a pair composed of i's endowment vector w_i^e and i's preference ordering $\gtrsim_{e,i}$ defined on commodity bundles. The characteristic e_i is observed by i. Let $P = (M, g, h)$ be a price process, privacy-preserving with respect to the characteristics e_i and covering E with regard to equilibria. M is an unbounded continuum; its typical element m is composed of prices and of feasible net trades, i.e. changes (positive or negative) in the n individual holdings, adding up to zero for every commodity. Thus M lies in a real space, say $\boldsymbol{R}^{nk+n-1}$. The function h is defined on M; it simply yields, for every m, the net-trades portion of m; h is a vector $(h_1, \ldots, h_m)$ and $h_i(m)$ is a change in i's holdings. For an equilibrium $\overline{m}$ [i.e. $g^i(e_i, \overline{m}) = 0$, all i], the vector $(h_1(\overline{m}) + w_1^e, \ldots, h_n(\overline{m}) + w_n^e)$ is Pareto-optimal for e. Consider $P^* = (M^*, g^*, h^*)$, an appropriate discrete approximation of the "surrogate" type illustrated in Section 4.3.4. The language M^* is a uniformly spaced and unbounded lattice in $\boldsymbol{R}^k$. The action set $h^*(M^*)$ corresponds to the net-trades portion of the messages in M^*; it is a lattice in $\boldsymbol{R}^k$. Suppose that $P^{**} = (M^{**}, g^{**}, h^{**})$ is any other discrete privacy-preserving process on E with $M^{**} \subseteq M^*$ and $h^{**}(M^{**}) \subseteq h^*(M^*)$.

The *conjecture* is then: for some $\bar{e} = (\bar{e}_1, \ldots, \bar{e}_n)$ in E, and every $m^* \in M^*$, $m^{**} \in M^{**}$, with $g_i^*(\bar{e}_i, m^*) = g_i^{**}(\bar{e}_i, m^{**}) = 0$, all i, if there is some individual j for whom

$$w_j^{\bar{e}} + h_j^{**}(m^{**}) >_{\bar{e},j} w_j^{\bar{e}} + h_j^*(m^*),$$

then there is some individual $t \neq j$ for whom

$$w_j^{\bar{e}} + h_t^*(m^*) >_{\bar{e},t} w_t^{\bar{e}} + h_t^{**}(m^{**}).$$

The conjecture says that the discrete price process is not dominated, with regard to all economies in the given class and the preferences of all members, by any other discrete process which generates trades while using no more informational resources, i.e. which uses a message lattice and an action lattice contained, respectively, in the lattices of the discrete price process.

This is perhaps the weakest general conjecture one can make about the informational efficiency of discrete price mechanisms. It would be reassuring if the conjecture could be verified when E is the class of classic economies.

4.4. *The pre-equilibrium study of adjustment processes*

We return now to the study of processes when they do *not* reach an equilibrium. They are not allowed to because it takes too long, or perhaps (given the initial message m_0 used) it never happens at all. Instead, a process is to terminate after T steps. Accordingly, a process must again be considered a quadruple: the initial message m_0 has to be added in order to determine what action is generated at the Tth step. Moreover, it is useful to return to the original formulation, where a process is a quadruple $\pi = (\mathscr{M}, m_0, f, h)$, whose functions $(f^1,\dots,f^n)$ imply specific message interchanges among the n members at each step. It is no longer appropriate to consider the "trial-message-announced-by-a-center" interpretation of a process; that was useful in the equilibrium analysis, where it suggested the compact form (M, g, h), which suppresses the fact that an element of M is an n-tuple of individual messages. As before, an environment set E and an action set A are given.

A central question for the pre-equilibrium study of a process is: When to stop? The question admits several interpretations. The simplest[43] is as follows. Suppose the process is to be started in response to the environment e, that is, $m_1^i = f^i(e, m_0)$, all i. An environment e prevails for *one time unit*. Suppose T is to be chosen while the process is operated – at each step it will be decided whether or not to stop there. Suppose the designer – the chooser of T – has a payoff function ρ on $A \times E$. In particular, if, during the period of one time unit during which e prevails, an action a^* is taken for q time units, $q \leqq 1$, then during that portion of the period, the total payoff $q\rho(a^*, e)$ is collected. The chooser of T wants the *total payoff collected in the entire period when e prevails*, denoted $\psi_T(e)$, to be high. One immediate proposition is helpful: If

it takes time $C < 1$ to complete one step of the process; (4.25a)

if we require that

the action $\hat{a}(T, e)$ has to be reached[44] before the instant at which the next environment occurs; (4.25b)

[43] Introduced in T. Marschak (1959) and pursued further in T. Marschak (1972).

[44] As before (Section 4.1), the symbol $\hat{a}(T, e)$ denotes the outcome (action) assigned by the outcome function h to the message reached at the Tth step, i.e. $\hat{a}(T, e) \equiv h(m_T, e)$.

a given constant action $\tilde{a} \in A$, called the *interim action*, has to be taken until $\hat{a}(T, e)$ is reached; (4.25c)

and if it is the case that for every $\bar{e}$ in E,

$$\begin{aligned} &\Delta\rho_t(e) > 0 \quad \text{for} \quad t > 1, \\ &\Delta^2\rho_t(e) < 0 \quad \text{for} \quad t > 2, \end{aligned} \tag{4.25d}$$

where

$$\begin{aligned} &\rho_t(e) \equiv \rho[\hat{a}(t, e), e], \\ &\Delta\rho_t(e) \equiv \rho_t(e) - \rho_{t-1}(e), \\ &\Delta^2\rho_t(e) \equiv \Delta\rho_t(e) - \Delta\rho_{t-1}(e); \end{aligned}$$

then there is a *unique best value of T for e*, i.e. a value which maximizes $\psi_T(e)$ under the constraints (4.25b) and (4.25c). There is also a unique best once-and-for-all value of T when successive environments have a stationary probability distribution, i.e. a value which maximizes $\mathscr{E}\psi_T(e)$ under the constraints (4.25b) and (4.25c), where $\mathscr{E}$ denotes expectation with respect to a given probability distribution on E.

For any T satisfying $CT \leqq 1$, the total payoff for the one-time-unit period when e prevails is

$$\psi_T(e) \equiv CT\rho(\tilde{a}, e) + (1 - CT)\rho_T(e).$$

So, under condition (4.25d), T^* is the unique maximizer of $\psi_T(e)$ on the interval $1 \leqq t < 1/C$ if $C(T^* + 1) < 1$ and T^* is the lowest positive integer for which

$$C[\rho_{T^*}(e) - \rho(\tilde{a}, e)] > (1 - CT^*)\Delta\rho_{T^*+1}(e). \tag{4.26}$$

At T^*, that is to say, the payoff which would be foregone if the process were stopped at $T^* + 1$ rather than at T^* (the expression on the left) exceeds the total payoff which would be gained by taking one more step beyond T^* and thereby improving the action which is reached when the process terminates and which is thereupon maintained as long as e continues to prevail (the expression on the right). Condition (4.25d) guarantees that the expression on the left is increasing in T and the expression on the right is decreasing. Hence $\psi_T(e)$ is increased until T^* is reached and decreases thereafter. To determine T^* requires a running check on *changes* in the payoff ρ from step to step, but it requires neither complete knowledge of e nor memory of the preceding values of $\hat{a}(t, e)$.

If there is no integer with the properties required for T^*, then, under (4.25d), the largest integer for which $\rho_T(e) > \rho(\tilde{a}, e)$ and $CT < 1$ is the unique maximizer of $\psi_T(e)$ on the interval $1 \leqq T < 1/C$. If $\rho_1(e) < \rho(\tilde{a}, e)$, then, under (4.25d), $\psi_T(e)$ is uniquely maximized at "$T = 0$", i.e. the process is not started at all.

Suppose next that "when to stop" is not to be decided afresh, following each repetition of the process, initiated by a fresh environment, but rather once and for all. Suppose successive environments are independently distributed according to a stationary probability distribution on E. A value of T which maximizes $\mathscr{E}\psi_T(e)$ is to be chosen [assume $\mathscr{E}\psi_T(e)$ to be finite]. Then if (4.25d) holds for every e, the lowest T satisfying both $C(T+1) < 1$ and the analogue of (4.26) (with expectations added) uniquely maximizes $\mathscr{E}\psi_T(e)$ on the interval $1 \leqq T < 1/C$.

The condition (4.25d) holds in certain cases which have been studied. Suppose that A is Euclidean n-space, so that a is an n-tuple $a = (a_1, \dots, a_n)$, where a_i can take any real value and is called "i's action"; e is an n-tuple of real-valued characteristics $(e_1, \dots, e_n)$ in $E = E_1 \times \cdots \times E_n$, with e_i in E_i; and for each e in E, $\rho(a, e) = \sum_{i=1}^{n} \alpha_i(e)\beta_i(a_i) + S(a)$, where α_i, β_i, and S are real-valued functions and for every i, β_i and S are differentiable with respect to a_i. The *gradient process* has received wide attention [Arrow and Hurwicz (1958, 1960)]. In this process, the language $\mathscr{M}$ is defined so that a message from i is a proposed value of his action.[45] Letting a_t^i denote i's proposal at step t (and a_0^i the initial proposal), the difference equations defining the process are

$$m_t^i = a_{t-1}^i + \lambda \cdot \left[\alpha_i(e_i)\beta_i'(a_{t-1}^i)\right] + S_i(a_{t-1}^1, \dots, a_{t-1}^n), \qquad t \geqq 1,$$

where the prime and the symbol S_i denote derivatives. The term λ is a positive constant; a small value of λ means a fine approximation to the differential-equation version of the gradient process. The process is privacy-preserving with respect to that n-tuple of partitionings on E which is defined when the member associated with a_i observes the characteristic e_i.

That is one appeal of the gradient process. The other has been its convergence properties. Though the focus of considerable work, these hold no direct implication for the good gross performance of the process in the sense of a high value of $\mathscr{E}\psi_T(e)$ for well-chosen T. Suppose that β_i is concave, $i = 1, \dots, n$, and that S is strictly concave, and that for each e some value of a, say $a^*(e)$, maximizes $\rho(a, e)$ on A. For any e there may be some step, say $\bar{t}(e)$, at which the proposed action $\hat{a}(t, e)$ first *overshoots* $a^*(e)$: while $\hat{a}_i(t, e) - a_i^*(e)$ has been of constant sign for every i until $\bar{t}(e) - 1$, this difference changes sign for some i at $\bar{t}(e)$. It *is* the case, however, that for $t < \bar{t}(e)$ the properties $\Delta\rho_t(e) > 0$, $\Delta^2\rho_t < 0$ hold (the

[45] Formally, if we want the message m^i to belong to the same set for all i, we have to define $\mathscr{M}$ as equal to the power set of A. Member i's message $m^i \in \mathscr{M}$ then has the form $m^i = \{a : a_i = \bar{a}_i\}$. The outcome function is $h(m^1, \dots, m^n) = \bigcap_i m^i$.

latter for $t>2$).[46] So, if the constraint $CT<1$ implies $T<\bar{t}(e)$, then there exists a T uniquely maximizing $\psi_T(e)$ subject to the constraint $CT<1$. If $CT<1$ implies $T<\max_{e\in E}\bar{t}(e)$, then there exists a T uniquely maximizing $\mathscr{E}\psi_T(e)$ subject to the constraint $CT<1$. In the special case where $\beta_i(a_i)=a_i$ and S is quadratic, i.e.

$$\rho(a,e)=\sum_{i=1}^{n}\alpha_i(e_i)a_i-a'Qa, \tag{4.27}$$

where Q is positive definite, there is never any overshooting at all and the conditions (4.25d) hold without modification.[47]

Turning back to general processes, note that the constant interim action $\tilde{a}$ is itself subject to choice. If for each e in E, some a in A maximizes $\rho(a,e)$, then there is a value of $\tilde{a}$ which maximizes $\mathscr{E}\psi_T(e)$. One could replace a constant interim action by an interim action chosen afresh at each repetition of the process, as a function of e (possibly a privacy-preserving function, with $\tilde{a}_i$ depending only on e_i). If one permits successive environments to be serially dependent – to be the realization of some stochastic process – then one could let the interim action be that action obtained in response to the preceding environment, and could measure the process's gross performance by the expected value of the suitably discounted stream of period payoffs[48] $\psi_T(e)$. [For the serially independent environments, choosing the interim action in this way can never be better, with regard to $\mathscr{E}\psi_T(e)$, then using the best constant interim action.] Finally, the vector of initial values could be chosen afresh in each repetition and could, in particular, equal the preceding terminal message. In these variants our

[46]As for convergence: for any $\varepsilon>0$ there is for any e an integer $\bar{t}(e)$ such that for all $t>\bar{t}(e)$ the distance $a^*(e)-\hat{a}(t,e)$ is as small as desired, provided λ is taken sufficiently small [Uzawa (1958)]. But for the steps following $\bar{t}(e)$, $\hat{a}(t,e)$ may fluctuate about $a^*(e)$. This property tells us nothing in itself about the existence of a T which uniquely maximizes $\psi_T(e)$ or $\mathscr{E}\psi_T(e)$ on given intervals.

[47]In most of the economic literature on gradient processes, the variables are constrained so as to guarantee that $\hat{a}(t,e)$ approaches, in a privacy-preserving manner, a *constrained* maximum. It is desired to maximize a function $F(a,e)$ subjected to the constraints $a_i\geqq 0$, $i=1,\ldots,n$, and $d_j(a)\geqq 0$, $j=1,\ldots,m$, d_j concave. The Lagrangean $L(a,\mu_1,\ldots,\mu_j,e)=F(a,e)+\Sigma_{j=1}^{m}\mu_j d_j(a,e)$ is formed and an associated gradient difference-equation system is defined in all the variables a_i and μ_j. The change in a_i at each step is proportional to $\partial L/\partial a_i$; the change in μ_j is proportional to $-\partial L/\partial\mu_j$; but neither a_i nor μ_j is permitted to become negative. Suppose that at the initial (non-negative) values of the variables, $d_j>0$ and $\mu_j=0$ for all j. If for subsequent steps $t\leqq t'$ that continues to be so, then for strictly concave F the condition $\Delta F(a_t,e)>0$, $\Delta^2F(a_t,e)<0$ holds for $2\leqq t\leqq t'$, provided that in that interval $a^*(e)$, the maximizer of F subject to the constraints d_j, is not overshot. But as soon as a t is reached for which $d_j(a_t,e)=0$ for some j, then the subsequent signs of $\Delta F,\Delta^2F$ may sometimes be positive and sometimes negative. So one cannot, in general, argue the existence of a unique best T for the given e, on an interval larger than $1\leqq T\leqq t'$.

[48]With serial dependence, moreover, one might drop the constraint $CT<1$: it might be desirable to delay the response to an environment until after the environment has ceased to prevail.

previous conditions no longer guarantee existence of a unique best once-and-for-all T. Computation of a once-and-for-all best T becomes complex, and so does the assessment of a process's performance for given costs.

The appeal of the classic price mechanisms, which inspired the study of abstract adjustment processes, has been not only their achievements at equilibrium but their temporal homogeneity as well as their privacy preservation with regard to natural partitionings of the environment. A constant interim action and a constant initial value preserve the homogeneity, whereas the other variants do not, since special "extra" steps have to be added to the typical repeated step of the process. For this reason, and because it implies the simplest gross performance measure, the preceding assumptions of a constant interim action and constant initial value seem to be a natural starting place, and $\mathscr{E}\psi_T(e)$ for well-chosen T a natural gross performance measure, at least for the case of serial independence.

Assume, then, that environments are serially independent and require again that the process be ended in the period in which it was started. Suppose the sets A and E are given and that two processes $\bar{\pi} = (\mathscr{M}, \bar{m}_0, \bar{f}, \bar{h})$ and $\bar{\bar{\pi}} = (\mathscr{M}, \bar{\bar{m}}_0, \bar{\bar{f}}, \bar{\bar{h}})$, each yielding an action in A in response to an environment in E, are to be compared. In our previous discussion of the cost of a one-step design, it was required only that the design's task be accomplished, and an action generated, some time in the period during which the initiating environment prevails. The cost of a design was the cost of this achievement, and further speeding up the performance of tasks had no purpose. Now, however, greater speed in completing a step of the process (a lower value of C) means that a given terminal step can be achieved sooner.

We may view completion of *one step* of a process as the operation of a one-step design, provided we first generalize the definition of task and design so as to add *memory*. Member i's (finite) task T^i is now a quintuple $(\bar{X}^i, X^i, S^i, \Lambda^i, \Delta^i)$. As before, $\bar{X}^i$ and X^i are i's input and output sets; S^i is the set of possible *states* of i's memory. Λ^i is a matrix of probabilities $\Pr(s^i|\bar{x}^i, s^{*i})$ and Δ^i a matrix of probabilities $\Pr(x^i|\bar{x}^i, s^{*i})$, where s^i, s^{*i} each run over all the elements in S^i, $\bar{x}^i$ over all the elements in $\bar{X}^i$, and x^i over all the elements in X^i. The task is noiseless if Δ^i, Λ^i contain only zeros and ones. So redefined, a noiseless task is, in another literature, a *finite-state machine* [Hartmanis and Stearns (1966)].

A *design with memory* for an n-member organization is, then, an n-tuple of tasks as just redefined. The one-step property is redefined in the obvious way to allow for the addition of memory. If every set S^i contains only one element, we have a *memoryless* design; this is a design as originally defined. Given a state of each member's memory, and given inputs received by members (observations, messages), a one-step design with memory yields a new memory state for every member and an n-tuple of outputs (messages, action-attribute values).

A noiseless one-step design with memory suffices to describe the efforts undertaken in one step of a temporally homogeneous adjustment process. Suppose in a process $\pi = (\mathcal{M}, m_0, f, h)$, $\mathcal{M}$ is finite. Suppose the process is privacy-preserving with respect to partitionings $\{\mathcal{P}^i\}_{i \in N}$ indexed by $\{e^i\}_{iN}$, where for each i, e_i can take a finite number of values comprising the set E_i. Let the typical element of i's memory be a pair – an element of $\mathcal{M}$, together with an element of E_i. At the start of the process's initial step, member i's memory is in an initial state $s_0^i = (\tilde{e}_i, m_0^i)$, where $\tilde{e}_i$ is an arbitrary element of E_i. Member i receives an input, which triggers the initial step. Its components are the new current value of e_i, say $\bar{e}_i$, and an arbitrary "dummy" message $\tilde{m}_j$, from every member $j \neq i$. In accordance with Λ_i, an output is formed and sent to all $j \neq i$, namely, the message m_0^i. In response to the input received, and in accordance with Δ^i, i's memory changes from the initial state to the state $(\bar{e}_i, m_i^0)$. The typical step t, $t \geqq 1$, is triggered by each member i's receipt of the messages $\{m_{t-1}^j\}_{j \neq 1}$ at the end of the preceding step. In response, member i, in step t, forms (in accordance with Λ_i), the output $m_t^i = f^i[\bar{e}_i, (m_{t-1}^1, \ldots, m_{t-1}^i, \ldots, m_{t-1}^n)]$, which is a function of the messages received *and* the state of i's memory [namely, $(\bar{e}_i, m_{t-1}^i)$] at the step's start; member i also, in response to the input received, updates his memory, giving it (in accordance with Λ^i) the new state $(\bar{e}_i, m_t^i)$, again a function of the same two arguments. At the tth step, the processes terminates. Formally, one has to add a "counter" to each member i's memory; the counter's reaching T causes i (in following Λ^i, Δ^i) to form the final output (the Tth) and to reset the memory (including the counter) to its starting position.

Modeling of the outcome function h depends on whom one gives the responsibility for transforming the terminal message into an action. If h is privacy-preserving – i.e., $h(m, e) = [h^1(m, e_1), \ldots, h^n(m, e_n)]$ – then one can add, for each member i, an action-taking device. This is a subsidiary memoryless task (machine) which takes as an input the main task's terminal output and terminal memory state and assigns to it an output, namely, an action fulfilling the function h^i. If h is not privacy-preserving, several formalisms are possible, including the addition of an $(n+1)$st task (machine) which generates actions and is triggered only at the Tth step; in each of them, an interchange of messages about e must take place at the terminal step in order to carry out h.

To compare two processes with regard to the best value of $\mathcal{E}\psi_T(e)$ achievable with a given cost, one needs knowledge of the cost of operating the associated one-step designs-with-memory *in C time units for alternative values of C*; the cost of alternative speeds of the final action-taking step is needed as well.[49] As in the

[49]It is easy to check that if $\Delta\rho_T(e) > 0$, $\Delta^2\rho_T(e) < 0$, then the T which maximizes $\mathcal{E}\psi_T(e)$ on a fixed interval $1 \leqq T \leqq \tilde{T}$ cannot become smaller if C is replaced by $C' < C$, and if C' is sufficiently small, then it becomes larger. But the effect of reducing C on the *clock* time CT [where T maximizes $\mathcal{E}\psi_T(e)$ on the given interval] is indeterminate [T. Marschak (1972)].

discussion of Section 2.2 above, models of observing, transmission, computing, and action-taking are required. The difficulties of existing models, surveyed there, arise again. In the case of transmission, the setting of the coding theorems is again inappropriate. If every *step* is to be completed in exactly C time units, no delay can be permitted.

As for the computing effort required in each task – the effort required to find the output and state to be associated with a given input/state pair – the theory of finite-state machines appears to hold some promise. It studies the decomposition of a given machine into a number of smaller ones which are linked together and realize the original machine. One might reasonably take as a unit of cost a certain "basic" machine (e.g. a two-input, two-output, two-state machine). It may be possible to find the smallest number of them which realize the original machine (algorithms for doing so are available for certain cases). If a given speed of the basic machine has a given cost, and if reasonable assumptions can be made about the way the speed of a machine built out of basic machines is obtained from the basic machine's speed, then the theory may eventually permit one to characterize usefully the costs of alternative speeds for the achievement of a given task.

If one is content to compare designs with regard to gross performance only, then the detailed study of $\mathscr{E}\psi_T(e)$ for fixed T and C may be fruitful. Thus, if ρ is taken to be quadratic, as in (4.27), and the process is the gradient process, then it is possible to find, from the solution to the difference-equation system, an explicit expression for $\mathscr{E}\psi_T(e)$. One can then study, for example, the question considered above at the end of Section 3.3: if Q is a matrix with ones on the diagonal and q's off it, how does increasing $|q|$ (the "strength of interaction") change the amount by which $\mathscr{E}\psi_T$ falls short of the expected period-payoff achieved in a "centralized" process which always yields (after an elapsed time equal to CT) a payoff-maximizing action? It turns out [T. Marschak (1972) and Simonovits (1976)], under certain assumptions about the way the "approximating" fineness λ is chosen, that increasing $|q|$ does *not* increase the penalty of "decentralizing" (operating the privacy-preserving gradient process rather than the centralized one).

The pre-equilibrium study of adjustment processes remains difficult and undeveloped. Yet theorists continue to produce additional adjustment processes[50] and claim informational (and incentival) virtues for them. It is difficult to see how those claims can be defended without pursuing a pre-equilibrium analysis of the sort we have outlined.[51]

[50] Some of which are described in Chapter 29.

[51] A major complicating step, which some might argue is essential for realism, is to permit the time required to complete one step of an adjustment process to vary and either to be determined uniquely by the initiating environment e or to be a random variable whose distribution depends on e. This may make relevant transmission models of the Shannon type, where transmission times differ from one

5. General concepts and issues in organization design

5.1. *Replacing "centralization versus decentralization" by "resources for coordination versus resources for local expertise"*

A one-step design is a complex object, and an adjustment process, requiring repeated operation of a design with memory, is still more so. Given the sets E and A, a rich variety of designs can be constructed. It is natural that discussions of organization design have struggled to introduce order by defining (loosely or otherwise) *categories* of designs, and even *orderings* of designs, and by making conjectures that some aspects of a design's performance might be deducible from its membership in a category or its place in an ordering. The most notorious attempts to categorize are associated with the terms "centralized" and decentralized", and the most notorious attempts to order with the term "more centralized than".

The terms have appeared so often that no survey can ignore them. "Decentralized" and "centralized" are typically used as though they denoted colors, understood by everyone in the same way without further elucidation. Yet in many discussions it is quite unclear to what objects the terms are being applied and how the terms are defined. We can attempt to catalogue some of the current usages, but only by first defining a class of objects. Let that class be noiseless one-step designs (possibly with memory) for a given set E and a given set A. Then the following usages (and others as well) have occurred:

(a) A design is "centralized" if there is one member, say i^*, whose output includes a message to every other member, and member i's output, $i \neq i^*$, is uniquely determined by the message received from i^* [see e.g. Camacho (1980)].

---

message to another. It also permits models in which the time required to perform a task – to assign outputs to inputs – varies from one input/output pair to another. The step completion time of a temporally homogeneous process with noiseless tasks then varies from one repetition of the process to another. If the tasks carried out in one step of the process are noisy, then for a given repetition of the process the completion time of a step is random – but the action generated at the terminal step, given the initiating environment, is random as well; we have not dealt with such "noisy" processes at all. Variable task-completion times are studied by Drenick (1977), who obtains some results about the probability distribution of completion times for certain tasks, given the distribution of inputs for each task.

Suppose, for a (noiseless) temporally homogeneous process, the step-completion time C becomes a function $C(e)$ of the initiating environment e (though a function choosable by the designer). Then the condition $CT<1$ has to be replaced. One possibility is the condition $T \max_{e \in E} C(e) < 1$. A second possibility is to adopt the convention that the process is stopped either at T, chosen once and for all, or at the end of the period wherein the initiating environment prevails, whichever comes first. Under either assumption, condition (4.21d) again implies the existence of a unique best once and for all choice of T.

(b) The same usage as (a), except that i's output is uniquely determined by the message received from i^* *and* the observation on the current e made by i.

(c) A design is "decentralized" if, for every member i, the action chosen [i.e. the value given to those components of i's output which are attributes of the organization's action] is independent of any message received from another member [Camacho (1980)].

(d) A design is "decentralized" if each member's environmental observing accords with a "natural" partitioning of E: a member who is made responsible for a given attribute of the organization's action is thereby naturally endowed with an associated partitioning on E. This association is given to the designer and defines the class of decentralized designs. In particular, a design with memory is "decentralized" if it achieves the typical step of a temporally homogeneous adjustment process which is privacy-preserving relative to the given n-tuple of natural partitionings on E [J. Marschak and R. Radner (1971, ch. 6), Hurwicz (1971, 1972), and many other writings].

(e) A design is "centralized" if there is one member for whom the partitioning on E induced by the input he receives (comprised of messages and observations) is at least as fine as the sum[52] of the observational partitionings of all other members [Hurwicz (1971, 1972), and many other writings].

All these usages define extreme categories. Many discussions loosely suggest that one can meaningfully speak of a greater or smaller distance from one of these extremes. Others explicitly use the terms "more centralized" or "more decentralized". Two possible definitions that appear to capture some of the intended usage follow. Let A and B be noiseless, memoryless one-step designs with the sets in i's observational partitionings $\mathscr{P}^i$ indexed by e_i, $i \in N$, and with the input and output sets $(\bar{X}^i, X^i)$ identical in both designs for every member i.

(a) Design A is "more centralized" than B if:

(i) In both designs there is a member i^* whose output x^{i^*} determines a subset of the output set X^k for every member $k \neq i$; in design A the subset is $\sigma_{Ax^{i^*}}(X^k)$, in design B it is $\sigma_{Bx^{i^*}}(X^k)$.

(ii) In design A, there is, for every $k \neq i$, for every value of e_k, and for every x^{i^*} in X^{i^*}, a function $\theta^A_{e_k x^{i^*}}$ on X^k such that k's output maximizes $\theta^A_{e_k x^{i^*}}$ on $\sigma_{Ax^{i^*}}(X^k)$; similarly, in B there is a function $\theta^B_{e_k x^{i^*}}$ on X^k such that k's output maximizes $\theta^B_{e_k x^{i^*}}$ on $\sigma_{Bx^{i^*}}(X^k)$.

(iii) For every $x^{i^*} \in X^{i^*}$ and every $k \neq i^*$, $\sigma_{Bx^{i^*}}(X^k) \supseteq \sigma_{Ax^{i^*}}(X^k)$ with strict inclusion for at least one $k \neq i^*$.

(iv) For at least one $x^{i^*} \in X^{i^*}$, at least one $k \neq i^*$, and at least one value of e_k, every maximizer of $\theta^B_{e_k x^{i^*}}$ on $\sigma_{Bx^{i^*}}(X^k)$ is distinct from all the maximizers of $\theta^A_{e_k x^{i^*}}$ on $\sigma_{Ax^{i^*}}(X^k)$.

[52] The sum of two partitionings on E, say $\mathscr{P} = \{S\}$ and $\mathscr{P}^* = \{S^*\}$ (where S, S^* denote typical sets) is the collection of sets $\{T: T$ is non-empty; for some $S \in \mathscr{P}$, $S^* \in \mathscr{P}^*$, $S \cap S^* = T\}$.

(v) There is no other member with the properties just attributed to i^*.

(b) Consider the partitioning on E induced by the n-tuple of *message* outputs which the design generates in response to an element of E. If the partitioning is finer for A than for B, then A is "more centralized" than B.

The first definition corresponds to the usage wherein the more a center (member i^*) constrains the sets (of messages and actions) out of which each other member chooses, the more "centralized" is the design. But it is not enough just to compare sizes of choice sets, because if, for some member, we have only specified a set of choosable outputs, then we have not specified a design. A rule for choice must be specified as well – and the definition uses a maximizing rule. The rule may be interpreted as the *normal*, *voluntary behavior* of a competent member confronted with an assigned choice set, a local observation, and some guidance from the center as to how the choice is to be made. One difficulty, however, is that there may in fact be *several* members such as i^*, so that if we did not add condition (v), then the "more centralized" ordering of A and B might be reversed as we go from one such "center" to another. It would be more satisfying, in such a case, to be able to select that one "center" who really matters, but there appear to be no appealing guidelines for such a selection. Another difficulty is that searching for a *unique* maximizer in a large set may be viewed by many as more "constraining" than searching for one of many tied maximizers (one of many equally acceptable outputs out of which the design arbitrarily selects one) in a small set. The definition masks this possibility.

The second definition attempts to capture the usage wherein "more centralized" means that the entire bundle of inter-member communications which follows a new environment is richer in information about that environment – as it would be, for example, if $n-1$ members told the remaining member a great deal about their local observations. Here a difficulty is that some rich message bundles may in fact have little influence on the actions generated by the design; the "centralization" of information may be to very little purpose and discarding some of the message bundle would define a new "less centralized" design which is only trivially distinct from the original one.

Both definitions are modest in that they provide only partial orderings. Nevertheless, a definition which combined *both* usages and which also met the difficulties just sketched would have to be complicated indeed. The outlook is not promising for propositions which use such a corrected definition and can answer, in interesting cases, the question "when is more (less) centralization good?"

If one despairs of sharpening the centralization (decentralization) concept so that it usefully classifies and orders designs, there is an alternative which may yet prove superior. It is suggested by the observation that in real organizations a very concrete choice facing a designer is the allocation of *informational resources* – e.g., administrative manpower – among various organizational activities, in particular, the allocation of resources *among a center and "local" members*. It may be useful, then, to group activities, and to compare designs with regard to the efforts devoted to each group and the resources required for these efforts.

To do so, one member, say n, is identified as the center. Consider a design in which each member i, $i=1,\ldots,n-1$, observes the environment (the observation is one of his input components) and forms one of his output components, namely, a message – a *report* – to n. The center's function is to coordinate – to use a common term – but he also makes an environmental observation of his own. In response to the observational component of his input *and* the "reports" component, the center forms his output, which is an $(n-1)$-tuple of messages called *instructions*, one to each member $i \neq n$. Member $i \neq n$ responding to the instruction received (an input component) *and* to his own observation, forms the other component of his output, namely, the value of one or more action attributes. The one-step requirement is satisfied (a "report" message is not based on messages received by the reporter which are influenced by the reporter himself; and similarly for "instruction messages").

Suppose we regard observing and action-taking by the members $i \neq n$ as *local-expertise* activities, and regard reporting by $i \neq n$ (and the reading of the reports by n), instructions by n, and observing by n as *coordination* activities. In a finite design a non-probabilistic indicator of coordination effort would be *fineness* of reporting, of central observing, and of instructing. A non-probabilistic indicator of local-expertise effort is fineness of local observing and action-taking. Fineness may be measured by the sizes of the corresponding finite sets (the number of possible reports, instructions, central observations, local observations, actions). Suppose there is a total cost function on these set sizes, namely, $C = C_1 + C_2$, where C_1 is increasing in the coordination set sizes and C_2 in the local-expertise set sizes. Consider two designs with equal total cost, i.e. with a common "budget". We can say that design I *devotes more of the budget to coordination and less to local expertise* than design II if C_1 is greater for I than for II. The issue to be studied then becomes: given the sets A and E, for what payoff functions, probability distributions on E, and functions C_1 and C_2, is it desirable (with regard to expected payoff) to choose a design in which a high proportion of the budget is devoted to coordination?[53] To define a design with given set sizes, the observational partitionings $\{\mathscr{P}^i\}_{i\in N}$ on E must be specified and each partitioning must contain the prescribed number of sets; and the permitted action-attribute values must be chosen for each $i \neq n$. Once this is done, the

[53] Camacho (1972) considers payoff functions which have an explicit "poor-coordination" penalty and an explicit "inadequate-local-expertise" penalty. The question studied is then the effect of changing those penalties on the relative performance of centralized and decentralized designs, where those terms are given the definitions (a) and (c) above. The approach is promising since it appears to permit replacing those extreme categories by an ordering and studying the effect of the two penalties on the desirability of "greater centralization".

choice of the best finite design among those for which the report and instruction sets have the required sizes becomes purely a combinatorial matter. Suppose, for example, that $n = 3$, and that members 1 and 2 each have three possible observations, two possible reports, and two possible actions (action-attribute values) while member 3 has two possible observations and two possible instructions to each other member. Then, given the three partitionings on E, and, for 1 and 2, the two possible actions, there are $9 \cdot (62)^2 R$ possible designs, where R is the number of ways eight objects can be partitioned into four non-empty sets.[54] It may turn turn out that for some highly symmetrical examples there are algorithms drastically cutting the number of alternatives to be examined in finding an expected-payoff-maximizing member of the given class of designs.[55,56]

5.2. *Hierarchies*

Hierarchies have been central to organization theories associated with the term "general systems" [Carzo and Yanouzas (1967), Lasdon (1970), Mesarovic, Macko and Takahara (1970), and Ruefli (1978)] and more recently in attempts to model centrally directed (Soviet-type) economies [Montias (1973, 1976)]. For a noiseless one-step design and given sets A and E, an *r-tier hierarchy* is simply an r-tuple $(G_1, \ldots, G_r)$ of disjoint groups, or *tiers*, of members; G_s is *higher* than G_t if $s < t$. An input for every member i of tier l consists of i's observations on E and messages received from one or more members in tier $l - 1$.

A hierarchy model may be natural where, for institutional reasons, the choosable designs are confined to those with the following property: many of the possible messages in the design can be viewed as determining, together with the recipient's own observations, both a permissible subset of the recipient's output set and a criterion for choosing an output out of that subset. Formally, one assumes a generalization, from two tiers to many, of properties (i) and (ii) in our first definition of "more centralized than". Those properties are now displayed

[54] There are three ways 1 or 2 can partition his three possible observations into two non-empty sets, corresponding to two reports; R ways 3 can partition the eight possible combinations of reports received and central observations made into four non-empty sets, corresponding to four instruction pairs; 62 ways 1 or 2 can each partition his six possible combinations of local observation and instruction received into two non-empty sets, corresponding to his two possible actions.

[55] Such an algorithm is sketched in T. Marschak (1980).

[56] The share of an organization's administrative resources to be given to "line" versus "staff" functions is discussed in general administrative literature. It could be modeled in the same spirit as "coordination versus local expertise". "Staff" resources are assigned to the center and permit greater fineness for his choices. (Or staff may be modeled as a group of members all of whom send messages to the center and adjust no action attributes.) "Line" resources are assigned to those members who may be arranged in hierarchies, the concept to be discussed next. Drenick (1977) studies aspects of the line-versus-staff issue for designs whose task-completion times are variable.

with respect to a member of a tier and another member, belonging to the tier immediately below. One might wish to add, in the definition of hierarchies, the further property that environmental information becomes poorer at higher tiers, i.e. the sum of the observational partitionings of the members of tier l is coarser than the sum of the partitionings of the members of tier $l+1$.

Some of the hierarchy theorizing has an incentive element. The output selected by a member in response to a message from above must be, in an appropriate sense, acceptable to that member, who has his own preferences over the possible choices [Mesarovic, Macko and Takahara (1970), Charnes, Clower and Kortanek (1967)].[57]

With regard to hierarchies, a further definition of "more centralized than" is possible. Consider a fixed set A and a fixed set E and two one-step designs which are both r-tier hierarchies. Consider the assignment of action attributes to members. Suppose that if an attribute is assigned to a member of tier s in design I and a member of tier t in design II, then $s \leqq t$, with "$<$" for at least one attribute. Then design I is called "more centralized" than design II.

One can certainly formulate general conjectures about the desirability, for given A, E, and n (the number of members) of many-tier versus few-tier hierarchies, and of more centralized hierarchies versus less centralized ones; and loose conjectures of this sort have been made. But such conjectures ought to specify a class of payoff functions, probability distributions on E, and functions yielding a design's cost, and they ought to interpret "desirability" as expected payoff attainable for a given cost. No conjectures of that form have yet been checked, not even for very simple cases.[58]

6. Concluding remarks

We have tried to view an assortment of recent economic studies in the same way: as steps toward characterizing those organization designs which do well, accord-

[57] Miyasawa (1973) considers the two-member team example (the "shipyard" example) with regard to which J. Marschak first illustrated team-theory concepts [an example treated further in Chapter 4 of J. Marschak and Radner (1972)]. Miyasawa makes the team a two-tier hierarchy by adding a center who gives commands specifying the sets out of which the two members can choose, and studies those commands for which the two members' Nash-equilibrium choices are good from the designer's point of view.

[58] Some discussions [e.g., Koopmans (1969)] have complicated the hierarchy concept, and perhaps made it more realistic, by introducing *activities* in which each member engages. One speaks, then, of a hierarchy with regard to a particular activity. In our terminology, one would divide the set of action attributes into subsets, each called an activity. A hierarchy with regard to an activity is then defined as before: each member of a tier receives messages from the tier above but now these messages constrain and guide him only with regard to his choice of a value for any action attribute in the given activity. Not every member, then, belongs to the hierarchy associated with a given activity.

ing to some measure of gross performance, with the informational and administrative resources they require. The steps turn out to be diverse and modest, but the problem is difficult. Piecing the assorted contributions together, one is still far indeed from a unified theory of efficient organization design.

The main stumbling block remains the modeling of technology and cost. Some elements of cost have been studied intensively and even elegantly. We have examined briefly the Shannon theory in connection with transmission, and the theory of finite-state machines in connection with the assignment of output/state pairs to input/state pairs in one-step designs with memory. We have not considered other well-developed efforts: the theory of pattern recognition in connection with observing [Fu (1974)], models from psychophysics and mathematical psychology in connection with various issues in the characterization of human information-processing capacities [e.g., Chevanny and Dickson (1974), Luce (1963), and Miller (1963)].

None of these topics were developed as part of a unified theory of organization design and so they appear difficult to fit into one. It may well be that only the *observing* and modeling of real organizations – something economists have been reluctant to do[59] – can lead to models of technology and cost which fit usefully into a unified theory of organization design.

On the other hand, techniques *have* been developed for study of a design's gross performance – e.g. the computation of expected payoff for a given team information structure – and these will remain useful in efficiency studies when good cost models do become available.

The theory, even in its present form, has already been useful in revealing how difficult it is (1) to define certain widely current terms sharply and agreeably to most usages; and (2) to verify certain widely held conjectures. The notion of supervision and hierarchy, boldly proclaimed in a thousand "organization charts", and the pervasive terms "centralization" and "decentralization" are examples of the first. An example of the second is the conjecture that "the stronger are externalities, the more centralized the organization ought to be". Theory-building shows us the many meanings one can give to "externalities", "stronger", "more centralized", and "ought", and leads us to study small examples, some supporting a version of the conjecture, others contradicting a version of the conjecture.[60]

A designer of a real human organization cannot in fact expect to choose designs as we have defined them, because the behavior of members is not in the designer's control. He may be able to specify the set of actions and messages out of which each member can choose and the administrative and informational

[59] Economists, of course, observe and model one organization of monstrous size and complexity – the market economy. Recently, theories of organization design have had some influence on the observing, modeling, and comparing of socialist economies [e.g., Duffy and Neuberger (1972), T. Marschak (1968), and Montias (1976)].

[60] See, e.g., Camacho (1980), J. Marschak and R. Radner (1970, ch. 7), and T. Marschak (1972).

resources to be placed at each member's disposal. He will then have to make behavioral *predictions* about the choices members will in fact make. These predictions determine an "expected" design, and the assignment of resources determines a cost. The expected-design/cost pair can then be ranked against others. If the designer is also free to choose the structure of members' *rewards*, then one guide to good predictions may possibly be norms for rational behavior in game situations. That approach is pursued in Chapter 28.

Which behaviorally robust designs – those wherein members are likely to behave in the way described by their assigned tasks – are also efficient designs? That is a complex question; it will have to be answered eventually by joint work of theorists *and* observers of the real behavior of the members of organizations.

References

Arrow, K. J. and L. Hurwicz (1958), "Gradient methods for concave programming", in: K. J. Arrow, L. Hurwicz and H. Uzawa, eds., Studies in Linear and Non-Linear Programming. Stanford.

Arrow, K. J. and L. Hurwicz (1960), "Decentralization and computation in resource allocation", in: R. W. Pfouts, ed., Essays in Economics and Econometrics, pp. 34–104. Chapel Hill: University of North Carolina Press.

Arrow, K. J. and R. Radner (1978), "Allocation of resources in large teams", Econometrica, 47:361–386.

Camacho, A. (1972), "Centralization and decentralization of decision-making mechanisms: A general model", Jahrbuch der Wirtschaft Osteuropas, 3:45–66.

Camacho, A. (1980), "Information, externality, and behavior in the performance of decision-making systems", Journal of Comparative Economics, 3:91–115.

Carzo, R. and J. N. Yanouzas (1967), Formal Organization: A Systems Approach. New York: Irwin–Dorsey Press.

Charnes, A., R. W. Clower and K. O. Kortanek (1967), "Effective control through coherent decentralization with preemptive goals", Econometrica, 35:294–320.

Chevanny, N. L. and G. H. Dickson (1974), "An experimental evaluation of information overload", Management Science, 20:1335–1344.

Chu, K. C. (1976a), "Designing information structures for quadratic decision problems", Report RC6191, IBM Thomas J. Watson Research Center.

Chu, K. C. (1976b), "Comparison of information structures in dynamic systems", in: Y. C. Ho and S. K. Mitter, eds., New Directions in Large-Scale Systems. New York: Plenum Press.

Drenick, R. (1977), "A mathematical theory of organization", Unpublished. Brooklyn: Polytechnic Institute of New York.

Duffy, W. and E. Neuberger (1972), "Toward a decision-theoretic approach to the study of economic systems", Jahrbuch der Wirtschaft Osteuropas, 3:67–86.

Feldman, A. (1973), "Bilateral trading processes, pairwise optimality, and Pareto optimality", Review of Economic Studies, 40:463–474.

Fu, K. S. (1974), Syntactic Methods in Pattern Recognition. New York: Academic Press.

Groves, T. and S. Hart (1982), "Efficiency of resource allocation by uninformed demand", Econometrica, 50:1453–1482.

Groves, T. and R. Radner (1972), "Allocation of resources in a team", Journal of Economic Theory, 3:415–441.

Hartmanis, J. and R. Stearns (1966), Algebraic Structure Theory of Sequential Machines. Englewood Cliffs: Prentice-Hall.

Huffman, D. A. (1962), "A method for the construction of minimum redundancy codes", Proceedings IRE, 40:1098–1101.

Hurwicz, L. (1960), "Optimality and informational efficiency in resource allocation mechanisms", in: K. Arrow, S. Karlin and P. Suppes, eds., Mathematical Methods in the Social Sciences. Stanford: Stanford University Press.

Hurwicz, L. (1971), "Centralization and decentralization in economic processes", in: A. Eckstein, ed., Comparison of Economic Systems. Berkeley: University of California Press.

Hurwicz, L. (1972), "On informationally decentralized systems", in: C. B. McGuire and R. Radner, eds., Decision and Organization. Amsterdam: North-Holland.

Hurwicz, L. (1973), "The design of mechanisms for resource allocation", American Economic Review, Papers and Proceedings, 63:1–30.

Hurwicz, L. (1977), "On the dimensional requirements of informationally decentralized Pareto-satisfactory processes", in: K. J. Arrow and L. Hurwicz, Studies in Resource Allocation Processes. Cambridge: Cambridge University Press.

Hurwicz, L. and T. Marschak, (1984), "Discrete allocation mechanisms", Working Papers in Economic Theory and Econometrics. Berkeley: Center for Research in Management, University of California.

Keeney, R. (1972), "Utility functions for multiattributed consequences", Management Science, 18:276–287.

Koopmans, T. C. (1969), "Notes on a social system composed of hierarchies with overlapping personnel", Orbis Economicus, July.

Lasdon, L. S. (1970), Optimation Theory for Large Systems. New York: Macmillan.

Luce, R. D. (1963), "Detection and recognition", in: R. Luce et al., eds., Handbook of Mathematical Psychology. New York: Wiley.

Marschak, J. (1971), "Economics of information systems", in: M. Intriligator, ed., Frontiers of Quantitative Economies. Amsterdam: North-Holland.

Marschak, J. (1977), "Optimal incentives in organizations", in: Quantitative Economics: Essays in Honor of William Krelle. Frankfurt: Anton Hain.

Marschak, J. (1979), "Efficient organizational design", in: Economic Theory for Economic Efficiency: Essays in Honor of A. P. Lerner. Cambridge: MIT Press.

Marschak, J. and R. Radner (1971), Economic Theory of Teams. New Haven: Yale University Press.

Marschak, T. (1959), "Centralization and decentralization in economic organizations", Econometrica, 27:399–430.

Marschak, T. (1968), "Centralized versus decentralized resource allocation: The Yugoslav 'Laboratory'", Quarterly Journal of Economics, 82:561–587.

Marschak, T. (1972), "Computation in organization: The comparison of price mechanisms and other adjustment processes", in: C. B. McGuire and R. Radner, eds., Decision and Organization. Amsterdam: North-Holland.

Marschak, T. (1980), "The best use of 'information budgets' in purposive organizations", International Journal of Policy Analysis and Information Systems, 4:37–46.

McGuire, C. B. (1963), "Some team models of a sales organization", Management Science, 7:101–130.

Mesarovic, M. D., D. Macko and Y. Takahara (1970), Theory of Hierarchical Multilevel Systems. New York: Academic Press.

Miller, G. A. (1963), "The magical number seven plus or minus two: Some limits in our capacity to process information", in: R. Luce et al., eds., Readings in Mathematical Psychology. New York: Wiley.

Miyasawa, K. (1973), "Decision processes in organizations", in: Proceedings of the IIASA Planning Conference on Design and Management of Large Organizations. Laxenburg: International Institute for Applied Systems Analysis.

Montias, J. M. (1973), "A framework for theoretical analysis of economic reforms in Soviet-type economies", in: M. Bornstein, ed., Plan and Market. New Haven: Yale University Press.

Montias, J. M. (1976), The Structure of Economic Systems. New Haven: Yale University Press.

Mount, K. and S. Reiter, "The informational size of message spaces", Journal of Economic Theory, 8:161–192.

Oniki, H. (1974a), "The cost of communication in economic organizations", Quarterly Journal of Economics, 58:529–550.

Oniki, H. (1974b), "The informational cost of economic decision-making: Some simple cases", Paper presented at 1974 Annual Meeting of the Econometric Society, New York.

Radner, R. (1972), "Allocation of a scarce resource under uncertainty: An example of a team", in: C. B. McGuire and R. Radner, eds., Decision and Organization, Chapter 11. Amsterdam: North-Holland.

Radner, R. (1975), "Satisficing", Journal of Mathematical Economics, 2:253–262.

Radner, R. and M. Rothschild (1976), "On the allocation of effort", Journal of Economic Theory, 10:358–376.

Reiter, S. (1974a), "The knowledge revealed by an allocation process and the informational size of message spaces", Journal of Economic Theory, 8:389–396.

Reiter, S. (1974b), "Informational efficiency of iterative processes and the size of message spaces", Journal of Economic Theory, 8:193–205.

Reiter, S. (1977), "The $(new)^2$ welfare economics", American Economic Review: Papers and Proceedings, 67:226–234.

Ruefli, T. (1971), "A generalized goal decomposition model", Management Science, 17:B505–B518.

Shannon, C. E. (1949), The Mathematical Theory of Communication. Urbana: University of Illinois Press.

Shannon, C. E. (1960), "Coding theorems for a discrete source with a fidelity criterion", in: R. E. Machol, ed., Information and Decision Processes. New York: McGraw-Hill.

H. Simon, (1972), "Theories of bounded rationality", in: C. B. McGuire and R. Radner, eds., Decision and Organization. Amsterdam: North-Holland.

Simonovits, A. (1976), "Some properties of a decentralized adjustment process", Management Science, 22:883–891.

Uzawa, H. (1958), "Iterative methods for concave programming", in: K. J. Arrow, L. Hurwicz and H. Uzawa, eds., Studies in Linear and Nonlinear Programming. Stanford: Stanford University Press.

Welch, R. (1980), "Vertical versus horizontal communication in economic processes", Review of Economic Studies, 47:733–746.

Chapter 28

INCENTIVE ASPECTS OF DECENTRALIZATION*

LEONID HURWICZ

University of Minnesota, Minneapolis

1. Resource allocation as information processing

A resource-allocation mechanism is sometimes viewed as a gigantic information processing system. Such a system utilizes the knowledge dispersed among economic agents concerning their preferences, technologies, and endowments in order to determine how resources should flow. Information is transmitted between economic units and processed by them through computations which result in allocative decisions. Alternative mechanisms can then be compared in terms of their efficiency in processing information adequate for optimal decisions. It is from this point of view that Hayek (1945) stressed the merits of the competitive-market mechanism. The informational aspects of resource-allocation mechanisms were formalized and analyzed in a number of papers in the last twenty-five years [Hurwicz (1960), Mount and Reiter (1974), Reiter (1974a, 1974b), Hurwicz (1972, 1977), Walker (1977), Osana (1978), Sato (1981), and Jordan (1982)].

From an informational point of view an economic mechanism may be thought of as an exchange of messages. In line with the tâtonnement idea, an outcome is determined when the exchange of messages is in a stationary position.

Denote by $\mathscr{M}^i$ the "language" of the messages to be used by agent i. Then the process of exchanging messages may be represented by a system of difference equations of the form

$$m^i_{t+1} = f^i(m^1_t, \ldots, m^n_t; e), \qquad t = 0, 1, \ldots, \quad i \in \{1, \ldots, n\} \equiv N,$$

where n is the number of agents, e represents the *economic environment* (preferences, endowments, technologies) and $m^i_t \in \mathscr{M}^i$ for all $t = 0, 1, \ldots,$ and all $i \in N = \{1, \ldots, n\}$.[1]

*This research was partially supported by National Science Foundation Grant No. SES-8208378.

[1] f^i is called agent i's *response* function.

Handbook of Mathematical Economics, vol. III, edited by K.J. Arrow and M.D. Intriligator
© 1986, Elsevier Science Publishers B.V. (North-Holland)

A message n-tuple $\overline{m}=(\overline{m}^1,\ldots,\overline{m}^n)$ is stationary if it satisfies the equation system

$$\overline{m}^i=f^i(\overline{m}^1,\ldots,\overline{m}^n;e),\qquad i\in N. \tag{1.1}$$

Denote by Z the space of conceivable resource allocations. Then, given the stationary message n-tuple $\overline{m}$, the resulting allocation $\bar{z}$ is determined by the outcome function $h:\mathscr{M}\to Z$, where[2] $\mathscr{M}=\mathscr{M}^1\times\cdots\times\mathscr{M}^n$, so that

$$\bar{z}=h(\overline{m}).$$

The response functions f^i are assumed to be defined for environments e which are elements of a class E of a priori admissible environments. An *adjustment process* for E is then defined as $\pi=(\mathscr{M}^1,\ldots,\mathscr{M}^n;f^1,\ldots,f^n;h)$.

Note that the ith equation of the system (1.1) may be interpreted as defining a correspondence $\mu^i:E\Rightarrow\mathscr{M}$ such that

$$m\in\mu^i(e),\qquad m=(m^1,\ldots,m^n),$$

if

$$m^i\in f^i(m^1,\ldots,m^n,e).$$

In turn, we also have the correspondence $\mu:E\Rightarrow\mathscr{M}$, defined by

$$m\in\mu(e)\quad\text{if and only if}\quad m\in\bigcap_{i\in N}\mu^i(e).$$

In this formulation a message n-tuple $\overline{m}$ is stationary (or: equilibrium) for e if and only if

$$\overline{m}\in\mu(e).$$

A mechanism is then defined as $(\mathscr{M},\mu,h)$ where $\mu=\bigcap_{i\in N}\mu^i$ and $\mathscr{M}$ may, but need not, be a Cartesian product of some $\mathscr{M}^i$'s.

Historically, the economists' interest has been focused on informationally decentralized mechanisms. An adjustment process is said to be *privacy-preserving* if each response function f^i depends only on the characteristic e^i of the ith agent. Typically, e^i is defined in terms of the endowments, preferences, and technologies of that agent; e.g. $e^i=(\omega^i,R^i,T^i)$ where ω^i is the ith agent's initial endowment, R^i his/her preference relation, and T^i his/her technology. We may

[2] $\mathscr{M}$ is called the *message space*.

then write

$$m_{t+1}^i = f^i(m_t^1,\dots,m_t^n; e^i), \qquad t=0,1,\dots, \quad i \in N.$$

It is understood that $(e^1,\dots,e^n) = e$.

A corresponding formulation of the privacy-preserving property in terms of the equilibrium correspondences is

$$\mu(e) = \bigcap_{i \in N} \mu^i(e^i),$$

where, this time, $\mu^i: E^i \Rightarrow \mathcal{M}$, and E^i is the a priori admissible domain of characteristics of the ith agent. Naturally, we have $E = E^1 \times \cdots \times E^n$.

A *message mechanism* $\chi = (\mathcal{M}, \mu, h; E, Z)$ on the class of environments E to an outcome space Z defines a correspondence, say $F: E \Rightarrow Z$, by

$$F(e) = \{z \in Z: z = h(m), m \in \mu(e) \text{ for some } m \in \mathcal{M}\},$$

for every e in E. F is said to be the (*static*) *performance correspondence* of the message mechanism.[3] Given a correspondence $G: E \Rightarrow Z$ and a message mechanism χ, we say that χ *realizes* G *over* E if $F(e) \neq \emptyset$ and $F(e) \subseteqq G(e)$ for all e in E. We say that χ *fully realizes* G *over* E if $F(e) \neq \emptyset$ and $F(e) = G(e)$ for all e in E.

It is important to note that the static aspects of performance depend on the response functions but only through the equilibrium message correspondence μ. I.e. if two n-tuples of response functions generate the same μ, then the resulting performance correspondence will be the same.

Much of the traditional welfare economics takes a mechanism (e.g. perfect competition) as given and investigates the properties (e.g. Pareto-optimality) of its (static) performance correspondence.

More recently, the reverse problem has come under investigation: given a correspondence $F: E \Rightarrow Z$, viewed as a social desideratum, are there mechanisms which (fully) realize it? In particular, one question has been whether there exist *decentralized* (privacy-preserving) mechanisms realizing the Pareto correspondence over "non-classical" sets of environments, such as those with indivisibilities, non-convexities, externalities, etc., where the competitive mechanism is known to fail. On the other hand, for the "classical" environments, in which the competitive mechanism is known to realize the Pareto correspondence, the problem has been whether there exist mechanisms that are informationally equally or more efficient (e.g. with message spaces of lower dimension) but that

[3] Mount and Reiter (1974) and Reiter (1977).

still realize the Pareto correspondence. The realization of other correspondences (e.g. those that are individually rational or envy free) has also been studied.

2. Decentralization in economies with public goods

The preceding analysis treats the economic agent as an information processing (communicating, computing) unit. But this, of course, is very inadequate, since economic agents also have motivations and preferences and their levels of satisfaction depend on the allocative decisions. Because of the dispersion of knowledge, it is usually possible for an economic agent to transmit false information. If this possibility is ignored in the design of a mechanism, the system is likely to malfunction. In early discussions, this danger became particularly obvious in two situations: the behavior of enterprise managers in Lange–Lerner socialist economies,[4] and in the Lindahl scheme of allocation of public goods.[5] For our purposes, the latter problem constitutes a particularly convenient point of departure.

Consider an economy with two goods (a public good Y and a private good X) and n agents, $n \geqq 1$.[6] The ith agent's initial endowments are denoted by $\omega^i_X, \omega^i_Y \in R_+$, and the preferences are represented by a utility function[7] $u^i(x^i, y)$ which is strictly increasing in each of its arguments. (x^i is the total amount of private good available to agent i; y is the total amount of public good available to all agents.) The private good can be used as an input to produce the public good. This technology is expressed by the input requirements function g: it takes $g(y)$ units of X to produce y units of Y.[8] For instance, we may think of X as leisure–labor, and y as measuring the width of a road to be constructed; then $g(y)$ is the number of units of labor required to construct the road of width y. (We ignore the costs of maintenance and crowding effects.) Denote by t^i the work contribution by the ith agent, so that the amount of good X available to this agent after the contribution is

$$x^i = \omega^i_x - t^i.$$

An allocation $(x^1, \ldots, x^n, y)$ is *feasible* if

$$g(y) \leqq \sum_{i=1}^{n} t^i, \qquad t^i = \omega^i_x - x^i, \qquad i = 1, \ldots, n.$$

[4] Weitzman (1974), Fan (1975), Bonin (1975) and Gindin (1970).

[5] Lindahl (1919), Samuelson (1954), Malinvaud (1971), Drèze and de la Vallée Poussin (1969), and Groves and Ledyard (1977).

[6] The public goods aspect is trivial for $n = 1$.

[7] The representability of preferences by utility functions is convenient but not always essential. The analysis can often be carried out in terms of preference relations. [See e.g. Hurwicz (1979).]

[8] The function g will be called the *input requirement function*. It is the inverse of the production function for the public good.

The well-known Samuelson condition states that an interior allocation $(x^1,\ldots,x^n, y)$ is Pareto-optimal for a differentiable utility function with $u^i_X > 0$ for all i and a differentiable input requirements function g only if

$$\sum_{i=1}^{n} \frac{u^i_y(x^i, y)}{u^i_x(x^i, y)} = g'(y),$$

where u^i_y, u^i_x are the partial derivatives with respect to x^i and y (marginal utilities) respectively. (Since $u^i_x > 0$ for all i, Pareto-optimality also requires that $\sum t^i = y$.)

A particularly simple case is obtained when constant returns prevail, so that $g(y) \equiv ky$ for some $k > 0$. It is then possible to choose the units of commodity measurement so that $k = 1$, so that $g(y) \equiv y$. In what follows we shall often confine ourselves to this case.

A number of allocation systems and equilibrium concepts for economies with public goods have been considered in the literature. These include the "equilibrium with subscription" [Malinvaud (1972, pp. 213–214)][9] and Foley's "public competitive equilibrium" [Foley (1970, p. 67) and Malinvaud (1972, pp. 215–218)]. The former is in general not Pareto-optimal. The latter is Pareto-optimal under classical assumptions, but it has been characterized by Milleron (1972, pp. 432, 453) as "an interpretation of Pareto optimum" rather than "a true definition of equilibrium"; as pointed out by Malinvaud, it is a partly cooperative solution, somewhat analogous to the concept of the core.

The best known concept of a decentralized Pareto-optimal equilibrium was proposed by Lindahl in 1919.[10]

Lindahl's "positive solution"[11] can be interpreted[12] in terms analogous to those involving the Walrasian auctioneer as follows. The auctioneer calls out a proposed price vector $(p_1,\ldots, p_n)$ where p_i is the price to be paid for each unit of the public good Y by agent i. The agent treats the price parametrically, and calculates his demand $(\tilde{x}^i, \tilde{y}^i)$ by maximizing $u^i(x^i, y)$ with respect to its arguments subject to the budget condition $x^i + yp_i \leqq \omega^i_x + \omega^i_y p_i$. Equilibrium is obtained when all agents' demands for the public good are equal, i.e. $\tilde{y}^1 = \cdots = \tilde{y}^n$. Under classical assumptions, such equilibrium is Pareto-optimal.[13] However, as stressed by Samuelson[14] there is the problem of incentives: "It is in the interest of each person to give *false* signals, to pretend to have less interest in a given

[9] See also Milleron (1972, pp. 451–453), Groves and Ledyard (1977, p. 789, ex. 2.1), and Roberts (1976).

[10] Independently proposed by Bowen (1943).

[11] Of the 'just taxation' problem.

[12] See Johansen (1963).

[13] Foley (1970); and Milleron (1972).

[14] Samuelson (1954).

collective consumption activity than he really has, etc." Furthermore, the objections he raised applied not only to the Lindahl solution but to *any* decentralized solution. (In fact, the relevant section of Samuelson's paper is entitled "Impossibility of Decentralized Spontaneous Solution".[15]) Was Samuelson's impossibility claim correct? And was he right in contrasting[16] the impossibility when public goods are present with the self-policing nature of markets for private goods?

To try to answer these questions we shall first formulate them in an analytically tractable manner, with sufficient generality to apply both to public and private goods, and to a variety of mechanisms.

3. Mechanisms (game forms) and implementation of social choice rules

Let E^i, $i \in \{1,\dots,n\}$, $n \geqq 1$, denote the class of a priori admissible characteristics[17] for agent i, and $E = E^1 \times \cdots \times E^n$ denote the class of a priori admissible environments (economies). It is assumed that the E^i's, and hence E, are known to the designer. The designer also knows a social choice rule F, i.e. a correspondence[18] from E to the set $\mathscr{A}$ of conceivable[19] outcomes (resource allocations). The designer's task is to find a mechanism whose outcomes would, in some sense, implement (or be acceptable for) F.

But what is meant by a mechanism in this context? In this exposition, we shall think of the (*game*) *mechanism or game form* as an ordered pair (S, h) where $S = S^1 \times \cdots \times S^n$, S^i is the strategy domain of the ith agent, and h an *outcome function*[20] $h: S \to \mathscr{A}$. (We shall see subsequently how this game mechanism can be related to the notion of a message mechanism [an adjustment process] as defined above in Section 1.) We shall assume that the agents participate in a non-cooperative game, with the S^i's as their strategy domains and with the ith "payoff function" $v^i: S \to R$ defined by $v^i = u^i \circ h$, where u^i is the ith utility function $u^i: \mathscr{A} \to R$.[21] [Note that the domain of the utility function is the set $\mathscr{A}$ of conceivable outcomes. For selfish preferences this can be reduced to the ith agent's component z^i of the outcome $z = (z^1,\dots,z^n)$, $z \in \mathscr{A}$.]

More generally, instead of a utility function, one may merely postulate for each agent a preference ordering on $\mathscr{A}$, to be denoted by $\textcircled{\geq}_i$ or R^i.

[15] See Samuelson (1955).
[16] Op. cit., p. 389.
[17] Typically, the characteristic e^i of the ith agent is given by $e^i = (C^i, \omega^i, R^i, Y^i)$ where C^i is the consumption set, ω^i the initial endowment, R^i the preference relation, and Y^i the production possibility description.
[18] By hypothesis, $F(e) \neq \emptyset$ for all $e \in E$.
[19] $\mathscr{A}$ includes all outcomes known to be possible, but may be broader.
[20] Called by Gibbard (1973) a "game form".
[21] R is the set of real numbers.

We define the *game* $(S; h, e)$ generated by the mechanism (S, h) in the environment e as the game $\Gamma = (S; \Pi)$ in which S^i is the strategy domain of the ith agent, $S = S^1 \times \cdots \times S^n$, $h: S \to R$ is an outcome function, and the *payoff relation* $\Pi_{h,e}$ of the ith player is given by

$$s' \Pi_{h,e} s'' \Leftrightarrow h(s') R^i(e) h(s'') \quad \text{for} \quad s', s'' \in S.$$

Here $R^i(e)$ is the weak preference relation of the ith agent in the environment e. E.g. in a pure exchange economy, if $\bar{e} = (\bar{e}^1, \dots, \bar{e}^n)$, $\bar{e}^i = (\bar{\omega}^i, \bar{R}^i)$, then $\bar{R}^i = R^i(\bar{e})$.[22]

Then a list (n-tuple) of strategies $s^* \in S$, $s^* = (s^{*1}, \dots, s^{*n})$, is called a *Nash equilibrium for the game* $(S; h, e)$ if, for each $i \in N$,[23]

$$h(s^*) R^i(e) h(s^*/s^i, i) \quad \text{for all} \quad s^i \in S^i,$$

where

$$(s^*/t, i) = (s^{*1}, \dots, t, \dots, s^{*n}),$$

with t in the ith place.[24] $h(s^*)$ is then called a *Nash equilibrium outcome* for $(S; h, e)$ (for a *Nash equilibrium allocation* when $\mathscr{A}$ is a space of allocations).

The set of all Nash equilibrium strategy lists for the mechanism (S, h) in the economy e is denoted by $\nu_{S,h}(e)$; it is a subset of S.

The set of all Nash equilibrium outcomes for the mechanism (S, h) in the economy e is denoted by $\mathscr{N}_{S,h}(e)$; it is a subset of $\mathscr{A}$. In fact, $\mathscr{N}_{S,h}(e) = h(\nu_{S,h}(e))$.

We say that a mechanism (S, h) *implements* or is *acceptable for*[25] the social choice rule F in the class of economies E, if, for all $e \in E$,

$$\mathscr{N}_{S,h}(e) \neq \emptyset, \tag{3.1}$$

and

$$\mathscr{N}_{S,h}(e) \subseteqq F(e). \tag{3.2}$$

[Recall that, by hypothesis, $F(e) \neq \emptyset$ for all $e \in E$.]

[22] $\bar{e}^i$ is called the ith agent's characteristic. $\bar{\omega}^i$ is that agent's initial endowment, and $\bar{R}^i$ his/her preference preordering (a total, reflexive, transitive binary relation). $a' R^i a''$ means that agent i either prefers a' to a'' or is indifferent between them.

[23] Where $N = \{1, \dots, n\}$.

[24] Equivalently, $s^* \Pi_{h,e} (s^*/s^i, i)$ for all $s^i \in S^i$ and all $i \in N$.

[25] At times, to avoid ambiguity we use the expression Nash-implementation. In Hurwicz and Schmeidler (1978) a mechanism is called acceptable if, in the sense of the above definition, it is acceptable for the Pareto correspondence.

We say that the mechanism (S, h) *fully implements* the social choice rule F for the class of economies E if, for all $e \in E$,

$$\mathcal{N}_{S,h}(e) \neq \emptyset, \tag{3.1'}$$

and

$$\mathcal{N}_{S,h}(e) = F(e). \tag{3.2'}$$

I.e. the inclusion in (3.2) of the definition of acceptability becomes an equality. [Since $F(e) \neq \emptyset$ by hypothesis, (3.1′) follows from (3.2′), but we have shown it explicitly for the sake of parallelism with the previous definition.]

We may note that when F is a (single-valued) function, the two definitions coincide.

Denote by $P^i(e)$ the strict preference agent i in e, and by $A(e) \subseteq \mathcal{A}$ the set of outcomes possible when the environment e prevails. As usual, an outcome $\hat{z} \in \mathcal{A}$ is *Pareto-optimal in e* if

$$\hat{z} \in A(e),$$

and there is no $z' \in A(e)$ such that $z' R^i(e) \hat{z}$ for all $i \in N$ and $z' P^j(e) \hat{z}$ for some $j \in N$.

In the earlier period of investigation, the question asked was relatively modest. Given the class of a priori admissible economies E, does there exist a mechanism (S, h) such that, for each $e \in E$, every Nash equilibrium outcome is Pareto-optimal in e? To avoid triviality, it is natural to add the requirement that the set of Nash equilibria, and hence Nash outcomes, be non-empty for all $e \in E$. As above, when these two conditions are satisfied, we say that the mechanism (S, h) is Pareto-*acceptable* for E.

Groves and Ledyard, in their path-breaking contribution (1977), constructed a Pareto-acceptable mechanism for a wide class of economies,[26] with three or more agents, containing public goods. Hurwicz and Schmeidler (1978) and, independently, Maskin (1977) undertook a systematic study of the existence of Pareto-acceptable mechanisms.

Paradoxically, it was discovered that, for the case of two agents, $n = 2$, if all strict preference orderings were a priori admissible in E, then only dictatorial mechanisms were Pareto-acceptable for E. [See Theorem 1 and Corollary 1 in Hurwicz and Schmeidler (1978, p. 1451) and Theorem 1 in Maskin (1977).] By contrast, many non-dictatorial Pareto-acceptable mechanisms could be found when there were three or more agents [see the kingmaker outcome function in Hurwicz and Schmeidler (1978, p. 1452) and Example 1 in Maskin (1977)].

[26] For the nature of this class, see Groves and Ledyard (1980).

Once it became clear that, for $n > 2$, Pareto-acceptable mechanisms do exist, ambitions expanded. In particular, it was noted that in the Groves–Ledyard mechanism, an agent's utility at equilibrium could be lower than at the initial endowment,[27] i.e. that the requirement of individual rationality was violated. This was in contrast to the properties of such traditional equilibria as Walras or Lindahl. Both of these give each agent the option of not trading (i.e. staying at the initial endowment), hence they do satisfy individual rationality.

Formally, when E is a class of pure exchange economies with initial endowment allocation $\omega(e)$, a social choice rule $F: E \Rightarrow \mathscr{A}$ is *individually rational in E* if, for all $e \in E$, and every $z \in F(e)$,

$$z R^i(e)\, \omega(e) \quad \text{for all} \quad i \in N.^{28} \tag{3.3}$$

The same definition can be used in a more general setting, with $\omega(e)$ interpreted as some type of reference allocation (e.g. status quo). We shall denote by $I(e)$ the set of all allocations z satisfying (3.3).

4. Revelation mechanisms and dominance equilibria

The new problem then is whether, for a given class of environments E, there exists a mechanism (S, h) such that, for any $e \in E$ and any $z \in \mathscr{N}_{S,h}(e)$,

$$z \in P(e) \cap I(e),$$

where $P(e)$ and $I(e)$ are, respectively, the sets of Pareto-optimal and individually rational allocations in e.

This issue was already studied in Hurwicz (1972) in the context of Edgeworth Box examples (pure exchange, two goods, two persons). However, the formulation there introduced an additional requirement, which in present-day terminology could be expressed as the requirement that (S, h) be a revelation mechanism. A mechanism (S, h) is said to be a *revelation mechanism for* $E = E^1 \times \cdots \times E^n$, if $S^i = E^i$ for each $i \in N$, where E^i is the class of a priori admissible characteristics for the ith agent.[29] Hence a revelation mechanism for E can be written (E, h). A

[27] See Groves and Ledyard (1980, p. 1487).

[28] It is not obvious how to define individual rationality in an economy with production. For a possible interpretation of this concept, see Hurwicz (1979b, p. 159).

[29] This usage of the term "revelation mechanism" corresponds to that of Green and Laffont (1979, p. 50, definition 4.3) except that we use characteristics where they use valuation functions. Dasgupta, Hammond and Maskin (1979, p. 188) use the term "direct mechanism" in the same sense as our "revelation mechanisms".

revelation mechanism (E, h) for E is said to be *compatible with* (or *natural for*) social choice rule F, if for every $e \in E$,

$$h(e) \in F(e),$$

where F is the social choice rule to be implemented. (Note, it is legitimate here to have e as an argument of h because in a revelation game $S = E$.)

In particular, if the social choice rule F is a (single-valued) function, then

$$h(e) = F(e) \quad \text{for each} \quad e \in E,$$

and the revelation mechanism compatible with F becomes (E, F), since here $S = E$ and $h = F$.

Let the ith agent's *true* characteristic be denoted by $\dot{e}^i$. In games of revelation to be considered it is understood that

$$\dot{e}^i \in E^i \quad \text{for all} \quad i \in N.$$

This means that the true characteristic is a priori admissible (otherwise the designer is misinformed!) and also, since $S^i = E^i$, $i \in N$, that each agent has the option of using the true characteristic as his strategy. On the other hand, unless E^i is a one-element set (which case we shall exclude), he also has the option of using as his strategy some element $\tilde{e}^i$ of E^i which is different from $\dot{e}^i$.

This raises the question whether the true profile $\dot{e} = (\dot{e}^1, \dots, \dot{e}^n)$ is a Nash equilibrium for the revelation mechanism (E, h). I.e. the question is whether, for each $i \in N$,

$$h(\dot{e}^1, \dots, \dot{e}^n) R^i(\dot{e}) h(\dot{e}^1, \dots, \tilde{e}^i, \dots, \dot{e}^n) \quad \text{for all} \quad \tilde{e}^i \in E^i.$$

[Note that $R^i(e)$ depends on e^i only. In fact, usually, $e^i = (C^i, \omega^i, R^i, Y^i)$.]

We say [see Hurwicz (1972)] that a direct mechanism (S, h) is *incentive-compatible on* E if truth is always a Nash equilibrium in E, i.e. if

$$\nu_{S,h}(e) \neq \emptyset \quad \text{and} \quad e \in \nu_{S,h}(e) \quad \text{for all} \quad e \in E.$$

[A more stringent requirement would be $\{e\} = \nu_{s,h}(e)$ for all $e \in E$, i.e. that truth be the only Nash equilibrium.]

Now denote by $L(e)$ the set of Lindahl allocations[30] in the economy e. Then L is a social choice rule (the *Lindahl Social Choice Rule*). We may note for later

[30] See, e.g. Johansen (1963) and Hurwicz (1979a, 1979b). In the two-good economy described in Section 2, $e = (C^i, \omega^i, R^i)_{i \in N}$ and $L(e) = \{(\bar{x}^i, \bar{y})_{i \in N}$: (1) $(\bar{x}^i, \bar{y}) \in C^i$ for all $i \in N$, and (2) for some $(p_1, \dots, p_n)$, and all $i \in N$, if $(x^i, y) \in C^i$ and $x^i + yp_i \leqq \omega^i_x + \omega^i_y p_i$, then $(\bar{x}^i, \bar{y}) R^i(e)(x^i, y)\}$.

reference that

$$L(e) \subseteq I(e),$$

i.e. a L is always individually rational,[31] and, under classical assumptions,

$$L(e) \subseteq P(e),$$

i.e. a Lindahl allocation is Pareto-optimal. To simplify exposition, let us suppose that, for each $e \in E$, there is a unique Lindahl allocation, also to be denoted by $L(e)$. Thus, now L is a function. Then the *Lindahl mechanism*, to be distinguished from the Lindahl social choice rule L, is the (unique) revelation mechanism natural for L, namely $(E; L)$.[32] I interpret Samuelson's critique as the claim that truth is not a Nash equilibrium for this game, i.e. that the Lindahl mechanism is not incentive compatible.

In a moment, we shall see that Samuelson's claim concerning the Lindahl mechanism was correct. But, somewhat surprisingly, it turned out that this was due not, as many thought, to the peculiar features of the public goods but rather to incentive problems that arise for private goods as well. This latter fact was shown in Hurwicz (1972). Consider a pure exchange economy with two *private* goods and two traders. First let the (revelation) mechanism require the agents to report their characteristics e^1, e^2; given these the outcome function dictates the Walrasian (competitive) allocation corresponding to $e = (e^1, e^2)$, to be denoted by $W(e)$.[33] Assume $\omega(e) \notin P(e)$, i.e. a non-optimal initial endowment. [For the sake of simplicity suppose that the competitive allocations are unique (i.e. that $W(e)$ is a singleton for all $e \in E$).] Then, if the a priori admissible class e is sufficiently rich, truth turns out not to be a Nash equilibrium. Thus the Walrasian (competitive) process, viewed as a revelation mechanism, was seen (in this special case) to be not incentive-compatible. But this fact turned out to be a special case of a more general phenomenon: for $n = 2$, pure exchange, two goods, any revelation mechanism guaranteeing Pareto-optimality and individual rationality[34] is not incentive-compatible if E is sufficiently rich.[35]

[31] Because it leaves each agent the freedom not to trade.

[32] This mechanism can be interpreted as follows. Each agent is asked to report his characteristic e^i. Given the reports $e^1, \dots, e^n$, the allocation prescribed by the mechanism is the Lindahl allocation for the economy in which $e^1, \dots, e^n$ are the agents' characteristics.

[33] I.e. we require the mechanism to be compatible with the Walrasian (competitive) social choice rule. In a pure exchange ("Edgeworth Box") economy where $e^i = (C^i, \omega^i, R^i)$, we have $W(e) = \{(\bar{z}', \dots, \bar{z}^n) : (1)\ \bar{z}^i \in C^i$ for each i, and (2) there exists $(p_1, \dots, p_n)$ such that, for all $i \in N$, if $z^i \in C^i$ and $p_1 z^i \leqq p_1 \omega^i$, then $\bar{z}^i R^i(e) z^i\}$.

[34] I.e. compatible with $P(e) \cap I(e)$.

[35] Although there are truthful equilibria when the initial endowment $\omega(e)$ is Pareto-optimal.

More precisely, for $n = 2$, pure exchange, two goods, if (S, h) is a revelation mechanism for E (i.e. $S = E$) and

$$\mathcal{N}_{S,h}(e) \subseteq P(e) \cap I(e) \quad \text{for all} \quad e \in E,$$

and if the class E is sufficiently rich, then

$$e \notin \nu_{S,h}(e),$$

unless $\omega(e) \in P(e)$.

Now, when there are two agents, the geometry of the Edgeworth Box so helpful for private goods has a close counterpart in the Kolm (equilateral) Triangle, using barycentric coordinates [Kolm (1964) and Malinvaud (1971)]. Using this fact, Ledyard and Roberts (1974) showed that the phenomenon just described for a private goods economy also occurs in a two-agent economy with two goods, one of which is public, and where constant returns prevail in producing the public good, using the private good as input.[36]

Thus they showed that for such public goods economies ($n = 2$), all revelation mechanisms guaranteeing Pareto-optimality and individual rationality over a sufficiently rich class of environments, are not incentive-compatible. In particular, since the Lindahl mechanism does have these two properties, it is not incentive-compatible.[37]

Thus the Samuelson claim turns out to be correct, although for reasons not related to the presence of public goods![38]

The results so far discussed are very specialized. A number of generalizations have been obtained. The impossibility results just cited for $n = 2$ were extended by Ledyard (1977) as applied to "core-seeking" mechanisms for arbitrary n. See also Satterthwaite (1976) and Dasgupta, Hammond and Maskin (1979, p. 198 thru Section 4.4.1).[39]

[36] See Section 2. An analytic example is given in Roberts (1979, p. 289).

[37] In fact, the geometry of Lindahl equilibria in the Kolm Triangle is completely analogous to the geometry of the Walrasian (competitive) equilibria in the Edgeworth Box (price line, double tangency, etc.).

[38] However, the similarity between the public and private goods economies appears to break down as the number of agents $n \to \infty$. For private goods, Postlewaite and Roberts (1976) showed the competitive mechanism (which has individually rational equilibria) to be, in a sense, asymptotically incentive compatible. By contrast, Roberts (1976) found that, for public goods, mechanisms acceptable for individually rational choice rules (such as Lindahl) are *not* asymptotically incentive compatible. [However, when the individual rationality requirement is abandoned, asymptotic incentive compatibility is possible. This is shown to be the case in certain economies for the pivotal mechanism with rebates in Green, Kohlberg and Laffont (1976).]

[39] This theorem drops the requirement of individual rationality but requires a broader class E than used in Hurwicz (1972) and an additional condition for $n \geq 2$. See also Hammond (1979) for related results in "large" economies, i.e. those with a non-atomic measure of space agents.

Furthermore, it was discovered that there is a close relationship between incentive-compatibility as defined above and dominance-equilibria.

Let $\Gamma = (S, \pi)$ be a game with the strategy space $S = S^1 \times \cdots \times S^n$ and a payoff function $\pi = (\pi^1, \ldots, \pi^n)$. Then s_i^* is said to be a *dominant strategy* for player i if

$$\pi^i\left(s_i^*, s_{)i(}\right) \geqq \pi^i\left(s_i, s_{)i(}\right) \quad \text{for all} \quad s_i \in S^i, \quad s_{)i(} \in S^{)i(},$$

where $S^{)i(} = S^1 \times \cdots \times S^{i-1} \times S^{i+1} \times \cdots \times S^n$. Similarly, when h is an outcome function, s_i^* is a dominant strategy for the game $(S; h, e)$ if

$$h\left(s_i^*, s_{)i(}\right) \; R^i(e) \; h\left(s_i, s_{)i(}\right) \quad \text{for all} \quad s_i \in S^i, \quad s_{)i(} \in S^{)i(}.$$

$s^* = (s_1^*, \ldots, s_n^*)$ is said to be a *dominance equilibrium* if s_i^* is a dominant strategy for each $i \in N$.

The discovery was the following: if, for a direct revelation mechanism (E, h), truth is a Nash equilibrium for all $e \in E$, then truth is a dominant strategy for every agent.[40] Thus such truthful Nash equilibria are dominance equilibria, and a search for incentive-compatibility is a search for dominance equilibria.

Moreover, as shown by Dasgupta, Hammond and Maskin (1979, theorem 4.1.1, p. 194),[41] given a mechanism with dominance equilibria, there exists an equivalent[42] *direct* mechanism in which truthtelling constitutes a dominance equilibrium.

One approach, pioneered[43] by Vickrey (1961), Clarke (1971), and Groves (1979), is focused on the design of mechanisms for which truthtelling is a dominant strategy. It turns out that such mechanisms can be designed, at least for the class of "parallel" preferences, i.e. preferences representable by utility[44]

[40] See d'Aspremont and Gérard-Varet (1979a, theorem 1, p. 31) and Dasgupta, Hammond and Maskin (1979, theorem 7.1.1, pp. 209–210). In Gibbard (1973, p. 595) a game form is called *straightforward* if every player always has a dominant strategy (so a dominance equilibrium exists).

[41] The construction is used in Gibbard (1973, p. 596) and in Green and Laffont (1977, p. 434, proof of theorem 5).

[42] See Dasgupta, Hammond and Maskin (1979, p. 189). Let (S, g) be any mechanism possessing dominance equilibria in games $(S; h, e)$ for e in E, and let $s^*: E \to S$ be a (single-valued) selection from the dominance equilibrium correspondence. Then the composition $h = g \circ s^*$ defines the outcome function of the corresponding revelation mechanism. This mechanism (E, h) is said to be equivalent to (S, g) if, for all $e \in E$, $e \in \mathscr{N}_h(e)$, i.e. truth is an equilibrium for all environments. However, it may happen that (S, h) is compatible with F while (E, h) is not. [See Dasgupta, Hammond and Maskin (1979, p. 195).]

[43] With Jacob Marschak as a precursor; see Groves (1979, p. 50, footnote 11) and Green and Laffont (1979, p. 36).

[44] Called quasi-linear or transferable.

functions linear in the private good (which can be thought of as numéraire). Let the ith agent's preferences be represented by the transferable utility function $u^i(x^i, y) = x^i + v_i(y)$, where x^i is the amount of private good retained (after taxes) by i, and y the level of the public good available to everyone. $v^i(\cdot)$ is called the ith *valuation function*. [Our terminology differs somewhat from that used by Green and Laffont (1979) because they subsume the costs of a project in the valuation function.] Equivalently,

$$u^i = \omega_x^i - t^i + v_i(y),$$

where t^i is the tax (contribution) paid by the ith agent. For any individual agent this tax may be positive, negative, or zero. But if no outside funds are available [assuming the input requirements function $g(y) \equiv y$, i.e. constant returns prevail and measurement units are normalized (see Section 1)], taxes must satisfy the feasibility requirement

$$\sum_{i=1}^{n} t^i \geqq y.$$

Furthermore, an allocation is not Pareto-optimal unless the equality holds, i.e.

$$\sum_{i=1}^{n} t^i = y.$$

[See Groves (1979, p. 47, proposition 2, condition (a)). The notion of Pareto-optimality used by Groves and in our text differs from that in Green and Laffont (1979, p. 33) where the utility of the public agency [an additional $(n+1)$st agent] permits Pareto-optimality even when $\sum t^i > y$.]

If the valuation functions are concave, at a Pareto-optimal allocation, the expression

$$-y + \sum_i v_i(y)$$

must be maximized. (For differentiable v_i's and an interior optimum this yields the Samuelson condition in the form

$$\sum_i v_i'(y) = 1.)$$

On the other hand, if $\sum t^i = y$ and the expression $-y + \sum_i v_i(y)$ is being maximized, then the allocation is Pareto optimal.[45] Thus, for transferable utilities, the maximization condition determines the optimal values of the public good, independently of the distribution of tax burdens. If $\hat{y}$ is an optimal value of the public good, then an n-tuple $(t^1,\ldots,t^n)$ of taxes is Pareto-optimal if and only if it satisfies the conditions

$$\sum_{i=1}^{n} t^i = \hat{y},$$

and[46]

$$t^i \geqq -\omega_x^i \quad \text{for all} \quad i \in N.$$

Consider now a revelation mechanism with the initial endowments known to the designer. Under the assumptions made, the only aspect of the ith characteristic unknown to the designer is the ith valuation function, v_i. Therefore, the space E^i may be identified with a set V^i of the a priori admissible valuation functions for agent i.

An outcome function for such a mechanism is of the form

$$h: V^1 \times \cdots \times V^n \to R^n \times R_+,$$

so that

$$h: (v_1,\ldots,v_n) \mapsto (x^1,\ldots,x^n, y).$$

We shall write $x^i = h_{x_i}(v_1,\ldots,v_n)$, $y = h_y(v_1,\ldots,v_n)$.

Revelation mechanisms independently proposed by Clarke (1971) and Groves (1970) have the remarkable property that truthtelling is a dominance equilibrium when preferences are representable by utility functions linear in the private good.[47] Furthermore, these mechanisms have the following additional property.

Let w_i denote the valuation function *reported* (whether truthfully or not) by agent i. Then, in the above mechanism,

$$h_y(w_1,\ldots,w_n)$$

maximizes the expression

$$-y + \sum_i w_i(y)$$

[45]Assuming that, for each $i \in N$, $C^i = R_+^2$ and $t^i \geq -\omega_x^i$ (C^i = the ith agent's consumption set). Cf. the corresponding theorem in Groves (1979, p. 47, proposition 2).

[46]Again assuming that each agent's consumption set $C^i = R_+^2$ (the non-negative quadrant).

[47]See Groves (1979, corollary, p. 51).

with respect to y. I.e. for all $(w_1,\dots,w_n)\in V^1\times\cdots\times V^n$,

$$h_y(w_1,\dots,w_n)=\max_y\left[-y+\sum_i w_i(y)\right].$$

But truth constitutes a dominant strategy for each agent. Hence we have, at dominance equilibrium,

$$w_i=\dot{v}_i \quad \text{for all} \quad i\in N,$$

where $\dot{v}_i$ is the true valuation function. So, there exists a dominance equilibrium such that the value of $\hat{y}$ maximizes the expression $[-y+\sum_i \dot{v}_i(y)]$, hence $\hat{y}$ is Pareto-optimal (with regard to the true preferences).

However, the fact that $\hat{y}$ is Pareto-optimal does not imply that the allocation $(x^1,\dots,x^n,\hat{y})$ produced by these mechanisms is Pareto-optimal. As seen above, for Pareto-optimality it is necessary that

$$\hat{y}=\sum t^i=\sum(\omega_X^i-x^i). \tag{4.1}$$

The latter condition is not satisfied by the "*pivotal*" (Clarke) mechanism. Now the pivotal mechanism is a special case of a class called the *Groves mechanisms*. It turns out that, under certain conditions on the families V^i of a priori admissible valuations, the search for mechanisms whose dominance equilibria yield Pareto-optimal outcomes can be confined to Groves mechanisms.[48] But, unfortunately, it has been found that every Groves mechanism will violate the balance condition on large classes of profiles.[49]

[48]See Green and Laffont (1977, corollary 3, p. 433; 1979, theorem 4.5, pp. 63–64), Walker (1978), and Holmström (1979). A sufficient condition [Holmström (1979, theorem 2, p. 1141)] is that $V\equiv V^1\times\cdots\times V^n$ be a convex set in the space of valuation n-tuples. This condition is satisfied in cases underlying the results due to Green and Laffont and to Walker where V_i consists of all continuous functions or all strictly concave (or convex) functions, or all concave quadratic functions on a convex subset of R^l.

Holmström (ibid., theorem 1, p. 1140) also shows a weaker condition (that V be "smoothly connected") to be sufficient.

On the relationship of the previous conditions, see Green and Laffont (1979, p. 65, footnote 15) and Holmström (1979, p. 1142).

[49]See Green and Laffont (1979, theorem 5.3, p. 90). Walker (1980, theorem 1, p. 1531) shows that the "failure set" is everywhere dense on a space of concave valuation functions when this space is endowed with any topology weaker than the strongest topology in which vector addition and scalar multiplications are continuous. Analogous results for private goods, pure exchange economies are presented in Hurwicz and Walker (1983). See also Hurwicz (1975a, 1975b, 1981) for impossibility results when $n=3$.

However, the balance condition may be satisfied on sufficiently small classes of profiles. See Groves and Loeb (1975) and Hurwicz (1975a, 1975b, 1981). These cases involve economies with three or more agents. For impossibility results for two agents, see Green and Laffont (1979, p. 94) and Hurwicz (1975a, 1975b, 1981).

To present these results more formally, it is convenient to switch to a model of public decision-making which subsumes the inputs (costs) of producing a public good under the more general rubric of a public decision (project).[50]

Formally, this reduces the problem to that of a "costless" project, but at the expense of increasing the dimensionality of the space in which the project is defined.

Thus, let c^i, $c^i \leqq \omega^i_X$, $i=1,\dots,n$, be such that $\sum_{i=1}^n c^i = y$, $y \geqq 0$. Then the "project" $z = (c^1,\dots,c^n; y)$ requiring the ith agent to contribute c^i units of the private good to the production of y units of the public good is feasible. Suppose that proposal p combines this project with a transfer scheme $(r^1,\dots,r^n)$ where, for each $i \in N$, r_i is the compensation (in terms of the private good) paid to agent i. Then the utility of the proposal to the ith individual is $u^i[p] = u^i(\omega^i_X - c^i + r^i, y)$. When u^i is linear in the private good, we may write this as

$$u^i[p] = \omega^i_X - c^i + r^i + \Psi_i(y).$$

Alternatively, we can write

$$u^i[p] = \omega^i_X + r^i + \varphi_i(z), \qquad i \in N,$$

where

$$z = (c^1,\dots,c^n; y) \quad \text{and} \quad \varphi_i(z) = -c^i + \Psi_i(y).$$

Since $x^i = \omega^i_X - c^i + r^i$, $i \in N$, the feasibility condition,

$$y + \sum x^i \leqq \sum \omega^i_X,$$

becomes

$$y + \sum_i (\omega^i_X - c^i + r^i) \leqq \sum \omega^i_X.$$

Hence

$$y - \sum c^i + \sum r^i \leqq 0 \quad \text{or} \quad \sum c^i - y \geqq \sum r^i.$$

But, by construction, $\sum c^i - y = 0$. Therefore, the feasibility condition becomes

$$\sum_{i=1}^n r^i \leqq 0.\text{[51]} \tag{4.2}$$

[50] See Green and Laffont (1979, pp. 29–31; p. 42, footnote 9; p. 52, footnote 5; pp. 74–75).

[51] Here and subsequently we shall ignore the implications of individual feasibility requirements, viz. that $x^i \equiv \omega^i_X - c^i + r^i \geqq 0$, $i \in N$.

Thus, in this formalization, the *public decision* (*project*) is a point $z = (c^1,\dots,c^n; y)$ in the $(n+1)$-dimensional Euclidean space R^{n+1} satisfying the condition

$$\sum_{i=1}^{n} c^i = y, \tag{4.3}$$

and a *proposal* is a point $p = (c^1,\dots,c^n; y; r^1,\dots,r^n)$ in the $(2n+1)$-dimensional space R^{2n+1} satisfying the conditions (4.2) and (4.3).

Somewhat more generally, and *reverting to the customary notation*,[52] a *proposal* is defined as a point $q = (r^1,\dots,r^n; y)$ where y is an element of a set Y of feasible projects and $\sum_{i+1}^{n} r^i \leqq 0$. As before, for preferences representable by utility functions of the form

$$u^i[q] = r^i + v^i(y), \qquad i \in N,$$

$v_i(\cdot)$ is again called agent i's *valuation function*.

The a priori admissible class of i's valuation function is denoted by V^i.

In a revelation game, an individual's strategy is an element w^i of V^i. A mechanism then is defined as a function

$$h: V^i \times \cdots \times V^n \to R^n \times Y,$$

so that

$$h(w_1,\dots,w_n) = (r^1,\dots,r^n, y) \quad \text{with} \quad \sum_{i=1}^{n} r^i \leqq 0.$$

Write $w = (w_1,\dots,w_n)$ and $r = (r^1,\dots,r^n)$. For $(r, y) = h(w)$, we shall use the notation $r^i = h^{ri}(w)$, $i \in N$, $r = h^r(y)$, and $y = h^y(w)$.

Example

Although we have come to this model through a re-formalization of a particular public goods problem, it has in fact much greater generality, as long as we retain the freedom of choosing a suitable set Y. In particular, a pure exchange economy with $k+1$ goods and selfish preferences can be modeled by choosing

$$Y = R^k \times \cdots \times R^k \quad (n \text{ times}).$$

Here, for $y \in Y$, we write $y = (y^1,\dots,y^n)$, $y^i \in R^k$, $i \in N$, and require that $\sum_{i=1}^{n} y^i \leqq 0$.

[52] Thus from now on y here corresponds to $z = (c^1,\dots,c^n; y)$ above (not to y above) and Y can be either multi- or one-dimensional; v_i here corresponds to Ψ_i above (not to v_i above).

Furthermore, the utility functions are made selfish by postulating that, for every $y \in Y$,

$$v_i(y) = \tilde{v}_i(y^i), \quad i \in N,$$

for some functions $\tilde{v}_1, \ldots, \tilde{n}_n$.

Clearly, this represents a pure exchange model with (r^i, y^i) as the net trade of agent i and the agent's selfish utility function (in terms of net trades),[53]

$$u^i(r^i, y^i) = r^i + v_i(y^i).$$

Going back now to the general model, we note first that[54] a proposal $(\hat{r}^1, \ldots, \hat{r}^n; \hat{y})$ is Pareto-optimal if and only if

$$\sum_{i=1}^{n} \hat{r}^i = 0, \tag{4.4}$$

and

$$\sum_{i=1}^{n} v_i(\hat{y}) \geqq \sum_{i=1}^{n} v^i(y) \quad \text{for all} \quad y \in Y. \tag{4.5}$$

Let $w_1, \ldots, w_n$, $w_i \in V^i$, $i \in N$, be the agents' *reported* (possibly false) valuation functions.

A Groves mechanism uses an outcome function $h: V^1 \times \cdots \times V^n \to R^n \times Y$, such that (i)[55]

$$\sum_{i=1}^{n} w_i(h^y(w)) \geqq \sum_{i=1}^{n} w_i(y) \quad \text{for all} \quad y \in Y,$$

and (ii) for each $i \in N$, there exists a function $g_i: V^1 \times \cdots \times V^{i-1} \times V^{i+1} \to R$ such that

$$h^{ri}(w) = \sum_{j \neq i} w_i(h^y(w)) - g_i(w_{)i(}).$$

[53]See Walker (1980, section 5, p. 1534) and Walker (1977). I have also benefited from private communications with Walker on this subject.

[54]See Groves (1979, p. 47, proposition 2). Note again that the "if" part of the statement ignores the lower bound restrictions (due to considerations of individual feasibility) on the x^i.

[55]Writing $w = (w_1, \ldots, w_n)$ and $w_{)i(} = (w_1, \ldots, w_{i-1}, w_{i+1}, \ldots, w_n)$.

I.e. the public decision $y^* = h^y(w)$ chosen by a Groves mechanism maximizes (over the set Y) the sum of *reported* valuations.[56] Clearly, if the reported valuations are truthful ($w_i = \dot{v}_i$ for all $i \in N$), then y^* will be a Pareto-optimal value of the public decision. Now,[57] truth is a dominant strategy for every agent. It follows that, for a Groves mechanism, there exist dominance equilibria yielding Pareto-optimal public decisions. I.e. if h is a Groves mechanism outcome function, then there exist some reported valuation lists $w^* \in V^1 \times \cdots \times V^n$, $\tilde{r} \in R^n$, $\sum_{i=1}^{n} \tilde{r}^i = 0$, such that (1) $w^* = \dot{v}$ (true valuation list), (2) w^* is a dominance equilibrium, and (3) $(\tilde{r}, h^y(w^*))$ is Pareto-optimal. However, this does not imply the optimality of $(r^*, y^*) = (h^r(w^*), h^y(w^*))$ because optimality requires that $\sum r^{*i} \leqq 0$, about which so far nothing has been said. (Note that $\tilde{r}$ above need not equal r^*.)

Example

As mentioned above, the pivotal (Clarke) mechanism is a Groves mechanism. To define it, it is sufficient to specify the "alien" component function g_i appearing in the definition of the compensation function h^{ri} for a Groves mechanism.

For each $i \in N$, let y_i^{**} denote the maximizer of $\sum_{j \neq i} w_j(y)$ over Y, i.e.

$$\sum_{j \neq i} w_j(y_i^{**}) \geqq \sum_{j \neq i} w_j(y) \quad \text{for all} \quad y \in Y.$$

Then, for the pivotal (Clarke) mechanism, we have

$$g_i(w_{)i(}) = -\sum_{j \neq i} w_j(y_i^{**}).$$

Thus, for this mechanism, the ith compensation function is given by

$$h^{ri}(w) = \sum_{j \neq i} w_j(y^*) - \sum_{j \neq i} w_j(y_i^{**}),$$

where y^* maximizes $\sum_{k=1}^{n} w_k(y)$ while y_i^{**} maximizes $\sum_{j \neq i} w_j(y)$; in both cases the maximization is over Y.

It is instructive to apply this in the special case where Y is a two-element set, say $Y = \{0, 1\}$.[59] Normalize $w_i(\cdot)$ so that $w_i(0) = 0$, and write $w_i(1) = w_i$. Then[60]

[56] Revelation mechanisms with this property are called *direct* revelation mechanisms in Green and Laffont (1979, p. 51, definition 4.4).
[57] See Groves and Loeb (1975), Green and Laffont (1979, pp. 56–57, theorem 4.1).
[58] See Green and Laffont (1979, pp. 42–43; p. 52, definition 4.7).
[59] The interpretation is that 1 represents undertaking a specific project and 0 not undertaking it.
[60] See Green and Laffont (1979, p. 42).

the function h_i for the pivotal mechanism becomes

$$g_i(w_{)i(}) = \max\left(\sum_{j \neq i} w_j, 0\right).$$

It turns out that the compensation r^i paid to agent i is either zero or $-\left|\sum_{j \neq i} w_j\right|$, and the latter can be the case for some agents[61] at a dominance (truthful) equilibrium. Hence it can happen at equilibrium that

$$\sum_{i=1}^{n} r^i < 0,$$

and this inequality implies absence of Pareto-optimality. Thus the pivotal mechanism does not guarantee the Pareto-optimality of the equilibrium allocation (r^*, y^*) it generates, although the choice of y^* itself is Pareto-optimal.[62]

Contrary to what one might have hoped, this difficulty is not remedied by broadening the horizon to the class of all Groves mechanisms. The fact that there exists no Groves mechanism such that the compensatory payments balance when the sets V^i are unrestricted, is demonstrated[63] in Green and Laffont (1979, p. 90, theorem 5.3). On the other hand, a balanced Groves mechanism[64] was constructed by Groves and Loeb (1975) for the case where $Y = \boldsymbol{R}_+$ and the valuation functions are quadratics,

$$v_i(y) = -\tfrac{1}{2}y^2 + \theta_i y, \qquad i \in N,$$

with $\theta_i \in \boldsymbol{R}_+$, provided $\# N \geqq 3$.

The requirement that $\# N \geqq 3$ is essential: when there are only two agents, balance cannot be achieved even for quadratic utility functions.[65]

Such results raise the question of conditions under which Groves mechanisms are balanced and hence Pareto-optimal. A necessary and sufficient differential

[61] It is the case for the "pivotal" agents. An agent i is "pivotal" if the sign of $\sum_{k=1}^{n} w_k$ is different from that of $\sum_{j \neq i} w_j$, namely where one of these sums is non-negative while the other is negative.

[62] In the sense that there exists some allocation $(\tilde{r}, y^*)$ which is Pareto-optimal. However, in general, $\tilde{r} \neq r^*$.

[63] Explicitly, for a two-element set Y.

[64] I.e. such that $h^r(w) = 0$ for all $w \in V^1 \times \cdots \times V^n$.

[65] Green and Laffont (1979, pp. 94–95).

condition for a class of differentiable valuation functions is given in Green and Laffont (1979, p. 96). It makes possible the direct testing of specific classes of environments.

So far we have been focusing on Groves mechanisms. But negative results concerning Groves mechanisms have broader implications. First, it has been shown[66] – under varying assumptions concerning the a priori admissible classes of valuations V^i, $i \in N$ – that Groves mechanisms are the only revelation mechanisms for which truth is always a dominant strategy. But even going beyond revelation mechanisms does not alter the situation materially as long as we insist on dominance equilibria. This is so because[67] given a mechanism (S, h), using an arbitrary strategy space $S = S^1 \times \cdots \times S^n$ and possessing a dominance equilibrium $s^* = (s_1^*, \ldots, s_n^*)$ over the environment $E = E^1 \times \cdots \times E^n$, there exists a corresponding revelation mechanism (E, g) in which truthtelling is dominant. Furthermore, the outcome generated by the truthtelling equilibrium corresponding to s^* yields the same outcome, hence preserves Pareto-optimality.[68]

Thus the difficulty arises as soon as one demands a mechanism with dominance equilibria. Such a mechanism will be subject to the same difficulties that would arise for Groves mechanisms. It has already been seen that there do exist[69] families of valuation functions for which balanced and truthtelling Groves mechanisms exist, hence Pareto-optimality can be guaranteed. But are such cases typical or exceptional? Answers to this question are provided in Walker (1980).

To formulate the relevant results one must specify the topology to be used on the space $V = V^1 \times \cdots \times V^n$ of valuation function profiles.

Walker requires V^i to be a subset of $C(\mathscr{A})$, the set of all continuous real-valued functions on $\mathscr{A}$ (the space of public decisions). Let $\bar{\bar{\mathscr{J}}}$ be the largest topology in which $C(\mathscr{A})$ is a topological vector space, and let $\bar{\mathscr{J}}$ be the topology on V^i inherited by V^i as a subspace of $C(\mathscr{A})$. ($\bar{\mathscr{J}}$ is called the vector space topology on V^i.)

Walker's Theorem 1 assumes $n \geq 2$ and $\mathscr{A}$ a convex open set in R^l and uses $V^1 = \cdots = V^n$, with V^1 the set of all strictly concave valuations on $\mathscr{A}$ which attain a maximum on $\mathscr{A}$. Then, for any Groves mechanism, the set of profiles in $V^1 \times \cdots \times V^n$ for which the mechanism yields non-Pareto-optimal outcomes is everywhere dense[70] in $V^1 \times \cdots \times V^n$. The same proposition is valid for any mechanism where truth is a dominant strategy.

[66] Green and Laffont (1977, theorem 3, corollary 3; 1979, theorem 4.5, pp. 63–64), Walker (1978), and Holmström (1979). See footnote 49.

[67] See footnote 41.

[68] See, however, Dasgupta, Hammond and Maskin (1979, pp. 189, 194–195) concerning complications due to additional, possibly non-optimal, equilibria.

[69] Quadratic in the public good, linear in the private good (see above).

[70] In the product topology $\bar{\mathscr{J}}^{(n)}$ or in any topology $\mathscr{J}^{(n)}$ such that $\mathscr{J} \subset \bar{\mathscr{J}}$.

Now, as mentioned above, mechanisms with dominance strategies can be transformed into revelation mechanisms where truthtelling is dominant. Given a mechanism (S, h) for E with a dominance equilibrium $(s_1^*,\dots,s_n^*)$, we define a mechanism (E, g) with truthtelling dominance equilibria, where

$$g(e) = h\left(s_1^*(e^1),\dots,s_n^*(e^n)\right),$$

and e^i is the characteristic of the ith agent.

This transformation yields a strengthening of Walker's Theorem 4 which, under the same assumptions on $\mathscr{A}$ and the V^i's, asserts that any mechanism yielding dominance equilibria will have an everywhere dense set where optimality fails.

However, a stronger conclusion is obtained if the mechanism yielding dominance equilibria is *continuous* and each V^i a compact subset of $C(Y)$. In that case, Walker's Theorem 5 states that the ("good") set of profiles with Pareto-optimal outcomes is closed and nowhere dense in $V^1 \times \cdots \times V^n$. Thus here the failure set is open as well as everywhere dense.

The results so far discussed show that, generally speaking, one cannot hope for both dominance equilibria and Pareto-optimality. Several alternative directions of research have been explored, including asymptotic properties of mechanisms as the number of agents tends to infinity,[71] sampling procedures,[72] and Bayesian specifications.[73]

These directions attempt to preserve, to the extent possible, the availability of dominance or truthful equilibria, possibly at the expense of other desirable properties of mechanisms.

5. Pareto-optimal Nash equilibria in economies with public goods

On the other hand, one may sacrifice the dominance equilibria and accept the weaker type, namely the non-cooperative Nash equilibria discussed above. This time, however, the mechanism is not one of the revelation type; hence, Nash equilibria do not become dominance equilibria.

The pioneering contribution in this direction is due to Groves and Ledyard (1977).

The Groves and Ledyard idea may be illustrated by a mechanism close to (but not identical with) theirs. Consider again a three-person economy with two goods (a private good X and public good Y), and constant returns in producing Y from

[71] Bowen (1943), Green, Kohlberg and Laffont (1976), Green and Laffont (1979, pp. 157–188). For private goods, see Roberts and Postlewaite (1976).

[72] Green and Laffont (1979, pp. 213–227).

[73] See, for instance, d'Aspremont and Gérard-Varet (1979a, 1979b) and Arrow (1977).

X, the input–output coefficient normalized to 1. Assume a differentiable (but not necessarily transferable) utility function, denoted for the ith agent by $u^i(x^i, y)$.

Let each agent's strategy domain be the set of reals. Denote the ith strategy domain by $\mathscr{M}^i$ (here $\mathscr{M}^i = R$, $i=1,2,3$) and its elements by m_i. Write $m = (m_1, m_2, m_3)$, and $\mathscr{M} = M^1 \times M^2 \times M^3$. The outcome function is defined by the tax functions T^i, $i=1,2,3$, and the public good function Y. Given the strategy triple $m=(m_1, m_2, m_3)$, agent i pays tax equal to $t^i = T^i(m)$, and the amount of public good produced is $y = Y(m)$. Now specify these functions as follows:

$$T^i(m) = m_i^2 + 2m_j m_k, \qquad i=1,2,3, \quad i \neq j \neq k \neq i,$$

$$Y(m) = M^2,$$

where

$$M = m_1 + m_2 + m_3.$$

Note first that the balance condition is satisfied because, for every $m \in M$,

$$Y(m) = M^2 = \sum_{i=1}^{3} \left(m_i^2 + 2m_j m_k\right) = \sum_{i=1}^{3} T^i(m).$$

Now examine the first-order necessary interior Nash equilibrium condition,

$$\partial v^i / \partial m_i = 0, \qquad i=1,2,3,$$

where v^i is the payoff (indirect utility) function defined by

$$v^i(m) = u^i\left(\omega_x^i - T^i(m), Y(m)\right).$$

Hence

$$0 = \partial v^i / \partial m_i = u_x^i \cdot \left(-T_i^i\right) + u_y^i \cdot Y_i, \qquad i=1,2,3,$$

where

$$T_i^i = \partial T^i / \partial m_i, \qquad \partial Y_i = \partial Y / m_i.$$

Using the outcome function specified above, this yields

$$u_x^i \cdot (-2m_i) + u_y^i \cdot 2M = 0, \qquad i=1,2,3,$$

and hence

$$u_y^i / u_x^i = m_i / M, \qquad i=1,2,3.$$

Summing, we obtain the Samuelson condition

$$\sum_{i=1}^{3} u_y^i / u_x^i = \sum_{i=1}^{3} m_i / M = \sum_{i=1}^{3} m_i \Big/ M = M/M = 1.$$

Thus, subject to the verification of second-order condition, we see that – locally at least – a Nash equilibrium allocation is Pareto-optimal.

For a precise version the reader is referred to Groves and Ledyard (1977, p. 796, esp. (4.3)–(4.4)), of which earlier version was circulated in 1974.

In a subsequent paper Groves and Ledyard (1980) (an early version was circulated in 1975) gave sufficient conditions for the existence of Nash equilibria in their model. As they point out, these conditions are slightly stronger than those required to prove the existence of Lindahl equilibria. Furthermore, the taxes may leave the consumers worse off than they had been with their initial endowments. I.e. the Groves–Ledyard mechanism is not *individually rational.*

6. Implementing the Lindahl correspondence

One is thus led to ask whether there exist alternative mechanisms whose Nash equilibrium allocations are not only Pareto-optimal but also individually rational. Also, it is of course desirable that equilibria exist for a wide class of economic environments.

All of these desiderata would be fulfilled if it were possible to design a mechanism whose Nash equilibria generate Lindahl allocations. We know already that this cannot be accomplished by a revelation game that is "natural for" (or "compatible with") the Lindahl correspondence. But there do exist mechanisms whose equilibrium allocations are precisely those of the Lindahl solution. Before describing them, a brief digression concerning *feasibility*.

In the public goods economies described above, there are two feasibility conditions: (a) "aggregate", $y = \sum_{i=1}^{n} t^i$, and (b) "individual", $t^i \leqq \omega_x^i$, for each $i = 1, \ldots, n$. Translated into properties of outcome functions, these become (a′) $Y(s) = \sum_{i=1}^{n} T^i(s)$ for all s in S, and (b′) $T^i(s) \leqq \omega_x^i$ for all i in $N = \{1, \ldots, n\}$ and all s in S (where $S = S^1 \times \cdots \times S^n$ is the Cartesian product of individual strategy domains). Condition (a′) is referred to as *balance*, and condition (b′) as *individual feasibility*.

Consider now the conventional scenario with the Lindahl analogue of a Walrasian auctioneer (Section 2). The auctioneer announces an n-tuple $p = (p_1, \ldots, p_n) \gg 0$, $\sum_{i=1}^{n} p_i = 1$, of personalized prices; for each $i = 1, \ldots, n$, the ith agent responds with the desired net trade commodity bundle (x^i, y^i) such that

$$x^i + p_i y^i \leqq 0,$$

and (x^i, y^i) is individually feasible.[74] In the usual interpretation, each agent treats the price parametrically and acts as if expecting to receive the requested bundle. Such a situation may be formalized as what has been called a *quasi-game*.[75] In a quasi-game there are two types of participants – not only the players (here, the n economic agents), but also non-players (here, the auctioneer). Each participant has a strategy domain. The ith agent's domain may be the set of individually feasible trades $Z^i = \{z^i : z^i \geqq -\omega^i\}$, and the auctioneer's domain may be the price space $P = \{(p_1,\ldots, p_n) \in R^n_{++} : \sum_{i=1}^n p_i = 1\}$. Thus, in conventional notation the $(n+1)$-tuple of strategies is $S = (S^1,\ldots, S^n, S^{n+1}) = (z^1,\ldots, z^n, p)$. Suppose that the outcome function for the ith agent is given by

$$\begin{aligned} h^i(s) &= z^i \quad \text{if} \quad (1, p^i)\cdot z^i \leqq 0, \quad i=1,\ldots,n, \\ &= 0 \quad \text{otherwise,} \end{aligned}$$

and the auctioneer's outcome function by

$$h^{n+1}(s) = \left(\sum_{j\in N'} x^j, - \sum_{j\in N'} x^j \right),$$

where j ranges over the set N' of those agents j for whom $(1, p^j)\cdot z^j \leqq 0$. (Thus the auctioneer is supplying $y = \sum_{j\in N'} x^j$ in exchange for the tax revenue of $x = \sum_{j\in N'} x^j$. In effect, the auctioneer is also the producer or, at least, the supplier of the public good or service.)

What we have described so far is not yet a quasi-game, but only a mechanism (outcome function) which is individually feasible but not balanced[76] and involving a non-player participant. To define a quasi-game we must specify the participants' behavior rules. For each player, as usual, we define a payoff (indirect utility) function,

$$v^i(s) = u^i(h^i(s)), \qquad i=1,\ldots,n,$$

where u^i is the utility function (assumed to be strictly increasing) of the ith agent in terms of net trades. The auctioneer's payoff function may be defined as

$$v^{n+1}(s) = \left(-\max_{i,k\in N'} |y_i - y_k|\right)\cdot \max_i p_i,$$

where $p_i > 0$, $\forall i$, and $\sum_{i=1}^n p_i = 1$.

[74] I.e. $(x^i, y^i) \geqq -\omega^i \equiv -(\omega^i_x, 0)$.

[75] Hurwicz (1979b).

[76] Since it is not in general the case that $y^k = \sum_{j\in N'} x^j$ for all k such that $x^k + p_k y^k \leqq 0$.

A *Nash equilibrium* s^* of this quasi-game is defined, as usual, by

$$v^r(s^*) \geqq v^r(r, s^r/s^{*\,)r(}) \quad \text{for all} \quad s^r \in S^r, \qquad r = 1, \ldots, n, n+1.$$[77]

It is clear that the Nash equilibrium s^* for the outcome and payoff functions specified above is characterized by: (1) $y^{*1} = \cdots = y^{*n}$ ($= y^*$ say) since otherwise v^{n+1} is not maximized with respect to $s^{n+1} = p$; (2) $x^{*i} + p_i^* y^* = 0$ for all $i = 1, \ldots, n$; and (3) $u^i(x^{*i}, y^*) \geqq u^i(x^i, y^i)$ for any (x^i, y^i) satisfying the budget constraint $x^i + p_i^* y^i \leqq 0$, since otherwise v^i is not being maximized with respect to $s^i = (x^i, y^i)$. Note also that the last condition implies that, at a Nash equilibrium, prices are treated parametrically.

Thus Nash equilibrium for this quasi-game yields a Lindahl equilibrium. But there are two features considered unsatisfactory by some. The first objection is to the introduction of a non-player participant, the auctioneer. This participant has a strategy variable but his payoff function is artificial – corresponding to the rule that the price vector should be chosen so as to equalize the agents' demands for public service. But it is not obvious how seriously to take this objection, since the role of the auctioneer could be programmed for a computer.

The second objection is to the lack of balance in the outcome function. When the agents' demands $y^1, \ldots, y^n$ for public good are not all equal, the actual supply y (here equal to $\sum_{j \in N'} x^j$) must be different from y^i for some i. The allocation $z^i = (x^i, y^i) = h(s)$ specified by the outcome function is therefore different from the allocation that would in fact be made, say $\tilde{z}^i = (x^i, \sum_{j \in N'} x^j)$ if $x^i + p_i y^i \leqq 0$ or $\tilde{z} = (0,0)$, otherwise. One would, in effect be assuming that the agents are either acting in ignorance of what the actual outcome would be, or willing to act on an "as if" basis. Such lack of realism can only be avoided by constructing a game that is balanced.

Thus it becomes of interest to know that a balanced outcome function can be constructed to implement the Lindahl correspondence. Indeed one need not use a separate auctioneer or any other non-player participant.[78]

7. Balanced outcome functions without an auctioneer

For $n = 3$, and without satisfying the individual feasibility condition, such outcome functions were constructed by Hurwicz (1979a) and Walker (1981). In

[77] Hurwicz (1979b) gives a somewhat more general definition of Nash equilibrium in a quasi-game.

[78] In fact, the issue of constructing balanced outcome functions implementing a Pareto-optimal individually rational social choice rule without an auctioneer arose first in the context of implementing the Walrasian correspondence and was accomplished for $n \geqq 3$, although without satisfying the individual feasibility condition, by Schmeidler (1976).

Hurwicz (1979a) one may think of agents as arranged in a circle, with each agent setting the price (acting in effect as an auctioneer) for his/her neighbors.[79]

At the same time each agent proposes the level of the public good. Thus agent i's message is of the form $m^i = (p_i, y_i)$, where y_i is the proposed level of the public good and $p_i \geqq 0$ will serve to determine the price to be paid by certain agents other than i. Specifically, the Lindahl price to be paid by agent j is[80]

$$R_j(m) = 1/n + p_{j+1} - p_{j+2},$$

where j is taken modulo n. (I.e., $n+1 \equiv_n 1$, $n+2 \equiv_n 2$, etc.) The outcome function is defined as follows:

$$Y(m) = \sum_{j=1}^{n} y_j/n,$$

and

$$X^i(m) = -R_i(m)\cdot Y(m) - p_i(y_i - y_{i+1})^2 + p_{i+1}(y_{i+1} - y_{i+2})^2, \quad i = 1,\dots,n.$$

As is necessary for the Lindahl prices in this normalized model, we have

$$\sum_{i=1}^{n} R_i(m) = 1 \quad \text{for all } m,$$

and hence

$$\sum_{i=1}^{n} X^i(m) = -Y(m) \sum_{j=1}^{n} R_i(m) - \sum_{i=1}^{n} p_i(y_i - y_{i+1})^2 + \sum_{i=1}^{n} p_{i+1}(y_{i+1} - y_{i+2})^2.$$

Now the last two terms cancel out and so

$$\sum_{i=1}^{n} X^i(m) = -Y(m)\cdot R_i(m) = -Y(m).$$

Hence the mechanism is balanced.[81]

[79] Thus every agent is forced to treat the price parametrically because, at a Nash equilibrium, the other players' strategies are taken as given.

[80] $m = (m^1,\dots,m^n)$, $m^i = (p_i, y_i)$, $i = 1,\dots,n$. All summations, unless otherwise indicated, are from 1 to n.

[81] The balance is essentially due to the presence of the term $p_{i+1}(y_{i+1} - y_{i+2})^2$ in $X^i(m)$. Note that this term cannot be influenced by agent i when $n \geqq 3$. (For $n = 2$, $y_{i+2} = y_i$.)

On the other hand, it is not the case that $(X^i(m), Y(m)) \geqq 0$ for all m. Hence the mechanism does not satisfy the condition of individual feasibility.[82]

It turns out that, at a Nash equilibrium $m^* = \{(y_i^*, p_i^*)\}_{i=1}^n$, we have, for every i, $y_i^* = y_{i+1}^*$ or $p_i^* = 0$, and hence

$$X^i(m^*) = -R_i(m^*)Y(m^*),$$

which is the Lindahl budget equation.

It was then shown that, for $n \geqq 3$, the mechanism so constructed (fully) implements the Lindahl correspondence.

A simpler mechanism, using a smaller message space,[83] was constructed in Walker (1981). Here each player has a one-dimensional strategy (message) space $M_i = R$. The outcome function is given [with $m = (m_1, \dots, m_n)$] by

$$Y(m) = \sum m_i,$$

and (in terms of net trades)

$$X^i(m) = -Y(m) \cdot q^i(m),$$

where

$$q^i(m) = 1/n + m_{i+2} - m_{i+1}$$

is the ith agent's Lindahl price [with $\sum q^i(m) = 1$], again outside the control of the ith agent. The balance property follows from $\sum X^i(m) = -Y(m) \sum q^i(m) = -Y(m)$ for all m. Again, however, individual feasibility may be violated.[84] For $n \geqq 3$, this mechanism is shown to (strongly) implement the Lindahl correspondence. It is obvious that[85] it does so with a message space of minimal dimension.[86]

Neither of the preceding two mechanisms implements the Lindahl correspondence when there are only two agents, and both violate the condition of individual feasibility. But we shall now see that either one of these defects may be avoided.

[82] In Hurwicz (1979a) the preference relation $\succcurlyeq_i$ was extended to all R^2, so formally the mechanism was well-defined, and individually feasible in terms of *extended* preferences. But there is no realistic reason why an agent would prefer one infeasible bundle to another.

[83] And having other advantages, see Walker (1981, p. 66, footnote 2). In addition, Walker shows how to generalize this mechanism to many agents and goods and more general production relations (ibid., pp. 68–69).

[84] See Walker (1981, p. 67, footnote 3) where, however, the point is made that an *interior* equilibrium has a neighborhood on which feasibility is assured. The same applies to Groves and Ledyard (1977) as well as Schmeidler (1976, 1980) and Hurwicz (1979a).

[85] For a sufficiently rich class of environments.

[86] Cf. Reichelstein (1983).

For an economy with two agents, a mechanism has been constructed by Miura (1982).[87] Here again $m^i = (p_i, y_i)$, $i = 1,\ldots, n$, and the outcome function is given by[88]

$$Y(m) = \sum y_j / n, \quad X^i(m) = -\left(p_i / \sum p_j\right) Y(m) \quad \text{if} \quad \prod p_j = 1,$$

while

$$Y(m) = p_j - 1, \quad X^i(m) = -\left(p_i / \sum p_j\right) Y(m) \quad \text{if} \quad \prod p_j \neq 1.$$

This outcome function is balanced under both regimes, but – like Hurwicz (1979c) – it is discontinuous and violates the individual feasibility condition; also, it uses a message space bigger than Walker's (1981).[89] It does implement the Lindahl correspondence for $n = 2$.[90]

Completely feasible implementation. The issue of implementation satisfying both the balance and individual feasibility conditions (we call this *completely feasible implementation*) is treated in Hurwicz, Maskin and Postlewaite (1980). It is shown there that, for $n \geqq 3$, the constrained[91] Lindahl correspondence can be Nash-

[87] This is a modification of the mechanism in Hurwicz (1979c) which Miura (1982) showed to contain an error. Hurwicz (1979) used the condition $\Sigma_{j=1}^{n} p_j = 1$ rather than $\Pi_{j=1}^{n} p_j = 1$ to distinguish between two regimes.

[88] All summations and products are from 1 to n.

[89] By an argument analogous to Hurwicz (1976, appendix 1) one can show that no smooth balanced Nash implementation of any Pareto-optimal correspondence is possible for $n = 2$ with a 2-dimensional message space.

[90] Note that the previously discussed mechanisms [Hurwicz (1979) and Walker (1981)], for $n \geqq 3$, are not only continuous but even smooth. For $n = 2$, it appears that the Lindahl correspondence cannot be Nash-implemented smoothly by a balanced outcome function [using an argument analogous to Reichelstein (1984)].

[91] In the setting of footnote 30, a *constrained Lindahl allocation* $L_c(e)$ for environment e, denoted by $L_c(e)$, differs from an ordinary Lindahl allocation $L(e)$ in that condition (2) is replaced by the following (2'): for some $(p_1,\ldots, p_n)$ and all $i \in N$, if $(x^i, y) \in C^i$, $x^i + yp_i \leqq \omega_x^i + \omega_y^i p_i$, and $x^i \leqq \Sigma_{i \in N} \omega_x^i$, then $(\bar{x}^i, \bar{y})\ R^i(e)\ (x^i, y)$.

Analogously, an allocation $(\bar{x}_i)_{i \in N}$, $\bar{x}_i \in R_+^l$ and a price vector p constitute a (pure exchange) *constrained Walrasian equilibrium* if

(i) $i \in N$, $p \cdot \bar{x}_i = p \cdot \omega_i$;

(ii) $i \in N$, $x_i R_i x$ for all $x \leqq \omega$, $\omega \equiv \Sigma \omega_i$, such that $p \cdot x \leqq p \cdot \omega_i$;

(iii) $\Sigma_{i \in N} \bar{x}_i = \omega$.

Note that $\bar{x}_i \in R_+^l$ represents total holdings (*not* the net trade) of the ith agent.

It is the condition "for all $x \leqq \omega$" that distinguishes this from the ordinary Walrasian equilibrium, since here the ith agent maximizes satisfaction subject not only to the individual budget constraint but also constrained by the aggregate availability of resources in the whole economy. (Similar remark applies to the condition $x^i \leqq \Sigma \omega_x^i$ in the definition of constrained Lindahl equilibrium.)

implemented by characteristic profile strategies over a class of environments E for which Lindahl allocations leave everyone with some private goods, i.e., such that for each e in E,

$$\omega_x^i + L_x^i(e) \geq 0,$$

where $L_x^i(e)$ is the net Lindahl increment of private good X obtained by agent i in the environment e.

In Hurwicz, Maskin, and Postlewaite (1980), a generic element of agent i's strategy space is of the form

$$s_i = (e_i^1, \dots, e_i^n, y_i), \qquad e_i^j = (w_i^j, R_i^j),$$

where y_i is i's proposal for the level of public goods, and e_i^j is agent i's statement concerning agent j's characteristic; w_i^j and R_i^j are, respectively, j's X-endowment and j's preference relation according to i. It is postulated that no agent can exaggerate his own endowment, i.e., $w_i^j \leqq \omega_i$ for all i. The outcome function is formulated in such a way that only truthful unanimity as to endowments can prevail at a Nash equilibrium. Once such unanimity as to endowments has taken place, a game of type considered in Maskin's Theorem 5 guarantees the Nash implementation of the (constrained) Lindahl correspondence through the use of preference profiles as strategies.

The result obtained [Hurwicz, Maskin and Postlewaite (1980, theorem VII)] applies to a class of performance correspondences broader than (constrained) Lindahl; it is sufficient that they be individually rational and monotone;[92] monotonicity is also necessary by Maskin's (1977) Theorem 2. This result presupposes that there are at least three agents, that the endowments are semi-positive, and preferences strictly increasing.

Every Walrasian equilibrium is a constrained Walrasian equilibrium. Every *interior* constrained Walrasian equilibrium is a Walrasian equilibrium.

When non-interior equilibrium allocations can occur, the (ordinary) Walrasian and Lindahl correspondences are not monotone, hence [by Theorem 2 in Maskin (1977)] are not Nash-implementable (see next page). The smallest monotone correspondences containing these are the corresponding constrained correspondences. Hence they are the smallest supercorrespondences that have a chance of being implementable.

[92]A performance correspondence $F: \mathscr{R} \Rightarrow A$ defined on the family $\mathscr{R}$ of preference profiles into the feasible set A is said to be *monotone* if, for any a in A, and any two profiles $\boldsymbol{R}, \boldsymbol{R}'$, the following holds: if (1) $a \in F(\boldsymbol{R})$, and (2) aR_ib implies $aR_i'b$ for all $i \in \{1, \dots, n\}$ and all $b \in A$, then $a \in F(\boldsymbol{R}')$. I.e. if $\underset{\sim}{a}$ is F-desirable for profile $\boldsymbol{R}$, and another profile $\boldsymbol{R}'$ is at least as favorable to $\underset{\sim}{a}$ as $\boldsymbol{R}$ was, then $\underset{\sim}{a}$ remains F-desirable for profile $\boldsymbol{R}'$. [Here $\boldsymbol{R} = (R_1, \dots, R_n)$, $\boldsymbol{R}' = (R_1', \dots, R_n')$. "$\underset{\sim}{a}$ is F-desirable for environment $\underset{\sim}{e}$" simply means $a \in F(e)$. Maskin (1977) uses the term "F-optimal". "At least as favorable" here permits replacing preference by indifference.]

It may be noted that the "no veto power"[93] condition used by Maskin in theorem 5 is not necessarily satisfied by the Lindahl or other performance correspondences to which Theorem VII of Hurwicz, Maskin, and Postlewaite (1980) applies. However, it is shown that, under the assumptions made, the "no-veto power" condition can be dispensed with.

Unfortunately, profile-using mechanisms require huge message spaces and are discontinuous. However, it may be possible to construct continuous mechanisms using smaller spaces, in a manner analogous to that used in Postlewaite and Wettstein (1983) for the implementation of Walrasian correspondence, but this question is still open. So is the problem of designing a feasible mechanism to implement the (constrained) Lindahl correspondence when there are only two agents.

8. Implications of Nash-implementability

We have already mentioned Maskin's result that only monotone performance correspondences are Nash-implementable. An example due to Postlewaite[94] shows that the Walrasian correspondence containing boundary allocations is not monotone; an analogous example could be constructed for the Lindahl correspondence. It is for this reason that the Nash-implementability results are for "constrained" Walrasian (respectively Lindahl) correspondences; these constrained correspondences are the smallest monotone ones containing the Walrasian (respectively Lindahl) correspondences.

Now it turns out that, in conjunction with other frequently made assumptions, monotonicity has very strong implications. In particular, suppose that a correspondence F is not only monotone, but also Pareto-optimal, individually rational, and continuous, over a sufficiently rich class of preferences. If we are dealing with pure exchange economies, it follows that

$$F(e) \supseteqq W_c(e) \quad \text{for all } e \text{ in } E, \tag{8.1}$$

where W_c is the constrained[95] Walrasian correspondence and $E \supseteq E_{L_c}$, the class of economies specified in Hurwicz (1979b).[96] Similarly, if we are dealing with a

[93] A performance correspondence $F: \mathscr{R} \Rightarrow A$ is said to have the "*no veto power*" property when the following is true: if, for any a in A, and $\boldsymbol{R}$ in $\mathscr{R}$, and for some i in $\{1,\dots,n\}$, we have aR_jb for all $b \in A$ and all $j \neq i$, then $a \in F(\boldsymbol{R})$. I.e. if the prevailing preferences are such that $\underline{a}$ is the most preferred alternative either for all players or for all players but one, then a is F-desirable.

[94] Described in Hurwicz, Maskin and Postlewaite (1980).

[95] See footnote 91.

[96] For $n \geqq 3$ this can be proved [see Hurwicz, Maskin and Postlewaite (1980)] by using Theorem 1 in Hurwicz (1979b) and Theorem 2 in Maskin (1977). However, a direct (unpublished) proof is available, valid for all $n \geqq 1$.

public goods economy,

$$F(e) \supseteqq L_c(e) \quad \text{for all } e \text{ in } E,$$

where L_c is the constrained Lindahl correspondence.

Now consider an arbitrary correspondence F which is continuous, Pareto-optimal, individually rational, and Nash-implementable over $E \supseteq E_{L_c}$. Since Nash-implementability implies monotonicity, it follows that $F \supseteqq W_c$ if we are in a private goods, pure exchange economy, or that $F \supseteqq L_c$ in a public goods economy. In other words a continuous, Pareto-optimal, individually rational correspondence, which does not contain the Walrasian (respectively Lindahl) performance correspondence, is not Nash-implementable. In particular, if F is singleton-valued, continuous, Pareto-optimal, and individually rational, it is Nash-implementable only if it is constrained Walrasian.[97]

A partial converse is obtained under additional (convexity or starlike) restrictions on the outcome function [Hurwicz (1979c) and Schmeidler (1982)].

If, in the above assumptions, *fairness* (in the sense of absence of envy) replaces individual rationality, it has been shown by Thomson (1979) that analogous conclusions obtain for a pure exchange economy, namely with formula (8.1) being replaced by

$$F(e) \supseteqq W_{c,I}(e),$$

where $W_{c,I}(e)$ is the constrained Walrasian allocation that would follow equal distribution of endowments.

Hence, if, in a private goods, pure exchange economy, F is continuous, Pareto-optimal, and envy-free, it is Nash-implementable only if it contains $W_{c,I}$. Analogous results were obtained by Thomson for other concepts of fairness, e.g. that of egalitarian equivalent [Pazner and Schmeidler (1978)]. Partial converses were also obtained.

9. Private goods, pure exchange economies

The preceding exposition has been focused on the public goods problem. But there are analogous results for private goods economies. The earliest balanced outcome function Nash-implementing the Walrasian correspondence for $n \geqq 3$ is due to Schmeidler (1976). A slightly modified form [Hurwicz (1979c)] works for all $n \geqq 2$. Here the ith strategy $m^i = (p_i, y_i)$, where both components are of dimension $l-1$, is an economy with goods X and Y [X one-dimensional, Y $(l-1)$-dimensional]. p_i is the price of Y while X is the numeraire.

[97] For a related result in "large" economies, see Theorem 5 in Hammond (1979).

Given an n-tuple $m=(m^1,\dots,m^n)$ of strategies, the set N of agents is partitioned so that (1) if agent i has proposed a price vector p_i that has not been proposed by anyone else, then i belongs to the subset T_0 of N, and only such "loners" belong to T_0; (2) if a price vector q has been proposed by two or more members of N, then all those proposing q belong to the subset $T(q)$.

Let $q^1,\dots,q^k$ be the complete list of price vectors each of which has been proposed by at least two agents, and write $T(q^r)=T_r$. Thus N is partitioned into subsets $T_0, T_1,\dots, T_k$, with some of these possibly empty. The outcome function is written

$$h(m)=\{X^i(m),Y^i(m)\}_{i=1}^{m}.$$

If $T_0\neq N$ and agent i belongs to the subset T_r, $r\in\{1,\dots,k\}$, then

$$Y^i(m)=y_i-\Big(\sum_{j\in T_r} y_j\Big)\Big/\#T_r,$$
$$X^i(m)=-q^r\cdot Y^i(m).$$

(This is precisely the Schmeidler rule.)

If T_0 is non-empty and i belongs to T_0, then

$$Y^i(m)=p_i-P,\qquad X^i(m)=-P\cdot Y^i(m),$$

where

$$P=\Big(\sum_{j\in T_0} p_j\Big)\Big/\#T_0.$$

(This rule is different from Schmeidler's rule for T_0.)

This outcome function is balanced but outcomes need not be non-negative; i.e., individual feasibility is not guaranteed. But with regard to extended preference relations, it (fully) Nash-implements the Walrasian correspondence for all $n\geqq 2$.

The preceding outçome function is discontinuous. A smooth balanced outcome function, also Nash-implementing the Walrasian correspondence (but again only for extended preferences, hence violating individual feasibility) is given, for $m=(m^1,\dots,m^n)$, $m^i=(p_i,y_i)$, p_i, y_i both $(l-1)$-dimensional vectors by $h(m)=\{X^i(m)=(X^i(m),\,Y^i(m))\}_{i=1}^{n}$, as follows: for each $i=1,\dots,n$,

$$Y^i(m)=y_i-y_{-i},$$

where

$$y_{-i}=\Big(\sum_{j\neq i} y_j\Big)\Big/(n-1),$$
$$X^i(m)=-p_{-i}\cdot Y^i(m)-L^i(m)+S^i(m),$$

where

$$p_{-i} = \left(\sum_{j \neq i} p_j \right) \Big/ (n-1),$$

$$L^i(m) = (p_i - p_{-i}) \cdot (p_i - p_{-i}),$$

and $S^i(m)$, which does not depend on m^i, is such that $\sum_{i=1}^n X^i(m) = 0$ for all m. Since $\sum_{i=1}^n Y^i(m) = 0$ for all m, the outcome function is balanced. The "penalty" term $L^i(m)$ provides an incentive to equalize all price proposals, and the "budget" term $-p_{-i} \cdot Y^i(m)$ prevents all agents from being their own price-setters.[98]

Completely feasible implementation. On the other hand, it is also possible to Nash-implement the Walrasian correspondence by a balanced outcome function without violating the individual feasibility condition, in a manner analogous to that outlined above for the Lindahl correspondence. In fact, a more general result is available. Assume $n \geq 3$, and consider a performance function[99] f (on a class E of pure exchange economies) which is individually rational and Nash-implementable by an outcome function g_v on strategy domain $D = D^1 \times \cdots \times D^n$ when v is a *known* endowment profile.

To implement f when the endowments are not known to the designer we give the ith agent the strategy domain S^i whose generic element is of the form

$$s_i = \left(w_i^1, \ldots, w_i^n, d_i \right),$$

where w_i^j is i's statement concerning the endowment of agent j,[100] and d_i an element of D^i. The outcome function on $S = S_1 \times \cdots \times S_n$ (for the game in which initial endowments are *not* known) is so designed that a Nash equilibrium can occur only when, for all i, $(w_i^1, \ldots, w_i^n) = (\dot{\omega}_1, \ldots, \dot{\omega}_n) =$ the true endowment profile. When such unanimity occurs, what remains is in effect the game form (D, g_v) implementing f on the assumption that the unanimously agreed endowment profile v is the correct one.

[98]An earlier version of Hurwicz (1979a) contained a "circular" variant, with $X^i(m) = -Y^i(m) \cdot p_{i-1} - L^i +$ a balancing term independent of r^i. Here, as in the above public goods mechanism, each agent is his neighbor's price-setter.

[99]With minor modifications of the outcome functions, these results can be extended to correspondences.

[100]It is assumed that $w^i \leq \dot{\omega}_i$ for all v, i.e. one cannot exaggerate one's endowment. See Hurwicz, Maskin and Postlewaite (1980).

Since f is assumed Nash-implementable for known endowments, it must [by Maskin (1977, theorem 2)] be monotone. We have also assumed $n \geqq 3$. If f has the "no-veto property", then Maskin's Theorems 4 and 5 provide us with a game form with a generic term of strategy domain D^i of the form

$$d_i = (R_i^1, \dots, R_i^n),$$

where R_i^j is the statement by i concerning the preference relation of agent j.

Maskin's Theorems 4 and 5 do more than construct a game form for a particular performance correspondence; these theorems constitute a recipe for constructing game forms Nash-implementing a large class of correspondences when there are at least three players ($n \geqq 3$). Let $F: E \Rightarrow A$ be a correspondence to be implemented in an economy where A is the set of feasible outcomes, $R = R_1 \times \cdots \times R_n$ is a class of preference profiles, and

$$E_A = \{e : e = (A; R_1, \dots, R_n), (R_1, \dots, R_n) \in R_1 \times \cdots \times R_n\}.$$

For each i, R_i is a preference preordering (a total, transitive and reflexive binary relation) on A. It is assumed that A and R are a priori known to the designer.

Agent i's strategy space is

$$S_F^i = \{(R_1, \dots, R_n, a) : (R_1, \dots, R_n) \in R,\ a \in F(A; R_1, \dots, R_n)\}.$$

A generic element of S^i is of the form

$$s_i = (R_i^1, \dots, R_i^n, a),$$

where R_i^j is i's statement about j's preference relation.

Note that when F is singleton-valued (i.e. a function), the a-component of s_i can be omitted. Theorem 4 states that an outcome function $h: S \to A$, $S = S^1 \times \cdots \times S^n$, fully implements F if it has the following three properties:

(i) If there is unanimity, i.e. if, for some $(\bar{R}_1, \dots, \bar{R}_n) \in R$ and $\bar{a} \in F(\bar{R}_1, \dots, \bar{R}_n)$, all agents' strategies satisfy $\bar{s}_1 = \cdots = \bar{s}_n = (\bar{R}_1, \dots, \bar{R}_n, \bar{a})$, then[101]

$$h(\bar{s}_1, \dots, \bar{s}_n) = \bar{a}.$$

(ii) For any agent $i \in \{1, \dots, n\}$, let $\bar{s}_1 = \cdots = \bar{s}_{i-1} = \bar{s}_{i+1} = \cdots \bar{s}_n = (\bar{R}_1, \dots, \bar{R}_n, \bar{a})$, with $\bar{a} \in F(\bar{R}_1, \dots, \bar{R}_n)$; then[102]

$$\{b \in A : b = h(\bar{s}_1, \dots, \bar{s}_{i-1}, s_i, \bar{s}_{i+1} = \cdots = \bar{s}_n), s_i \in S^i\}$$
$$= \{c \in A : \bar{a}\bar{R}_i c\}$$
$$= \text{the lower contour set of } \bar{a} \text{ under } \bar{R}_i.$$

[101] Note that when F is a function this relation becomes $h(\bar{s}) = F(\bar{s})$.

[102] I.e. when all agents but i are unanimous, every outcome in A can be reached by i through unilateral choice.

(iii) For $\bar{s} \in S$, let there exist $i \in \{1,\dots,n\}$ such that "$\bar{s}_1 = \cdots = \bar{s}_{i-1} = \bar{s}_{i+1} = \bar{s}_n$" is false; then[103]

$$\{b \in A : b = h(\bar{s}_1,\dots,\bar{s}_{i-1},\bar{s}_i,\bar{s}_{i+1},\dots,\bar{s}_n), s_i \in S^i\} = A.$$

Maskin's Theorem 5 shows by construction that when $n \geqq 3$, and F is monotone and has the "no-veto power" property, then there exists an outcome function h satisfying the above conditions (i), (ii), (iii). [See, however, Williams (1984a, 1984b) and Saijo (1984).]

However, the strategy domain used in the above example is extremely large, and the outcome function discontinuous. For the special case of a Walrasian correspondence, $n \geqq 3$, Postlewaite and Wettstein (1983) have designed a balanced *continuous* outcome function, with a generic element of D^i of the form

$$d_i = (z^i, p^i, r^i),$$

where z^i and p^i are, respectively, a net trade vector and a price vector announced by individual i, and r^i a positive real number. Thus the dimension of D^i is $1+2l$, where l is the total number of goods.

Reichelstein (1982) has constructed, for $n \geqq 3$, smooth mechanisms Nash-implementing the Walrasian correspondence with strategic domains of smaller dimensions, but these do not satisfy the individual feasibility condition. It is not known at present what the minimal required dimension is when we insist on individual feasibility as well as balance and continuity or smoothness. [But see Williams (1984c).]

10. Informational aspects of Nash-implementability

An exciting aspect of recent research is the convergence of informational and incentive aspects of economic mechanisms. One example of this is the investigation of minimal dimensional requirements[104] on the strategy spaces of mechanisms implementing the Pareto (and, in particular, Walras or Lindahl) correspondence. Another is a study of the relationship between message mechanisms and game forms giving rise to Nash equilibrium outcomes.[105]

It is clear that such a game form may be viewed as arising from an adjustment process or a message mechanism. Given a game form (S, h), let $\nu_{S,h}$ be its Nash equilibrium correspondence over E; i.e. for each e in E,

$$\nu_{S,h}(e) = \{s \in S : s \text{ is a Nash equilibrium for the game } (S; h, e)\},$$

[103] I.e. when there is no unanimity among agents other than i, agent i can reach every feasible point by unilateral choice.

[104] Reichelstein (1982) and Hurwicz (1976).

[105] Williams (1984b).

where (h, e) is the payoff function defined by the composition of the outcome function h with the preferences in e. E.g. if $e = (e^1, \dots, e^n)$, $e^i = (u^i, \dots)$, where u^i is the ith utility function, then the ith payoff function is $v^i = u^i \circ h$.

Furthermore the resulting process is privacy-preserving. This is seen most easily when the preferences are represented by concave differentiable functions u^i and the outcome function h is also concave differentiable. Then the ith payoff function is $v^i = u^i \circ h$. At a Nash equilibrium, $\partial v^i / \partial s_i = 0$ for all i. More explicitly, in an l-dimensional commodity space, and with a k_i-dimensional strategy vector s_i, the condition becomes

$$\sum_{j=1}^{l} \left(\partial u^i / \partial x_j^i \right) \left(\partial h_j^i / \partial s_{i, r_i} \right) = 0, \qquad r_i = 1, \dots, k_i.$$

It is clear that each agent can verify this condition knowing only the values of the strategy variables s and his/her own utility function. So the process is privacy-preserving.

Given a game form (S, h) which Nash-implements the correspondence F, we can easily construct a privacy-preserving message mechanism (S, ν_{Sh}, h) which realizes F. But the converse problem is much more difficult: given a privacy-preserving message mechanism $(\mathscr{M}, \mu, h)$ realizing F, to construct a game form (S, h) which Nash-implements F. Sufficient conditions for such construction are given by Williams (1984b).

In particular, let $(\mathscr{M}, \mu, h)$ be the natural direct revelation mechanism for a social choice *function* $f: E \to Z$. Thus $\mathscr{M} = \mathscr{M}^1 \times \cdots \times \mathscr{M}^n = E^1 \times \cdots \times E^n = E$, $h = f$, and μ is given by

$$m_i = e^i \quad \text{for all } i.$$

Then the Williams construction (which, however, covers a much broader class of mechanisms) yields a Maskin-type game form, with $S = E$.

The present paper fails to cover a number of important topics: the Harsanyi–Bayes type mechanism, the dynamics of allocation processes (e.g. Malinvaud, Drèze and de la Vallée Poussin), performance of mechanisms in large economies, and many others.

Several excellent surveys cover some of the topics neglected here as well as many that are discussed in the present essay [e.g. Dasgupta, Hammond and Maskin (1979), Groves (1982), Laffont and Maskin (1982), Postlewaite (1983)].

References

Arrow, K. J. (1977), "The Property Rights Doctrine and demand revelation under incomplete information", IMSSS Technical Report No. 243. Stanford: Stanford University.

d'Aspremont, C. and L.-A. Gérard-Varet (1979a), "On Bayesian incentive compatible mechanisms", in: J.-J. Laffont, ed., Aggregation and Revelation of Preferences, pp. 269–288. Amsterdam: North-Holland.

d'Aspremont, C. and L.-A. Gérard-Varet (1979b), "Incentives and incomplete information", Journal of Public Economics, 11:25–45.

Bonin, John P. (1975), "On the design of managerial incentive structures in a decentralized planning environment", Discussion Paper No. 75-9. San Diego: Department of Economics, University of California.

Bowen, H. (1943), "The interpretation of voting in the allocation of economic resources", Quarterly Journal of Economics, 58:27–48.

Clarke, E. H. (1971), "Multipart pricing of public goods", Public Choice, 2:19–33.

Dasgupta, Partha, Peter Hammond and Eric Maskin (1979), "The implementation of social choice rules: Some general results on incentive compatibility", Review of Economic Studies, 46:181–216.

Drèze, J. and D. de la Vallée Poussin (1969), "A tâtonnement process for guiding and financing an efficient production for public goods". Louvain-la-Neuve: C.O.R.E.

Fan, L. S. (1975), "On the reward system", American Economic Review, 65:226–229.

Foley, D. K. (1970), "Lindahl's solution and the core of an economy with public goods", Econometrica, 38:66–72.

Gibbard, A. (1973), "Manipulation of voting schemes: A general result", Econometrica, 61:587–602.

Gindin, S. (1970), "A model of the Soviet firm", Economic Planning, 10:145–157.

Green, J. and J.-J. Laffont (1977), "Characterization of satisfactory mechanisms for the revelation of preferences for public goods", Econometrica, 45:427–438.

Green, J. and J.-J. Laffont (1979), Incentives in Public Decision-Making. Amsterdam: North-Holland.

Green, J., E. Kohlberg and J.-J. Laffont (1976), "Partial equilibrium approach to the free rider problem", Journal of Public Economics, 6:375–394.

Groves, T. (1970), "The allocation of resources under uncertainty", Ph.D. Dissertation. Berkeley: University of California.

Groves, T. (1979), "Efficient collective choice with compensation", in: J. J. Laffont, ed., Aggregation and Revelation of Preferences, pp. 37–59. Amsterdam: North-Holland.

Groves, T. (1982), "On the theories of incentive compatible compensation", in: W. Hildenbrand, ed., Advances in Economic Theory, Ch. 1. Cambridge: Cambridge University Press.

Groves, T. and J. Ledyard (1977), "Optimal allocation of public goods: A solution to the 'free rider' problem", Econometrica, 45:783–811.

Groves, T. and J. Ledyard (1980), "The existence of efficient and incentive compatible equilibria with public goods", Econometrica, 48:1487–1506.

Groves, T. and M. Loeb (1975), "Incentives and public inputs", Journal of Public Economies, 4:211–226.

Hammond, P. J. (1979), "Straightforward individual incentive compatibility in large economies", Review of Economic Studies, 46:263–282.

von Hayek, F. A. (1945), "The use of knowledge in society", American Economic Review, 35:519–530.

Holmström, B. (1979), "Groves scheme on restricted domain", Econometrica, 47:1137–1144.

Hurwicz, L. (1960), "Optimality and informational efficiency in resource allocation processes", in: K. J. Arrow, S. Karlin and P. Suppes, ed., Mathematical Methods in the Social Sciences, pp. 27–46. Stanford: Stanford University Press.

Hurwicz, L. (1972), "On informationally decentralized systems", in: R. Radner and C. B. McGuire, eds., Decision and Organization (Volume in Honor of J. Marschak), pp. 297–336. Amsterdam: North-Holland.

Hurwicz, L. (1975a), "On the Pareto-optimality of manipulative Nash equilibria", presented at the International Econometric Society World Congress, August 1975, Toronto.

Hurwicz, L. (1975b), "Incentives in resource allocation systems with public goods", Mimeographed. Boulder, CO.

Hurwicz, L. (1976), "On informational requirements for nonwasteful resource allocation systems", in: Sydney Shulman, ed. (NBER), Mathematical Models in Economics: Papers and Proceedings of a U.S.–U.S.S.R. Seminar, Moscow.

Hurwicz, L. (1977), "On the dimensional requirements of informationally decentralized Pareto-satisfactory processes", in: Studies in Resource Allocation Processes, pp. 413–424. Cambridge: Cambridge University Press.

Hurwicz, L. (1979a), "Outcome functions yielding Walrasian and Lindahl allocations at Nash equilibrium points", Review of Economic Studies, 46:217–225.

Hurwicz, L. (1979b), "On allocations attainable through Nash equilibria", Journal of Economic Theory, 21:140–165. Also in: J.-J. Laffont, ed., Aggregation and Revelation of Preferences, pp. 397–419. Amsterdam: North-Holland.

Hurwicz, L. (1979c), "Balanced outcome functions yielding Walrasian and Lindahl allocations at Nash equilibrium points for two or more agents", in: Jerry R. Green and José A. Scheinkman, eds., General Equilibrium, Growth, and Trade. New York: Academic Press.

Hurwicz, L. (1981), "On incentive problems in the design of non-wasteful resource allocation systems", in: J. Łoś, ed., Studies in Economic Theory and Practice (in Honor of Edward Lipiński), pp. 93–106. Amsterdam: North-Holland.

Hurwicz, L. and D. Schmeidler (1978), "Construction of outcome functions guaranteeing existence and Pareto-optimality of Nash equilibria", Econometrica, 46:1447–1474.

Hurwicz, L. and M. Walker (1983), "On the generic non-optimality of dominant-strategy mechanisms with an application to pure exchange", Research Paper No. 250, circulated by the Economic Research Bureau. Stony Brook: S.U.N.Y.

Hurwicz, L., E. Maskin and A. Postlewaite (1980), "Feasible implementation of social choice correspondences by Nash equilibria", Mimeographed.

Jordan, J. S. (1982), "The competitive allocation process is informationally efficient uniquely", Journal of Economic Theory, 28:1–18.

Johansen, L. (1963), "Some notes on the Lindahl theory of determination of public expenditure", International Economic Review, 4:346–358.

Kolm, S. Ch. (1964), "Introduction à la théorie du ròle économique de l'état: Les fondements de l'économie publique". Paris: IFP.

Laffont, J.-J. (1979), Aggregation and Revelation of Preferences. Amsterdam: North-Holland.

Laffont, J.-J. and E. Maskin (1982), "The theory of incentives: An overview", in: W. H. Hildenbrand, ed., Advances in Economic Theory, Ch. 2. Cambridge: Cambridge University Press.

Ledyard, J. (1977), "Incentive compatible behavior in core-selecting organizations", Econometrica, 45:1607–1621.

Ledyard, J. and J. Roberts (1974), "On the incentive problems with public goods", Discussion Paper No. 116. Evanston: CMSEMS, Northwestern University.

Lindahl, E. (1919), "Just taxation–A positive solution", in: 1967, Musgrave and Peacock, eds., Classics in the Theory of Public Finance, pp. 168–176.

Malinvaud, E. (1971), "A planning approach to the public goods problem", Swedish Journal of Economics, 72:96–112.

Malinvaud, E. (1972), Lectures on Microeconomic Theory.

Maskin, E. (1977), "Nash equilibrium and welfare optimality", Working Paper. Cambridge: M.I.T.

Milleron, J.-C. (1972), "Theory of value with public goods: A survey article", Journal of Economic Theory, 5:419–477.

Miura, Rei (1982), "Counter-example and revised outcome function yielding the Nash–Lindahl equivalence for two or more agents", privately circulated paper. Tokyo: Keio University.

Mount, K. and S. Reiter (1974), "The informational size of message spaces", Journal of Economic Theory, 8:161–192.

Nakayama, M. (1980), "Optimal provision of public goods through Nash equilibria", Journal of Economic Theory, 23:334–347.

Osana, H. (1978), "On the informational size of message spaces for resource allocation processes", Journal of Economic Theory, 17:66–78.

Postlewaite, A. (1983), "Implementation via Nash equilibria in economic environments", CARESS Working Paper No. 83-18. Philadelphia: University of Pennsylvania.

Reichelstein, S. (1982), "On the informational requirements for the implementation of social choice rules", Handout for the Decentralized Conference.
Reichelstein, S. (1983), Presentation at a seminar. Minneapolis: Institute for Mathematics and Its Applications, University of Minnesota.
Reichelstein, S. (1984), "Smooth versus discontinuous mechanisms", privately circulated paper.
Reiter, S. (1974), "The knowledge revealed by an allocation process and the informational size of the message space", Journal of Economic Theory, 8:389–396.
Reiter, S. (1974), "Informational efficiency of iterative processes and the size of message spaces", Journal of Economic Theory, 8:193–205.
Reiter, S. (1977), "Information and performance in the $(\text{new})^2$ welfare economics", 67:226–234.
Roberts, J. (1976), "The incentives for correct revelation of preferences and the number of consumers", Journal of Public Economics, 6:359–374.
Roberts, J. (1979), "Incentives in planning procedures for the provision of public goods", Review of Economic Studies, 46:283–292.
Roberts, J. and A. Postlewaite (1976), "The incentives for price-taking behavior in large economies", Econometrica, 44:115–128.
Saijo, T. (1984), "The Maskin–Williams theorems: Sufficient conditions for Nash implementation", Mimeographed. California Institute of Technology.
Samuelson, P. (1954), "The pure theory of public expenditure", Review of Economics and Statistics, 36:387–389.
Samuelson, P. (1955), "Diagrammatic exposition of a theory of public expenditure", Review of Economics and Statistics, 37:350–356.
Sato, F. (1981), "On the informational size of message spaces for resource allocation processes in economies with public goods", Journal of Economic Theory, 24:48–69.
Satterthwaite, M. (1975), "Strategy-proofness and Arrow's conditions: existence and correspondence theorems for voting procedures and social welfare functions", Journal of Economic Theory, 10:187–217.
Satterthwaite, M. (1976), "Straightforward revelation mechanism", Mimeographed. Evanston: Northwestern University; cited in Dasgupta et al. (1979).
Schmeidler, D. (1976), "A remark on a game theoretic interpretation of Walras equilibria", Mimeographed. Minneapolis: University of Minnesota.
Schmeidler, D. (1980), "Walrasian analysis via strategic outcome functions", Econometrica, 48:1585–1594.
Schmeidler, D. (1982), "A condition guaranteeing that the Nash allocation is Walrasian", Journal of Economic Theory, 28:376–378.
Schmeidler, D. and H. Sonnenschein (1978), "Two proofs of the Gibbard–Satterthwaite theorem on the possibility of a strategy proof social choice function", in: H. Gottinger and W. Leinfellner, eds., Decision Theory and Social Ethics, Issues in Social Choice, pp. 227–234. Dordrecht: Reidel.
Thomson, W. (1979), "Comment on L. Hurwicz: On allocations attainable through Nash equilibria", in: J. J. Laffont, ed., Aggregation and Revelation of Preferences, pp. 420–431. Amsterdam: North-Holland.
Vickrey, W. (1961), "Counterspeculation, auctions and competitive sealed tenders", Journal of Finance, 16:8–37.
Walker, M. (1977), "On the informational size of message spaces", Journal of Economic Theory, 15:366–375.
Walker, M. (1978), "A note on the characterization of mechanisms for the revelation of preferences", Econometrica, 46:147–152.
Walker, M. (1980), "On the nonexistence of a dominant strategy mechanism for making optimal public decisions", Econometrica, 48:1521–1540.
Walker, M. (1981), "A simple incentive compatible scheme for attaining Lindahl allocations", Econometrica, 49:65–73.
Williams, S. (1984a), "Sufficient conditions for Nash implementation", Preprint. Minneapolis: Institute for Mathematics and Its Applications, University of Minnesota.
Williams, S. (1984b), "Realization and Nash implementation: Two aspects of mechanism design", Preliminary draft. Minneapolis: Institute for Mathematics and Its Applications, University of Minnesota.

Williams, S. (1984c), "Implementing a generic smooth function", Preprint. Minneapolis: Institute for Mathematics and Its Applications, University of Minnesota.

Weitzman, M. (1974), "The new Soviet incentive model", Working Paper No. 141. Cambridge: M.I.T.

Chapter 29

PLANNING

GEOFFREY HEAL

University of Sussex

1. Introduction

Economists have been discussing the way in which a planned economic system might be run for about three-quarters of a century now, though progress towards the clarification of the issues involved has certainly not been uniform during this period. Not surprisingly, the literature was given considerable impetus by the Russian revolution, with a number of major contributions following within a decade or so [Lange (1936) and Taylor (1929)]. But further advances had to await the absorption into economics of the mathematical techniques of constrained optimisation (see Chapter 24) and occurred only in the 1960s [Arrow and Hurwicz (1960) and Malinvaud (1967)].

It has generally been accepted in the literature that the problem of economic planning is best viewed as one of solving an extremely large constrained maximisation problem. [For a dissenting view, see Kornai (1967), and for a general discussion, see Heal (1973, ch. 1).] In such a formulation, the objective function is identified with a measure of economic welfare, which has to be maximised subject to a variety of constraints, including those imposed by the details of available production processes and by the economy's endowment of economic resources.

A problem of this sort naturally involves very large numbers of variables and constraints – "millions of equations in millions of unknowns", according to Barone (1935) – and it had been recognised that because of this it would not be either desirable or feasible to assemble a complete description of the problem under the auspices of a single agency. Such a procedure would be undesirable because, as observed by Hayek (1945), the collection and transmission of information could lead to errors: and it would anyway be infeasible simply because of the scale of the problem. Much attention has, therefore, been devoted to the analysis of informationally decentralised planning procedures: the precise definition of informational decentralisation is still open to some debate [an interesting contribution is Hurwicz (1969), and there is a survey in Heal (1973, ch. 3); see also Chapter 28], but in essence the phrase is used to describe a way of breaking

Handbook of Mathematical Economics, vol. III, edited by K.J. Arrow and M.D. Intriligator
© 1986, Elsevier Science Publishers B.V. (North-Holland)

the planning problem down into a number of independent operations, each of manageable size and each performed by a separate agency. In a typical decentralised planning scheme, one might find that a central authority was responsible for ensuring that for each good supply and demand were in balance, taken across the whole economy, and that this authority was provided with the minimum information sufficient for execution of this task. Responsibility for ensuring that technological constraints were satisfied would be delegated to firms in whose processes the constraints were embodied.

In addition to being decentralised, in the above sense, most planning procedures discussed have been iterative in the sense that they view the planning problem as being solved by a trial and error process, in which information exchanges between the participants in the decentralised process lead to the construction of better and better approximations to the solution of the planning problem. Formally, therefore, the literature is concerned with procedures for solving constrained maximisation problems which are iterative (involve taking successive approximations) and which have the distinctive feature that the information available to the participants about the economy available is never pooled: there is never any one agent having a complete specification of the problem.

Within this framework, it is conventional to distinguish between price-guided and quantity-guided planning procedures, though not all contributions fit happily into this framework. The former are procedures within which the information flowing from the central planning authority to the firms or sectors takes the form of prices for the goods and services that they consume and produce: the latter are, as their name would suggest, approaches within which this information takes the form of quantitative input–output targets. In general, the information flows from firms to the central agency are dual to those in the reverse direction, in that price information flowing from the centre to the sectors elicits quantitative responses, and vice versa.

Unfortunately, there are no general theorems available about the relative merits of the different approaches: the literature takes the form of studies of individual processes, supplemented by pairwise comparisons of their properties. The conclusions that seem to be emerging are that quantity-guided approaches are more demanding than price-guided alternatives in terms of the amounts of information collected and transmitted by the central authorities (though the measures of information on which such statements rest are extraordinarily crude) but they do have the substantial advantage of superior convergence properties in economies where production possibility sets are not convex. In view of the fact that much of heavy industry has non-convex production possibilities and that it is with the development of large-scale industry that planning is primarily concerned, superior performance under these conditions is of importance.

A significant omission from the literature on planning, is any discussion of the objective function for the planning problem. This is typically taken to be a real-valued, quasi-concave function of the outputs of goods and services which is known to, or chosen by, planners. If planning is to be in some sense democratic, then this function must be related to the preferences of individuals in the society. Of course, preferences typically differ from individual to individual, so that we need a representative or aggregate preference. There is an extensive literature on social choice theory (see Chapter 22) which makes it clear how elusive such a concept is. In view of this, it seems reasonable to conjecture that if a planning procedure is to take adequate account of individual preferences, it will not be sufficient merely to work with an aggregate objective function, but there will be a need to decentralise the consumption as well as the production side of the operation. This is an issue which, with the exception of the literature on planning with public goods (see Section 6), has not been considered. In conclusion, it is probably worth noting that obtaining adequate information about preferences is a major problem in many of the planned economies of eastern Europe. There it is widely felt that an inadequate amount of information is available to planners about individual preferences, and the resulting problems are becoming more acute with the growing range of goods and services available [see, for example, Dyker (1976)].

2. The Lange–Arrow–Hurwicz approach

The earliest carefully articulated contribution to the subject was that of Lange (1936) and his contribution was subsequently formalised by Arrow and Hurwicz (1960). Their formalisation concentrated on that aspect of Lange's work that was concerned with the use of prices in an iterative and decentralised resource-allocation procedure and did not pursue his interesting discussion of distributional issues. In this respect we shall follow Arrow and Hurwicz, as these issues have continued to be largely neglected in the post-war literature.

The present approach can be illustrated by a fairly straightforward model: suppose that there are n firms and s commodities, with each firm using a range of inputs to produce several different outputs. The key variable in describing a firm's production activities is its scale of operation: at a given scale, the vectors of inputs used and outputs produced are uniquely determined, with no scope for substitution. Let the scale at which firm j operates be X_j, with $g_{ij}(X_j)$ standing for the amount of good i produced by firm j at scale j. (Inputs will be represented by negative numbers.) The object of the planning exercise is to find an allocation of resources that maximises a function $U(y_1,\dots, y_s)$ of the amounts of various goods allocated to final demand. The total net output of good i is $\sum_{j=1}^{n} g_{ij}(x_j)$, and if e_i is the economy's initial endowment of good i, then the overall planning problem can be stated as

$$\left.\begin{array}{l} \text{maximise } U(y_1,\ldots,y_s) \\ \text{subject to} \quad -y_i+\sum_{j=1}^{n} g_{ij}(x_j)+e_i \geqq 0, \quad i=1,\ldots,s \end{array}\right\}. \tag{2.1}$$

If the functions $U(\ )$ and $g_{ij}(\ )$ are concave and satisfy a constraint qualification, the Kuhn–Tucker Theorem can be used to characterise a solution to (2.1): under the postulated assumptions, necessary and sufficient conditions for $(y_1,\ldots,y_s)$ and $(x_1,\ldots,x_n)$ to solve (2.1) are that

$$U_i-\lambda_i \leqq 0, \quad = \text{if } y_i>0, \qquad i=1,\ldots,s, \tag{2.2}$$

where $U_i=\partial U(Y_1,\ldots,Y_s)/\partial y_i$,

$$\sum_{i=1}^{s} \lambda_i(\partial g_{ij}/x_j) \leqq 0, \quad = \text{if } x_j>0, \qquad j=1,\ldots,n, \tag{2.3}$$

$$\sum_{j=1}^{n} g_{ij}(x_j)+e_i-y_i \geqq 0, \quad = \text{if } \lambda_i>0, \qquad i=1,\ldots,s, \tag{2.4}$$

where $(\lambda_1,\ldots,\lambda_s)$ are the dual variables associated with the constraints in (2.1). Now suppose that this simple economy is run by managers of firms, who seek to maximise profits, and by a distributor, who controls the allocations of goods to final uses and who determines these allocations so as to maximize the difference between the value of the objective function and the cost of the final demand vector. Formally if λ_i is the price of good i, then managers maximise

$$\left.\begin{array}{l} \sum_{j=1}^{s} \lambda_i g_{ij}(x_j) \\ \text{subject to} \quad x_j \geqq 0, \qquad j \geqq 1,\ldots,n \end{array}\right\}. \tag{2.5}$$

The distributor maximises

$$\left.\begin{array}{l} U(y_1,\ldots,y_s)-\sum_{i=1}^{s} \lambda_i y_i \\ \text{subject to} \quad y_i \geqq 0. \end{array}\right\}. \tag{2.6}$$

Necessary and sufficient conditions for solutions to these problems are

$$\sum_{i=1}^{s} \lambda_i(\partial g_{ij}/\partial x_j) \leqq 0, \quad = \text{if } x_j > 0, \qquad j=1,\dots,n, \tag{2.7}$$

and

$$U_i - \lambda_i \leqq 0, \quad = \text{if } y_i > 0, \qquad i=1,\dots,s. \tag{2.8}$$

Obviously (2.7) and (2.8) are identical in form to (2.3) and (2.2), and if the functions involved are strictly concave, (2.7) and (2.8) will have unique solutions. In such a situation, it is clear that if managers and the distributor are faced with market prices equal to the dual variables associated with a solution to (2.1), they will choose consumption and production vectors which solve (2.1). The solution to the resource-allocation problem (2.1) can, therefore, be attained simply by the centre quoting the appropriate prices, if all the functions involved are strictly concave.

How are these "appropriate prices" to be calculated? Arrow and Hurwicz show that they may be found by an iterative procedure which in a very direct sense imitates a competitive market. Consider the following equations:

$$\begin{aligned} \dot{y}_i &= 0 && \text{if} \quad y_i = 0, \quad U_i - \lambda_i \leqq 0, \\ &= a(U_i - \lambda_i) && \text{otherwise,} \end{aligned} \tag{2.9}$$

$$\begin{aligned} \dot{x}_j &= 0 && \text{if} \quad x_j = 0, \quad \sum_{i=1}^{s} \lambda_i(\partial g_{ij}/\partial x_j) < 0, \\ &= a\sum_{i=1}^{s} \lambda_i(\partial g_{ij}/\partial x_j) && \text{otherwise,} \end{aligned} \tag{2.10}$$

$$\begin{aligned} \dot{\lambda}_i &= 0 && \text{if} \quad \lambda_i = 0, \quad e_i + \sum_{j=1}^{n} g_{ij}(x_j) - y_i > 0, \\ &= a\left(y_i - \sum_{j=1}^{n} g_{ij}(x_j) - e_i\right) && \text{otherwise.} \end{aligned} \tag{2.11}$$

These have very simple interpretations: (2.9) requires that the final demand for good i should be adjusted at a rate dependent on the difference between its marginal contribution to the objective function and its price, (2.10) requires that each firm should alter its operations in such a way as to raise its profits, and (2.11) simply implies that the price of a good is raised if demand exceeds supply, and vice versa. All three statements are complicated by the need to respect non-negativity constraints. The information flows involved in this are clear: the centre, which sets prices, has to inform firms and the distributor of these, so that they can calculate the expressions on the right-hand sides in (2.9) and (2.10). In exchange, they inform the centre of demands and supplies, so that it can calculate the price adjustment in (2.11). Arrow and Hurwicz (1958) showed that the process described by (2.9) to (2.11) converges to a solution to problem (2.1), thus enabling the central authority to find an optimal resource allocation without at any stage receiving information about the production possibilities open to firms. In mathematical terms, (2.9) to (2.11) can be seen as defining a gradient process which is applied to locate the saddle-point of the Lagrangian corresponding to (2.1): it is interesting that this mathematical procedure has such a simple economic interpretation. It should be noted that (2.9), (2.10) and (2.11) have discontinuous right-hand sides: this is a feature also of the adjustment equations of Sections 4, 5 and 7 below. In these cases it is not straightforward to establish the existence of a solution. This problem is discussed by Henry (1972, 1973).

3. The contribution of Malinvaud

Malinvaud (1967) has analysed a procedure which is also price-guided, but which is very different in spirit from that just discussed. Pose the planning problem in the following terms:

$$\left.\begin{array}{l}\text{find a vector } \boldsymbol{y} \text{ which maximises } U(\boldsymbol{y}) \\ \text{subject to } \quad 0 \leqq \boldsymbol{y} \leqq \sum_{j=1}^{n} \boldsymbol{x}_j + \boldsymbol{e} \quad \text{and} \quad \boldsymbol{x}_j \in X_j \\ \text{for } \quad j = 1, \ldots, n\end{array}\right\}, \tag{3.1}$$

where $\boldsymbol{y}$ is once again a vector of amounts allocated to final demand, $\boldsymbol{e}$ is the vector of the economy's endowments, $\boldsymbol{x}_j$ is firm j's production programme, and X_j is the set of all such programmes feasible for j. At the tth iteration of the planning process, the central authority solves this problem with the constraint $\boldsymbol{x}_j \in X_j$ replaced by the constraint $\boldsymbol{x}_j \in X_j^t$, where $X_j^t \subset X_j$ is an approximation to X_j, constructed as follows. At each iteration of the planning procedure, the central authority announces a set of prices: firms are required to calculate their profit-maximising production programmes at these prices, and inform the centre of these. At the tth iteration, therefore, the centre knows of t feasible plans for each firm. For firm j, let these be $\boldsymbol{x}_j^1, \boldsymbol{x}_j^2, \ldots, \boldsymbol{x}_j^t$: firms' production possibility sets X_j are assumed to be convex, so that the approximation X_j^t, defined by

$$X_j^t = \left[x/x = \sum_{i=1}^{t} \lambda_i \boldsymbol{x}_j^i, \quad \lambda_i \geqq 0, \quad \sum_{i=1}^{t} \lambda_i = 1 \right],$$

is contained in X_j. In order to complete a description of the process, it is only necessary to define the rule by which the centre chooses the prices announced at each step. This is straight-forward: they are simply the dual variables associated with a solution to problem (3.1) with X_j replaced by X_j^t. Hence a typical iteration runs as follows:

(i) The centre announces as prices the dual variables of the solution to the problem:

$$\text{maximise } U(y)$$

$$\text{subject to} \quad 0 \leqq y \leqq \sum_{j=1}^{n} x_j^t + e, \qquad x_j^t \in X_j^{t-1},$$

where

$$X_j^{t-1} = \left[x/x = \sum_{i=1}^{t-1} \lambda_i X_j^i, \quad \lambda_i \geqq 0, \quad \sum_{i=1}^{t} \lambda_i = 1 \right],$$

and the x_j^i are firm j's responses at earlier iterations.

(ii) Firms inform the centre of the input–output plans that maximise their profits at the new prices.

(iii) The centre constructs new approximations X_j^t to the X_j by incorporating this new observation.

(iv) Step (i) is repeated with X_j^{t-1} replaced by X_j^t.

Malinvaud (1967) proves that this process converges to a solution to the problem (3.1): perhaps equally importantly, he establishes that it does so via a sequence of feasible plans y^t having the property that $U(y^t) \geqq U(y^{t-1}) \geqq U(y^{t-2}), \dots$. In practical terms this is very important, as no procedure can be iterated through to convergence: it is thus of great significance that a finite number of iterations can be guaranteed to produce a feasible plan superior to the initial proposal. And, as Weitzman (1974) has observed, the fact that Malinvaud's procedure is a discrete-step and not a gradient process may well enable it to converge more rapidly. On the other hand, as Heal (1973, ch. 6) has emphasised, it does require the central authority to process very large quantities of information.

4. Non-price approaches

All planned economies have relied heavily on forms of organisation within which the central authorities dispense quantitative input and output targets, supplemented only minimally by prices. During the last decade, several writers have turned to an analysis of such approaches, to enquire whether they are fundamentally misguided or whether they have real advantages in resource-allocation terms. It turns out, as mentioned in the introduction, that they do have such advantages, though they are bought at the cost of extra complexity in the field of information-processing. The simplest non-price approach is that of Heal (1969), which may be formalised as follows. Consider an economy with n different firms, each producing a good used only to supply final demand. The output of firm i is described by $Y_i = f_i(X_{i1}, X_{i2},\dots, X_{im})$, where X_{ij}, $j = 1,2,\dots, m$, is the amount of resource j used as an input by firm i. The economy has endowments of m different resources, X_j being that of the jth, so the X_{ij} must satisfy $\sum_{i=1}^{n} X_{ij} = X_j$ for all j. The planning problem is

$$\left.\begin{array}{l} \text{maximise } U(Y_1, Y_2,\dots, Y_n) \\ \text{subject to} \quad Y_i = f_i(X_{i1},\dots, X_{im}), \quad \sum_{i=1}^{n} X_{ij} \leqq X_j, \quad X_{ij} \geqq 0 \end{array}\right\}. \tag{4.1}$$

The procedure suggested for solving this is as follows: the centre proposes an initial allocation X_{ij}^0, $i = 1,\dots, n$, $j = 1,\dots, m$, of all inputs in all firms and in return is informed of the values of the derivative of $\partial Y_i / \partial X_{ij}$ for all i and j. It then alters the initial allocation according to the equations

$$\begin{aligned} \dot{X}_{ij} &= U_i f_{ij} - \mathrm{Av}(K_j) U_i f_{ij} \quad \text{for} \quad i \in K_j, \\ &= 0 \qquad\qquad\qquad\qquad\quad\; \text{otherwise}, \end{aligned} \tag{4.2}$$

where $\mathrm{Av}(K_j)U_i f_{ij}$ is to be read as "the average of the f_{ij} for all i in K_j" and K_j, a set of indices, satisfies $K_j = (i / X_{ij} > 0$ or $X_{ij} = 0$ and $U_i f_{ij} > \mathrm{Av}(K_j)U_i f_{ij})$; its construction is described in Heal (1969; 1973, ch. 7).

The basic idea behind (4.2) is that each input is reallocated in a way which directs more to firms where its productivity on the margin is above-average, and vice versa: the obvious procedure has to be complicated to avoid violating non-negativity constraints. It is relatively easy to prove that this gradient-like process generates a sequence of feasible plans associated with steadily increasing values of the objective function, and that it converges to a critical point of (4.1) that is not a local minimum, independently of any convexity assumptions about the feasible sets. Hori (1975) discusses in some detail the nature of the critical points to which the process might converge.

These are stronger results than can be established for either of the previous approaches without a convexity assumption. The general strategy of the proof runs as follows. The rate at which the objective function changes can be expressed as

$$\dot{U} = \sum_{i=1}^{n} U_i \dot{Y}_i = \sum_{i=1}^{n} U_i \sum_{j=1}^{m} f_{ij} \dot{X}_{ij}$$

$$= \sum_{j=1}^{m} \sum_{i \in K_j} U_i f_{ij} \left(U_i j_{ij} - \mathrm{Av}(K_j) U_i j_{ij} \right)$$

$$= \sum_{j=1}^{m} \sum_{i \in K_j} \left(U_i f_{ij} - \mathrm{Av}(K_j) U_i f_{ij} \right)^2 \geqq 0,$$

with equality if and only if $U_i f_{ij}$ is a constant for all $i \in K_j$, each j, and less than or equal to this common value for $i \notin k_j$. By letting $\overline{U}$ be the maximum value of U feasible for problem (4.1), one can construct a Lyapunov norm [see, for example, La Salle and Lefschetz (1961)] equal to $\overline{U} - U$ and use this to prove convergence.

The approach just discussed can, of course, be extended to more complex and realistic models of an economy, and details of these extensions are given in the references cited.

In view of the informational advantages of a price-guided approach, and the superior performance of non-price approaches in non-convex environments, there is an obvious interest in attempting to combine the attractive aspects of the two. A certain amount of progress in this direction has been made by Aoki (1971) and Heal (1971): both discuss mixed price-and-command planning where some resource-allocation decisions are made centrally and dictated for firms, while other decisions are left to firms that, subject to the constraints imposed by centrally-made decisions, are free to act in a profit-seeking manner. These procedures bear an interesting similarity to the mixed command-and-market systems that have been developed lately in some eastern European economies, and are discussed in the following section.

5. Price and quantity approaches

As just mentioned, one can in some measure synthesise the approaches outlined in the previous sections, and produce an amalgam which has several attractive features. This approach is necessarily somewhat more complex, and so can be presented only in outline and in its simplest interpretation here: in Heal (1971) the basic idea is shown to be open to several quite different institutional interpretations.

The model within which we shall consider the planning procedure may be formalised as follows. The only inputs to the production process are resources:

these are used exclusively as inputs to production, and are not themselves produced. They are indexed by $j \in M$, $M = (1, \ldots, m)$. There are n firms, indexed by $i \in N$, $N = (1, \ldots, n)$, and p distinct produced goods, indexed by $g \in P$, $P = (1, \ldots, p)$.

Our notation is as follows: X_{ij} is the amount of resource j allocated to firm k, Y_{ig} is the amount of good g produced by firm i, $R_j > 0$ is the total amount of resource j available to the economy, $\boldsymbol{X}_i$ is the vector of inputs to firm i, and $\boldsymbol{Y}_i$ is the vector of outputs of firm i.

The production possibilities of firm i are represented by an implicit function,

$$T_i(\boldsymbol{Y}_i, \boldsymbol{X}_i) \leqq 0,$$

where $\boldsymbol{Y}_i$ and $\boldsymbol{X}_i$ are the vectors defined above. It will be assumed that the set of efficient production programmes open to firm i can be represented by

$$T_i(\boldsymbol{Y}_i, \boldsymbol{X}_i) = 0.$$

It is also assumed that the T_i are once continuously differentiable.

In the ensuing argument we shall make frequent use of a slightly unconventional derivative: we shall use the symbol F^i_{gj} to stand for the rate at which firm i's output of good g changes, as the input of good j to firm i is varied, assuming that the quantities of the firm's various outputs are maintained in their existing proportions to each other. It is fairly easy to derive an expression for this derivative in terms of the conventional partials of T_i,

$$F^i_{gj} = -\left(Y_{ig}\left(\partial T_i / \partial X_{ij}\right)\right) \Big/ \left(\sum_{g \in P} Y_{ig}\left(\partial T_i / \partial Y_{ig}\right)\right).$$

This equality is well-defined as long as the Y_{ig} are not zero for all g: in such a case, the Y_{ig} can be assigned arbitrary values, though the derivatives must still be evaluated at $\boldsymbol{Y}_i = 0$.

We need to make one further, fairly innocuous, assumption about firms' production possibilities – a "finite input, finite output" assumption. Formally, we assume that if $\|\boldsymbol{X}_i'\| < A$, where A is finite, then any $\boldsymbol{Y}_i$ satisfying $T_i(\boldsymbol{Y}_i', \boldsymbol{X}_i') = 0$, also satisfies $\|\boldsymbol{Y}_i'\| < B$, for some finite B.

The symbol $\|\boldsymbol{X}_i\|$ denotes the Euclidean norm of the vector $\boldsymbol{X}_i$. Let $Y_g = \sum_{i \in n} Y_{ig}$, etc. Then the objective of the planning procedure can be specified as follows:

$$\text{maximize } U(Y_1, \ldots, Y_p)$$

$$\text{subject to} \quad T_i(\boldsymbol{Y}_i, \boldsymbol{X}_i) \leqq 0 \quad \text{for all} \quad i \in N, \tag{5.1a}$$

$$\sum_{i \in N} X_{ij} \leqq R_j \quad \text{for all} \quad j \in M, \tag{5.1b}$$

$$X_{ij}, Y_{ig} \geqq 0 \quad \text{for all} \quad i, g, j. \tag{5.1c}$$

U is a function of class C^1 from R^p to R^1, and has finite first partial derivatives. We denote

$$\partial U/\partial Y_g \text{ by } U_g, \text{ etc.}$$

Before describing the details of the planning procedure, it is necessary to introduce one additional concept – the value of a resource in a particular use. The value of resource j in firm i, V_{ij}, is given by

$$V_{ij} = \sum_{g \in P} U_g F^i_{gj},$$

and thus gives the rate at which U would change if X_{ij} were changed marginally, and firm i maintained its existing output proportions. It is, in a sense, a "shadow price" for the variable X_{ij}.

As mentioned in the introduction, there are a number of different institutional interpretations that can be given to the planning procedure under consideration. However, in all variants, the differential equations governing the reallocation of resources are the same: this provides the justification for speaking of different interpretations of one planning procedure, rather than about three distinct procedures. We consider here a simple price-and-command interpretation, where economic activity is controlled partly by input quotas set by the central planning board (CPB), and partly by the use of output prices, also set by the CPB.

There are two main elements in the planning procedure:

(1) The re-allocation of resources amongst firms. This is carried out by the CPB in the light of the V_{ij}: it increases the allocation of a resource to a firm where its value is above average and vice versa.

(2) The substitution of one output for another. This is carried out by firms: at each stage of the process, the CPB announces "prices" for each produced good – the price of good g is U_g, the derivative of the objective function w.r.t. the output of that good at the current output levels. Taking these prices, and its inputs of resources, as given, each firm then adjusts its output mix so as to increase the value of its output.

Details of the planning process are as follows. Starting from an arbitrary feasible plan satisfying (5.1a) and (5.1b) with equality:

(1) Firms inform the CPB of their outputs of the various produced goods.

(2) The centre computes the totals Y_g, and the prices, U_g, for $g \in P$, and informs firms of the latter.

(3) Each firm now calculates a value for every resource in its productive processes, and informs the CPB of these. (Alternatively, firms may inform the CPB of the quantities F^i_{gj} and leave the CPB to calculate the V_{ij}.)

(4) The centre now changes the allocation of inputs amongst firms according to the following rules:

$$\begin{aligned} \dot{X}_{ij} &= V_{ij} - \text{Av}(K_j)V_{ij} \quad \text{for} \quad i \in K_j, \\ &= 0 \qquad\qquad\qquad\quad \text{otherwise,} \end{aligned} \tag{5.2}$$

where a dot over a variable denotes its time derivative, and the notation $\text{Av}(K_j)V_{ij}$ denotes the average of the values of V_{ij} over the subscripts i contained in the set K_j. The set K_j is constructed iteratively as in Heal (1969). It is defined by the following property:

$$K_j = \left[i: X_{ij} > 0 \text{ or } X_{ij} = 0 \text{ but } V_{ij} > \text{Av}(K_j)V_{ij}\right],$$

and contains only firms whose allocation of resource j is positive, or those whose allocation is zero but where the value is above the average over K_j. Hence application of (5.2) will never violate the non-negativity constraints.

(5) At the same time, each firm, remaining on the efficient surface given by the current input vector, substitutes between outputs so as to increase the total value of its output. That is if $\dot{Y}^s_{ig}$ is the rate of change of i's output of g due to substitution between outputs, then the $\dot{Y}^s_{ig}$ are chosen so that the Y_{ig} vary continuously and

$$\sum_{g \in P} U_g \dot{Y}^s_{ig} \geqq 0,$$

with equality if and only if the necessary conditions for a maximum of the value of output at prices U_g are satisfied.

This completes one step of the process: we now return to item (1).

The necessary conditions referred to in item (5) can easily be derived. The relevant maximisation problem is

$$\text{maximize} \sum_{g \in P} U_g Y_{ig}$$

$$\text{subject to} \quad T_i(\mathbf{Y}_i, \mathbf{X}_i) = 0, \quad \mathbf{Y}_i \geqq 0, \quad U_g, \mathbf{X}_i \text{ given,}$$

which yields as necessary conditions

$$U_g - \mu_i(\partial T_i/\partial Y_{ig}) \leqq 0, \quad = \text{ if } Y_{ig} > 0,$$

which must hold for all $g \in P$, each $i \in N$.

Note that the total change in firm i's output of good g is the sum of any effect due to substitution between outputs and any effect due to changes in inputs, governed by (5.2). Hence the total is

$$\dot{Y}_{ig} = \sum_{j \in M} \dot{X}_{ij} F^i_{gj} + \dot{Y}^s_{ig}.$$

It is now relatively straightforward, using the techniques outlined in the previous section, to prove the following:

Theorem

If the production relations T_i and the objective function satisfy the assumptions specified and if the initial allocation satisfies constraints (5.1a) to (5.1c) and is not a local minimum, then:

(a) every limit point of the re-allocation process is a critical point: such limit points exist and are not local minima;
(b) along the paths produced by the process, the objective function increases monotonically;
(c) every proposed allocation satisfies the constraints (5.1a) to (5.1c).

6. Cremer's quantity–quantity model

A paper by Cremer develops the idea of quantity-guided procedures in a different direction. The procedure that he proposes has a very strong intuitive resemblance to that of Malinvaud, in that it is one in which the central authority receives output proposals from firms, accumulates these in its memory, and uses them to construct progressively more accurate approximations to their production possibility sets. There are however two important differences:

(i) The information flows from the centre to firms take the form of proposed production vectors, rather than price vectors.

(ii) Production possibility sets are approximated from the outside rather than from the inside. That is, if X_j is firm j's true production set and X_j^t is the tth approximation to it, $X_j \subseteq X_j^t$ (whereas in Malinvaud's model, $X_j^t \subseteq X_j$).

It immediately follows from these two points that this procedure requires very substantial quantities of information to be transmitted – the loads on the participants exceed those in Malinvaud's model – and that not all plans proposed short of the optimum will be feasible. However, the procedure is still of interest because Cremer establishes that it will lead to a globally optimal allocation of resources even if the production sets are non-convex.

The details of Cremer's model can be appreciated with the help of the model used in Section 3 to discuss Malinvaud's contribution: recall that in this case the overall problem is posed as

$$\left.\begin{array}{l}\text{find a vector } y \text{ which maximises } U(y)\\ \text{subject to} \quad 0 \leqq y \leqq \sum_{j=1}^{n} x_j + e \quad \text{and} \quad x_j \in X_j\\ \text{for} \quad j = 1, 2, \ldots, n\end{array}\right\}, \tag{6.1}$$

where the X_j are the individual production possibility sets. In Malinvaud's model, as in the present one, the centre constructs approximations to these, the approximation at the tth iteration being X_j^t, and in both cases the centre at the tth iteration solves the problem

$$\left.\begin{array}{l}\text{maximise } U(y)\\ \text{subject to} \quad 0 \leqq y \leqq \sum_{j=1}^{n} x_j^t + e \quad \text{and} \quad x_j^t \in X_j^t\end{array}\right\}. \tag{6.2}$$

The differences, as mentioned, lie in the ways in which the X_j^t are constructed, and in the information the centre sends to firms. Let x_j^* be production vectors which solve (6.1): it is assumed that the centre knows for each firm a vector $x_j^0 > x_j^*$ which forms an upper bound for its optimal production plan. We then set

$$X_j^0 = \{x | x \leqq x_j^0\},$$

ask firms to specify an efficient feasible production vector $x_j^1 < x_j^0$, and set

$$X_j^1 = \{x | x \in X_j^0, x \not> x_j^1\},$$

which is just X_j^0 minus that portion strictly greater than x_j^t. At this point the centre solves problem (6.2) with $X_j^t = X_j^1$, leading to solution vectors $\boldsymbol{x}_j^2$ for the firms. Firms are then asked if they can produce $\boldsymbol{x}_j^2$ and, if not, to specify an efficient feasible production vector $\boldsymbol{x}_j^2$ strictly less than $\boldsymbol{x}_j^2$. In this case,

$$X_j^2 = \{ \boldsymbol{x} | \boldsymbol{x} \in X_j^1, \boldsymbol{x} \not> \boldsymbol{x}_j^2 \},$$

and so the procedure continues.

Cremer shows that, under certain mild assumptions on the function $U(\ )$ and the sets X_j, and provided that the responses of the firms satisfy technical conditions not described in the brief outline above, this process converges to a global solution of the problem (6.1). He also shows that, although the problem (6.2) to be solved by the centre at each step is typically a highly non-convex problem, the feasible set has in fact a special structure which might make the problem tractable.

In summary, then, we see that Cremer has proposed a process which has stronger convergence properties than any of the others mentioned, but at the cost of substantial information flows and the imposition of a more demanding task on the central authority. This confirms the nature of the trade-offs suggested by earlier work.

7. Planning with public goods

It has been known since the work of Samuelson (1954) that, because of the free-rider problem, we cannot rely on a competitive market to produce an efficient allocation of public goods. In recent years an extensive literature has developed on this subject. An excellent survey is Tulkens (1976); important contributions have been made by Malinvaud (1971, 1972) and Dreze and Poussin (1971), who have developed a planning procedure for use in an economy with public and private goods which is now generally known as the M.D.P. procedure. In this brief treatment, we shall be concerned to present the essentials of this in a simple context, drawing heavily on Tulkens (1976).

We consider a set of N consumers, indexed by $i = 1,\dots,n$, each possessing a differentiable utility function $U_i(y_i, Z)$, where y_i is his allocation of a single consumption good and Z is the output of the single public good. Individual i enters the picture with an endowment e_i of the private good. Production of the public good uses as an input the private good, according to the formula $w = g(Z)$, where w is the input needed to produce Z. The overall problem with which we are concerned can thus be stated as follows:

$$\left.\begin{aligned} &\text{maximize} \sum_i U_i(y_i, z) \\ &\text{subject to} \quad \sum_i y_i + w = \sum_i e_i, \quad w = g(z) \end{aligned}\right\}. \tag{7.1}$$

Letting

$$\pi_i(y_i, z) = \frac{\partial U_i(y_i, z)/\partial z}{\partial U_i(y_i, z)/\partial y_i},$$

one can readily verify that the first-order conditions for a solution to (7.1) are that

$$\sum_i \pi_i(y_i, z) = g'(z), \tag{7.2}$$

which is of course just the familiar condition [see Samuelson (1954)] that the sum of the consumers' marginal rates of substitution between public and private consumption, must equal the marginal cost of the public good in terms of the private.

In intuitive terms, the essence of the M.D.P. approach is that, starting from an arbitrary but feasible initial allocation ($y_i, i \in N, Z$), a planning board (C.P.B.) revises the y_i and Z so as to eliminate the discrepancies between the left- and right-hand sides of (7.2). In particular, it raises the output of the public good if $\sum_i \pi_i$ – the sum over all consumers of their willingness to pay for an extra unit – exceeds $g'(z)$, the marginal cost of that unit, and vice versa.

More formally, the procedure runs as follows. At each point, the C.P.B. proposes an output of public good, Z, and a set of private consumption levels Y_i, which are feasible. Consumers respond by reporting to the centre the associated marginal rates of substitution π_i, and the producer responds with $g'(z)$, the marginal cost of the public good. The C.P.B. then revises the proposed allocation according to the following equations, and the procedure continues:

$$\dot{z}(t) = a\left(\sum_i \pi_i(t) - g'[z(t)]\right), \tag{7.3}$$

$$\dot{w}(t) = g'(z)\dot{z}(t), \tag{7.4}$$

$$\dot{y}_i(t) = -\pi_i(t)\dot{z}(t) + \delta_i a\left[\sum_j \pi_j(t) - g'(z)\right]^2. \tag{7.5}$$

In (7.3) $0 < a < \infty$ is an adjustment coefficient, and in (7.5) $\delta_i \geqq 0$, $\sum_i \delta_i = 1$ are a set of distributive weights. Clearly (7.3) is merely a formalisation of the intuitively appealing idea mentioned above and (7.4) is the consequential adjustment to the inputs allocated to producing the public good. As the production of the public good changes, so of course does that of the private good. (7.5) specifies that i's allocation y_i changes at a rate which depends partly on her expressed willingness to pay for the public good, and partly on the distributive weight δ_i and the squared error in the first-order conditions.

Although the rationale for the presence of this second term may not be immediately clear, it is in fact essential, as it enables one to establish that along any path satisfying (7.3), (7.4) and (7.5), each individual's utility is non-decreasing. The proof is as follows:

$$\begin{aligned} \dot{U}_i &= (\partial U_i/\partial y_i)\dot{y}_i + (\partial U_i/\partial z)\dot{z} \\ &= (\partial U_i/\partial y_i)\dot{y}_i + \pi_i\dot{z} \\ &= (\partial U_i/\partial y_i)\delta_i a\Big(\sum_j \pi_j - g'(z)\Big)^2 \geqq 0. \end{aligned} \tag{7.6}$$

The fact that utilities are non-decreasing is clearly an appealing feature of the procedure, and is also of significance in a discussion of whether participants have an incentive to misrepresent their preferences for the public good. This issue is discussed in Section 8.

It is evident by inspection of (7.3) and (7.5) that any equilibrium of the M.D.P. procedure must satisfy the first-order conditions (7.2). Malinvaud and Dreze and Poussin show that if for each i, $U_i(y_i, z)$ is strictly concave, and if $g(z)$ is convex, and if certain technical assumptions are satisfied, then the M.D.P. procedure converges to a solution to the problem (7.1) – and hence converges to a Pareto optimum. Given the monotonicity property established above, this optimum is clearly individually rational with respect to the initial allocation. For any initial allocation, there are typically many Pareto optima which are unanimously preferred and hence individually rational: Champsaur (1976) has established that to any such optimum there corresponds a set of δ_i, $i \in N$, such that the M.D.P. procedure had this optimum as its limit point. The procedure is thus in a certain sense distributionally neutral: it does not restrict the set of Pareto optima attainable, other than by the condition of individual rationality relative to the initial allocation.

8. Revelation of information

8.1. *With public goods*

All planning procedures rely on the revelation of information by participants to the central authority. This may take the form of information about outputs and marginal productivities at certain input configurations, or of information about preferences for public goods. The need to ensure accurate revelation of this latter sort of information raises particularly acute problems, for it has long been recognised in the theory of the allocation of public goods that individual consumers have an incentive to misrepresent their preferences [Samuelson (1954)]. This problem has recently given rise to some interesting literature by, amongst others, Groves and Ledyard (1977) and Green and Laffont (1977). These authors have shown that under certain circumstances one can devise ways of ensuring that it is in an individual's interest to reveal his true preference. The basic idea is very straightforward: by raising my demand for a public good, I raise the output and hence everyone else's consumption, conferring on them a positive externality. The procedures of Groves and Ledyard and Green and Laffont can be viewed as ingenious ways of internalising this externality.

What then is the state of play in the field of planning with public goods? Clearly the relevant procedures need preferences to be revealed to the centre: can it be assumed that they will be revealed accurately? This is a field in which much work remains to be done, but there are indications that the M.D.P. procedure is to an encouraging degree cheatproof. There have been two approaches to this issue, which have been termed local and global. The former asks the question: suppose that at each instant of the planning procedure, a consumer is concerned only with the change in his utility that results from the re-allocation carried out at that instant, and gives a reply which will maximise this, taking due account of the behaviour of others. Depending on whether or not he collaborates with the others, we have a cooperative or non-cooperative local game, and we can ask whether the solution to this involves telling the truth. The global approach to this issue involves assuming that at each instant a consumer attempts to anticipate the effects of his actions on the allocation that will eventually result when the procedure converges, and behaves so as to optimise this, again either with or without cooperating with the other individuals.

The most extensive results are available for local games. Dreze and Poussin (1971), working in a non-cooperative framework, showed that telling the truth is a minimax strategy in the non-cooperative local game, and is in general the only such strategy. A proof of this is easily outlined. Let $\psi_j(t)$ be j's declared marginal rate of substitution between public and private goods at time t: this need not equal the true value, still denoted by $\pi_j(t)$. Then from (7.6),

$$\dot{U}_i = \frac{\partial U_i}{\partial y_i}\left(\delta_i a\left[\sum_j \psi_j - g'(z)\right]^2 + \dot{z}(\pi_i - \psi_i]\right)$$

$$= \frac{\partial U_i}{\partial y_i}\delta_i a\left[\sum_j \psi_j - g'(z)\right]^2 + (\pi_i - \psi_i)\left[\sum_j \psi_j - g'(z)\right].$$

It is clear from this that setting $\psi_i = \pi_i$ guarantees i a non-negative change in his utility, and it can readily be shown that any other strategy will allow other players to choose ψ_j, $j \neq i$, which can ensure that $\dot{U}_i$ is non-positive. Hence the result. Dreze and Poussin also show that at an equilibrium of the procedure, and only at such a point, telling the truth and setting $\pi_i = \psi_i$ for all i is a Nash equilibrium.

Roberts (1976) considers the Nash equilibrium of a local non-cooperative game away from an equilibrium and shows that although truthful revelation of preferences is not in general a Nash equilibrium (it is in the special case when there are two consumers with $\delta_1 = \delta_2 = \frac{1}{2}$), it is still the case that if all consumers play their Nash strategies at each instant, the process converges to a Pareto optimum. This may be a different optimum from that which would have been reached if preferences had been revealed correctly (so that some consumers may gain from misrepresentation), and convergence may be slower, but the outcome is still efficient. Henry (1977) has extended this result to the case when there are constraints on the numbers ψ_i that each consumer feels able to present as his marginal rate of substitution – constraints imposed for example by the need to lie in a manner which is plausible a priori and is consistent with previous replies. Roberts has also studied global games associated with the M.D.P. procedure, and has shown that a Nash equilibrium in a global non-cooperative game cannot in general involve accurate revelation of preferences.

It can be seen from this review that our understanding of the problems of eliciting accurate information about preferences for public goods has progressed considerably, and indeed that the M.D.P. procedure for determining the allocation of such goods has encouragingly strong properties in this respect.

8.2. *The production of private goods*

As mentioned earlier, information has to be revealed by agents to the centre even when planning is entirely in the context of private goods. In general it is simply assumed that this information will be accurate. There is very little literature on the problem of ensuring accuracy, though Heal (1971) does discuss this problem in the context of the mixed price-and-command procedure of Section 5. The essentials of the argument can be appreciated in a one-firm model. Let the economy consist of a single firm: when using all the resources available (these are not consumable) it can produce an output vector $\boldsymbol{Y} = (Y_1, \ldots, Y_p)$ according to the equation $F(\boldsymbol{Y}) = 0$. The social welfare function is $U(\boldsymbol{Y})$. Suppose now that the firm is told that the use of resources is free, that it will be paid an amount U_g for each unit of good g it produces, and that the price vector $(U_1, \ldots, U_p)$ will vary as the output vector varies. The firm is then instructed to maximize its profits (which in this case equal its revenue): this puts it in the position of a monopolist, and it has to solve the problem

$$\left.\begin{aligned} &\text{maximise} \sum_{g \in P} Y_g U_g \\ &\text{subject to} \quad F(\boldsymbol{Y}) = 0 \end{aligned}\right\}, \tag{8.1}$$

where the vector $(U_1,\ldots,U_g)$ depends on $(Y_1,\ldots,Y_g)$ in a way that may or may not be known to the firm. Assume the social welfare function to have the following property:

$$\sum_{g \in P} Y_g U_g = \phi[U(Y)] \quad \text{for some } \phi \text{ with } \phi' > 0, \tag{8.2}$$

then the solution to the problem (5.2) is identical to the solution to the planning problem

$$\left.\begin{array}{l} \text{maximise } U(\boldsymbol{Y}) \\ \text{subject to} \quad F(\boldsymbol{Y}) = 0 \end{array}\right\}, \tag{8.3}$$

and the social optimum is the monopolist's profit-maximum. Condition (8.1) is by no means completely unacceptable: the class of positively homogeneous functions is a subclass of those satisfying it. Three interesting implications result from this identity of the solutions to (8.1) and (8.3):

(i) During the application of any of the planning processes described above to this simple economy, the monopolist's profits will rise monotonically as the process continues.

(ii) Any departure from the socially optimal production plan will lower the firm's profits.

(iii) No false reporting of outputs during the procedure leading to the optimum could increase the total profit that accrues to the firm at the equilibrium.

These conclusions have to be modified for the many-firm model set out in Section 5. In that case one has corresponding to point (i) the fact that the total revenue from the sale of the outputs of all firms is rising during the planning process; in points (ii) and (iii) the "total revenue of all firms" must likewise be substituted for "firm's profits".

It is also possible to make the following statement about this more general case. Consider an arrangement whereby once the optimum had been attained, firms were permitted to trade resources with each other and the auctioneers. Then the sum of the profits earned by all firms and auctioneers would be maximised, over all possible allocations, at the social optimum. The proof of this assertion is trivial: if P_j is the price of resource j, and π the sum of the profits,

$$\begin{aligned} \pi &= \sum_{i \in N} \sum_{g \in P} Y_{ig} U_g - \sum_{i \in N} \sum_{j \in M} X_{ij} P_j + \sum_{j \in M} \sum_{i \in N} X_{ij} P_j \\ &= \phi\left[U(Y_1,\ldots,Y_p)\right] \quad \text{with } \phi' > 0 \text{ if (8.2) is satisfied.} \end{aligned}$$

It is therefore true that if in such a situation any agent departs from his socially optimal action the resulting gain to him must be less than the losses to others: the losers could therefore bribe the gainers not to make such a departure.

In summary, it is clear that although the social optimum located by the planning processes discussed cannot be supported by prices in the normal way, there is implicit in the rules of the economy a structure of incentives that, if the objective function satisfies (8.2), makes a departure from the social optimum against the interests of all agents taken together. By the same reasoning, false reporting during the planning process done with the intention of diverting the process to a point other than the social optimum, though it may be in the interests of a subset of agents, cannot be in the interests of all. Those who lose from such a diversion to a non-optimal point could profitably bribe those who gain from it not to cause it. The idea which underlies this analysis, that of supporting an optimum in a non-convex technology by demand functions rather than prices, has recently been extended to a general equilibrium context by Brown and Heal (1976, 1979, 1980). For a related discussion of non-convex optimisation problems in economics, see Chichilnisky and Kalman (1978).

Another contribution of interest in the present context, is that of Weitzman (1976), who gives a theoretical development of recent Soviet innovations in planning practice. The background to his analysis is as follows: in planning procedures in the Soviet Union, the central authorities will typically assign inputs to firms and request information about the output that these can produce. The response is used (typically together with other information available to the centre) to set an output target for the firm, and various bonus payments to the firm will then depend on the amount by which it manages to overshoot this target. The firm clearly has a strong incentive to understate its output possibilities, so that the target is easily surpassed and the bonus earned. This is of course disadvantageous to the central authorities, who need accurate information about the likely output in order to assign to other firms outputs which are intermediate goods.

The basic idea behind the new system that Weitzman discusses, is that firms are invited to suggest output targets for themselves, and are then penalised for departing from these in either direction. This clearly gives the firm an interest in suggesting as a target, the output that it actually expects to produce. Formally, if B is the bonus, q is the output that will be produced from the inputs assigned and q^* is the output target that the firm suggests, then a suitable scheme would be:

$$B = \alpha + \beta(q - q^*)^2, \quad \alpha > 0, \quad \beta < 0. \tag{8.4}$$

In this expression, q is assumed to be determined by the available inputs, so that q^* is the firm's only choice variable. Evidently B is maximised by setting $q^* = q$ – i.e. by telling the truth. This point is of relevance to our earlier theoretical discussions, because in the planning procedures of Sections 4 and 5 it was assumed that the central authorities could ask for, and receive, accurate information about the outputs proceeded by any given set of inputs. It is also true that they assume information about marginal productivities to be obtainable, and fortunately the present idea can be extended to cover this need also. The basic point is that the marginal productivity of an amount x_{ih} of input h to firm i can be accurately approximated from information about the output levels associated with inputs $x_{ih}+\Delta x_{ih}$ and $x_{ih}-\Delta x_{ih}$ which means that marginal productivities can be approximated by asking the sort of question to which we know how to obtain an accurate answer. Indeed, in the discrete-step reformulations of the procedures of Sections 4, 5 and 6, to be discussed in the next section, the only information about outputs associated with input levels slightly different from the present ones.

In the analysis of the previous paragraphs, it was assumed that a firm's output was uniquely specified by the inputs assigned to it by the centre. There will of course be occasions when this is not true, because there are certain other local inputs controlled by the firm's management. Important amongst these might be what could loosely be called managerial effort. In such situations, the centre will of course wish to ensure that the management is provided with an incentive to supply this local input, and the formulation (8.4) fails to do this. It could however easily be adapted:

$$B=\alpha q+\beta(q-q^*)^2,\quad \alpha>0,\quad \beta<0,$$

provides an incentive to raise output, and to announce as a target the output it is intended to produce.

Weitzman's formulation of the 1971 Soviet incentive reforms is clearly of interest, but these proposals suffer from the drawbacks that they are in essence partial equilibrium in approach – that is, they are based on an analysis of an individual firm in isolation, and do not take account of the interactions between firms that may arise because one firm's proposals affect the allocations that the centre makes both to that firm and to others. A recent paper by Loeb and Magat (1978) has analysed this dimension of the problem, bringing to bear on it the results and techniques already developed by Groves, Ledyard, Green, Laffont and others for the analysis of problems of preference revelation with public goods. They show that in a multi-firm situation there may indeed be circumstances in which the Weitzman scheme leads firms to report inaccurate information to the centre, and go on to develop an incentive system which ensures that revealing accurate information on production possibilities is a dominant strategy for each firm.

The basic idea of the Loeb–Margat model can be explained very simply. Suppose that each firm's output Y_i, $i=l,z,\dots,n$, is a function of the amount of material inputs M_i allocated to it by the centre, and of the work effort l_i supplied by its staff:

$$Y_i = Y_i(M_i, l_i)$$

where M_i is fixed centrally and l_i is chosen locally, subject to the constraint $l_i \in L_i$, to maximise the firm's bonus receipts. The overall resource allocation problem is then to

$$\left.\begin{array}{ll} \text{maximise} & \sum_{i=1}^{n} Y_i(M_i, l_i) \\ \text{subject to} & \sum_{i=1}^{n} M_i \leqq M, \quad M_i \geqq 0, \quad l_i \in L_i, \quad \text{each } i \end{array}\right\}. \tag{8.5}$$

The firm's problem is to

$$\left.\begin{array}{l} \text{maximise } B_i(\tilde{M}_i, l_i) \\ \text{subject to } \quad l_i \in L_i \end{array}\right\}, \tag{8.6}$$

where B_i is the bonus payment scheme and $\tilde{M}_i$ is the amount of material allocated to the firm by the centre. The centre is not of course in a position to solve problem (8.5), as it is unaware of the true production functions Y_i ( ). It has to obtain information about these, and asks each firm for a forecast $y_i(M_i)$ of the output it would produce from any given material allocation. The rational firm will choose $y_i(M_i)$ to equal the output which would maximise its bonus B_i, given M_i and the constraint $l_i \in L_i$. That is,

$$\begin{array}{l} y_i(M_i) = Y_i(M_i, l_i^*(M_i)) \\ \text{where } l_i^*(M_i) \text{ maximises } B_i(M_i, l_i) \\ \text{subject to } \quad l_i \in L_i. \end{array}$$

Now consider the following bonus payment scheme:

$$B_i(\tilde{M}_i, l_i) = Y_i(\tilde{M}_i, l_i) + \sum_{j \neq i} y_j(\tilde{M}_j) - C_i,$$

where C_i is a number independent of firm i's forecast. The ith firm's bonus thus equals the value of the output it actually produces, plus the forecast outputs of other firms at allocated input levels, minus an amount which is independent of firm i's forecast. The central authority will choose $\tilde{M}_i$ to solve the problem

$$\begin{aligned} &\text{maximise} \quad \sum_{i=1}^{n} y_i(M_i) \\ &\text{subject to} \quad \sum_i M_i \leqq M, \quad M_i \geqq 0, \end{aligned}$$

and if firm i sets

$$y_i(M_i) = Y_i(M_i, l_i^*(M_i)),$$

then the centre is in fact choosing the $\tilde{M}_i$ to maximise

$$Y_i(M_i, l_i(M_i)) + \sum_{j \neq i} y_j(M_j),$$

which is of course identical to the part of $B_i(M_i, l_i)$ which depends on i's output forecast to the centre. Hence by giving an accurate forecast, the firm can ensure that the materials allocation chosen by the centre is so chosen as to maximise its bonus function. In other words, reporting accurate information to the centre becomes a dominant strategy for the firm.

The Loeb–Magat contribution is interesting because it stresses the applicability of the literature on incentive compatibility, which originated with the free rider problem for public goods, to the general question of incentives in economic planning. At the same time, their particular formulation obviously has many limitations. It deals with firms producing a single good, and using a single input. It thus avoids the problems of aggregation which are particularly difficult in a planned economy because of the absence of market prices. It also assumes sufficient separability and linearity that the social objective is the sum of the individual objectives. Such strong assumptions are probably necessary to ensure that truth-telling is a dominant strategy. If one were to be content with the weaker result that telling the truth was a Nash equilibrium, then recent work [see, for example, Dasgupta, Hammond and Maskin (1979) and Chichilnisky and Heal (1980)] suggests that considerably weaker assumptions could be used. But it would still be necessary to model explicitly the interactions and interdependencies between firms before really convincing results could be established.

9. Discrete steps

It will no doubt not have escaped the reader's attention that whereas most of the planning procedures presented so far have been continuous processes, any actual planning exercise must consist of only a finite number of discrete steps. The procedures are usually presented in continuous form because the mathematics is far more tractable, but it is clearly essential to confirm the existence of discrete analogues. Although this may appear a straightforward matter, such an appearance would be most misleading: workers in this area have needed great ingenuity to surmount considerable technical problems.

The first to tackle these problems was Uzawa (1958), who considered what is in essence a discrete step reformulation of the Lange–Arrow–Hurwicz procedure. Instead of (2.9) one has

$$\begin{aligned} y_i^{t+1} - y_i^t &= 0 \qquad \text{if} \quad y_i^t = 0 \quad \text{and} \quad U_i - \lambda_i < 0, \\ &= \alpha(U_i - \lambda_i) \quad \text{otherwise,} \end{aligned} \tag{9.1}$$

with corresponding alterations to (2.10) and (2.11). A difficulty arises in proving convergence for such an approach. This is that if a fixed value of the adjustment coefficient α is maintained, the procedure cannot be shown to converge to an optimum, but merely to within a certain neighbourhood of an optimum. Once that neighbourhood is entered, it may overshoot and oscillate around the optimum. Naturally, the smaller the coefficient α, the smaller the neighbourhood, but of course the slower the rate of adjustment towards that neighbourhood. Consequently there is a trade-off between the speed of convergence and the asymptotic precision of that convergence.

It is reasonable to suppose that the conflict that gives rise to this trade-off could be overcome if the parameter α were systematically reduced during the procedure, and this is indeed an ingredient of a discrete reformulation of Heal's planning-without-prices procedure that has been analysed by Henry and Zylberberg (1976). The essence of their proposal is that at each step the centre proposes to a typical firm i both an input vector $x_i = (x_{i1}, \dots, x_{im})$ and a "pitch" $\boldsymbol{a} \in R^m$ which specifies the amount by which it considers changing each component of the allocation. The firm then has to respond with the outputs from the $2m$ input vectors $(x_{i1} \pm a_1, x_{i2}, \dots, x_{im})$, $(x_{i1}, x_{i2} \pm a_2, x_{i3}, \dots, x_{im})$, ... which of course form a basis for computing discrete analogues of marginal productivities. Henry and Zylberberg specify a rule by which the centre can use the information just described to decide whether to alter the pitch a: they also specify a rather distinctive reallocation rule according to which at each step the centre changes the allocation of only one good between one pair of firms – contrast this with equation (4.2), which may require all inputs to all firms to be changed simultaneously. The reallocation rule is as follows: if the indices $i, j \in N$ refer to firms, and $h \in M$ refers to goods, then the centre has to choose a triple (i, j, h) so as to maximise the difference

$$U_i V_{ih}^{+}(x_i^t, a^t) - U_j V_{jh}^{-}(x_j^t, a^t),$$

where U_i and U_j are the partial derivatives of the objective function (assumed for convenience to be linear) and

$$V_{ih}^{+}\left(x_i^t, a^t\right) = \frac{1}{a_h^t} f_i\left(x_{i1}^t, \ldots, x_{ih}^t + a_h, \ldots, x_{im}^t\right),$$

and so on. The centre thus performs that transfer of an amount a_h which maximises the increment to the objective function. For this revised procedure, Henry and Zylberberg establish monotone convergence, via a sequence of feasible plans, to a critical point, thus establishing that all of the properties of the continuous procedure hold for the discrete case. Cremer (1983) generalises these results of Henry and Zylberberg so that the procedure requires less information and may have faster rates of convergence.

It is worth noting that Champsaur, Dreze and Henry (1977) have completed a similar analysis for the M.D.P. procedure with public goods, again demonstrating the existence of a discrete version enjoying the same properties as the original continuous formulation.

Postscript

This entry was written in 1978. Between then and the publication date there have been a number of mathematical developments in the analysis of convergence rates of algorithms for solving optimization problems. It appears likely that these results could provide a basis for a more systematic study of the properties of alternative planning procedures, and for an investigation of the trade-off between the informational requirements of a procedure and its convergence properties. These mathematical results are summarized in Nemirovsky and Yudin (1984).

References

Aoki, M. (1971), "Investment planning in an economy with increasing returns", Review of Economic Studies, 273–281.

Arrow, K. J. and L. Hurwicz (1958), "Gradient methods for concave programming, I, II and III", in: K. J. Arrow, L. Hurwicz and H. Uzawa, eds., Studies in Linear and Non-Linear Programming, pp. 117–146. Stanford.

Arrow, K. J. and L. Hurwicz (1960), "Decentralisation and computation in resource-allocation", in: R. Pfouts, ed., Essays in Economics and Econometrics in Honour of Harold Hotelling, pp. 34–103. Chapel Hill: University of North Carolina Press.

Barone, E. (1935), "The Ministry of Production in a collectivist state", in: F. A. von Hayek, ed., Collectivist Economic Planning.

Brown, D. J. and G. M. Heal (1977), "The existence and efficiency of equilibrium with increasing returns in production", Cowles Foundation Discussion Paper. New Haven: Yale University.

Brown, D. J. and G. M. Heal (1979), "Equity, efficiency and increasing returns", Review of Economic Studies, Oct.

Brown, D. J. and G. M. Heal (1980), "Two-part tariffs, marginal cost pricing and increasing returns in a general equilibrium model", Journal of Public Economics, 13:25–49.

Champsaur, P. (1976), "Neutrality of planning procedures in an economy with public goods", Review of Economic Studies, 43:293–300.

Champsaur, P., J. H. Dreze and Cl. Henry (1977), "Stability theorems with economic applications", Econometrica, 45:273–295.

Chichilnisky, G. and P. Kalman (1979), "Comparative statics of less neoclassical agents", International Economic Review, Feb.

Chichilnisky, G. and G. M. Heal (1980), "Incentives to reveal preferences: The case of restricted domains", University of Essex Discussion Paper 19.

Cremer, Jacques (1977), "A quantity-quantity algorithm for planning under increasing returns to scale", Econometrica, 45.

Cremer, Jacques (1983), "The discrete Heal algorithm with intermediate goods", Review of Economic Studies, 50:383–391.

Dasgupta, P. S., P. J. Hammond and E. Maskin (1979), "The implementation of social choice rules: Some general results on incentive compatibility", Review of Economic Studies, 46:185–216.

Dreze, J. H. and D. della Vallee Poussin (1971), "A tatonnement process for public goods", Review of Economic Studies, 38:133–150.

Dyker, D. A. (1976), The Soviet Economy. London: Crosby, Lockwood, Staples.

Green J. and J. J. Laffont (1977), "Characterisation of satisfactory mechanisms for the revelation of preferences for public goods", Econometrica, 45:427–439.

Groves, T. and J. Ledyard (1977), "Optimal allocation of public goods: A solution to the free rider problem", Econometrica, 45:783–811.

Hayek, F. A. von (1945), "The use of knowledge in society", American Economic Review, 35:519–530.

Heal, G. M. (1969), "Planning without prices", Review of Economic Studies, 347–363.

Heal, G. M. (1971), "Planning, prices and increasing returns", Review of Economic Studies, 281–295.

Heal, G. M. (1973), The Theory of Economic Planning. Amsterdam: North-Holland.

Henry, Cl. (1972), "Non linear evolution equations on mathematical economics", Techniques of Optimisation.

Henry, Cl. (1973), "An existence theorem for a class of differential equations with multivalued right hand side", Journal of Mathematical Analysis and Application.

Henry, Cl. (1977), "The free rider problem in the M.D.P. procedure", Discussion Paper. Paris: Laboratoire d'Econometrie, Ecole Polytechnique.

Henry, Cl. and A. Zylberberg (1976), "Planning algorithms to deal with increasing returns", Review of Economic Studies, forthcoming.

Hori, H. (1975), "The structure of equilibrium of points of Heal's process", Review of Economic Studies, 42:457–467.

Hurwicz, L. (1969), "On the concept and possibility of informational decentralisation", American Economic Review, Papers and Proceedings.

Kornai, J. (1967), Mathematical Programming of Structural Decisions. Amsterdam: North Holland.

Lange, O. (1936), "On the economic theory of socialism", Review of Economic Studies, 4:53–71, 123–142.

Loeb, M. and W. A. Magat (1978), "Success indicators in the Soviet Union: The problem of incentives and efficient allocations", American Economic Review, 68.

Malinvaud, E. (1967), "Decentralised procedures for planning", in: E. Malinvaud and M. O. L. Bacharach, eds., Activity Analysis of Growth and Planning, pp. 170–211. London: Macmillan.

Malinvaud, E. (1971), "A planning approach to the public good problem", Swedish Journal of Economics, 73:96–112.

Malinvaud, E. (1972), "Prices for individual consumption, quantity indicators for collective consumption", Review of Economic Studies, 384–407.

Nemirovsky, A. S. and D. B. Yudin (1984), Problem Complexity and Method Efficiency in Optimization. New York: Wiley-Interscience.

Roberts, D. J. (1976), "Incentives in planning procedures for the provision of public goods", C.O.R.E. Discussion Paper 7611. Louvain: Université Catholique de Louvain.

Samuelson, P. A. (1954), "The pure theory of public expenditure", Review of Economics and Statistics, 26:387–389.

Taylor, F. M. (1929), "The guidance of production in a socialist state", American Economic Review, 19.

Tulkens, H. (1976), "Dynamic procedures for allocating public goods: An institution-orientated survey", Mimeographed. Louvain: C.O.R.E.

Uzawa, H. (1958), "Iterative methods for concave programming", in: K. J. Arrow, L. Hurwicz and H. Uzawa, eds., Studies in Linear and Non-Linear Programming. Stanford: Stanford University Press.

Weitzman, M. L. (1974), Review of G. M. Heal, "The theory of economic planning", Journal of Economic Literature, 12:499–500.

Weitzman, M. L. (1976), "The new Soviet incentive model", Bell Journal of Economics, 7:251–258.

LIST OF THEOREMS

INDEX